ORGANIC REACTIONS

MECHANISM WITH PROBLEMS

ORGANIC REACTIONS
MECHANISM WITH PROBLEMS

By

Dr. Rajpal Tyagi
Dept. of Chemistry
M.M.H. Post Graduate College
Ghazaibad (U.P.)

D P H

DISCOVERY PUBLISHING HOUSE
NEW DELHI-110002

Edition - 2015

ISBN: 978-81-8356-012-2

Organic Reactions
Mechanism with Problems

Published by:
DISCOVERY PUBLISHING HOUSE PVT. LTD.
4383/4B, Ansari Road, Darya Ganj
New Delhi-110 002 (India)
Phone: +91-11-23279245, 43596064-65
Fax: +91-11-23253475
E-mail: discoverypublishinghouse@gmail.com
sales@discoverypublishinggroup.com
web: www.discoverypublishinggroup.com

Printed at:
Infinity Imaging Systems
Delhi

Preface

The present title "Organic Reaction" has been designed for undergraduate and post-graduate student of all Indian Universities. We live and breed in a world that owes to organic chemistry many times more than organic chemistry owes to it. The domain of organic chemistry is so enormous that it defies the imagination of any individual, let alone mastering it in entirety. This is not a text book, but a reference book supplement to the text of organic chemistry meant for University students. However some advanced students may find the book inadequate.

It has been constant endeavour of the authors to furnish maximum substance, keeping in view the limitations of size of the volume. Efforts have been made to condense the matter as far as practicable. It is hoped that this book will not only meet the requirement of Indian students but will also be useful as a guideline to the teachers in their teaching.

The authors is deeply grateful to for his expertise and advice during the preparation of the present title.

The author expresses his gratitude to Mr. Wasan and staff of M/s Discovery Publishing House for their whole hearted co-operation in the publication of this book.

—Author

Contents

1

ACYLOIN REACTION OR ACYLOIN CONDENSATION

This reaction is given by ester in presence of sodium metal to give α-hydroxy carbonyl compound known as acyloin and the reaction is known as Acyloin condensation.

Reaction takes place between two moles of ester (intermolecular condensation) or one mole of diaster (intramolecular condensation).

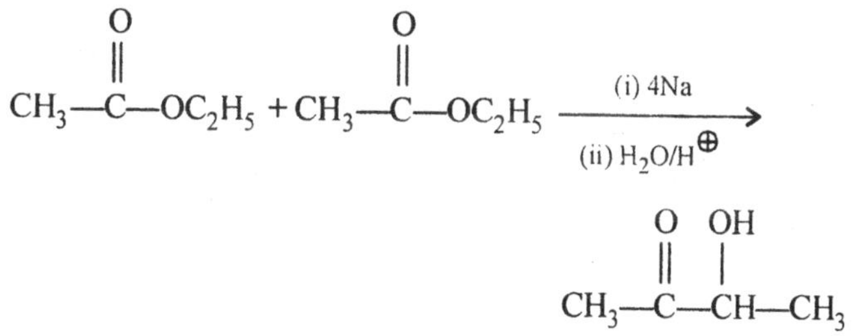

Acetoin (Acyloin)

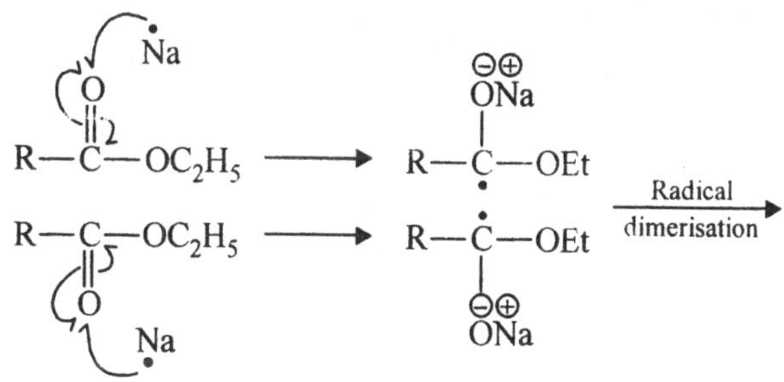

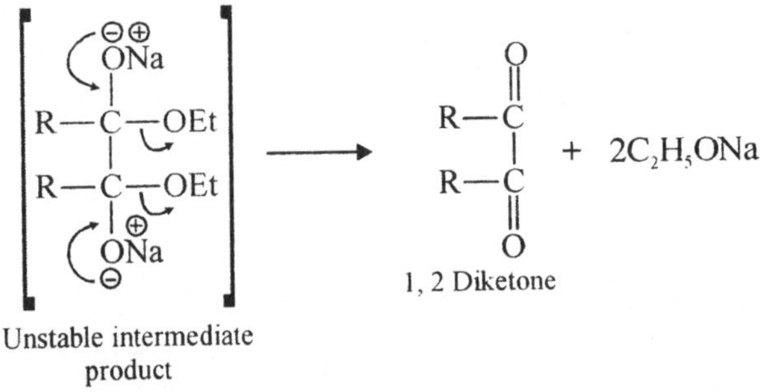

1,2-diketons is highly reductive and reactive, and undergoes reduction with Na.

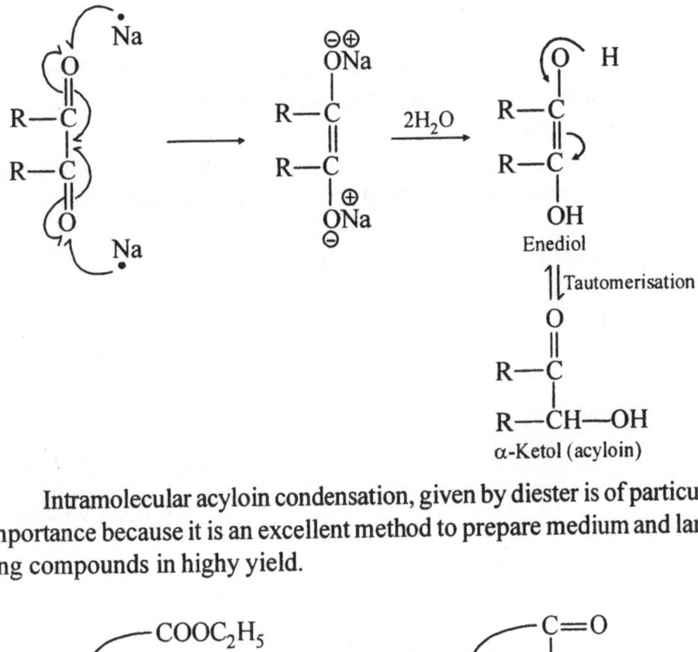

Intramolecular acyloin condensation, given by diester is of particular importance because it is an excellent method to prepare medium and large ring compounds in highy yield.

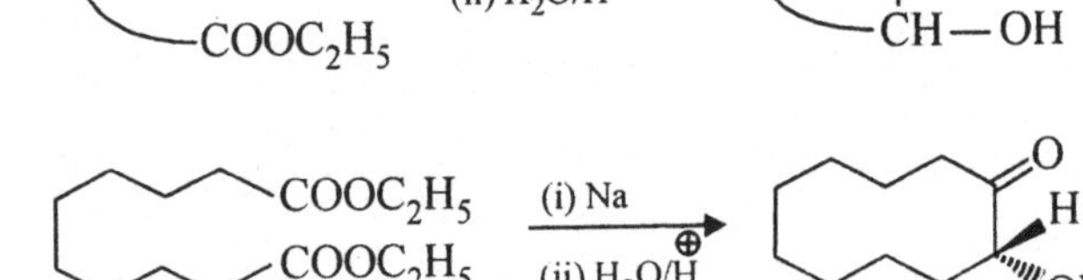

2

ALDOL CONDENSATION

Condensation between two molecules of an aldehyde or a ketone to a β-hydroxyketone is known as the **aldol condensation.**

1. This reaction takes place between two molecules of carbonyl compounds (same or different). One molecule should have at least one α-hydrogen.
2. The reaction is catalysed by acid as well as base but the best result is obtained when reaction is carried out in presence of dil base below room temp. The aldehyde reaction works well when one drop of dil. NaOH is added to aldehyde. The ketone reaction is best done with insoluble $Ba(OH)_2$.
3. Product of the reaction is β-hydroxy carbonyl compound known as Aldol.

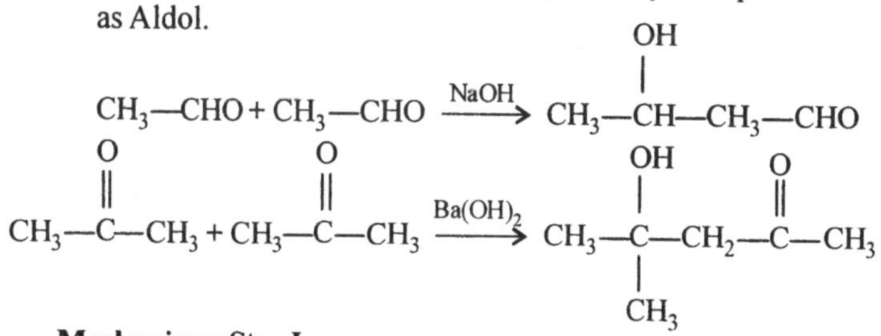

Mechanism : Step I.

In this step the base removes a proton from α-carbon of one molecule of aldehyde to give a resonance stabilised enolate anion.

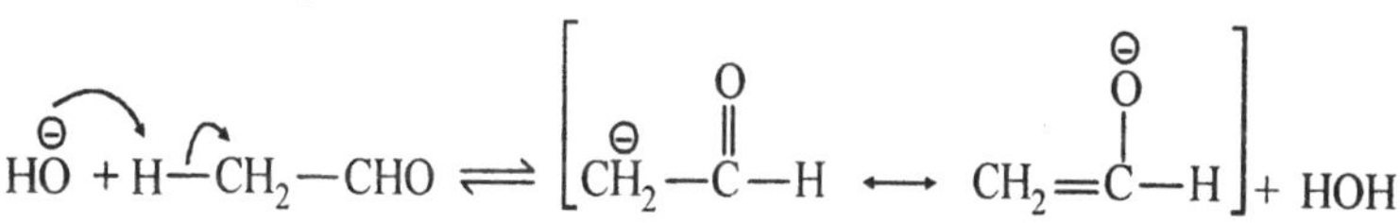

Step II.

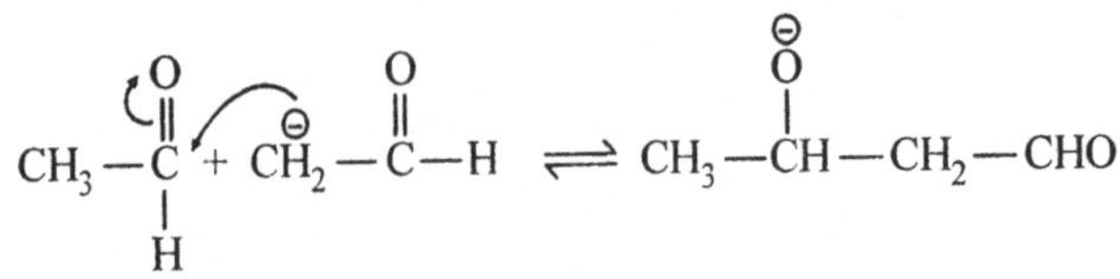

The enolate anion acts as a nucleophile-as a carbanion-and attacks the carbonyl carbon of second molecule of aldehye, producing an alkoxide anion.

Note. If the aldehyde or ketone is not having an α-hydrogen, a simple aldol condensation will not take place. For example,

$$\left.\begin{array}{l} C_6H_5CHO \\ HCHO \\ (CH_3)_3CCHO \\ C_6H_5COC_6H_5 \\ C_6H_5COCK_3 \end{array}\right] \xrightarrow{\text{dil.}OH^-} \text{no reaction}$$

(NO α-Hydrogen)

In concentrated base, these may undergo the Cannizzaro's reaction.

Step III. The alkoxide ion now removes a proton from water molecule to form aldol.

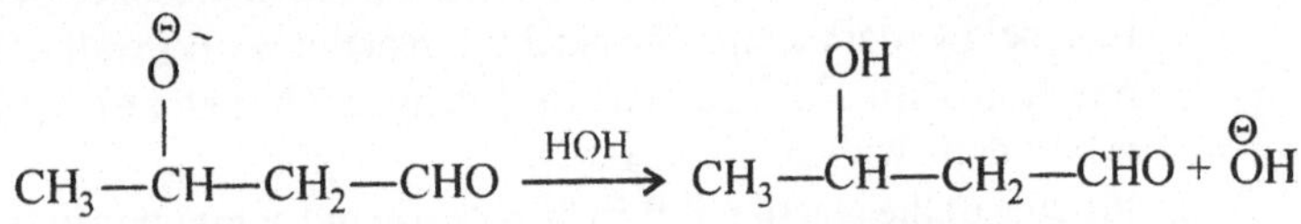

Thus aldol condensation is nucleophilic addition reaction between carbonyl group of one molecule and α-carbon of other molecule having at least one hydrogen. Reaction in short, is represented as :

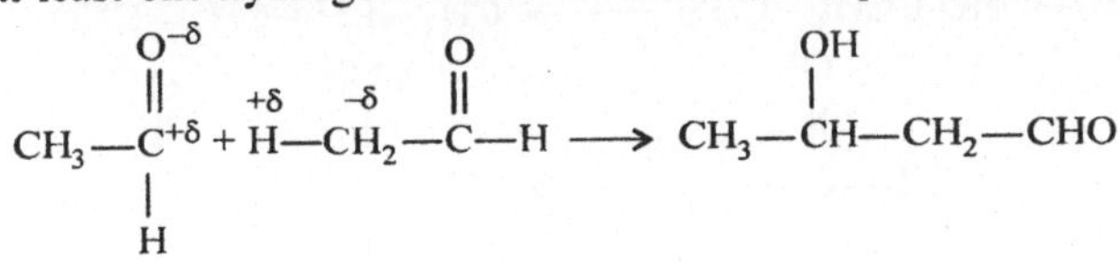

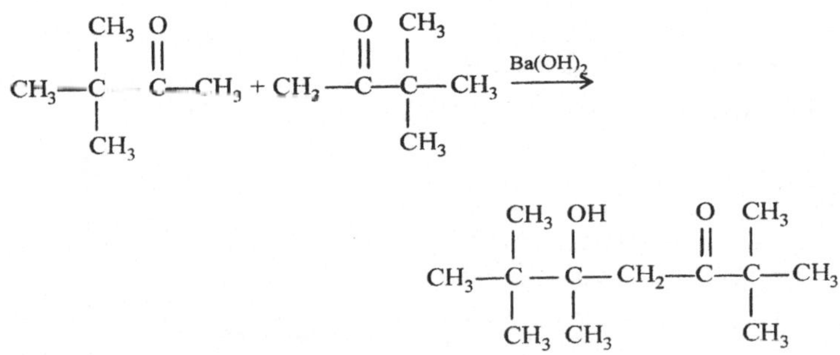

Intra-molecular aldol condensation : Intramolecular aldol condensation is given by dialdehydes, diketones or keto aldehydes. But best result is given by diketones. The diketones should be 2,5 2,6 2,7 and 2,8- diketones. Intramolecular aldol condensation offers a convenient method for the preparation of five and six membered rings.

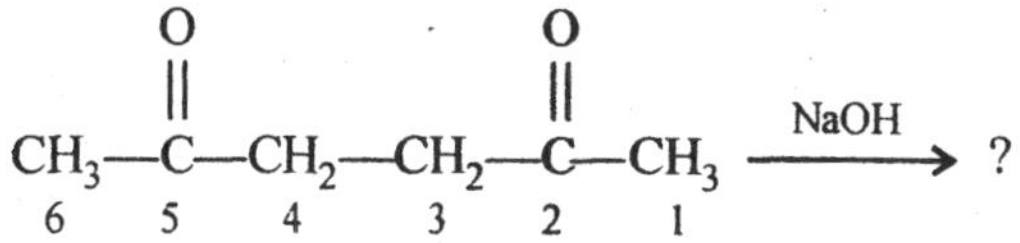

Reaction should take place between C_1 and C_5 because this will give five membered cyclic compound.

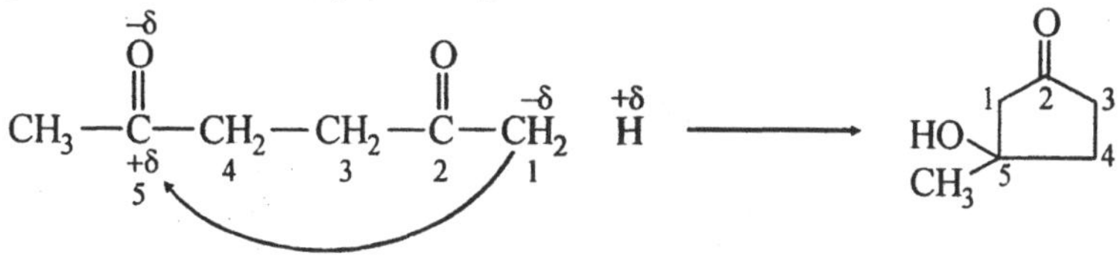

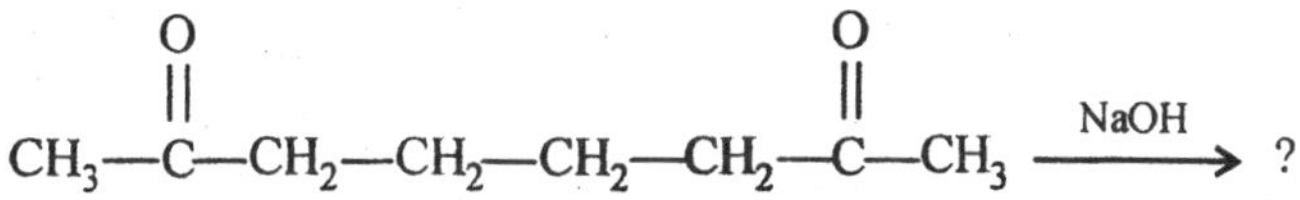

Note : If diketone is 2,7 2,8 or 2,9 then inner α-carbon converts into carbanion in the presence of base.

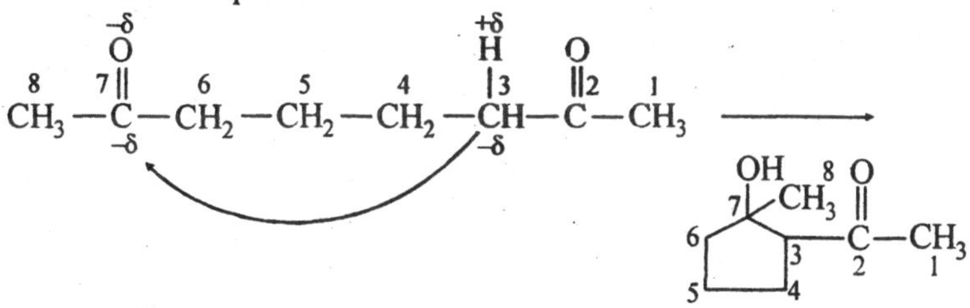

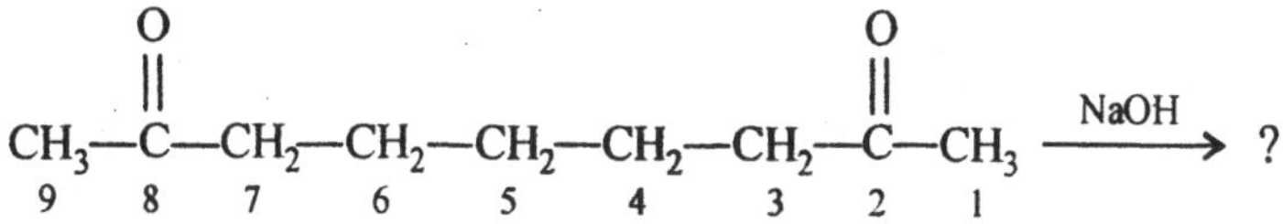

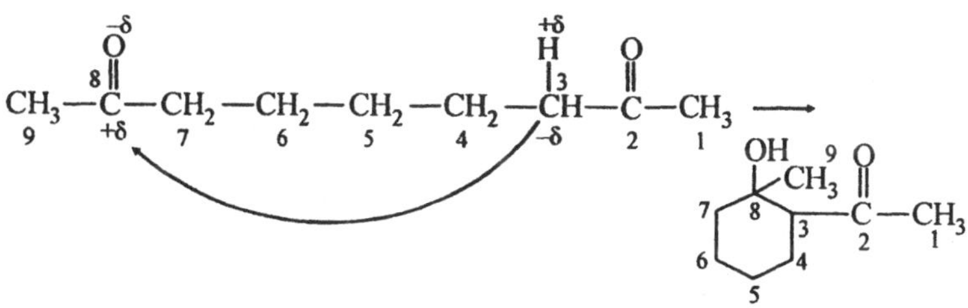

In case of keto aldehydes, the reaction almost certainly involves the formation of at least three different enolates. However, it is the enolate from the ketone side of the molecule that adds aldehydic group leading to the main product.

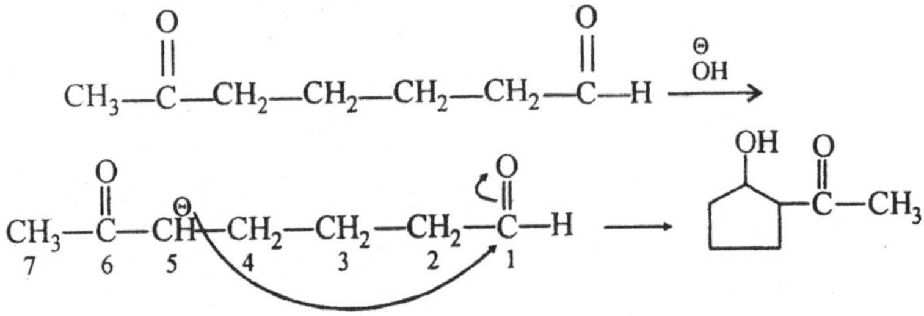

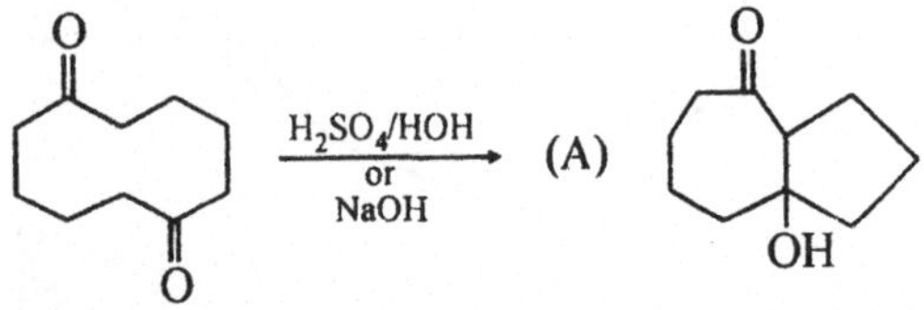

CROSSED AIDOL CONDENSATION

If the condensation takes place between two different carbonyl compounds, it is known as crossed aldol condensation.

1. Crossed aldol condensation between two different aldehydes

If both the aldehydes are having α-hydrogen (s), both can prouduce carbonions and also can act as carbanion acceptors. Hence there occurs the formation of a mixture of four products which is of little synthetic importance. If one of the aldehydes lacks α-hydrogen, then it can act only as a carbanion acceptor. In such cases there occurs the formation of two products, *e.g.*,

(a) $R_3C{-}CHO + CH_3CHO \xrightarrow{\overset{\ominus}{O}H} R_3C{-}CH(OH) + CH_2{-}CHO$
Crossed product

(b) $CH_3CHO + CH_3CHO \xrightarrow{\overset{\ominus}{O}H} CH_3CH(OH)CH_2CHO$
Simple product (normal)

However, it is possible to obtain a good yield of the crossed product by slowly adding the aldehyde having α-hydrogen to a mixture of the aldehyde lacking α-hydrogen and the catalyst, *e.g.*,

$$C_6H_5CHO + \overset{\ominus}{O}H \xrightarrow[\text{Slow addition}]{CH_3CHO} C_6H_5CH(OH)CH_2CHO \xrightarrow{-H_2O} C_6H_5CH{=}CH{-}CHO$$

Cinnamaldelyde

As formaldehyde lacks α-hydrogen, it acts as a reactive carbanion acceptor because of the absence of steric hindrance and +I effect. Hence when acetaldehyde is reacted with excess of formaldehyde in the presence of $Ca(OH)_2$, crossed aldol condensation starts which will continue three times until trihydroxy methyl acetaldehyde (I) gets formed. The product (I) lacks α-hydrogen and it undergoes crossed Cannizzaro's reaction with formaldehyde to form pentaerythritol which is a medicine for heart trouble.

$$CH_3CHO + \overset{\ominus}{O}H \rightleftharpoons \overset{\ominus}{C}H_2CHO + H_2O$$

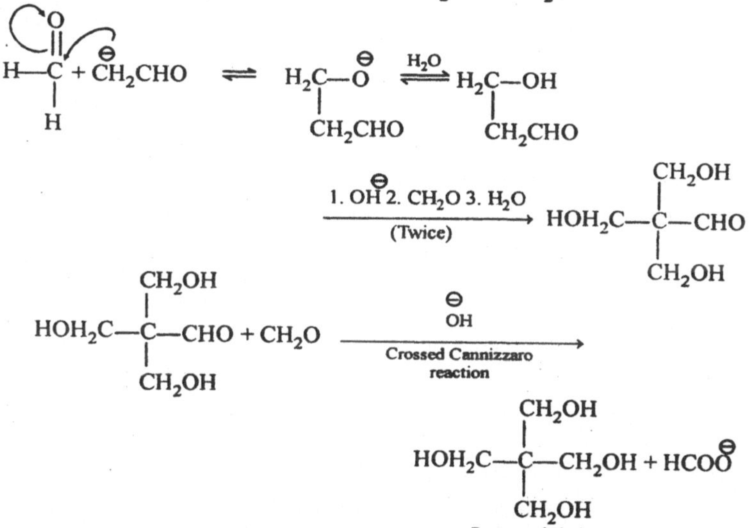

Pentaerythritol

2. Crossed aldol condensation between two different ketones

As the reactivity of ketones is quite low, a poor yield of condensation products is obtained. Therefore, it is rarely attempted.

3. Crossed aldol condensation between an aldehyde and an ketone

(a) If both an aldehyde and a ketone are having α-hydrogens, they undergo aldol condensation to form two products. The reason it that ketones are poor carbanion acceptors and so cannot undergo selfcondensation. Aldehydes, which are more reactive than ketones, act as carbanion acceptors and the ketones provide the carbanions.

$$CH_3CHO + CH_3COCH_3 \xrightleftharpoons{\overset{\ominus}{O}H} CH_3\text{—}CH(OH)\text{—}CH_3\text{—}COCH_3$$

4-Hydroxypentan-2-one(crossed product)

$$CH_3CHO + CH_3CHO \xrightleftharpoons{\overset{\ominus}{O}H} CH_3\text{—}CH(OH)\text{—}CH_2\text{—}CHO$$

Acetaldol (normal product)

Generally, the predominant product is the crossed product. However, it is possible to minimise the formation of acetaldol by slowly adding the aldehyde to the mixture of ketone and the catalyst base.

(b) If the addition takes place between a ketone and an aldehyde lacking α-hydrogen, only one product is obtained. The reason is that the ketones are less reactive and cannot undergo self-condensation while the aldehyde does not undergo cannizzaro's reaction which is slower than aldol condensation. Hence, the ketone provides the carbanion whereas the aldehyde acts as the carbanion acceptor. For instance,

(i) $C_6H_5CHO + CH_3COCH_3 \xrightleftharpoons{\overset{\ominus}{O}H} C_6H_5\text{—}CH{=}CH\text{—}CO\text{—}CH_3$

Benzylideneacetone

(ii) $C_6H_5CHO + CH_3COC_6H_5 \xrightarrow{\overset{\ominus}{O}H} C_6H_5\text{—}CH{=}CH\text{—}CO\text{—}C_6H_5$

Benzylideneacetophenone

(iii) $CH_3COCH_3 + HCHO \xrightarrow{\overset{\ominus}{O}H} CH_3\text{—}CO\text{—}CH_2\text{—}CH_2OH$

4-Hydroxybutanone-2

It is difficult to stop the reaction (*iii*) at this stage and continues until all the hydrogens of the ketone are replaced by hydroxy methyl group to form $(HOCH_2)_3C\text{—}CO\text{—}(CH_2OH)_3$.

DEHYDRATION OF ALDOLS

Both in the acidic or basic medium, aldols are easily dehydrated to α,β-unsaturated compounds. The dehydration can also take place on simple heating.

Acid catalysed

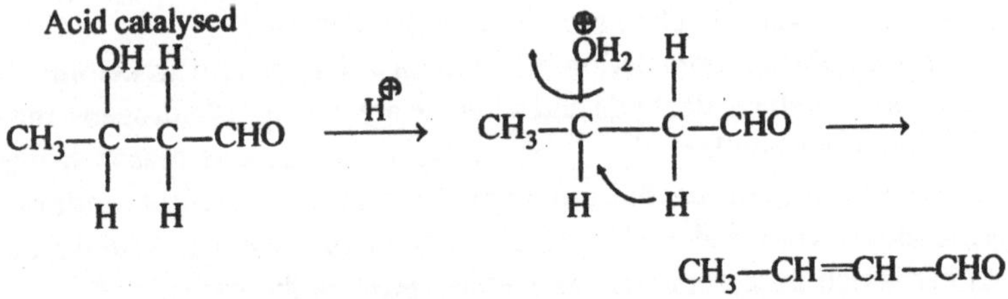

Base catalysed

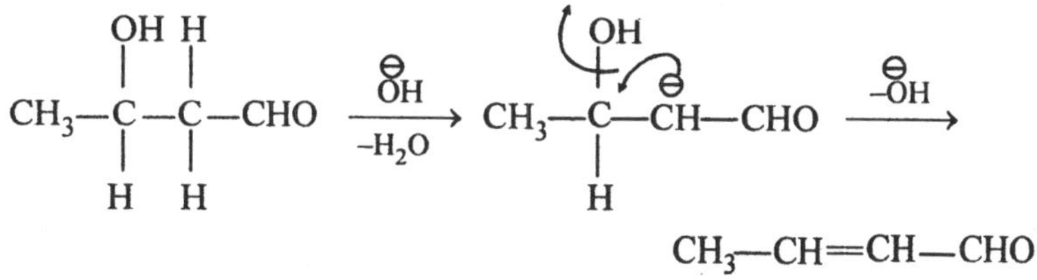

Applications

(i) Both simple and crossed aldol condensation are used for synthesizing saturated and unsaturated aldehydes and alcohols of synthetic importance, *e.g.*,

$$2CH_3CHO \xrightarrow{\overset{\ominus}{OH}} CH_3CH(OH)CH_2CHO \xrightarrow[-H_2O]{\Delta}$$

$$CH_3CH{=}CHCHO \xrightarrow{H_2/Ni} \underset{\text{n-Butanol}}{CH_3CH_2CH_2CH_2OH}$$

$$2\,C_3H_7CH_2OH \xrightarrow[250°C]{Cu} 2C_3H_7CHO \xrightarrow{\overset{\ominus}{OH}}$$

$$C_3H_7CH(OH)CH(C_2H_5)CHO \xrightarrow[\Delta]{H^{\oplus}} C_3H_7CH{=}C(C_2H_5)CHO$$

$$\xrightarrow{H_2/Ni} \underset{\text{2-Ethylhexanol-1}}{C_3H_7CH_2CH(C_2H_5)CH_2OH}$$

(*ii*) For the synthesis of vitamin A, the intermediate β-ionone is required which is synthesized by the condensation of citral with acetone and subsequent treatment with boron trifluoride.

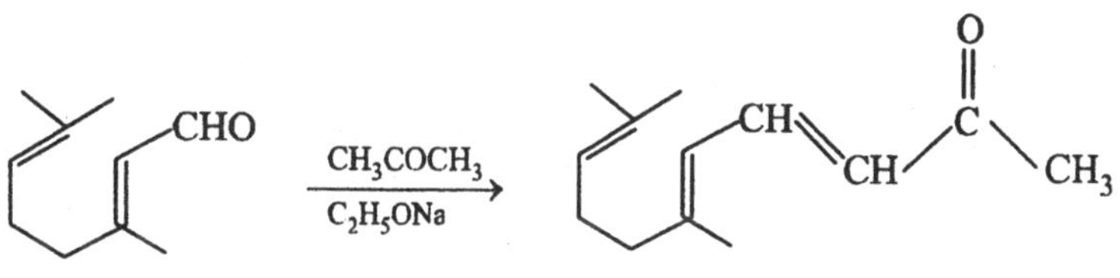

Citral Ψ-Ionone

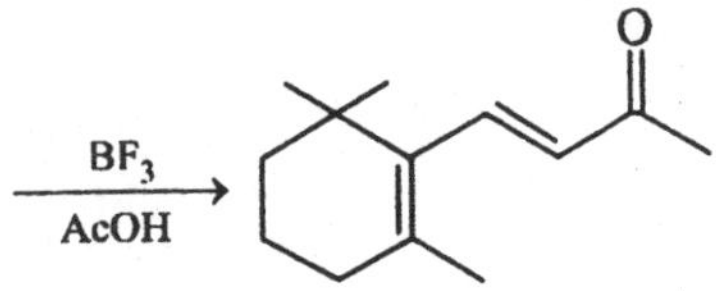

β-Ionone

(iii) It is also possible to carry out aldol condensation in the presence of acid.For example, acetone with dry hydrochloric acid slowly results in the formation of mesityl oxide and phorone. When acetone is distilled with sulphuric acid, mesitylene is obtained.

$$2CH_3COCH_3 \xrightarrow{\text{Dry HClgas}} \underset{\text{Mesityl oxide}}{(CH_3)_2C{=}CHCOCH_3} \xrightarrow[\text{HCl gas}]{CH_3COCH_3}$$

$$\underset{\text{Phorone}}{(CH_3)_2C{=}CH{-}CO{-}CH{=}C(CH_3)_2}$$

(*iv*) The pyridine and quinoline having methyl groups at positions - 2 and - 4 undergo aldol type of condensation. These reactions find use in the synthesis of an important class of dyes, *i.e.*, cyanine dyes (photographic sensitizers).

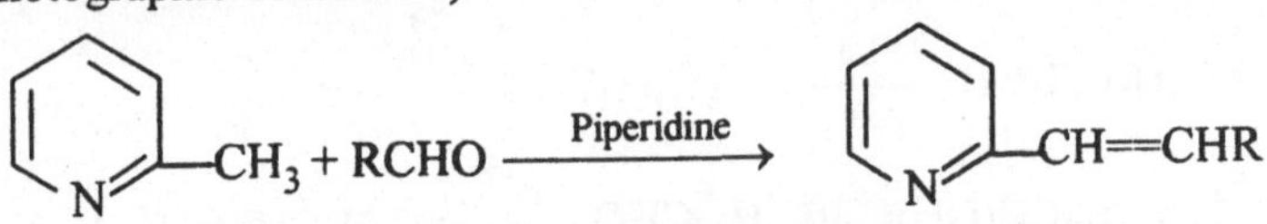

(v) A large number of compounds having acidic C—H bonds undergo aldol type condensation with aldehydes and ketones to form compounds of valuables synthetic importance. For example, condensation of acetylene with aldehydes or ketones yields acetylenic alcohol, which is a valuable commercial product.

$$2CH_2O + HC{\equiv}CH \xrightarrow{Cu_2C_2} \underset{\text{2-Butyne-1, 4-diol.}}{HOH_2C{-}C{\equiv}C{-}CH_2OH}$$

The process is called **ethynylation.**

Pentaerythritol (see above) is a valuable medicine for heart trouble. Its nitroderivative, penterythritol tetra nitrate (PENT) is a valuable explosive and is also used as a surface-coating agent.

(vi) Between two different aldehydes (or ketones), when all the four possible condensation products are obtained. However, by using different catalysts one of the products can be made to predominate in the mixture, *e.g.*,

$$\underset{\text{Ethanal}}{CH_3CHO} + \underset{\text{Propanal}}{CH_3CH_2CHO} \xrightleftharpoons{\text{NaOH}} \underset{\text{3-Hydroxypentanal}}{CH_3CH_2CH(OH)CH_2CHO}$$

$$CH_3CHO + CH_3CH_2CHO \xrightleftharpoons{\text{HCl}} \underset{\text{3-Hydorxy-2-methylbutanal}}{CH_3CH(OH)\overset{\overset{CH_3}{|}}{C}HCHO}$$

(vii) Between one molecule of an aldehyde and a molecule of a ketone, when it is α-hydrogen atom of the ketone which gets involed in the condensation, *e.g.*,

$$CH_3CHO + HCH_2COCH_3 \xrightleftharpoons{\text{NaOH}} \underset{\text{4-Hydroxypentan-2-one}}{CH_3CH(OH)CH_2COCH_3}$$

When the aldols are heated, they eliminate molecule of water, to form the corresponding unsaturated compounds, for example, acetaldol on heating yields 2-butenal (crotonaldehyde).

$$\underset{\text{Acetaldol}}{CH_3CH(OH)CH_2CHO} \xrightarrow{\Delta} \underset{\text{2-Butenal}}{CH_3CH{=}CHCHO} + H_2O$$

PROBLEM BASED ON ALDOL AND RELATED REACTION

Problem 1. You are provided with two compound ethanol and acetone. Aldol reaction is carried out in D_2O.

(i) Will the reaction occur in D_2O?

(ii) In which compound isotopic substitution occur?

Ans. When aldol reaction is carried out in D_2O, deuterium may be incorporated in acetone but not in acetaldehyde because in case of acetaldehyde second step which include attack of carbanion on carbonyl carbon is very fast. So carbanion does not have time to interact with D_2O.

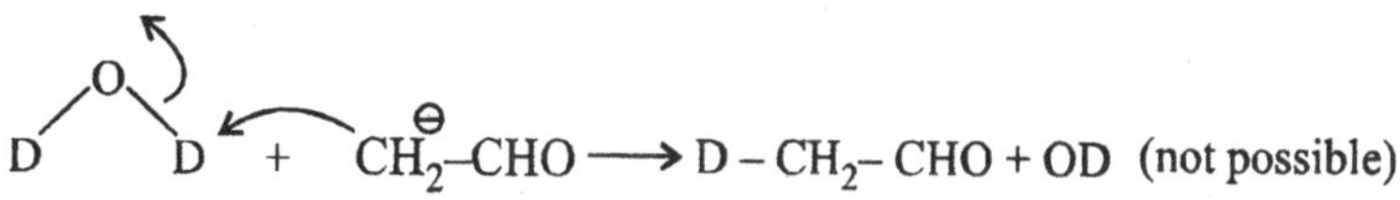

But in case of acetone, carbanion attack slowly on carbonyl carbon so it has enough time to interact with D_2O.

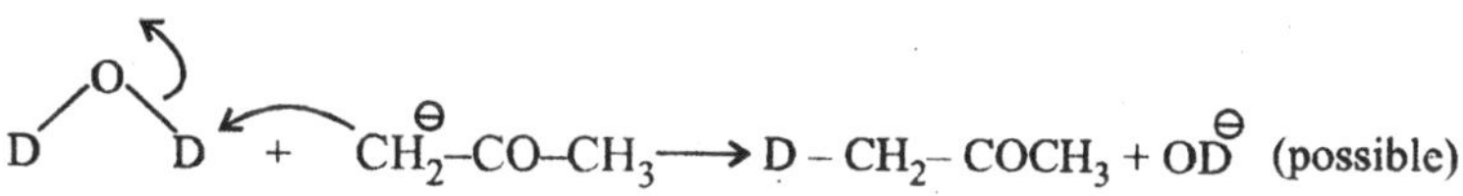

Problem 2. Write the mechanism of acid-catalysed Aldol reaction.

$$Me_2C=O \underset{}{\overset{H^{\oplus}}{\rightleftharpoons}} Me_2\overset{\oplus}{C}-OH$$

Ans.

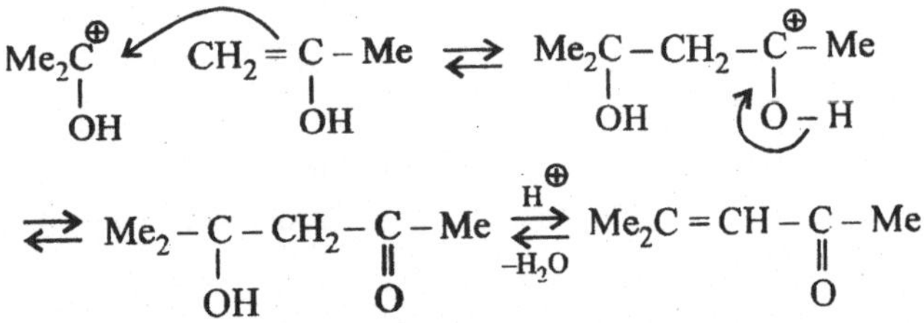

Problem 3. Write the mechanism.

(i) 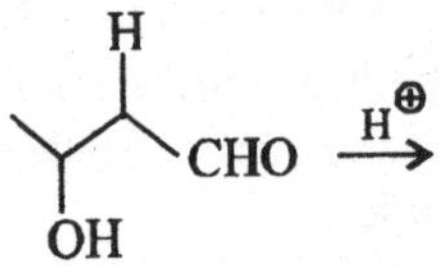 *(ii)* 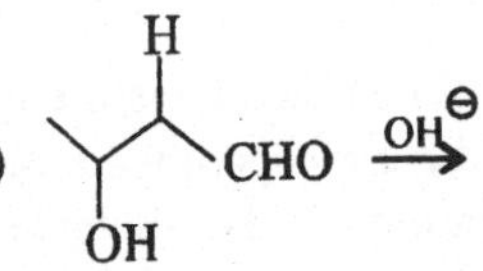

Ans. *(i)* 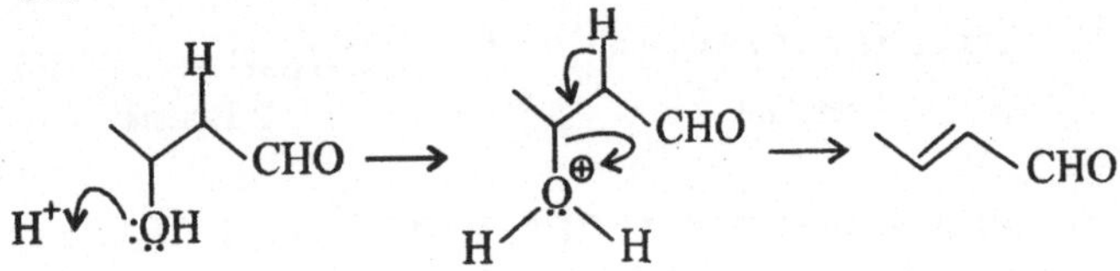

(ii) 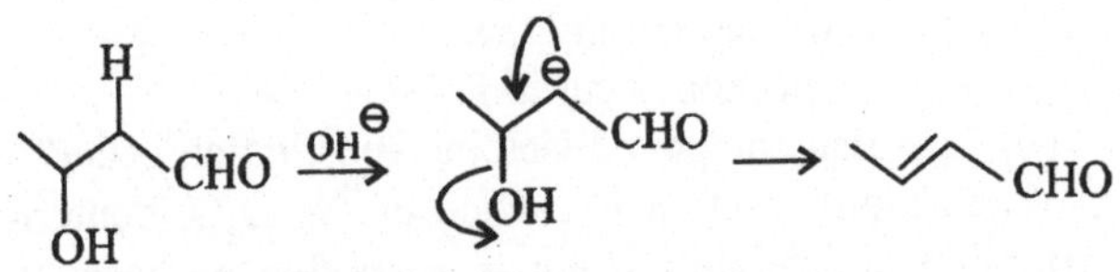

Problem 4. What is the necessary condition for aldol reaction?

Ans. Aldehyde or ketone must have an α - H. For example, CH_3 CHO undergoes aldol reaction easily but

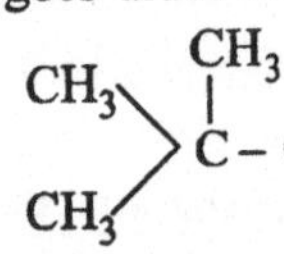 will not undergo this reaction.

Problem 5. Which of the following react fast with base?

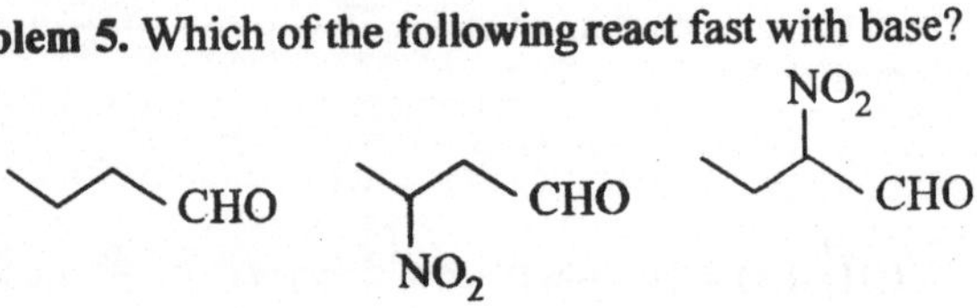

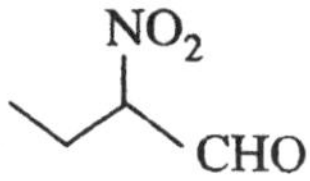

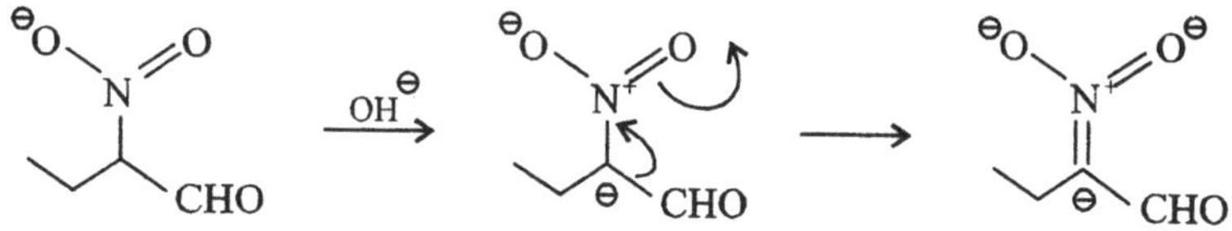

$$\mathbf{CH_3CHO + CH_3NO_2 \xrightarrow{OH^{\ominus}}}$$

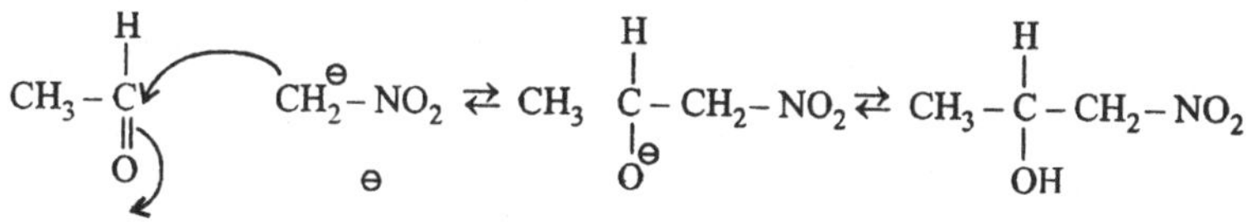

CH_3CHO

$CH_2^{\ominus}NO_2$

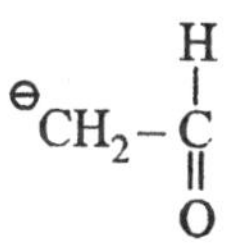

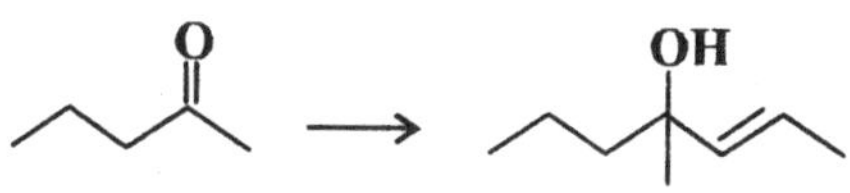

$$CH \equiv C - CH_3 \xrightarrow{NH_2^{\ominus}} \overset{\ominus}{C} \equiv C - CH_3$$

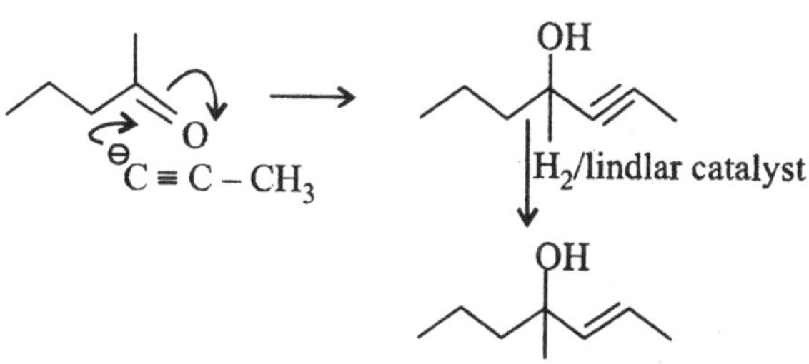

3

ALLYLIC REARRANGEMENT

The compounds which have a functional group X other than an unsaturated linkage on a carbon atom α to an olefinic bond are known as allylic compounds.

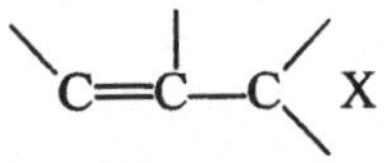

These compounds (I), when subjected to acid or base catalysis, undergo functional group (X) migration to yield new compounds (II).

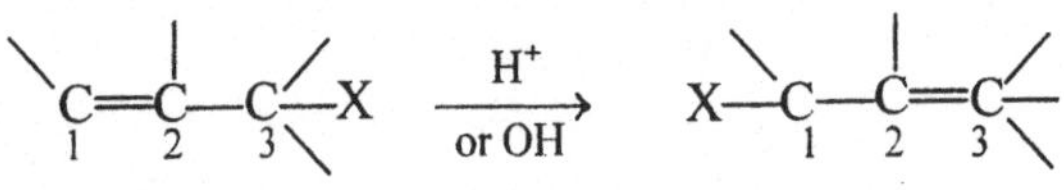

Allylic compound (I) New compound (II)

Some examples of allylic rearrangements are as follows :

(i) $CH_3CH{=}CH{-}CH_2OH \xrightarrow{H^+} CH_2.\overset{OH}{\overset{|}{C}H}.CH{=}CH_2$

Crotyl alcohol → Methyl vinyl carbinol (rearranged product)

(ii)

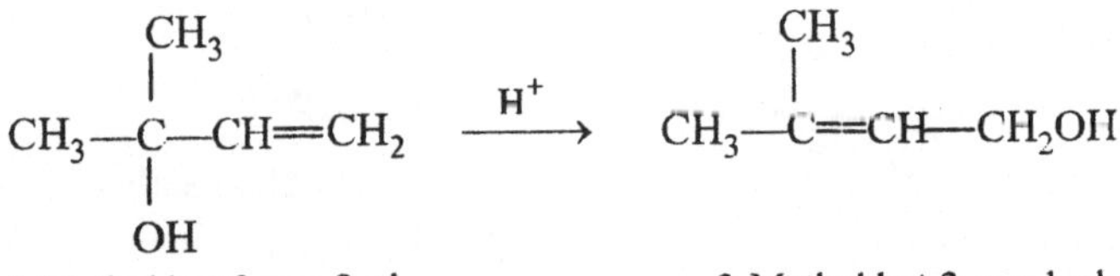

2-Methyl but-3-ene-2-ol (a tertiary alcohol) → 3-Methyl but-2-ene-1-ol (a primary alcohol)

(iii) The incoming and outgoing nucleophiles may not be the same.

$$CH_3—\underset{\underset{Cl}{|}}{CH}—CH{=}CH_2 \xrightarrow{EtOH}$$

α-Methylallylchloride

$$CH_3—\underset{\underset{OEt}{|}}{CH}—CH{=}CH_2 \; + \; CH_3—CH—CH_2OEt$$

α-Ethoxybutene-1 1-Ethoxybutene-2

The reactions at the allylic positions which are involving the migration of the double bond (and the functional group, if present) from their original position to other carbon atom are known as allylic rearrangemnts.

MECHANISM

As allylic rearrangements are more common in subsititution reactions, their mechanism implies the mechanism of the following four different types of substitution reactions in the allylic compounds.

(a) Unimolecular nucleophilic substitution reactions (S_N1)
(b) Bimolecular nucleophilic substitution reactions (S_N2)
(c) Unimolecular electrophilic substitution reactions (S_E1)
(d) Bimolecular electrophilic substitution reactions (S_E2)

We will now dicuss the mechanism of the four types of substitution reactions in the allylic system one by one.

(a) *Unimolecular nucleophilic substitution reaction* (S_N1). Unimolecular nucleophilic substitution (S_N1) reaction in allylic system always yields a mixture of two substituted products, *viz.* the normal (expected) product and the rearranged product, as intermediate carbonium ion gets stabilised by delocalization of the p electrons of the olefinic bond between two structures either of which may be attackd to form two products.Thus, there is partial positive charge on both C_1 and C_3. The nucleophilic attack can thus take place at both C_1 and C_3 to yield two products which are indeed obtained. For example, the hydrolysis of 3-chlorobut 1-ene in ethanol takes place in the following way to yield a mixutre of two products.

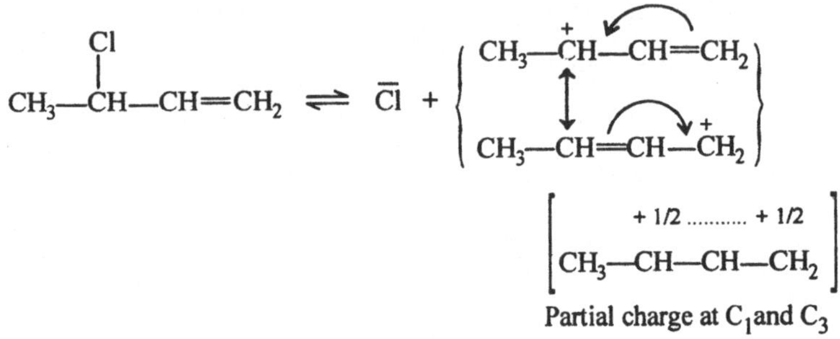

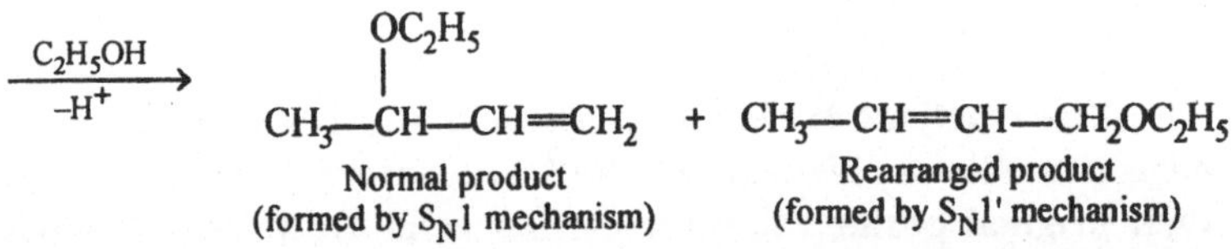

As the reaction usually yields the normal product, the rearrangement product is formed by the special case of S_N1 mechanism which is generally designated as S_N1' mechanism.

(b) *Bimolecular nucleophilic substitution* (S_N2) *reactions* : As in S_N2 reactions the powerful nucleophilic reagent, such as sodium ethoxide, displaces directly the chloride ion without forming any carbonium ion, these reactions ordinarily yield the normal (expected) products. For example, solvolysis of 3-chloro-but-I-ene in ethanol with sodium ethoxide yields the normal product.

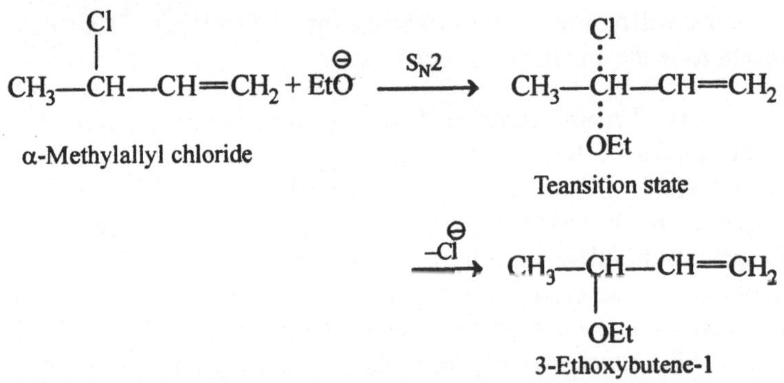

However. allylic rearrangements in bimolecular substitution reactions are also observed in some cases.Such reactions are designated as S_N2'

In S_N2' reaction, the nucleophile attacks the unsaturated γ-carbon rather the α-carbon to yield the rearranged product, *e.g.*,

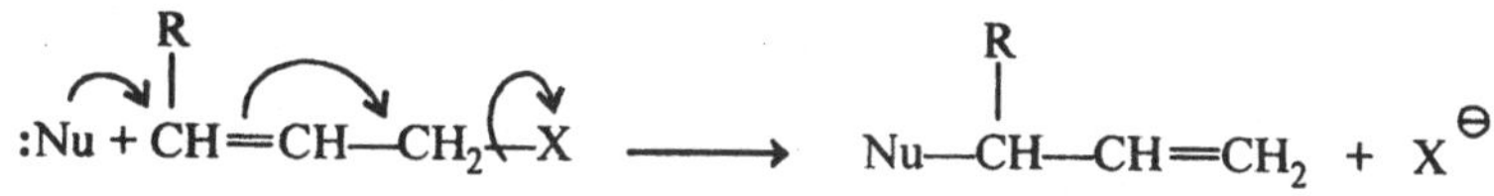

Generally the S_N2 reaction has been much faster than the S_N2 reaction. However, the S_N2' reaction may take place concurrently with S_N2 or even to exclusion of S_N2 if the α-carbon is crowded with substituents. The crowding in the α-carbon restricts the direct approach of the nucleophile to the α-carbon so that S_N2 reaction becomes significant.

Hence, α-chloroallyl chloride with sodium ethoxide in ethanol yields a normal S_N2 product and a rearranged S_N2' product as well.

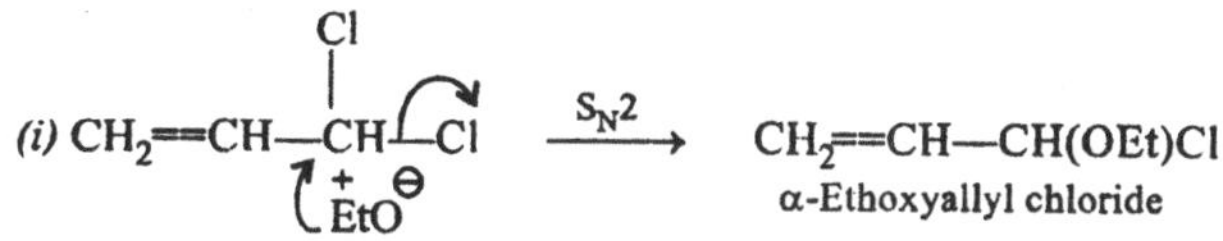

α-Ethoxyallyl chloride

(Normal S_N2 product)
30%

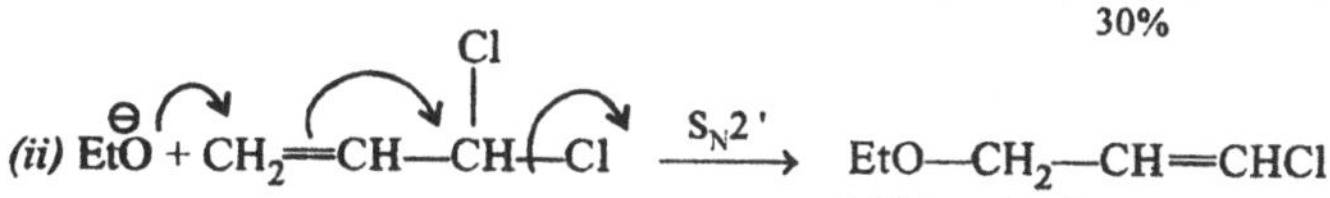

1-Chloro-3-ethoxypropene-1

(Rearranged S_N2' product)
60%

When having substituted α-carbon, the S_N2 attack is completely inhibited and only the rearranged product would be obtained (S_N2')

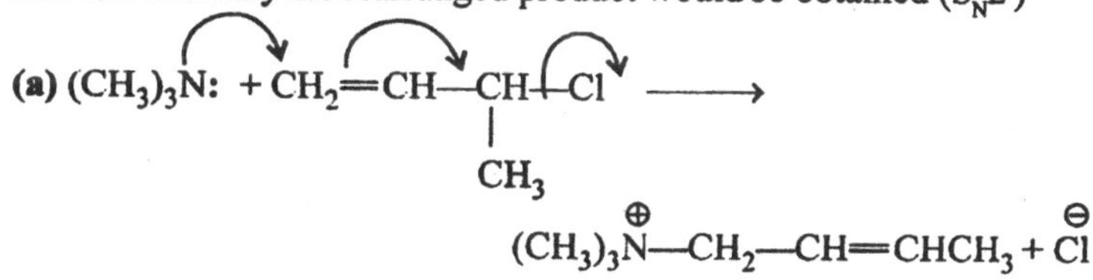

(b) $C_6H_5SNa + CH_2{=}CH{-}C(Cl)(CH_3)_2 \longrightarrow C_6H_5S{-}CH_2{-}CH{=}C(CH_3)_2 + NaCl$

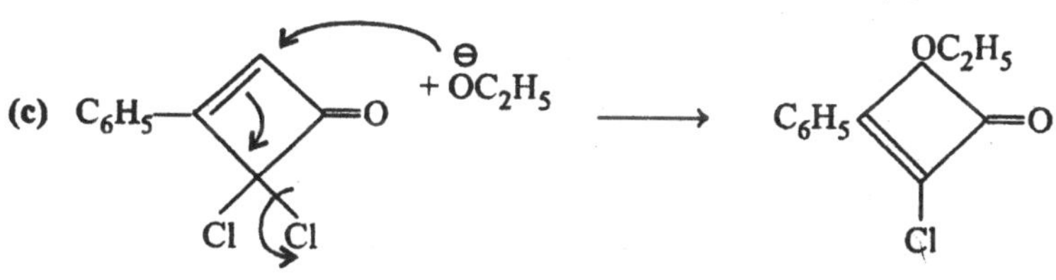

The S_N2' reaction has been found to be stereospecific. The nucleophile attacks from the same side of the molecule as the leaving group yielding the same steroisomer.

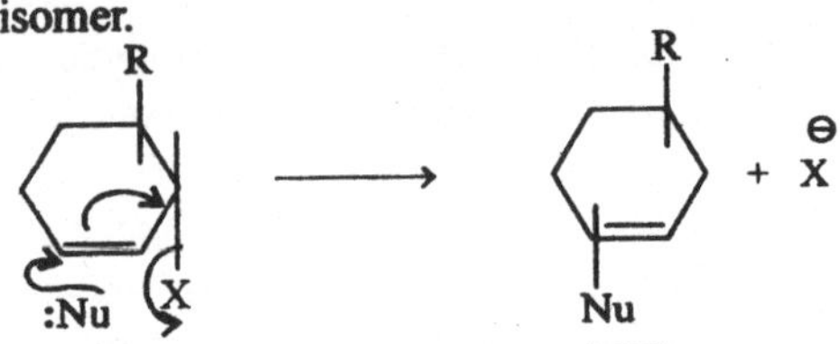

trans trans

S_Ni **type :** When treated with $SOCl_2$, a suitably substituted allyl alchol yields two halides.

$$CH_3-\underset{\substack{|\\OH}}{CH}-CH=CH_2 \xrightarrow[HCl-SO_2]{SOCl_2}$$

α-Methylallyl alcohol
(2°alcohol)

$$CH_3-\underset{\substack{|\\Cl}}{CH}-CH=CH_2 + CH_3-CH=CH-CH_2Cl$$

α-Methylallyl chloride (2°halide) Crotyl chloride (1°halide)

The formation of 2° halide with retention of configuration reveals 2 normal S_N i reaction. The formation of 1° halide is explained on the basis of the internal attack of Cl^- on the γ-carbon with consequent shift of the double bond.

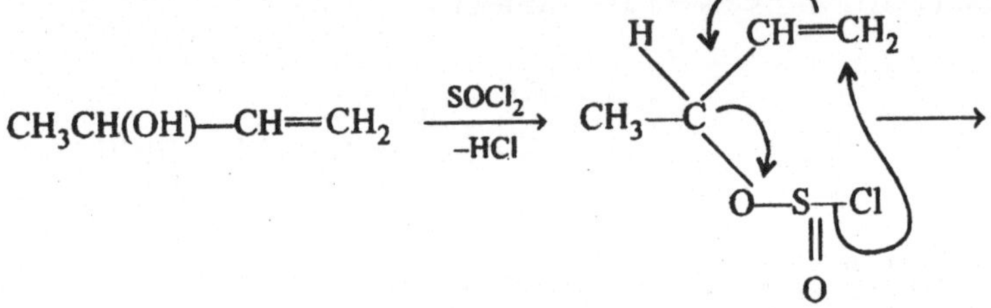

$$CH_3CH=CH-CH_2Cl + SO_2$$

As the reaction is analogous to S_Ni, but the attack taking place at the γ-carbon instead of the α-carbon, the reaction has been designated S_Ni'.

(c) Unimolecular electrophilic substitution (S_E1) *reactions* : When compounds are treated with strong bases like KOH, NaOH, etc. they undergo rearrangement in the following way.

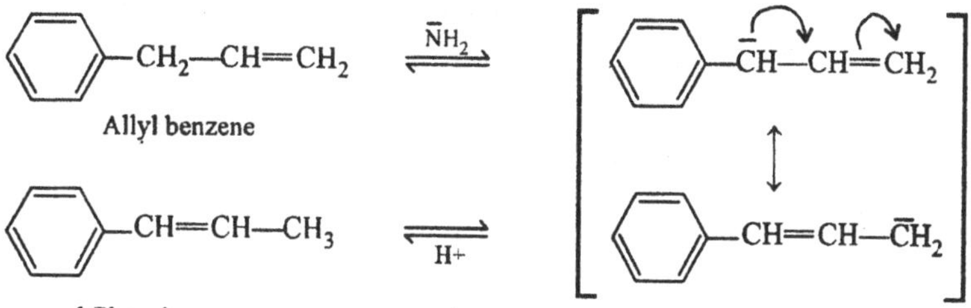

Allyl benzene

1-Phenylpropene
(formed by S_E1' mechanism)

$$CH_2{=}CH{-}CH_2{-}COOH \xrightarrow[\text{reflux}]{\text{dil. NaOH}} CH_3{-}CH{=}CH.COOH$$

Vinylacetic acid; Crotonic acid (formed by S_E1' mechanism)

$$CH_2{=}CH{-}CH_2{-}CH_2{-}CH{=}CH_2 \xrightarrow{\text{KOH}} CH_3.CH{=}CH.CH{=}CH.CH_3$$

Biallyl; 2,4-Hexadiene (formed by S_E1' mechanism)

(d) Bimolecular electrophilic substitution (S_N2) *reactions* : S_E2 reactions yield rearranged product if the allylic compound or the electrophilic reagent has bulky group, *viz.*

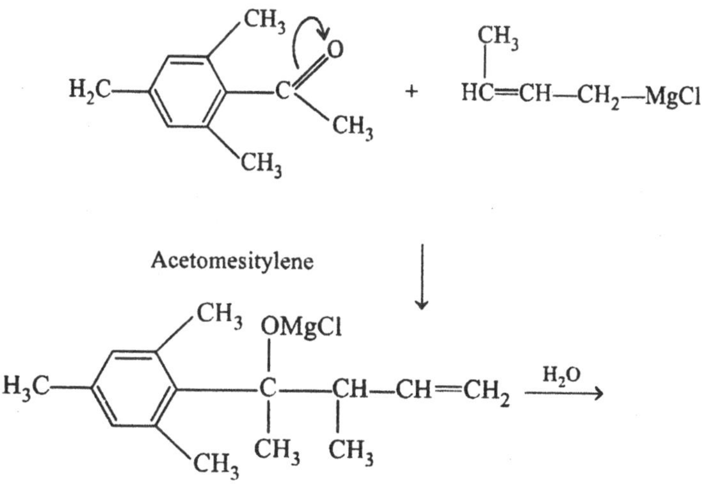

Acetomesitylene

γ-Methylallyl magnesium chloride

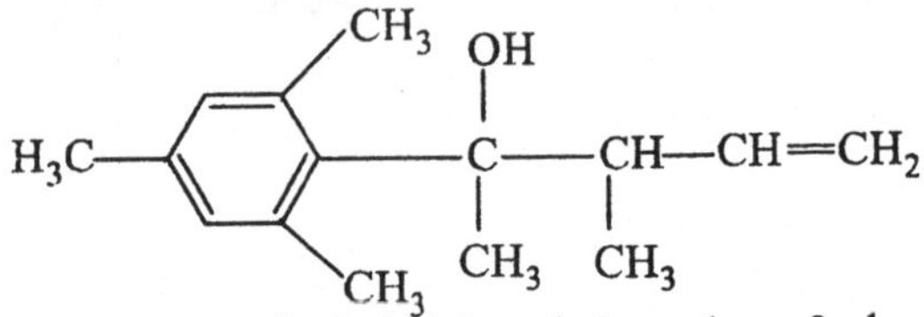

2-Mesityl-3-methylpent-4-ene-2-ol

Application : Allylic rearrangements are mainly used for converting tertiary alcohol into primary alcohol during the synthesis of several natural products, *viz.* citral, farnesol, irones, phytol, vitamin A, etc.

4

ARNDT-EISTER REACTION

This reaction is a method for converting an acid RCOOH to a homologous acid R—CH_2—COOH in which the Wolff rearrangement is the key step. In this reaction R—CO—Cl from RCOOH is converted to R—CH_2—COOH by diazomethane via formation of ∝-diazoketone.

Mechanism : The Mechanism of Arndt—Eister synthesis involves two principal steps-

(1) Wolff Rearrangement
(2) Hydrolysis

Mechanism :

Diazomethane, CH_2N_2,

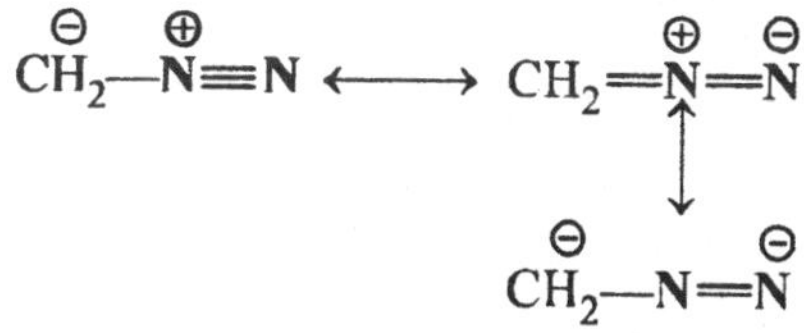

Step I.

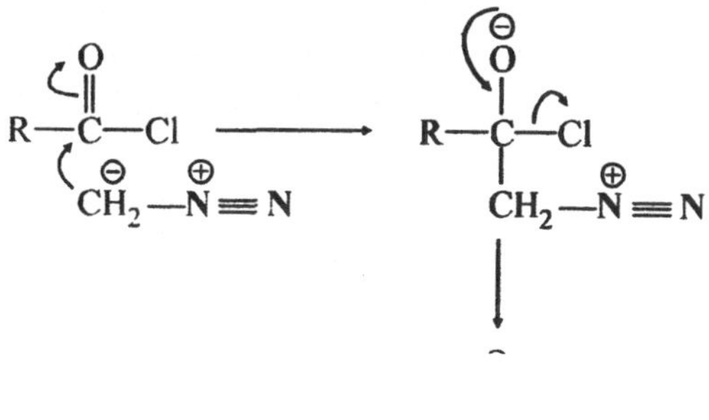

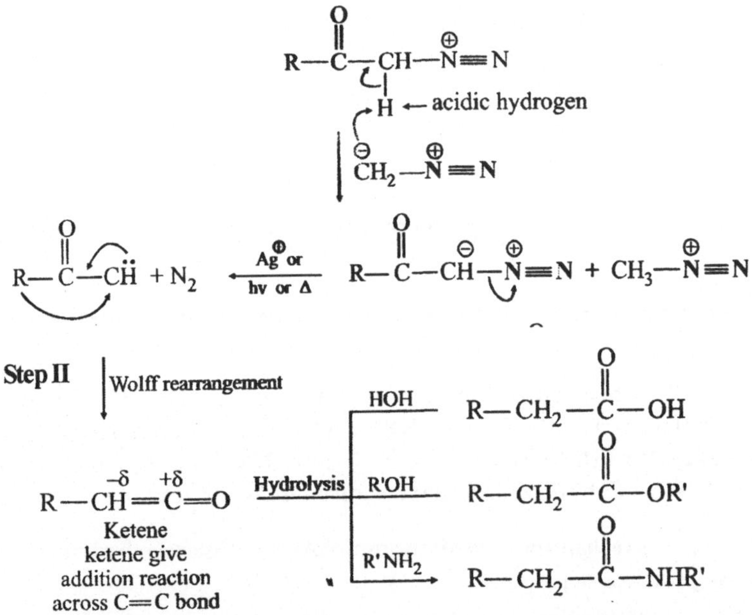

In this reaction, Wolff rearrangement is intramolecular rearrangement and carbonyl group of the product comes from the carbonyl group of the reactant.

SYNTHETIC APPLICATIONS

Arndt-Eistert homologation has been used in the synthesis of many compounds.

1. *Synthesis of acids and their derivatives* : Arndt-Eistert homologation has been used for synthesising higher acids from lower ones. This method has certain advantages over the other methods :

(i) Only a small amount of material is required.

(ii) Yields of this method are generally 50–80% of the theoretical amount.

(iii) As this method does not require reduction, it is especially useful for the compounds having reducible groups such as nitro, cyano, quinone, keto, lactone, ester, etc.

The reaction is applicable for the homologation of almost all types of acyclic, alicyclic, aromatic, carbocyclic and heterocarboxylic acids, *e.g.*,

$$R.COOH \longrightarrow R.CH_2COOH$$

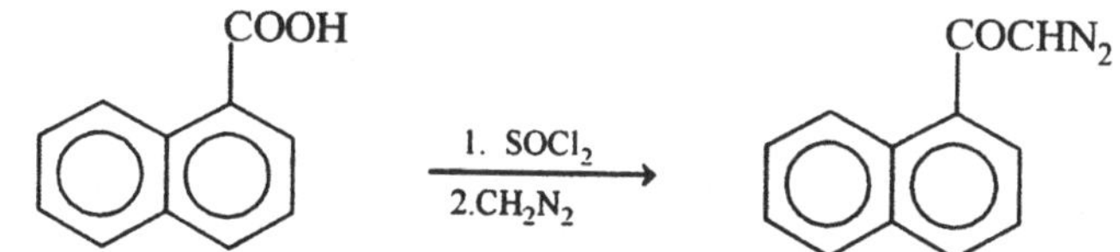

α-Naphthoic acid

Diazo-α-acetonaphthone

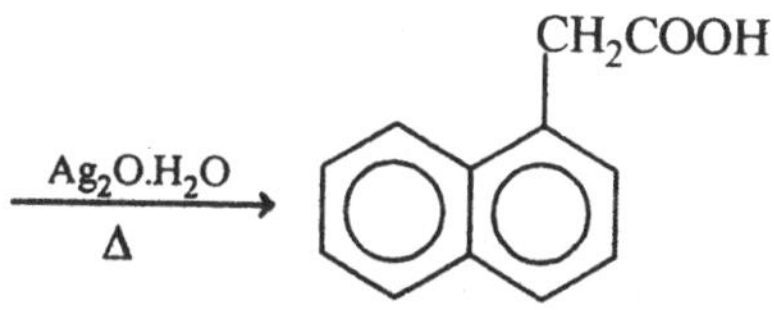

α-Naphthylacetic acid
(80%)

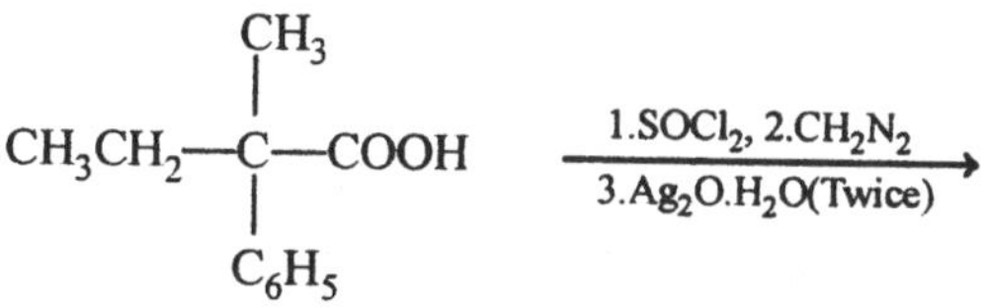

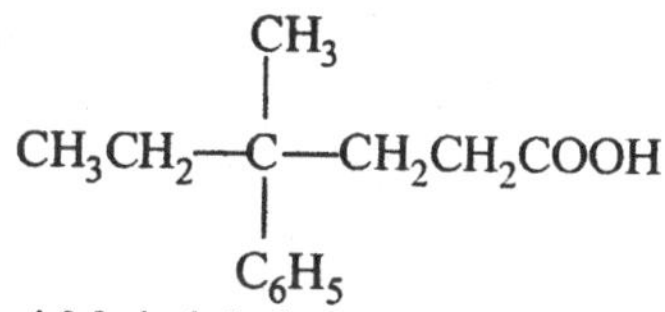

4-Methyl-4-phenylcaproic acid

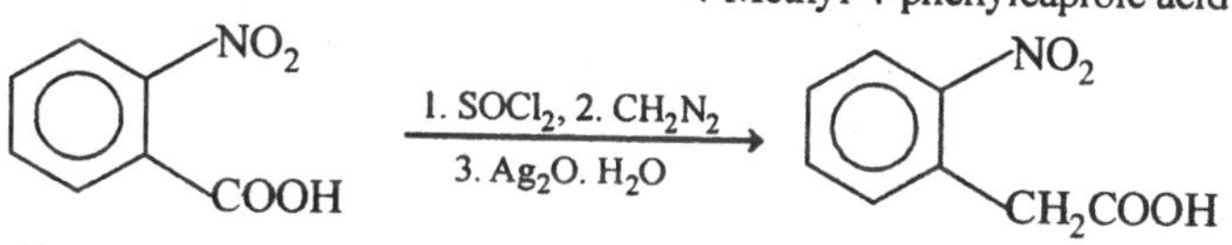

o-Nitrobenzoic acid

2-Nitrophenylacetic acid

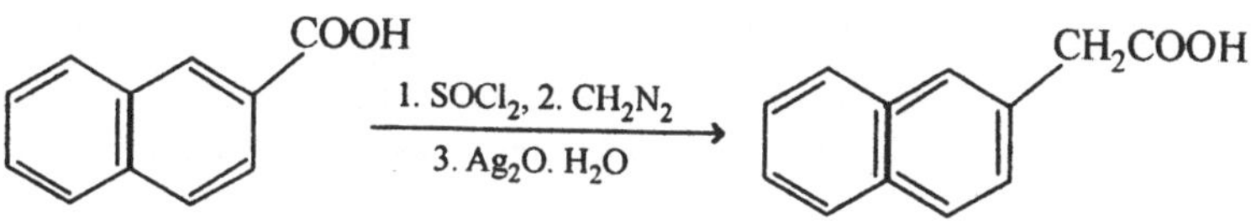

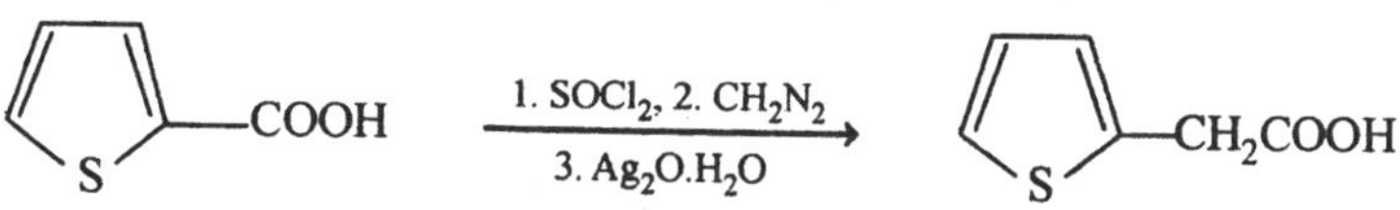

2-Thienylacetic acid

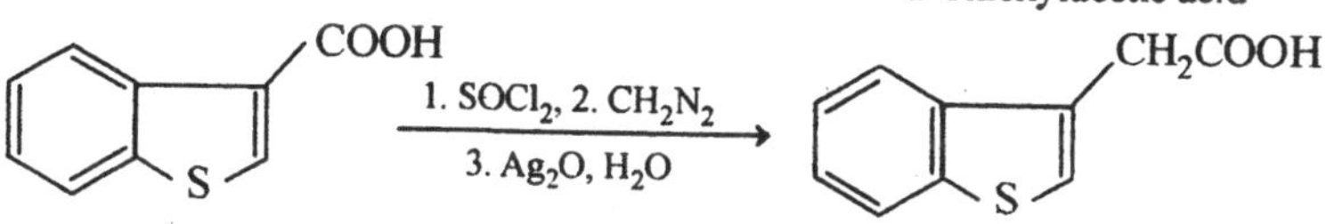

Homologation of dicarboxylic acids can also be carried out which is difficult otherwise *e.g.*

Glutaric acid — Adipic acid

It is also a method of bis-homologation :

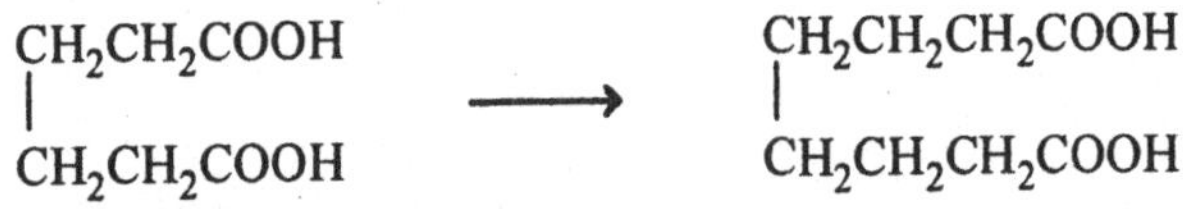

It is also applicable to unsaturated molecules. Double bond in the substrate remains unaffected. *e.g.*,

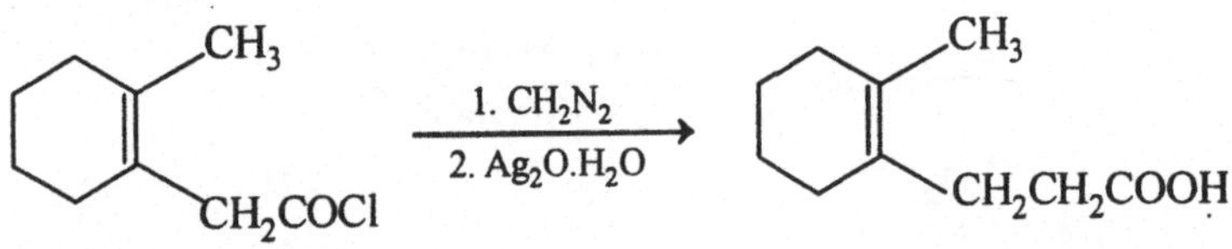

2-Methyl cyclohexenyl acetic acid chloride — 2-Methyl cyclohexenyl propionic acid

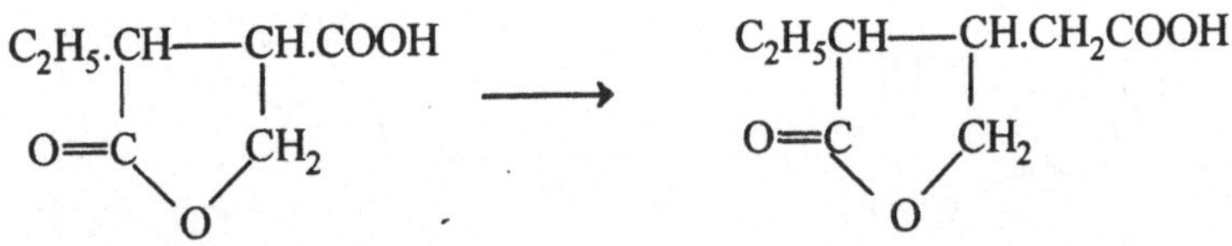

2. *Synthesis of natural product* : Arndt-Eistert homologation has been used for synthesising various natural products such as oestrone, corticoids, *e.g.* deoxycorticosterone, substituted acetophenone (starting material for the synthesis of anthocyanidins), mescaline (alkaloid), etc.

(a) Arndt-Eistert homologation is used in the synthesis of deoxycorticosterone.

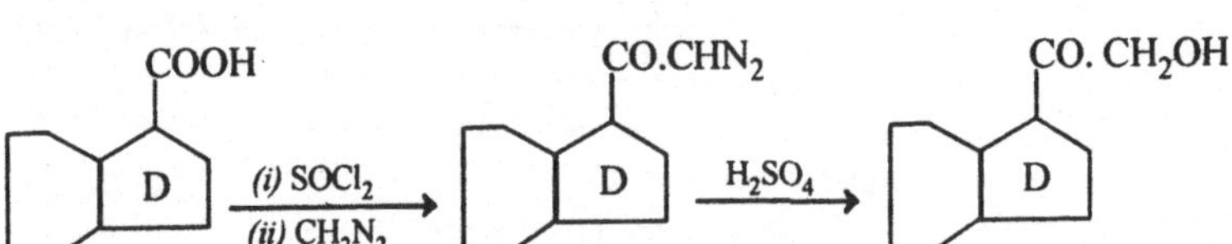

(b) Synthesis of ω, 3, 4-trimethoxyacetophenone by Arndt-Eistert homologation is done as follows :

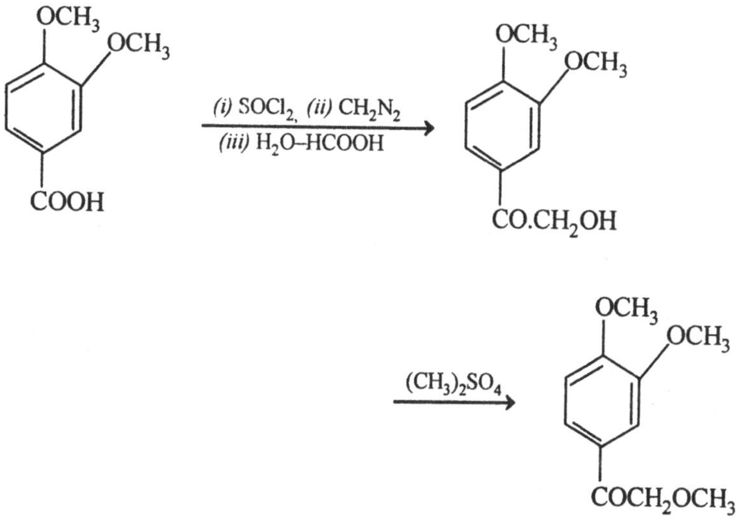

(c) Synthesis of oestrone by Arndt-Eistert homologation is done as follows :

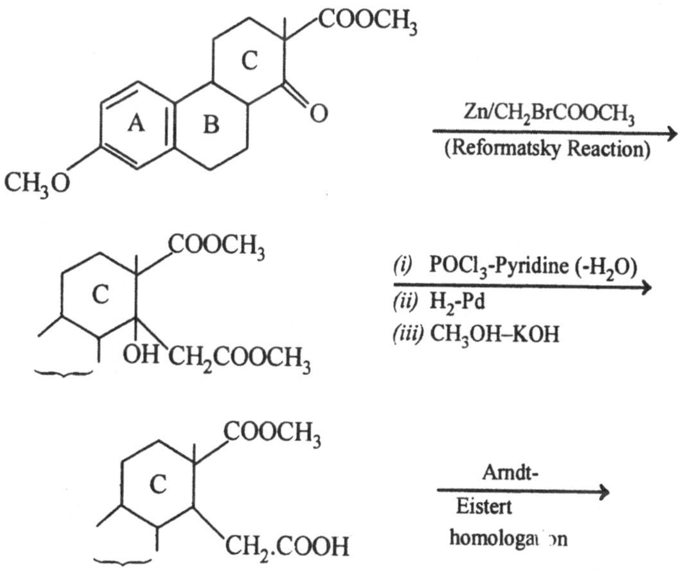

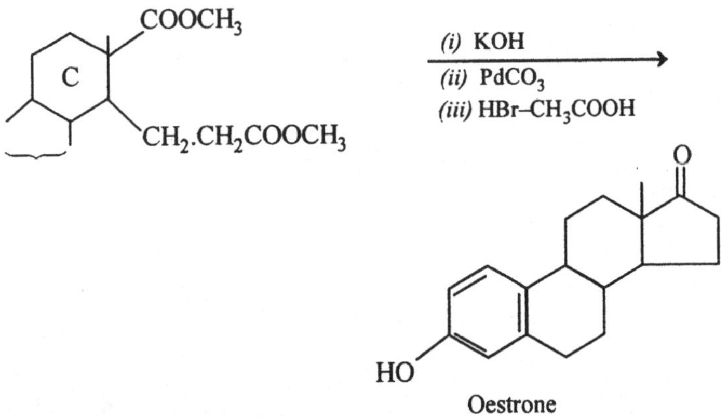

(d) Synthesis of mescaline (mezcaline) by Arndt-Eistert homologation is done as follows :

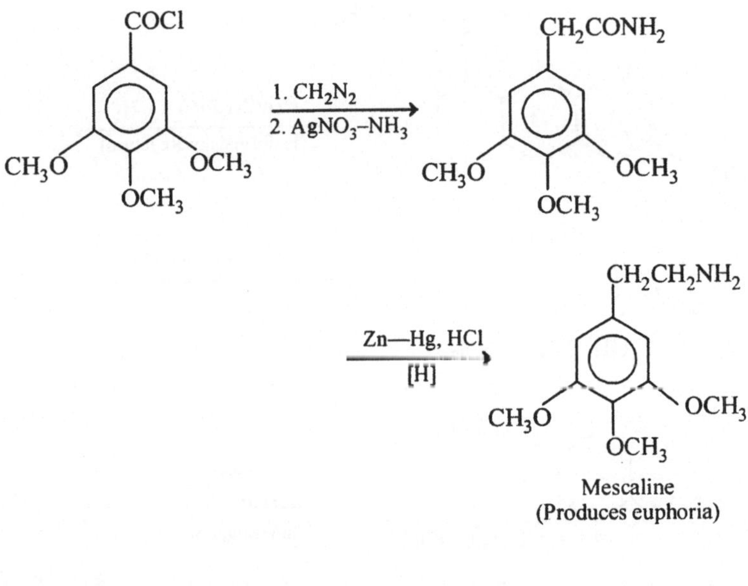

(e) Synthesis of homoveratroy! **chloride, an intermediate for** papaverine synthesis :

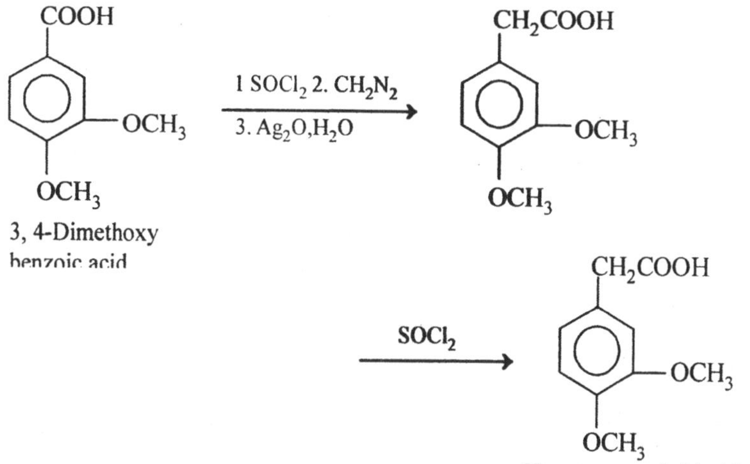

3, 4-Dimethoxy benzoic acid

Homoveratroyl chloride

3. *Synthesis of hydroxy ketones* : **Diazoketones on treatment with** aqueous formic acid give hydroxy **ketones.**

$$RCOCHN_2 \xrightarrow{HCOOH + H_2O} RCOCH_2OH_2$$

PROBLEMS BASED ON ARNDT-EISTER REACTION

Problem-1

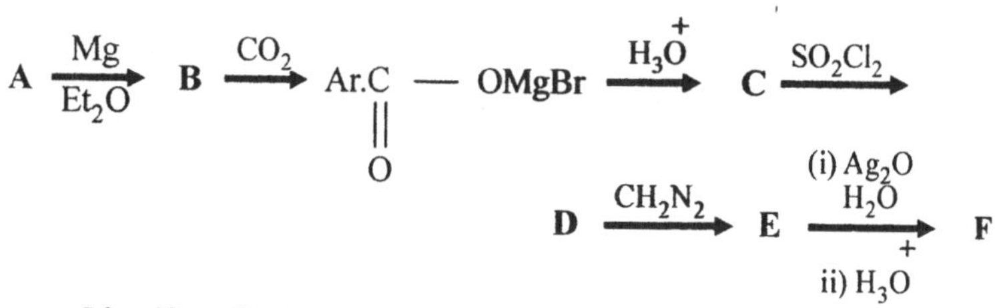

Identify **A, B, C, D, E & F.**

Answer	**A**	Ar. Br	**Phenyl bromide**
	D	Ar. COCI1	**Benzoyl chloride**
	B	Ar. MgBr	**Phenylmagnesium bromide**

E	Ar. CO. CHN_2	Phenyldiazoketone
F	Ar. CH_2CO_2H	Phenylacetic acid
C	Ar. CO_2H	Benzoic acid

Problem-2

$$\underset{[A]}{C_4H_6O_4} \xrightarrow{PCl_5} \underset{[B]}{C_4H_4O_2Cl_2} \xrightarrow{CH_2N_2} \xrightarrow[H_2O]{Ag_2O} \xrightarrow{H_3O^+} \underset{[C]}{C_4H_{10}O_4}$$

C on heating prouduces cyclopentanone. Identify **A, B, C.**

Answer

$$\underset{[A]}{[C_2H_4]\begin{matrix}-CO_2H\\-CO_2H\end{matrix}} \xrightarrow{PCl_5} \underset{[B]}{[C_2H_4]\begin{matrix}-COCl\\-COCl\end{matrix}} \xrightarrow{CH_2N_2} [C_2H_4]\begin{matrix}-CO.CHN_2\\-CO.CHN_2\end{matrix}$$

$$\xrightarrow{Ag_2O\ /\ H_2O} \xrightarrow{H_3O^+} \underset{[C]}{[C_2H_4]\begin{matrix}-CH_2.CO_2H\\-CH_2.CO_2H\end{matrix}}$$

A	Succinic acid
B	Succinyl chloride
C	Adipic acid

Problem-3

$$A \xrightarrow{Br_2} B \xrightarrow[Et_2O]{Mg} C \xrightarrow{CO_2} D \xrightarrow{H_3O^+} E \xrightarrow{SOCl_2} F \xrightarrow{CH_2N_2} G$$

$$\xrightarrow[H_2O]{Ag_2O} \xrightarrow{H_3O^+} H \xrightarrow[\Delta]{Ac_2O} \text{Cyclopentanone}$$

A absorbs one molar equivalent of bromine

Answer

HINTS

A absorbs 1 mole Br_2 per mole of **A.**

Therefore **A** contains one ethylenic bond [C=C] which adds to bromine to form a dibromide that reacts with Mg/Et_2O to form a Grignard complex.

Answers

A $CH_2=CH_2$ Ethylene

B CH_2——CH_2 Ethylene dibromide
| |
Br Br

C CH_2——CH_2 Ethylene magnesium dibromide
| |
MgBr MgBr

D CH_2——CH_2 Ethylene carbonyl magnesium dibromide
| | CO_2MgBr
CO_2MgBr

E CH_2——CH_2 Succinic acid
| |
CO_2H CO_2H

F CH_2——CH_2 Succinyl chloride
| |
COCI COCI

G CH_2——CH_2 Ethane-1,2 - diazoketone
| | $CO.CHN_2$
$CO.CHN_2$

H CH_2——CH_2
| | $CH_2.CO_2H$
$CH_2.CH_2H$

5

BECKMANN REARRANGEMENT

A unique product of 1:2-shift, Beckmann Rearrange ment has its good many synthesis applications.

REACTION

The transformation of oximes into acid amides in presence of strong acids

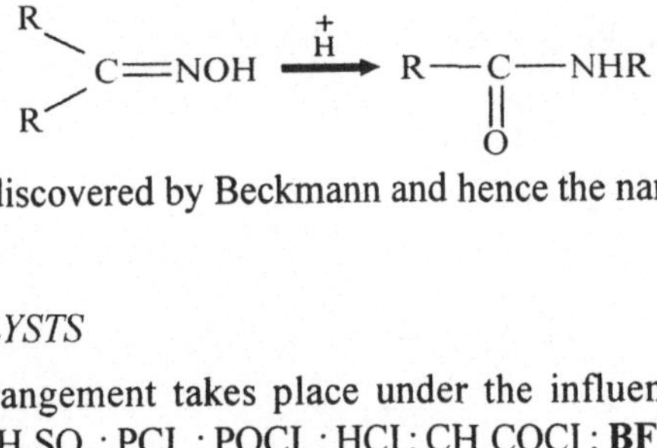

was first discovered by Beckmann and hence the name Beckmann Rearrangement.

CATALYSTS

The rearrangement takes place under the influence of certain catalysts : conc. H_2SO_4 ; PCI_5 ; $POCI_3$; HCI ; CH_3COCI ; $\mathbf{BF_3}$; $SbCl_5$, HCl–HOAc–Ac_2O ; polyphosphoric acid etc.

However, the presence of catalyst is not absolutely essential. For example, oxime picrates rearrange on heating.

MECHANISM

[I] *Generation of Electrophile*

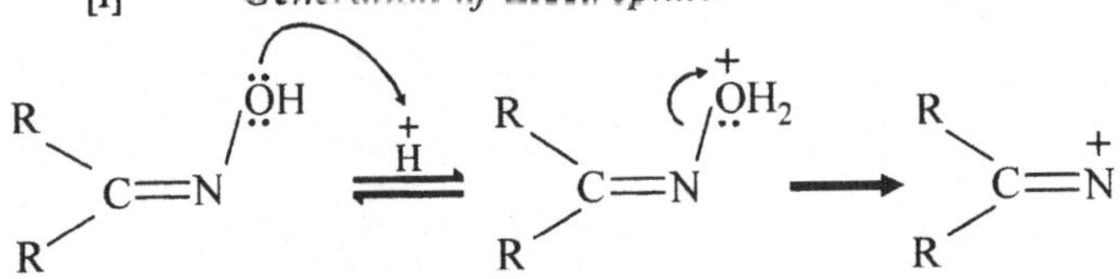

[II] *1:2-Shift*

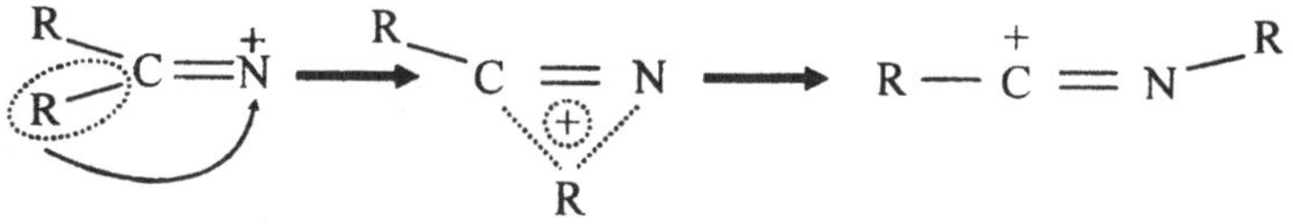

[III] *Hydration*

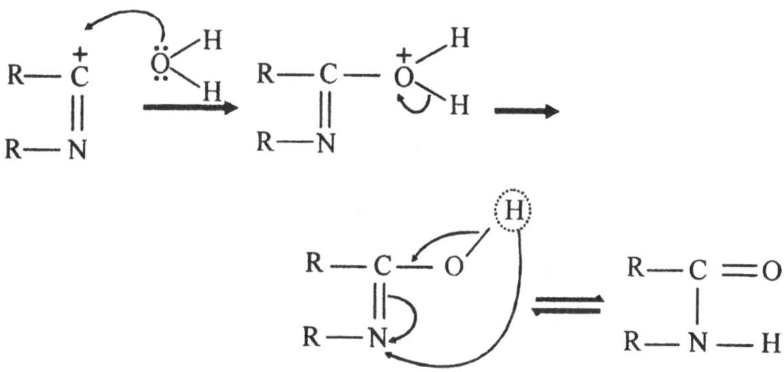

Isotopic Substitution Method : Simple interchange of R and OH does not take place, it was confirmed by carrying out Beckmann Reaction in $H_2\,^{18}O$. It is found that had there been intramolecular exchange of R and OH group $Ph\overset{\parallel}{\underset{^{18}O}{C}}NH\,Ph$ would have not been isolated. This means that in one step OH is removed and in other $H_2\,^{18}O$ is added. Actually, it is found that resultant product contain same proportion of ^{18}O as the original water. So the rearrangement must involve reintroduction of oxygen from water molecule. The main function of the acidic catalyst is to convert OH of oxime to better leaving group by protonation.

Beckmann reaction is highly stereospecifie in that the migrating group always approaches the nitrogen atom on the side opposite to the oxygen atom.

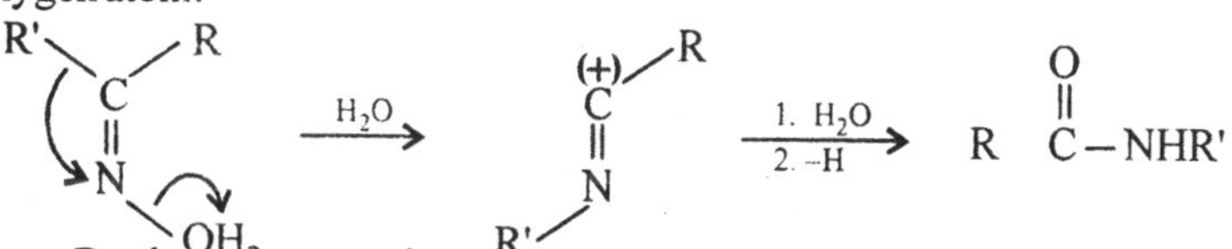

Beckmann reaction is used to synthesise Nylon-6 from cyclohexanone oxime via the cyclic amide.

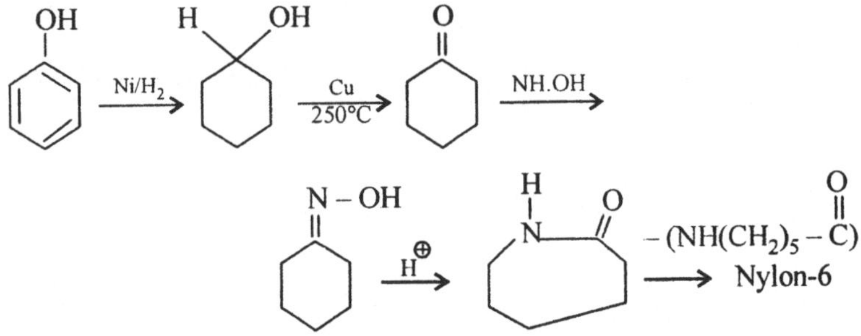

Mechanism:

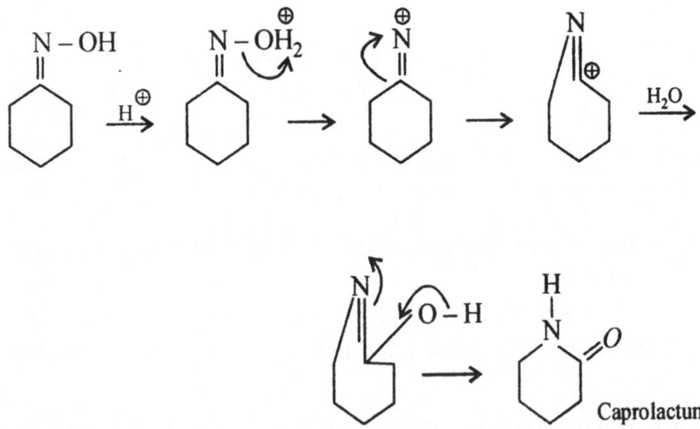

Caprolactum

Beckmann reaction normally required acid which react with hydroxyl group of oxime, makes $OH_2^{\oplus}$ which is better leaving group than–OH. But there are some oxime which undergo this reaction without acid. Such an example was studied by Kuhera et. all (1914) and Chapman (1934). Kuhera prepare the benzene sulphonate of benzophenone oxime and he showed that this oxime undergoes reaction very rapidly in the neutral solvents.

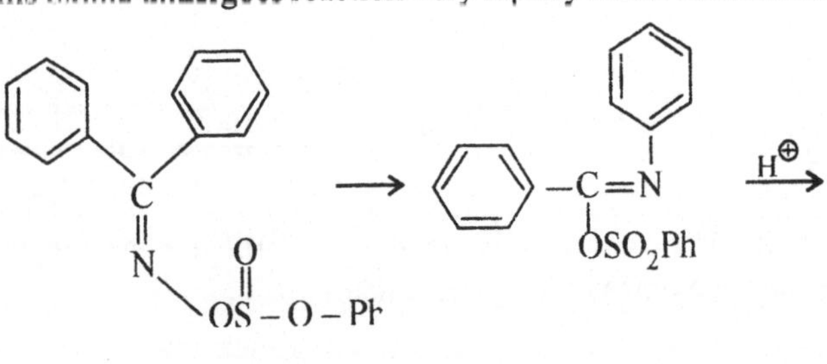

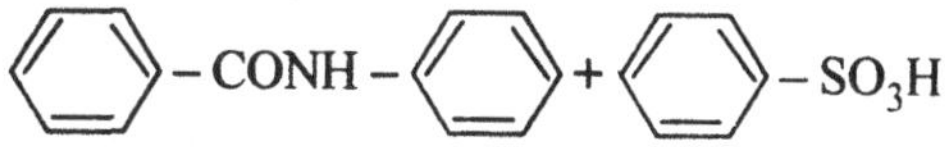

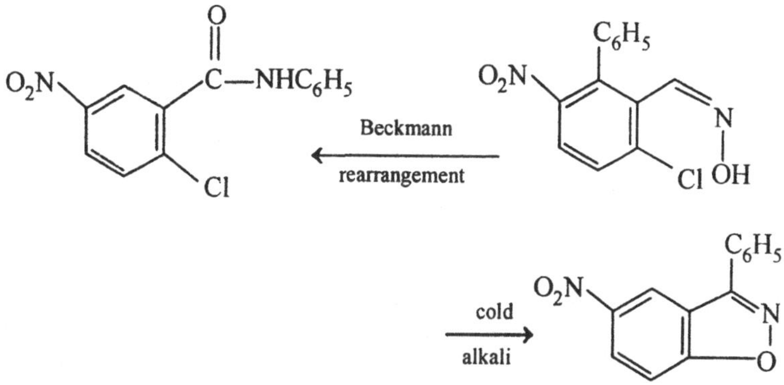

As the *anti* group always migrates, it is evident that either the breakage of the N–O bond and the group migration are synchronous or these two steps may be following each other quickly. There is so much certainty in the migration of the *anti* group in the Beckmann rearrangement that it is normally the configuration of a particular oxime which can be established readily by identifying its rearrangement products.

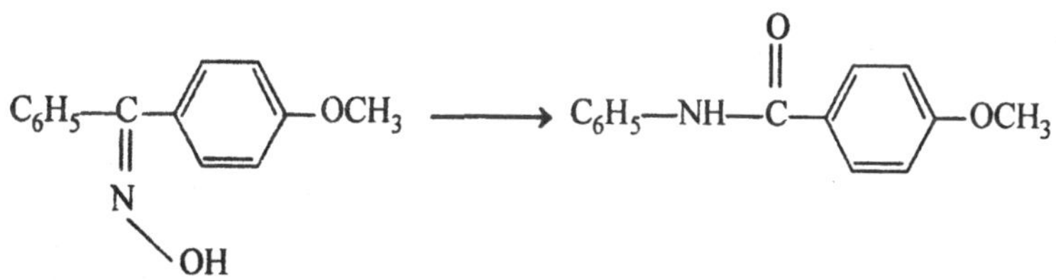

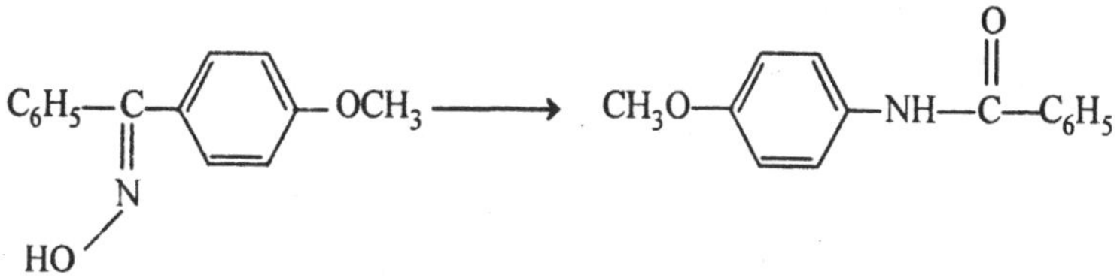

In the Beckmann rearrangement the carbon atom of the migrating group *retains its comfiguration.* By carrying out the various series of reactions, Kenyon proved retention of configuration as well as correlated Beckmann and Hofmann rearrangements. In the series of experiments optically active 3-heptylcarboxylic acid after converting into amide was subjected to Hofmann rearrangement. In another series of experiments the same acid was first converted into 3-heptylmethyl ketone and then the oxime of this ketone was subjected to the Beckmann method. Both these rearrangements gave the same optical isomer of 3-heptylamine.

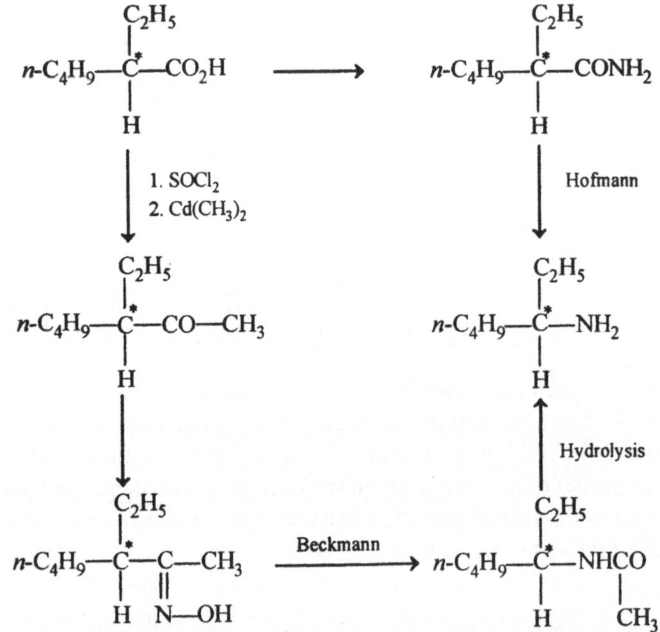

But the retention of configuration at the migrating carbon is well ascertained in the Hofmann rearrangement. Therefore, it means that the Beckmann rearrangement also takes place with *retention of configuration.* Furthermore the migrating group does not become free from the remainder of the molecule and it is evident that the breakage of C–C bond and the formation of the new C–N bond occur synchronously.

Mechanism :

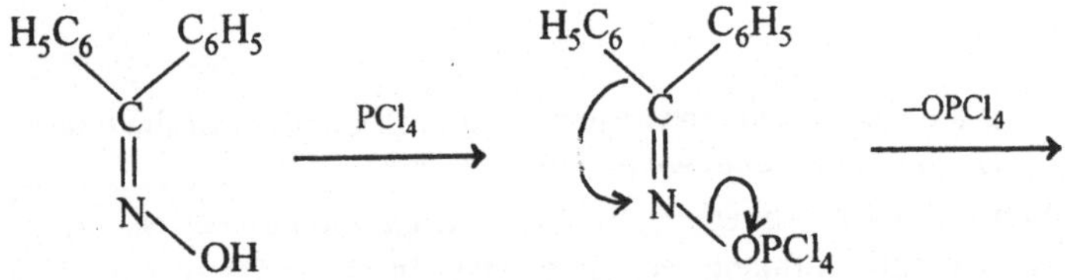

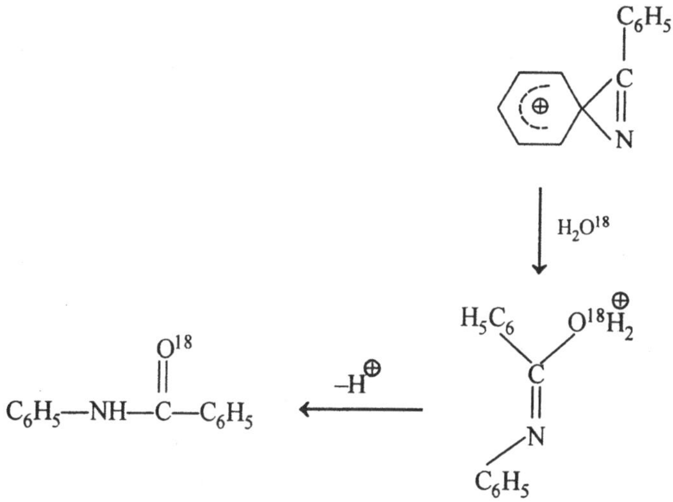

When the Beckmann rearrangement is carried out in the presence of strong aids, the reaction gets initiated by protonation of the oxime followed by loss of water to form the same species as formed by the decomposition of ester.

Aldoximes have also been shown to undergo this rearrangement. For example, *syn*-and *anti*-benzaldoximes undergo Beckmann rearrangement in the presence of PPA.

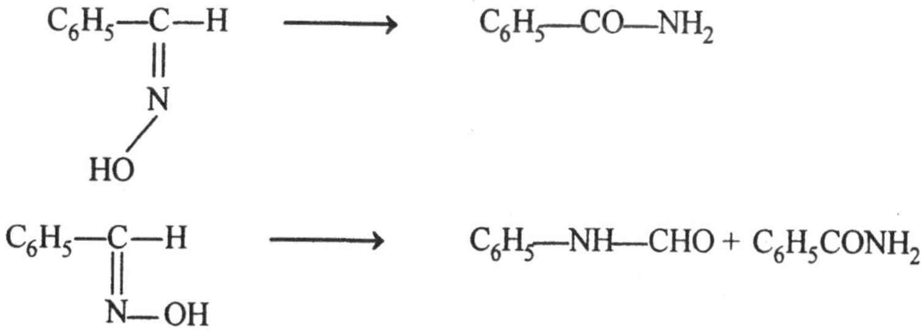

The reason for the formation of benzamide in the rearrangement of the *syn*-isomer is assigned to the ready conversion of some of the *syn*-form into the *anti* which then gets rearranged with a hydrogen migration.

APPLICATIONS

(i) *Determination of configuration of ketoximes* : It is the most useful and general application of the Beckmann rearrangement. Its principle is based on the fact that the two isomeric ketoximes, *viz*. *syn* and *anti*, give different amides when undergo Beckmann rearrangement. The amides so obtained in turn can be characterised by their hydrolysis products.

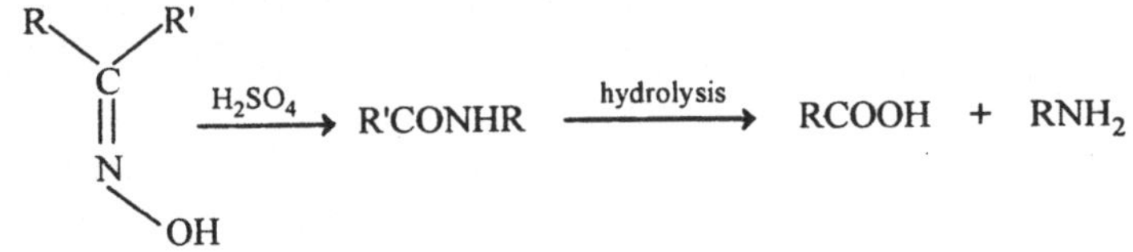

(ii) *Synthesis of isoquinoline* : The conversion of cinnamaldoxime into isoquinoline by heating with phosphorus pentoxide may be assumed to involve the Beckmann rearrangment which is followed by cyclisation

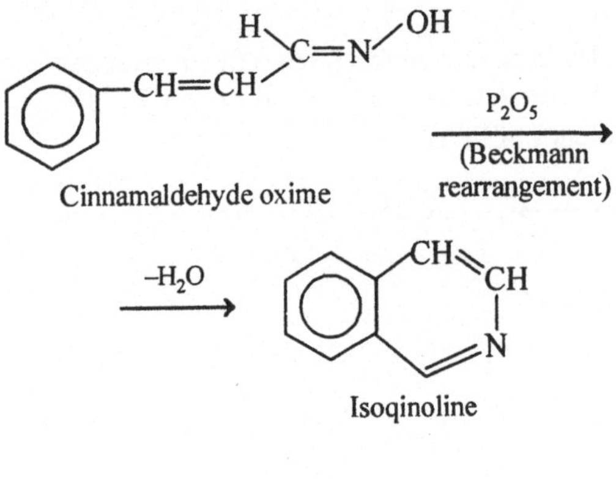

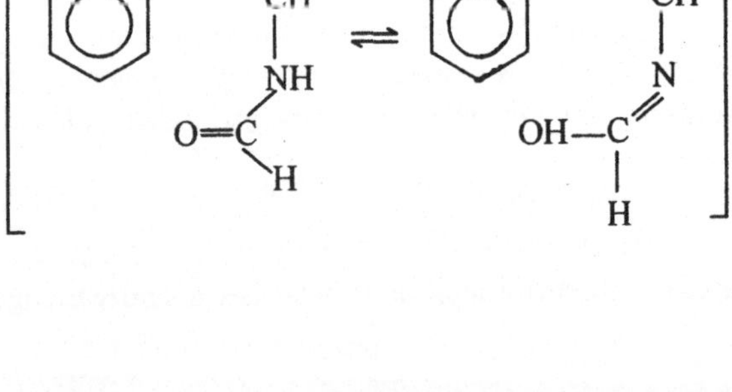

(iii) *Synthesis of ε-caprolactam* : It is an interesting application of the rearrangement by which ε-caprolactam is synthesised from cyclohexanone oxime and concentrated sulphuric acid. ε-Caprolactam is used as the raw material for nylon type polymers and on heating it can be polymerised.

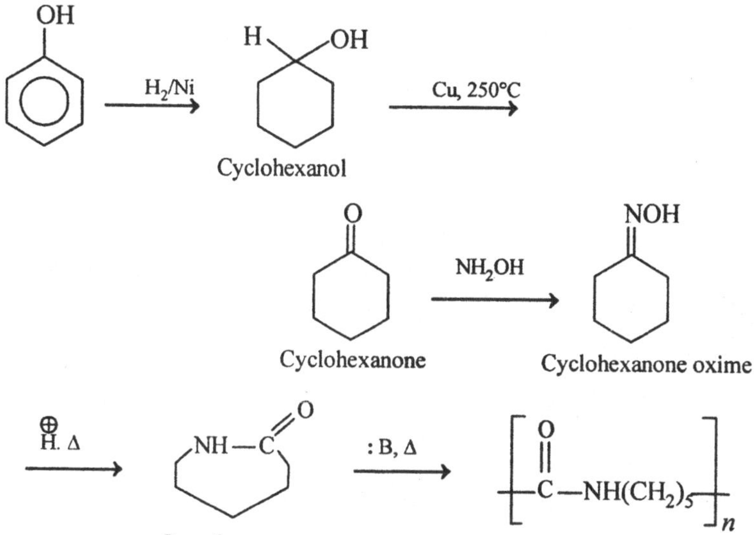

Cyclohexanol

Cyclohexanone

Cyclohexanone oxime

ε-Caprolactam

Perlon

Similarly, cyclopentanone oxime yield 2-piperidone under Beckmann condition. Aldoximes under the Beckmann reaction conditions undergo dehydration to yield nitriles.

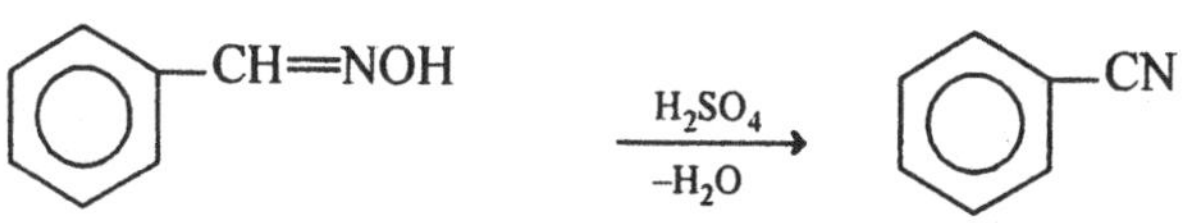

Benzaldoxime

Benzonitrile (Phenyl cyanide)

(iv) 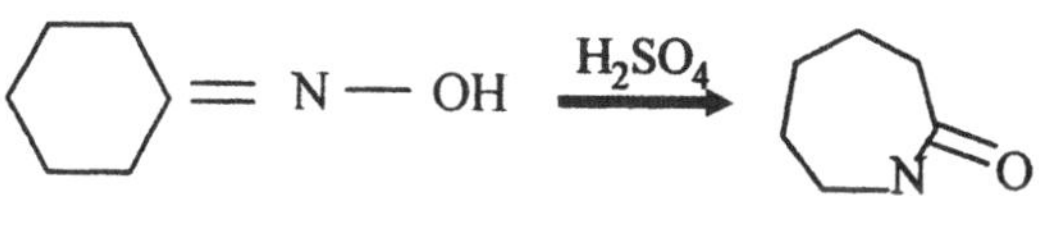

cyclohexanone oxime

ε - caprolactam

(v) 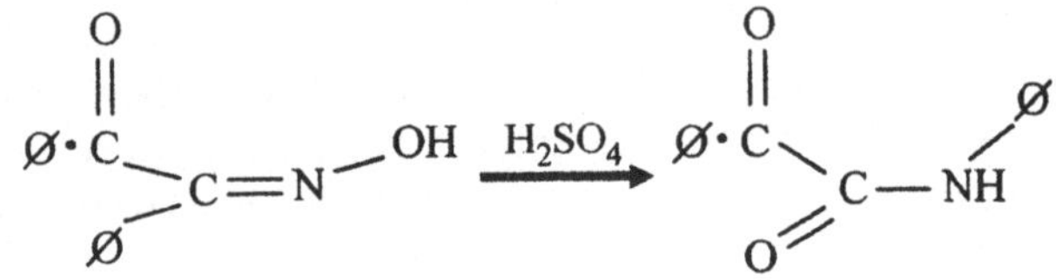

PROBLEMS BASED ON BECKMANN REACTION

Problem 1. Rate of Beckmann reaction is fast when strong acids are employed. Write following in order of preference for using in the Beckmann reaction.

$ClCH_2COOH$, $PhSO_3H$, CH_3COOH

Ans. Order of acidic strengths.

$CH_3COOH < ClCH_2COOH < PhSO_3H$ stronger the acid more capable is the anion for independent existance, the order of capability of independent existance is $CH_3COO^{\ominus} < ClCH_2COO^{\ominus} < PhSO_3^{\ominus}$. So $PhSO_3H$ will easily do the Beckmann reaction.

Problem 2. What is nitrilium ion?

Ans. $R-C\equiv N^{\oplus}-R$ ion is called as nitrilium ion.

Problem 3. Carry out following transformation :

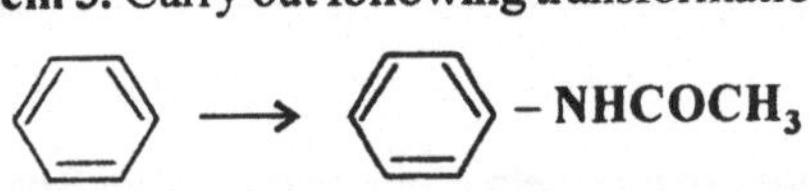

Ans.

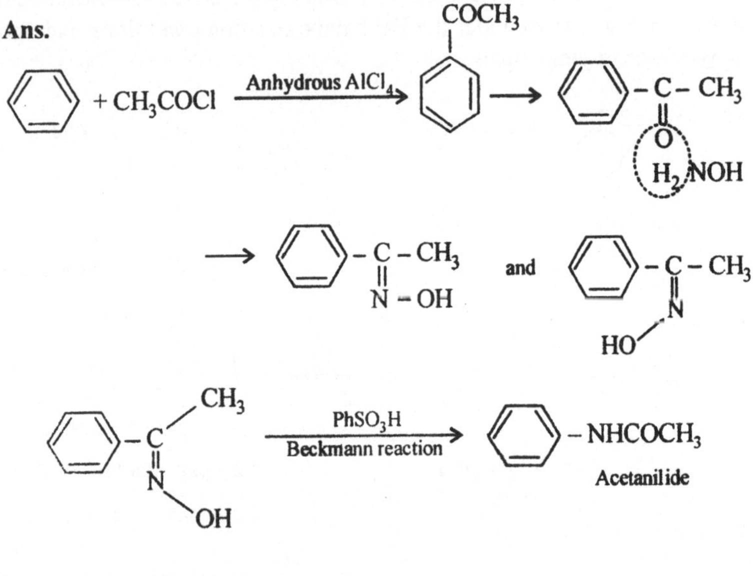

Problem 4. In Beckmann reaction, anti-migration take place. Elaborate this statement.

Ans. Anti-migration keeps the migrating and departing group well apart so that repulsion are minimised on the contrary the approach to the transition state for syn migration forces the migrating and departing group into close cantact causing repulsion and raising energy of the transition state.

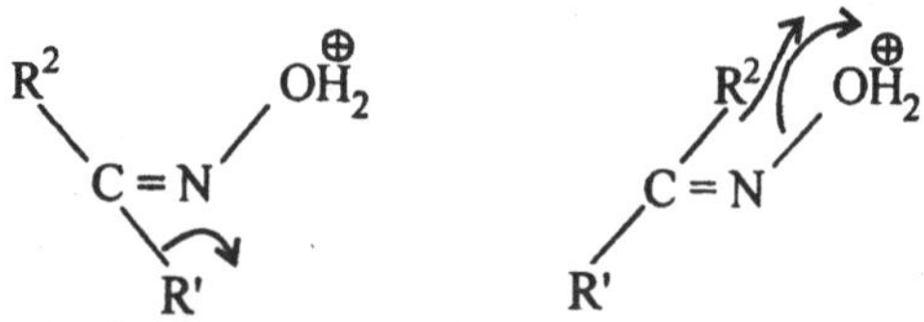

Anti-migration Syn migration

Problem 5. In the Beckmann reaction, migrating group is chiral. Will it retain or change its configuration while migration?

Ans. It will retain its configuration while migration.

Problem 6. Complete the following reaction:

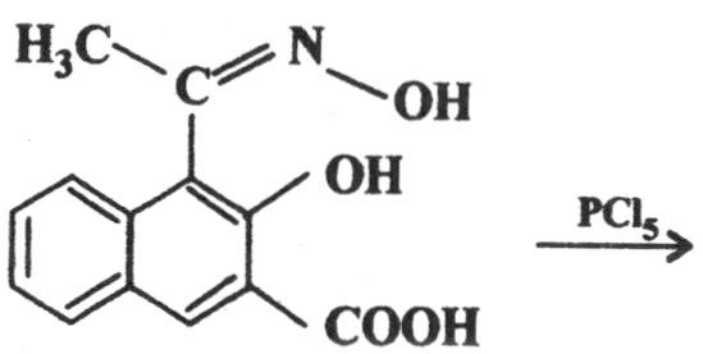

Ans.

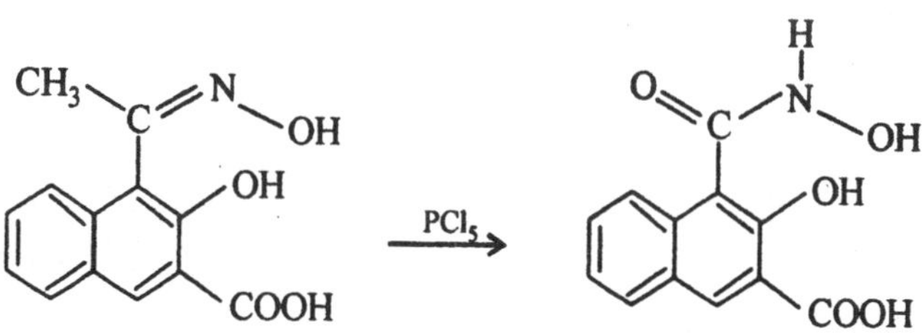

Beckmann reaction

In the above equation CH_3 migrates as it is present anti to–OH group.

Problem 7. Write a chemical reaction for reduction of caprolactum by LAH.

Ans.

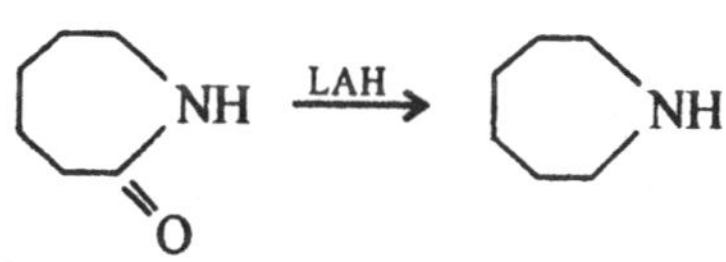

Caprolactum

Problem 8. Show how Beckmann reaction can be used to degrade an acid by one carbon atom.

Ans.

1. $RCH_2CH_2CH_2COOH \xrightarrow{SOCl_2} RCH_2CH_2CH_2COCl + SO_2 + HCl$

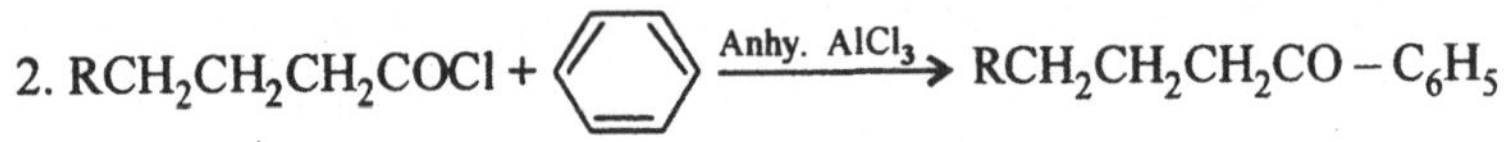

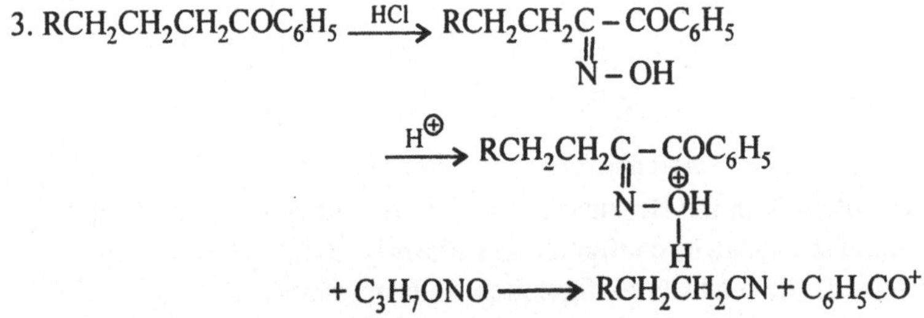

4. $C_6H_5CO^{\oplus} \xrightarrow{H_2O} C_6H_5COOH + H^{\oplus}$

5. $RCH_2CH_2CN \xrightarrow{H_3O^{\oplus}} RCH_2CH_2COOH$

Third step include second order Beckman reaction. The mechanism of second order Beckmann reaction may be :

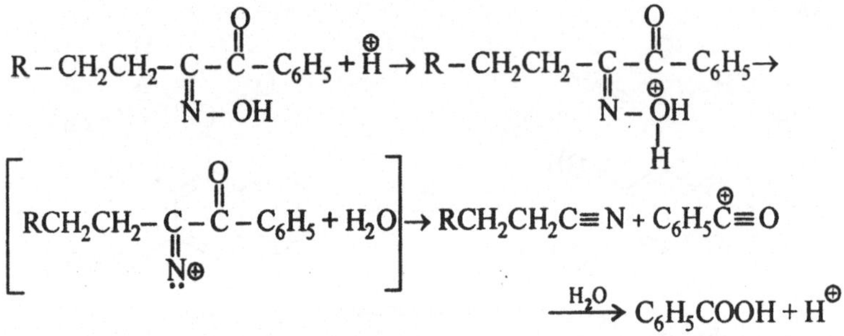

Problem 9. Complete the following :

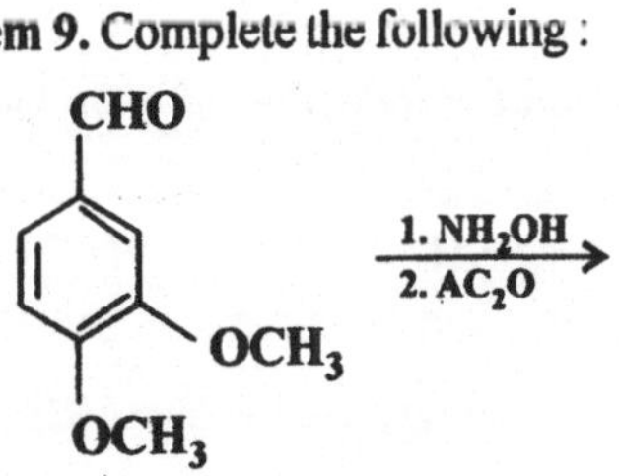

Ans.

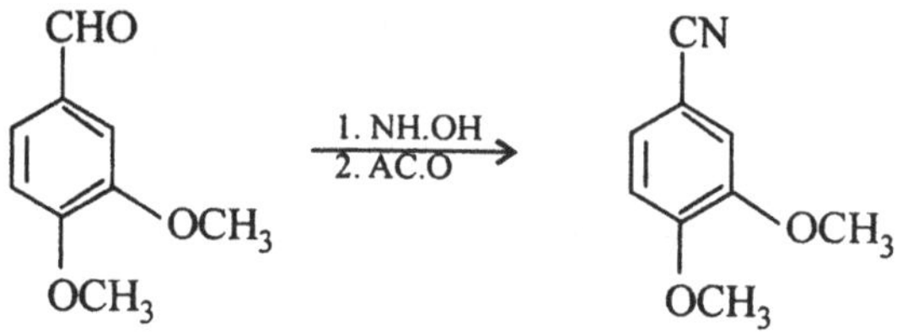

Problem 10. Write the mechanism of Beckmann reaction when PCl_5 is used as Lewis acid.

Ans.

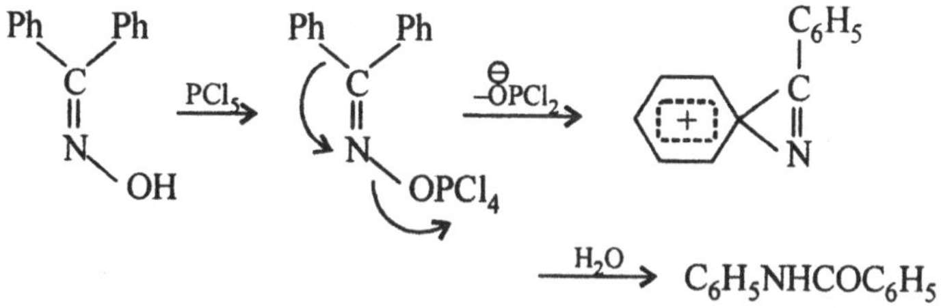

Problem 11. In Beckmann reaction, carbon atom of migrating group retains its configuration. Explain by taking suitable example.

Ans. Optically active 2-ethyl hexanoic acid was converted to 3-ethyl 2-heptanone and then the oxime of this ketone was rearranged by Beckmann reaction.

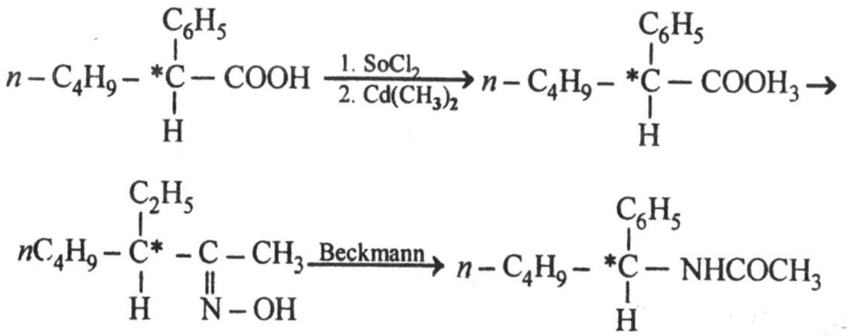

Problem 12. Suggest a mechanism for the conversion of oximes to alpha amino ketones.

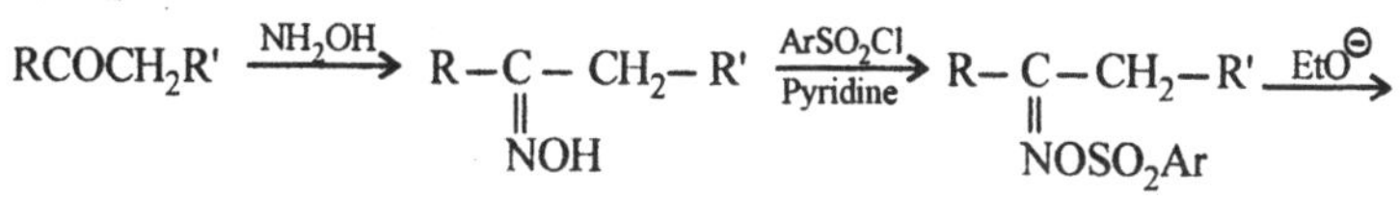

Ans.

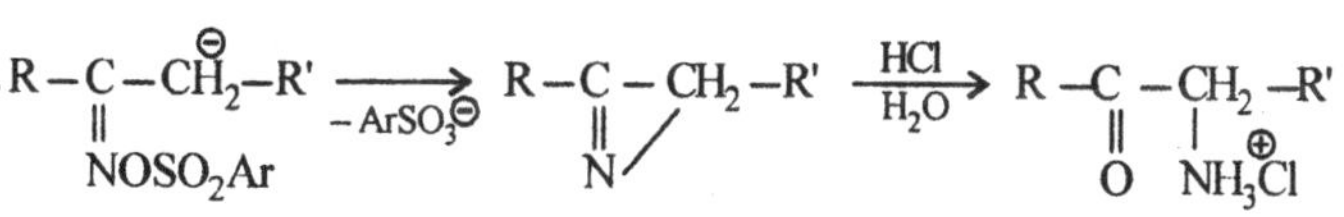

This reaction is an example of neber rearrangement which is used to prepare α-amino ketones.

Problem 13. Write the product of the reaction :

$$CH_3-CH_2-\underset{\underset{N-OTS}{\|}}{C}-CH_3 \xrightarrow{OEt^{\ominus}}$$

Ans. $CH_3CH_2\underset{\underset{N-OTs}{\|}}{C}-CH_3 \xrightarrow{OEt^{\ominus}} CH_3\underset{\underset{NH_2}{|}}{CH}-\underset{\underset{O}{\|}}{C}-CH_3$

Problem 14. Draw the structure of azirine intermediates. Give one example of reaction in which these are formed.

Ans. 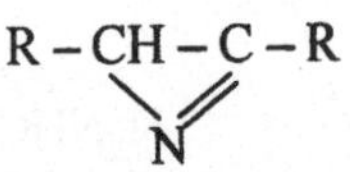

is the azirine intermediate. These are formed in neber rearrangement.

Problem 15. How Neber rearrangement differ from Beckmann rearrangement?

Ans. In neber rearrangement both anti as well as syn keto-oxime gives the same product.

Problem 16. Discuss the synthesis of amino ketones by the Baumgarten modification of the neber reaction.

Ans.

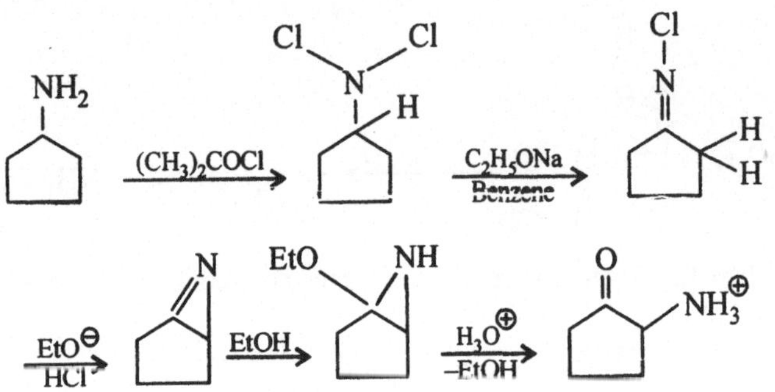

6

BENZIDINE REARRANGEMENT

INTRODUCTION

Benzidine rearrangement includes the whole family of acid-catalysed rearrangements of aromatic hydrazocompounds. The name of the rearrangement has been derived from the major product, benzidine (4,4′-diaminodiphenyl) which is formed when hydrazobenzene (1, 2-diphenylhydrazine) is warmed with dilute HCl or H_2SO_4. The other product of the rearrangement is the isomeric diphenylene (2, 4′-diaminodiphenyl).

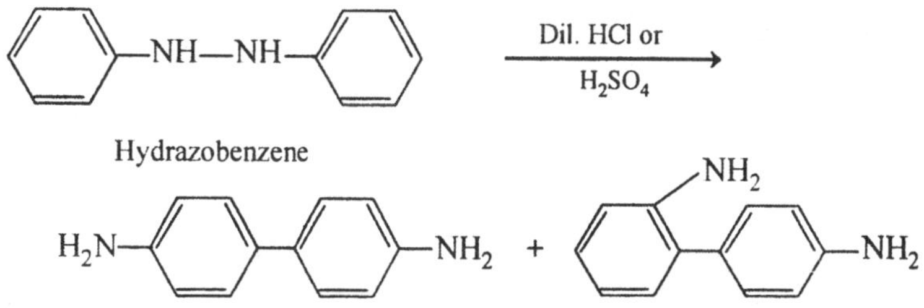

Benzidine (70%) Diphenylene (30%)

In addition to these two products, less frequently, three other products, namely *o*-benzidine and *o*-semidine and *p*-semidine may also be formed in smaller amounts. For example, when solid hydrazobenzene is treated with dry hydrogen chloride, all the possible products are obtained. Thus, the types of products obtained by benzidine rearrangement may be summarised as follows :

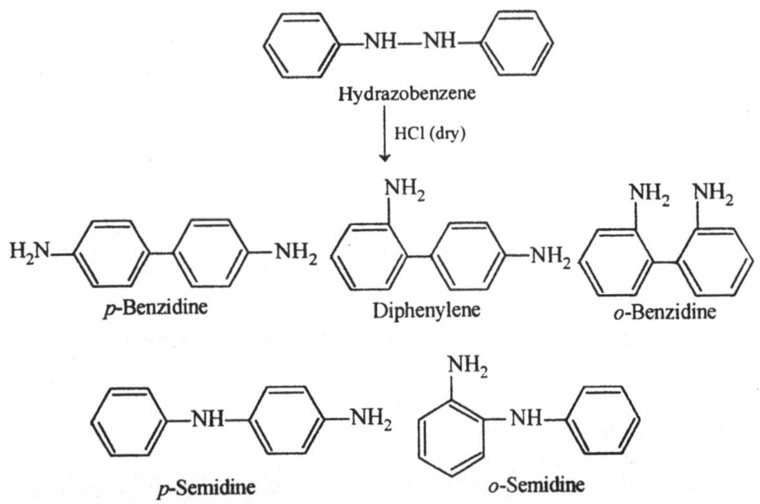

When 1-hydrazonaphthalene is treated with $HClO_4$, it undergoes benzidine rearrangement forming *p*-benzidine type compound as well *o*-benzidine type compound (major product), the latter of which may cyclise to form pyrrole derivative.

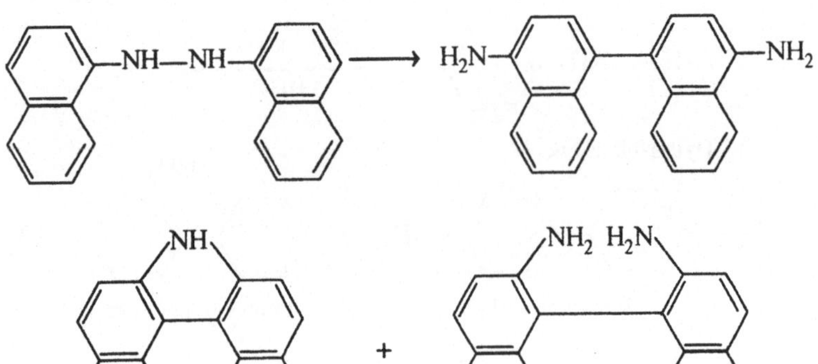

EFFECT OF SUBSTITUENTS

The nature of the product formed during bezidine rearrangement is largely determined by the nature of the substituent, if any, especially in the *p*-position in the bezene nucleus of hydrazobenzene. It is evident from the following example :

(a) If either of the two *p*-positions of the hydrazobenaene molecule is substituted, the major product formed might be diphenylene (2, 4-diaminobiphenyl), *o*-semidine (2-aminodiphenylamine) or *p*-semidine (4-aminodiphenylamine), depending upon the nature of the substituent.

(i) If the *para*-position is blocked by a methyl, methoxy or an ethoxy group, the main product has been found to be *o*-semidine accompanied by a small amount of *p*-semidine, *e.g.*,

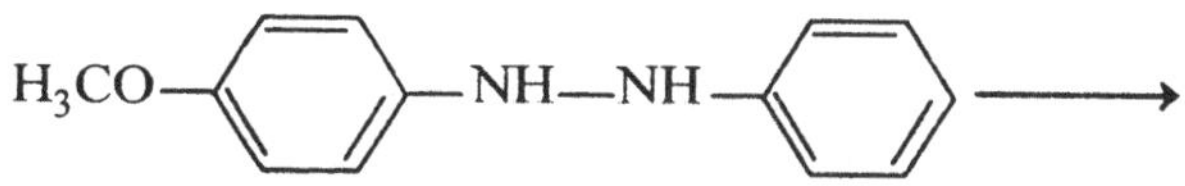

p-Methoxy hydrazobenzene

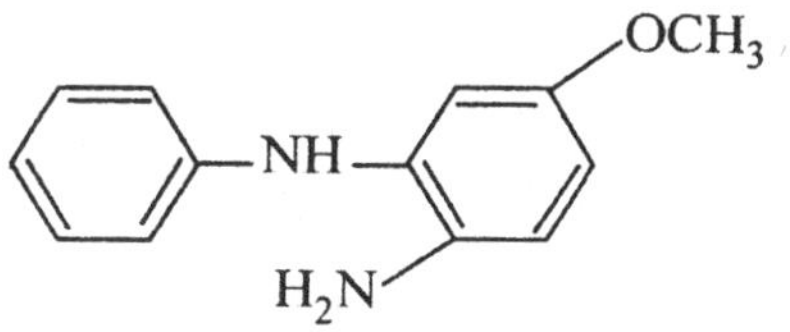

Anilinomethoxyaniline
(An *o*-semidine)

(ii) If the *para*-position is blocked by Cl, Br, I or $N(CH_3)_2$, the major product has been found to be a diphenylene accompanied by a small amount of *o*- and *p*-semidines, *e.g.*

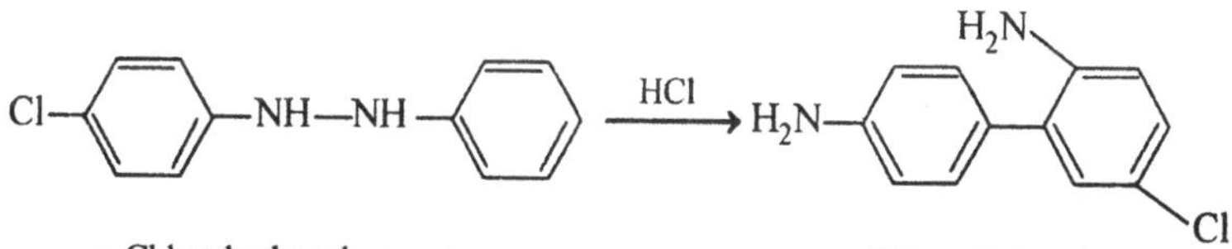

p-Chlorohydrazobenzene

Chlorodiphenylene
(a diphenylene)

(iii) If the para position has been blocked by NH_2 or $NHCOCH_3$, the main product is *p*-semidine, *e.g.*,

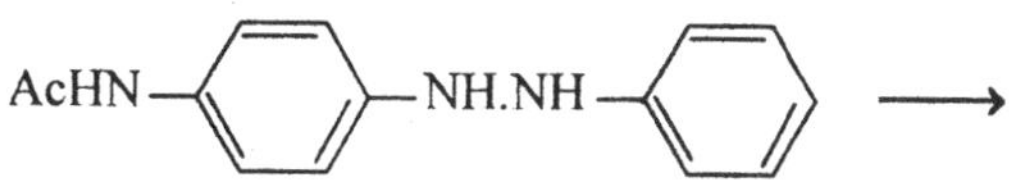

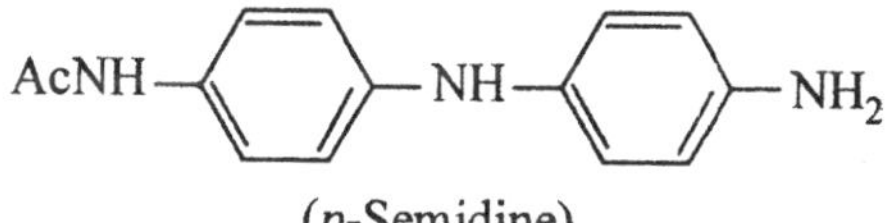

(*p*-Semidine)

(iv) If both the *para* positions of the hydrazobenzene are substituted, that only *o*-semidine is formed.

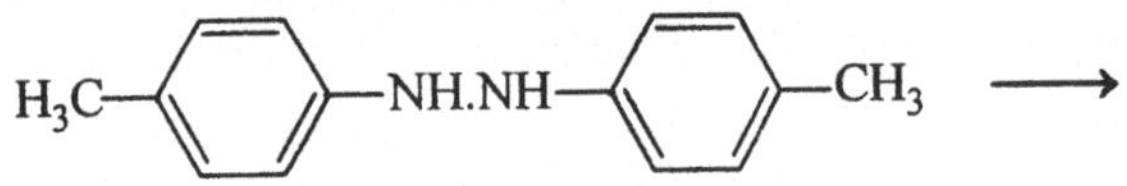

p-Hydrazotoltuene

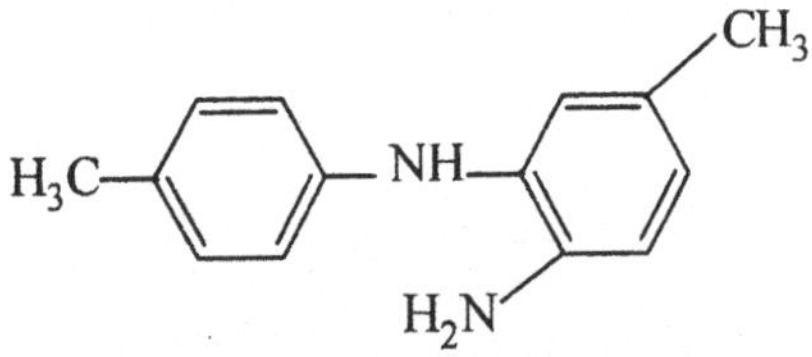

o-Amino-(4, 3')-ditolamine

(v) Some groups especially carboxyl and sulphonic acid get eliminated during the benzidine rearrange
ment.

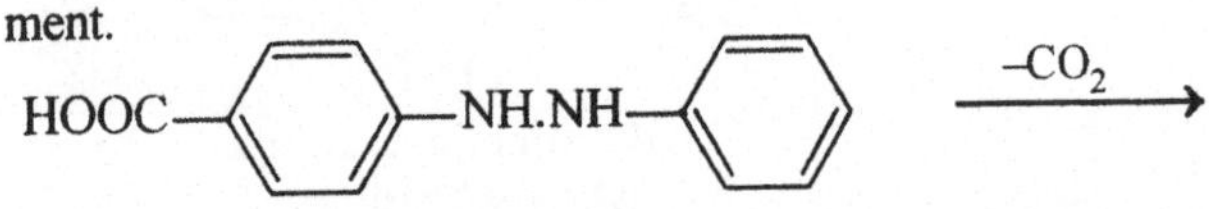

Hydrazobenzene-4-carboxylic acid

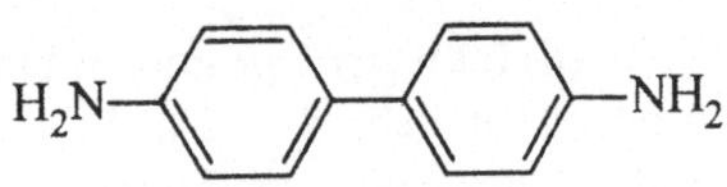

Benzidine

MECHANISM

Although the mechanism of benzidine rearrangement is still uncertain, the intramolecularity of the reaction has been established by a number of experiments.

(i) When a mixture of *o,o'*-dimethoxy hydrazobenzene and *o,o'*-diethoxybenzidine is subjected to benzidine rearrangement, they rearrange to form *m, m'*-diethoxybenzidine. However, no cross product, ethoxymethoxybenzidine, is detected.

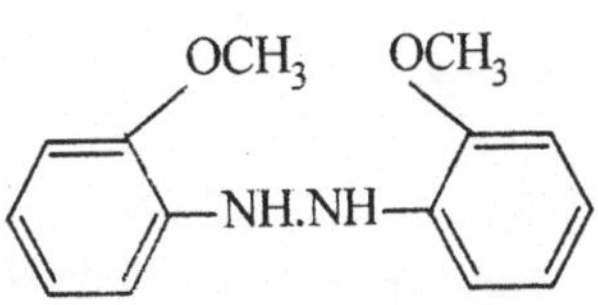

o, o'-Dimethoxy hydrazobenzene

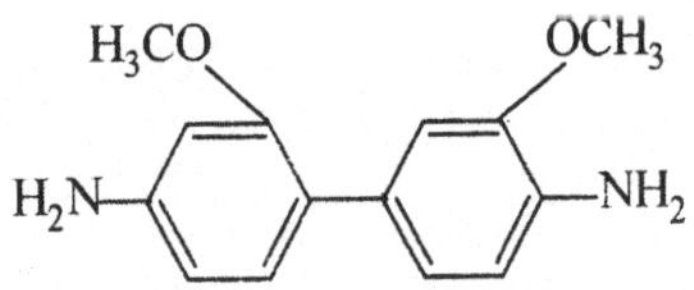

m, m'-Dimethoxy benzidine

\+ $\longrightarrow$ +

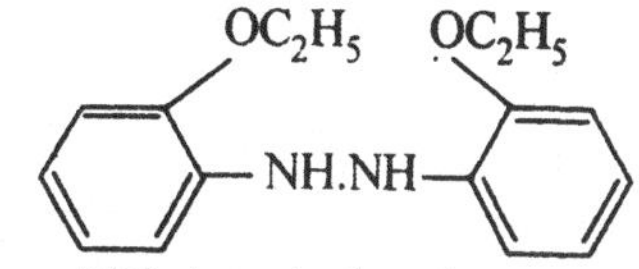

o, o'-Diethoxy hydrazobenzene

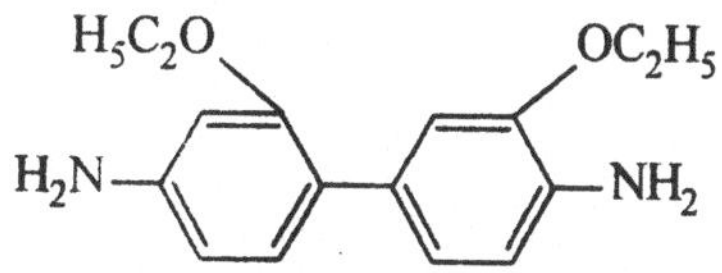

m, m'-Dimethoxy benzidine

(ii) When a mixture of two hydrazobenzenes, one having isotopically labelled methyl group ^{14}C as a substituent, is subjected to benzidine rearrangement, it is found that the whole of the tracer atom ^{14}C gets incorporated into the product derived from the labelled hydrazobenzene.

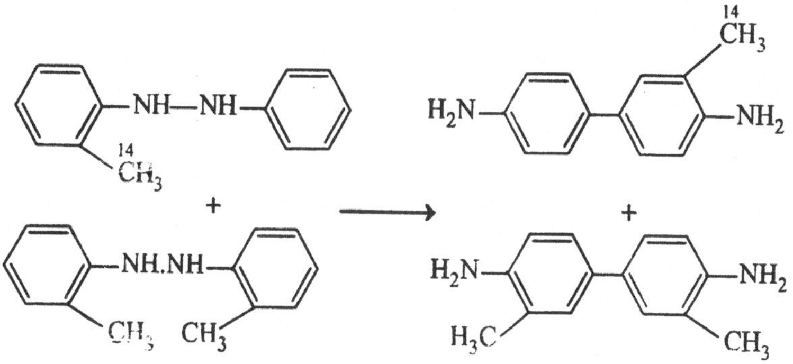

(iii) It was also shown that the benzidine rearrangement involves specific acid catalysis. It is of first order in the hydrazo compound and may be of both first and second order in the acid.

$$\frac{-d[\text{Hyd}]}{dt} = k_2\,[\text{Hyd}]\,[H^+] + k_3\,[\text{Hyd}]\,[H^+]^2$$

A number of *ortho* hydrazo compounds such as hydrazotoluene have been found to rearrange under a mixed first and second order catalysis. However, the overall result is near the mean 1.5.

(*iv*) By comparing the rate of rearrangements in H_2O (where the acid is H_3O^+) and D_2O (where the acid is d_3O^+), it was proved that the protonation at one or both of the nitrogens is not a slow, determining step. If the protonation is the rate determining step of the rearrangement, it is expected that there should occur a retardation in the rate of reaction in D_2O. This is expected because D_3O^+ in D_2O is a stronger acid than H_3O^+ in H_2O.

(*v*) As the replacement of the ring protons in the hydrazo compounds by deuterium or tritium could not slow down the rate of their rearrangements, it menas that the loss of two ring protons required for the formation of benzidine and diphenylene is also not a slow rate-determining step.

A number of theories have been postulated to explain the mechanism of the benzidine rearrangement. Some of these theories are described as follows :

1. Caged radical theroy

According to this theroy, benzidine rearrangement involves a two-proton mechanism which operates through the formation of two cation radicals in the rate-determining step.

$$Ar.\overset{+}{N}H_2—\overset{+}{N}H_2Ar \longrightarrow Ar.\overset{+\cdot}{N}H_2/H_2\overset{+\ \cdot}{N}Ar$$

Each of the cation radical is stabilised by reasonance and is prevented from diffusing away from the other by a cage of solvent. All types of products obtained in benzidine rearrangement are assumed to be formed by recombination of fragments within the solvent cage.

The caged-radical theory could not be accepted because ESR spectroscope failed to detect any radical in the rearrangement of hydrazobenzene. Also, it is difficult to visualise the homolytic dissociations of monocation into the unsymmetrical radical ion $Ar.\overset{+\cdot}{N}H_2/H\overset{+\ \cdot}{N}Ar$ in the absense of repulsion between like charges.

2. π-Complex mechanism

As the benzidine rearrangement is kinetically of first order in acid, Dewar postulated that the monoprotonated hydrazo compound undergoes a rate-determining heterolytic fission to form a π-complex. This complex does not represent transition state but represents a definite compound in which the covalent bond between the nitrogen atoms has been replaced by a delocalised π-bond between the aromatic rings.

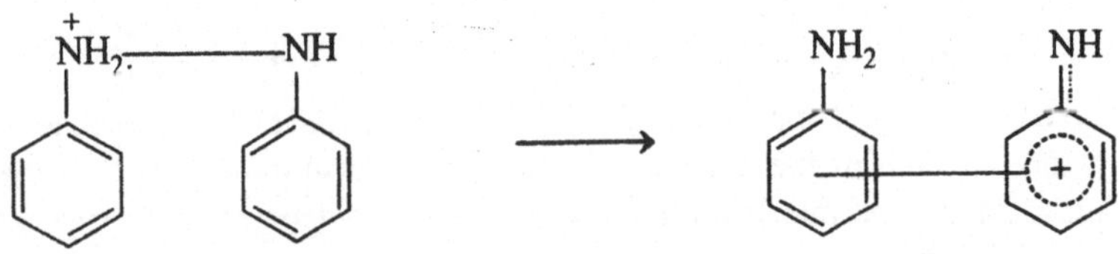

(I)

The π-bonds (I) can hold the aromatic rings in parallel planes. These rings can rotate and collapse to form the various products.

When it was found that the rearrangement is kinetically of second order in acid, the original theory was modified and it was postulated that the attack of second proton on (I) was rate determining.

3. Polar transition state theory

This theory was proposed by Banthorpe, Hughes and Ingold. According to this theory, it was assumed that a ploar transition state is developed in the rate-determining step of the rearrangement. We shall illustrate this theory by considering the arrangement of hydrazobenzene (II) *via*. a two-proton mechanism. As the diprotonated molecule of (II) has two adjacent positively charged nitrogen atomse, the repulsion between them facilitates the cleavage of the N—N bond. As a result of heterolysis two loosely bound aromatic rings are formed. One of these rings carries two units of positive charge and thus behaves like a dication. However, the other ring does not have charge but has the character of an aryl amine. This possesses fractional dipolar charge which is distributed in the same manner as in an aniline molecule.

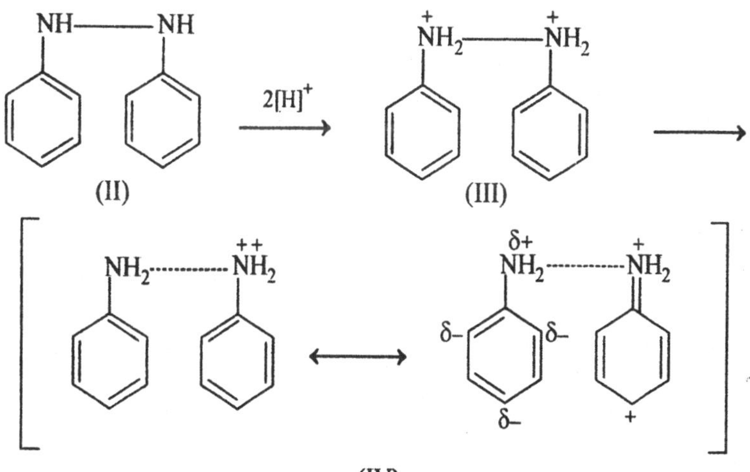

(IV)

When the inspection of the charge distribution in the transition state (IV) is done carefully, it can be understood very easily that the bond formation could only take place at 4, 4′,-positions. Thus, the rearrangement of (III) involves a transition state in which the heterolysis of the N-NO bond is concerted with the formation of new bonds at the most favourable positions in the transition state. On the basis of these assumptions, a two-proton mechanism for the formation of (V) and (VI) from the rearrangement of (II) can be postulated as follows :

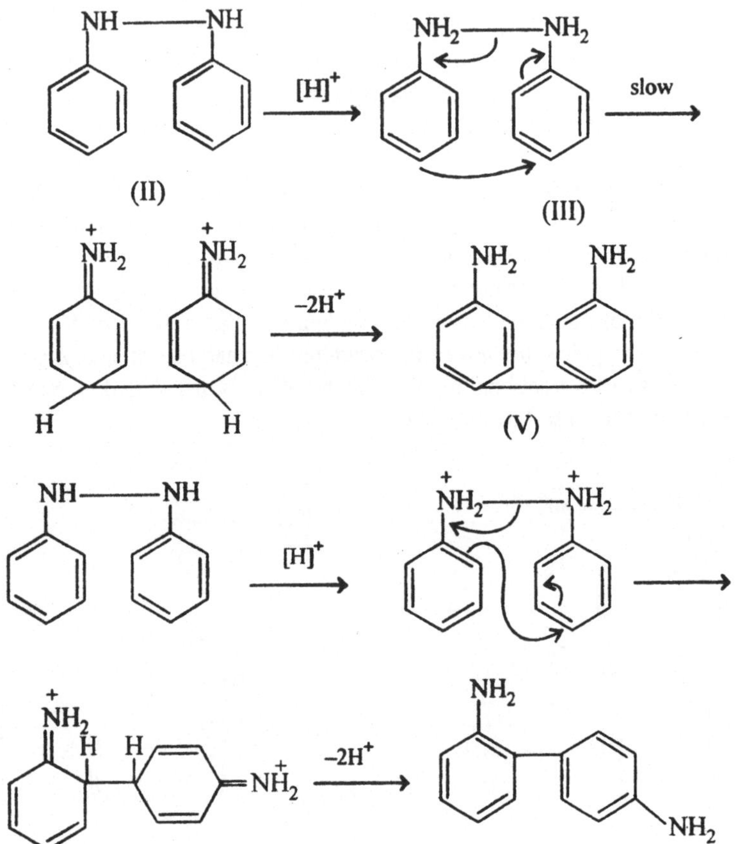

(II) (III) (V) (VI)

APPLICATIONS

As the alkaline reduction of the nitro compounds yields hydrazobenzenes very readily, benzidine rearrangement provides an important route for the preparation of 4, 4′-disubsitituted biphenyls.

The 4, 4′-diaminodiphenyl (benzindine) and its derivatives are especially used in the synthesis of dyes, especially for cotton. An example is benzopurpurin 4B which is an important red dye and is obtained from *o*-toluidine as a starting material

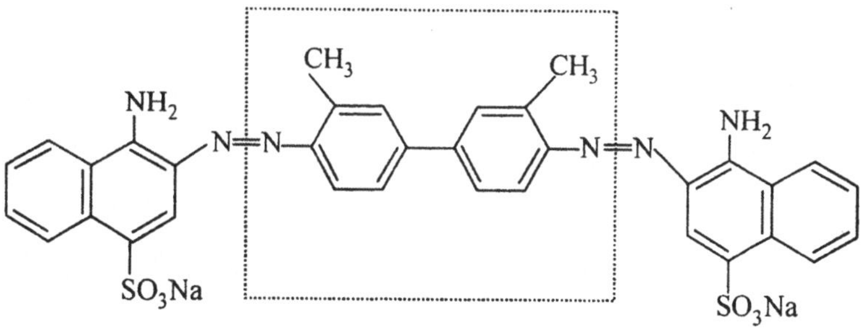

Benzoupurpurin 4B

The dotted rectangle encloses the protein of the molecule which is formed from, a benzidine rearrangement.

7

BENZOIN CONDENSATION

This reaction is give by aromatic aldehydes in the presence of NaCN/ C_2H_5OH/HOH or KCN/C_2H_5OH/HOH. Product of the reaction is α-hydroxy ketone known as Benzoin. In Benzoin, hydroxy group is always secondary.

$$C_6H_5\text{—}CHO + C_6H_5\text{—}CHO \xrightarrow{KCN/C_2H_5OH/HOH/\Delta} C_6H_5\text{—}\underset{\substack{|\\OH}}{CH}\text{—}\overset{\substack{O\\||}}{C}\text{—}C_6H_5$$

α-hydroxy ketone or Benzoin

Mechanism : First step :

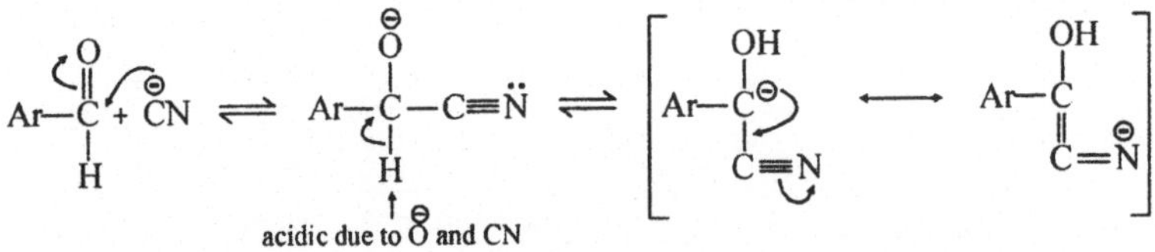

Second step :

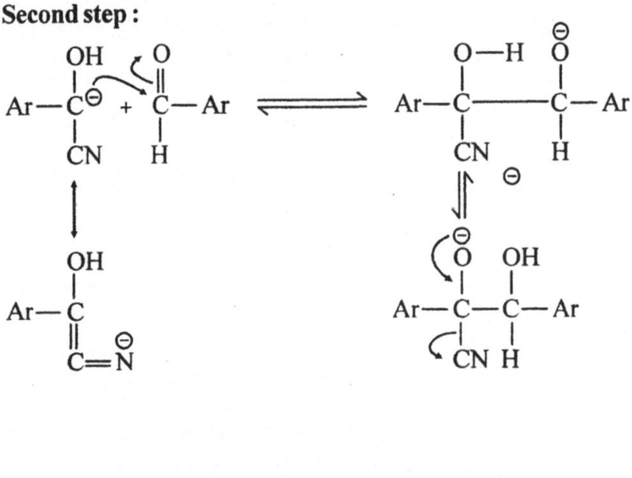

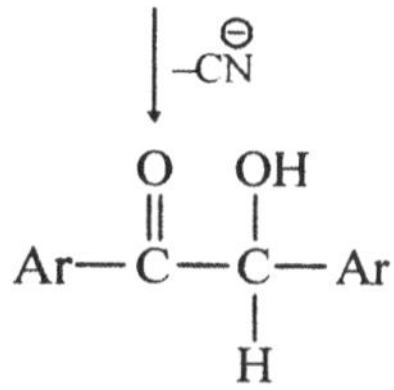

Aldehyde on which CN adds is **proton donor** and the other aldehyde is **proton acceptor** or aldehyde which gives addition with CN is known as donor and aldehyde which gives addition with carbanion is known as acceptor.

Acceptor should be highly reactive for nucleophilic addition reaction.

The unique catalytic property of cyanide ion is due to its.

(*i*) Good nucleophilicity

(*ii*) Property of stabilising the carbanion which reacts in the slow step

(*iii*) Property of being a good leaving group as nucleophile.

Electron attracting groups retard the reaction because they decrease nucleophilicity of carbanion which gives nucleophilic addition with the other molecule of aldehyde. Electron donating groups at *o* or *p*-position also reatard the reaction because such groups decrease reactivity of carbonyl compounds for nucleophilic addition reaction. *p*-nitro benzaldehyde, *p-N, N* -dimethyl amino benzaldehyde do not give this reaction.

This reaction is also given by glyoxal R—C(=O)—CHO

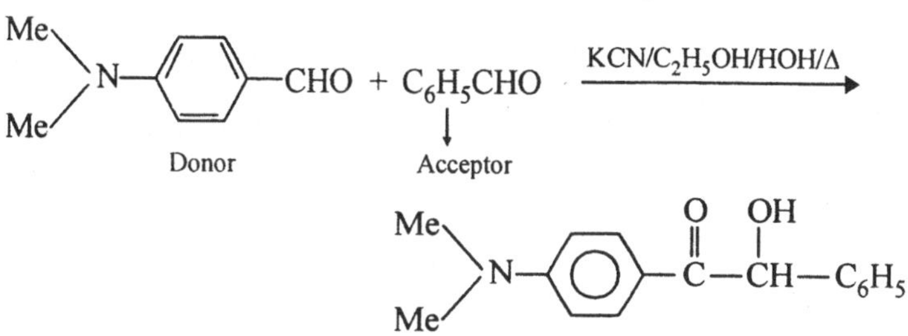

$Me_2\ddot{N}$ gp will increase nucleophilicity of carbanion.

8

BENZILIC ACID REARRANGEMENT

INTRODUCTION

It is a base-catalysed reaction whereby α-diketones are converted into α-hydroxy acids. The best known example is that benzil on refluxing with alcoholic potassium hydroxide gets rearranged into benzilic acid. This conversion is about 89%.

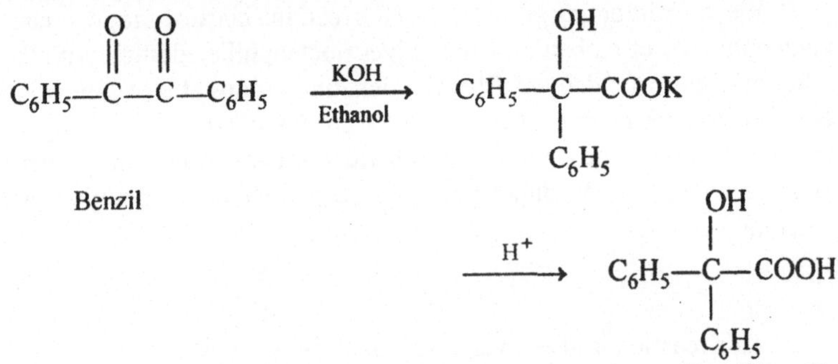

Benzil

Benzilic acid

This reaction is similar to the intramolecular Cannizzaro reaction of phenyl-glyoxal except that phenyl has taken the place of hydrogen.

Another example of benzilic acid rerrangement is the rearrangement of furil into furilic acid.

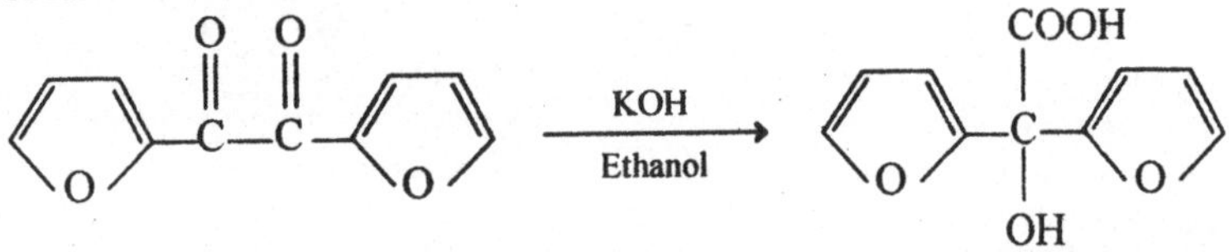

Furil

Furilic acid

The aliphatic diketones also undergo benzilic acid rearrangement but condensation involving α-hydrogen atoms usually competes with the rearrangement. Thus, the yields are low. For instance, ketopinic acid rearranges to citric acid

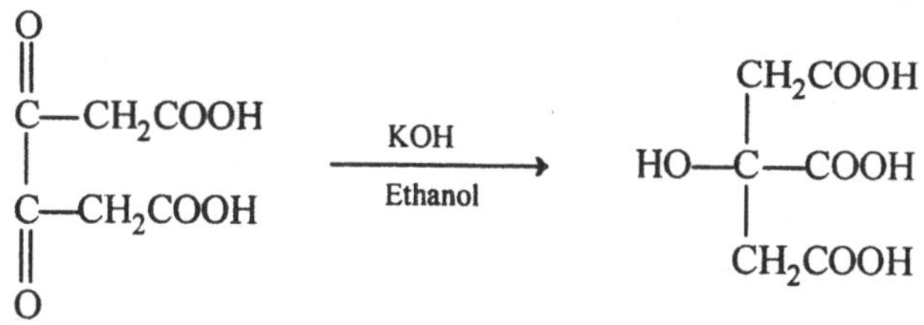

Ketopinic acid   Citric acid

It was proved by Pfeif in 1956 that barium and thallous hydroxides are more effective than the commonly used KOH or NaOH.

MECHANISM

Benzilic acid rearrangement is not subject to general base catalysis and when it is carried out in $H_2\,^{18}O$, benzil exchanges ^{18}O at a faster rate than it undergoes the rearrangement. All these facts show that the mechanism of this rearrangement involves a nucleophilic attack at a carbonyl group by the hydroxide ion, this step is a reversible and fast step. The intermediate so obtained then undergoes a rate-determining rearrangement, *i.e.*, it is actually a 1, 2-shift of the phenyl group to the adjacent carbonyl carbon.

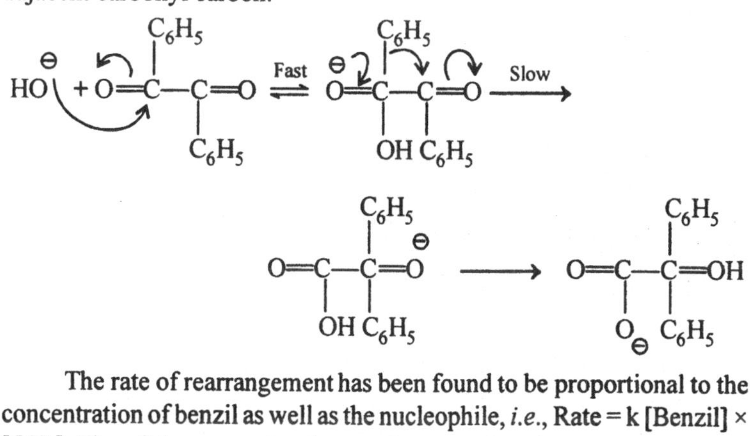

The rate of rearrangement has been found to be proportional to the concentration of benzil as well as the nucleophile, *i.e.*, Rate = k [Benzil] × [OH^-]. Thus, it is a second order reaction.

Further the rate of rearrangement in D_2O-dioxane has been found to be almost twice that of in H_2O-dioxane. This reveals that the proton transfer occurs after the migration and not simultaneously. Otherwise the rate ratio would have been reversed.

In the case of substituted compounds the aryl group migrates in preference to the aryl group having electron releasing groups. The reason is that the electron releasing group will tend to neutralize the positive charge on the carbonyl carbon and thus decrease its electrophilic character. Hence, the nucleophile adds on to the other carbonyl group. This has been proved experimentally using labelled *para*-methyl as well as *para*-methoxybenzil. In both cases the phenyl group gets migrated in preference to *p*-methylphenyl or *p*-methoxyphenyl group.

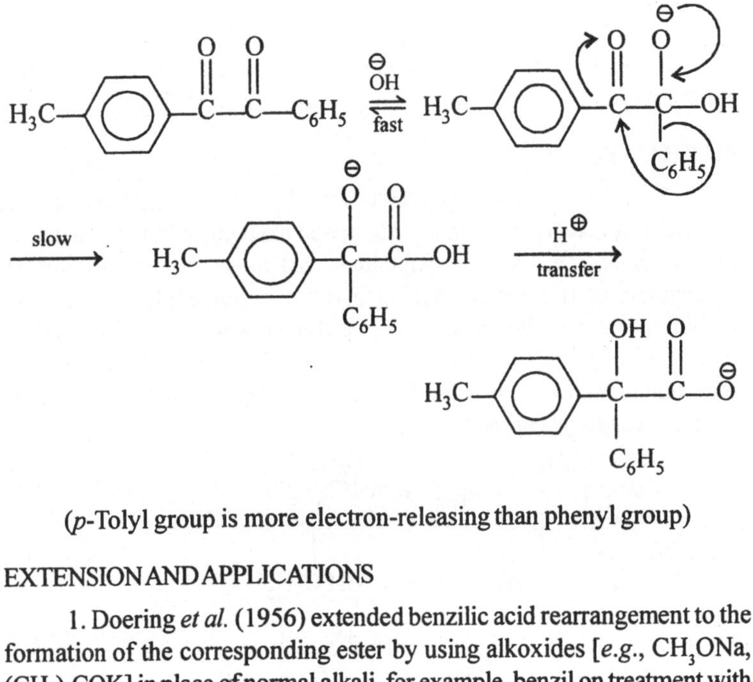

(*p*-Tolyl group is more electron-releasing than phenyl group)

EXTENSION AND APPLICATIONS

1. Doering *et al.* (1956) extended benzilic acid rearrangement to the formation of the corresponding ester by using alkoxides [*e.g.*, CH_3ONa, $(CH_3)_3COK$] in place of normal alkali, for example, benzil on treatment with sodium alkoxide is converted into alkyl benzilate.

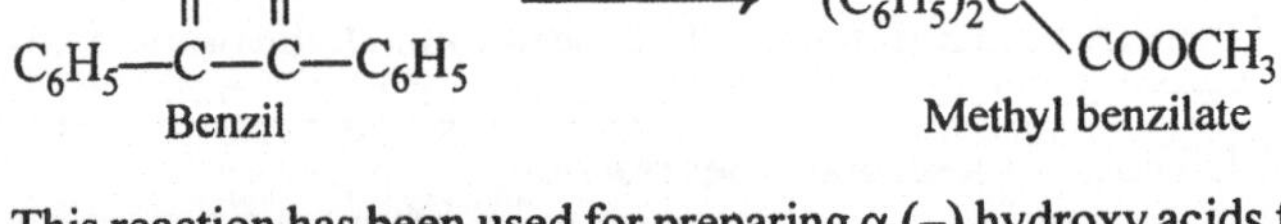

This reaction has been used for preparing α (–) hydroxy acids from the easily avaiblable starting substances.

(a)

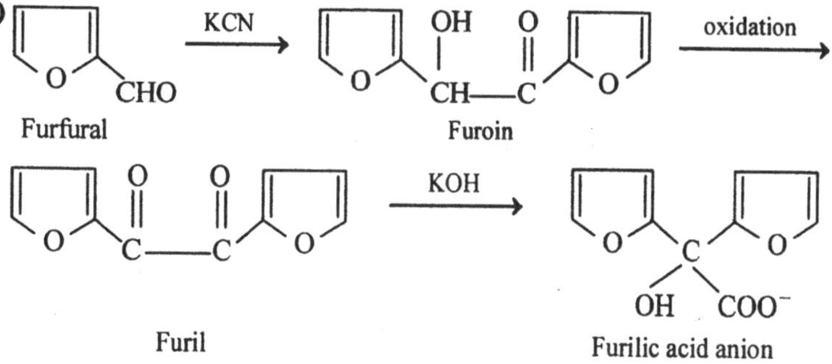
KCN
OH
O
CHO
CH—C
Furfural
Furoin
oxidation
O
KOH
C—C
OH
COO⁻
Furil
Furilic acid anion

(b)

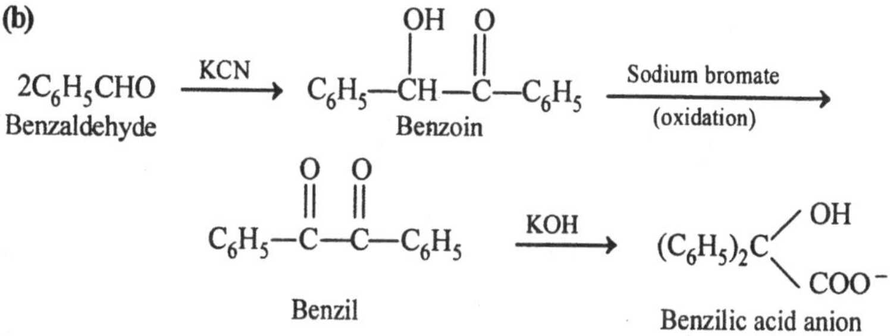
2C6H5CHO
Benzaldehyde
KCN
OH O
C6H5—CH—C—C6H5
Benzoin
Sodium bromate
(oxidation)
O O
C6H5—C—C—C6H5
Benzil
KOH
(C6H5)2C
OH
COO⁻
Benzilic acid anion

(c)

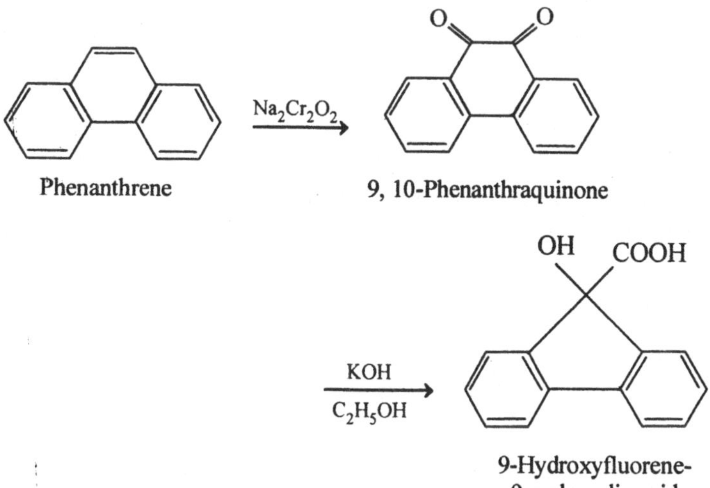
Phenanthrene
Na2Cr2O2
O
O
9, 10-Phenanthraquinone
KOH
C2H5OH
OH
COOH
9-Hydroxyfluorene-
9-carboxylic acid

Diaryl diketones are better starting materials than aliphatic diketones because the latter with enolizable hydrogen usually allow side reactions such as aldol condensation.

In benzilic acid rearrangement, electron withdrawing substituents as the migrating groups, retard the rearrangement. This is expected because the migration occurs to an electron-poor carbonyl carbon atom.

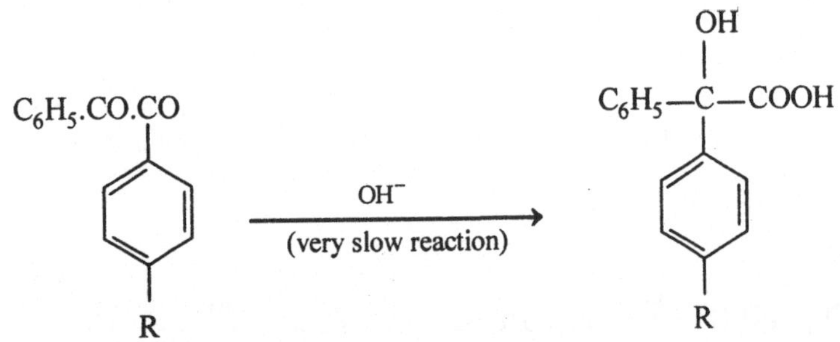

where R is an electron-withdrawing group

(d) Benzilic acid rearrangement is generally applicable to aromatic ketones. However, ketonpinic acid, an aliphatic ketone, undergoes the rearrangement to form citric acid.

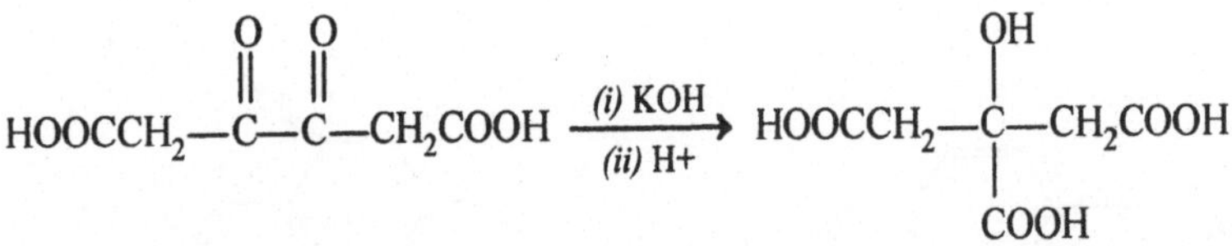

Ketopinic acid　　　　Citric acid

(e) In the case of steriods, benzilic acid rearrangement is used as a route to the nor-steroids. For example, cholestan-2, 3-dione on reaction with potassion hydroxide in refluxing propanol gives an almost quantitative yicld of the α-hydroxy acid (I) which an oxidation yields the norketone (II).

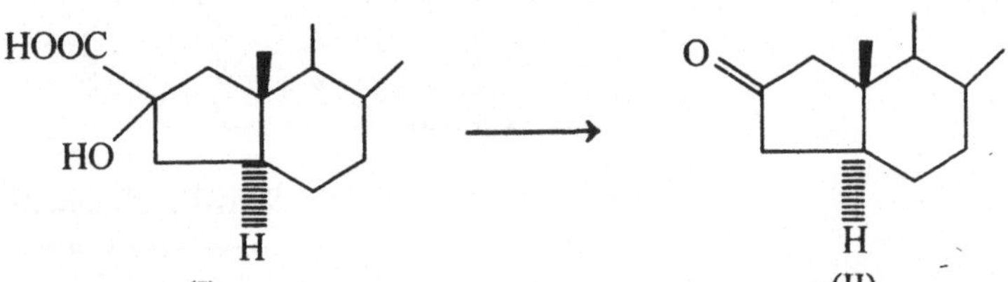

(I)　　　　(II)

(f) The reaction is also useful for the synthesis of following compounds.

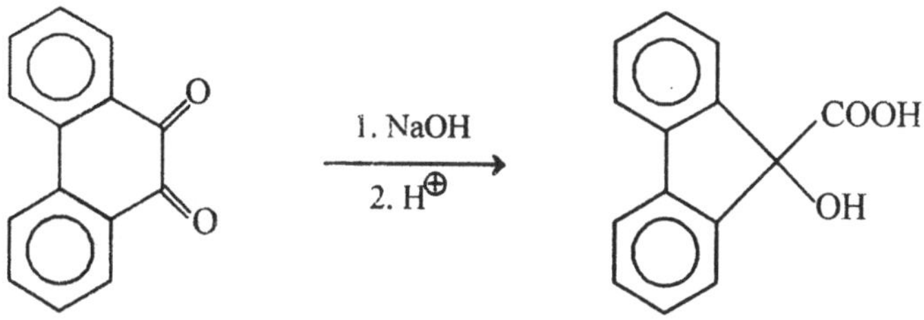

Phenanthraquinone

9-Hydroxyfluorene-9-carboxylic acid.

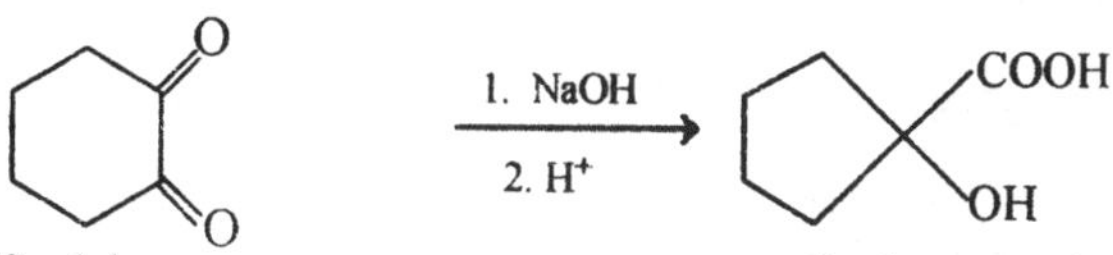

Cyclohexane -1, 2-dione

Cyclopentanol -1-carboxylic acid

(g) Similar benzilic acid rearrangement is observed when α-haloketones not having α-hydrogens are treated with alkoxides.

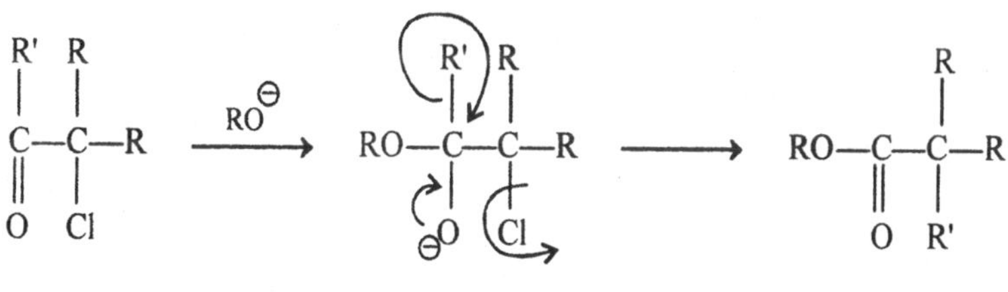

This reaction is known as semibenzilic rearrangement.

9

BISCHLER-NAPIERALSKI SYNTHESIS

INTRODUCTION

This reaction constitutes an elegant method of preparing certain isoquinoline derivatives. This reaction involves the following steps :

(*i*) The acyl derivatives of β-aryl ethylamine (β-aryl ethylamide) are heated with phosphorus pentoxide, anhydrous zinc chloride, etc, when dehydration occurs with the formation of following compound :

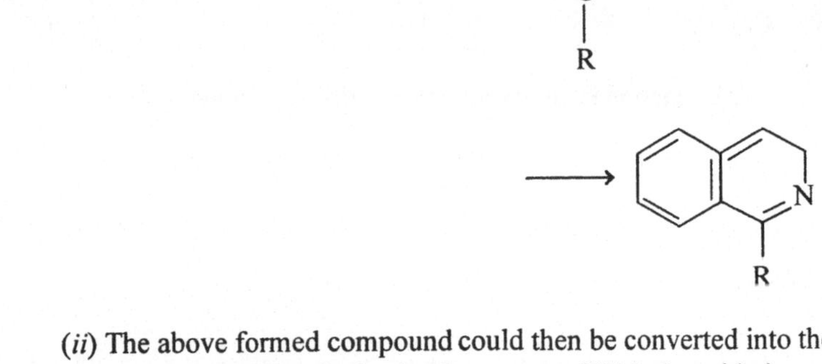

(*ii*) The above formed compound could then be converted into the corresponding isoquinoline derivative by means of $KMnO_4$ oxidation or by dehydrogenation with S, Se or Pd black.

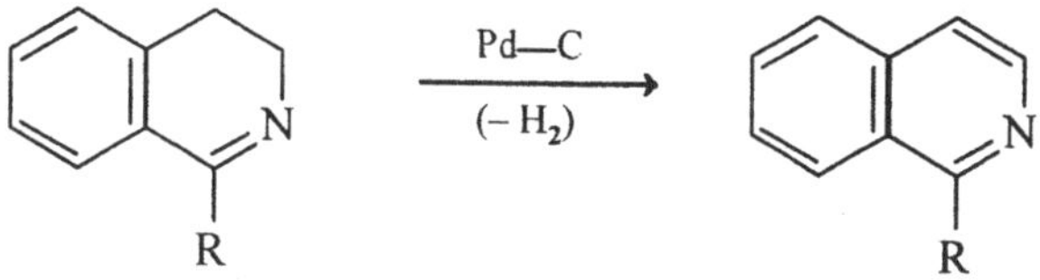

Yields of the product were either poor or almost negligible. However, yields can be increased by using lower temperature and a milder condensing agent ($POCI_3$, PCI_5). Later, Pictet and Gams prepared isoquinoline directly from β-hydroxyl-β-phenlyethylamide without involving the dehydrogenating step.

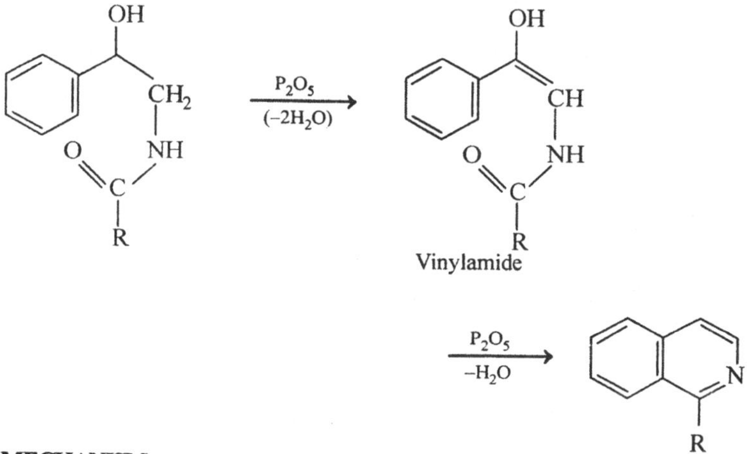

MECHANISM

The mechanism of this reaction proceeds through the intial formation of a carbonium ion I which then undergoes an eletrophilic intramolecular attack on the aromatic double bond to form the 3, 4-dihydroisoquinoline II which on dehydrogenation catalytically yields isoquinoline III.

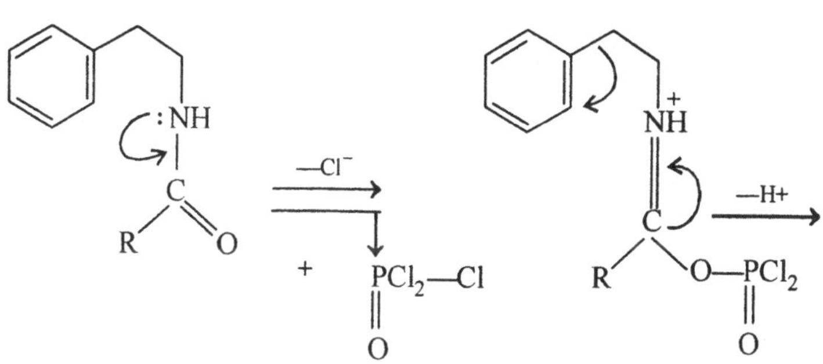

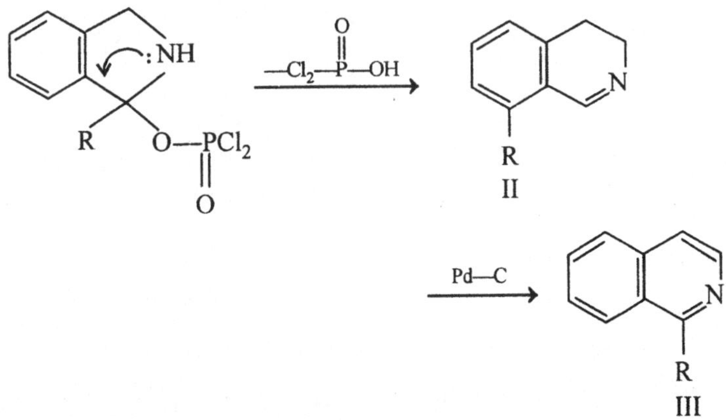

SCOPE OF THE REACTION

This reaction has been used for synthesising many alkaloids such as cotarnine, harman, papaverine, etc.

(a) Synthesis of Papaverine:

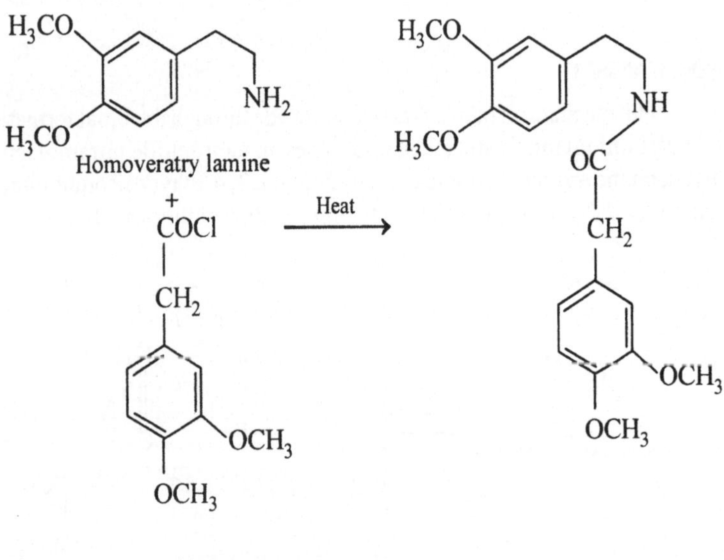

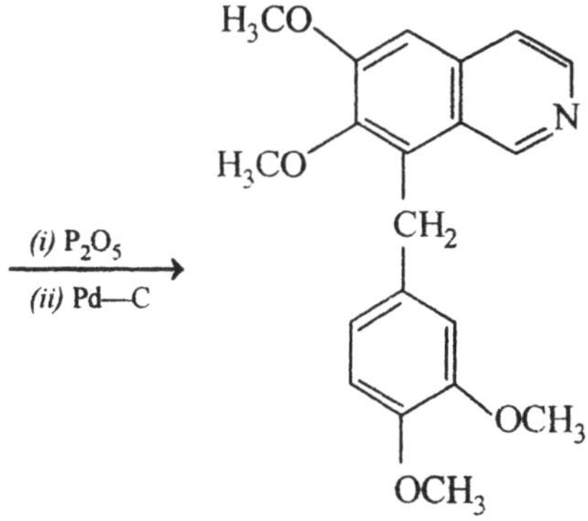

Papaverine

(b)

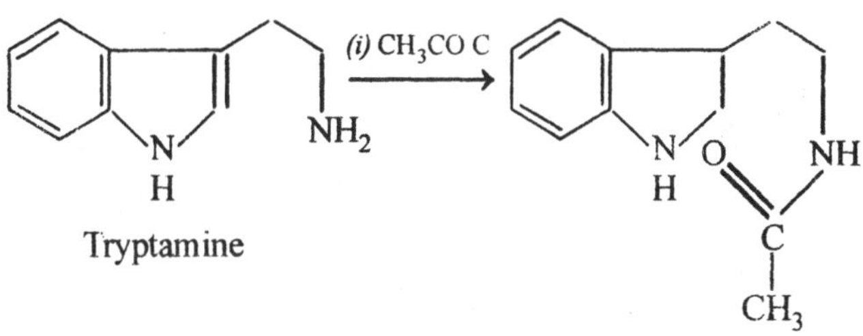

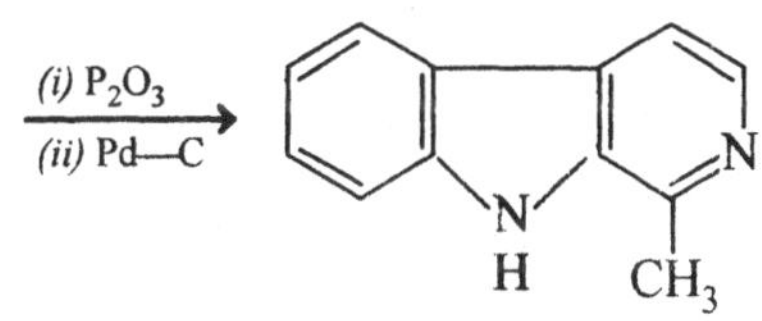

Harman

10

BAYER-VILLIGER OXIDATION

With the insertion of oxygen in the carbonyl compounds subjected to bayer-villiger oxidation, a new vista in organic synthesis was opened.

REACTION

Bayer-Villiger oxidation is essentially the oxidation of aldehydes into carboxylic acids and ketones into esters. The oxidizing agent is a peracid

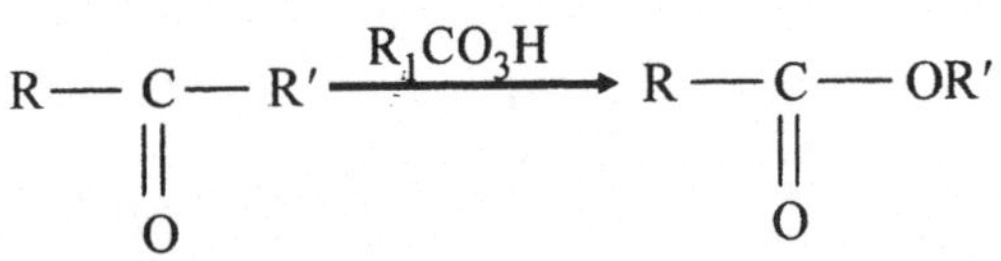

ketone ester

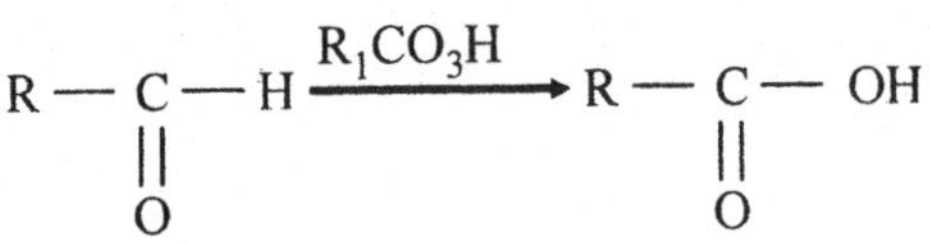

aldehyde Carboxylic acid

MECHANISM

The entire mechanism can be broken into three steps:

Step-[I] ***Formation of Carbonium Ion***

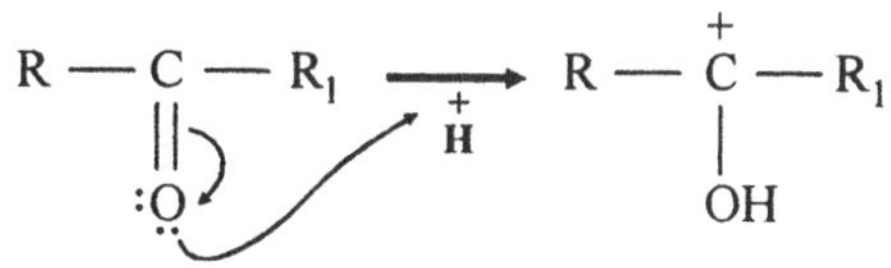

Step-[II] ***Nucleophilic attack by Peroxy Acid***

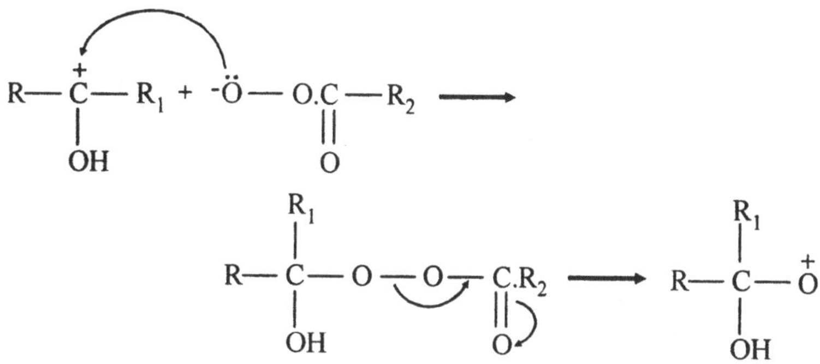

Step-[III] ***Migration***

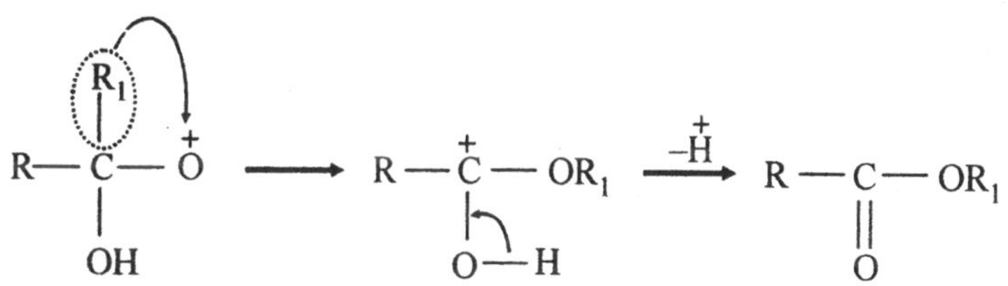

For aldehydes, it is the hydrogen atom of the CHO group that migrates.

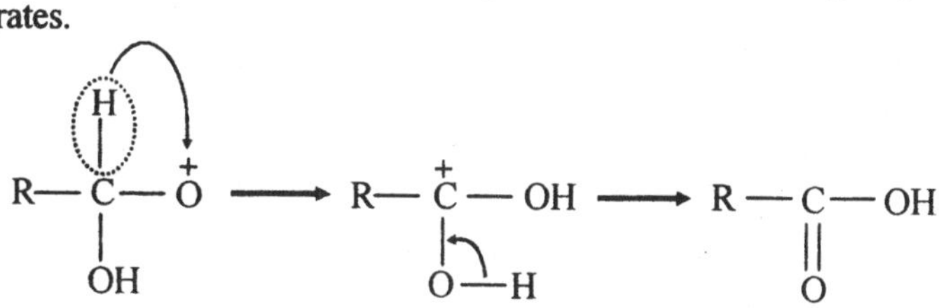

It is important to note that in the presence of a strong acid there may be addition of peroxy-acid to the protonated ketone, but additional acid is not needed and in its absence addition may take place to the ketone itself. The general mechanism is supported by the fact that the reaction is catalysed by acid and is accelerated *by electron-releasing groups in the ketone and by electron with-drawing groups in the peracid.* Further that the oxygen atom of the carbonyl group of the ketone is the same one as in the carbonyl group of the ester was clearly demonstrated in the following experiment with the isotope O[18] present in the molecule.

$$C_6H_5-\overset{\overset{O^{18}}{\parallel}}{C}-C_6H_5 \xrightarrow{C_6H_5COOOH} C_6H_5-\overset{\overset{O^{18}}{\parallel}}{C}-OC_6H_5$$

During the mechanism we see that an alkyl group migrates from carbon to oxygen ; the order of this migratory aptitude in the alkyl series is tertiary > secondary > primary [Fries ; 1950], *e*, g.

$$\underset{\text{Pinacolone}}{Me_3C-CO-CH_3} \xrightarrow{RCO_3H} \underset{t\text{-butyl acetate}}{Me_3C-O-CO-CH_3}$$

Among the aryl series the order is : *p*-anisyl > *p*-toly > phenyl > *p*-chlorophenyl > *p*-nitrophenyl > *p*-aminophenyl, etc. [Doering ; 1950]. When aryl and alkyl, both groups, are present, the migratory aptitude of various groups is : tertiary alkyl > sec. alkyl, aryl, benzyl > primary alkyl > methyl. Thus in acetophenone, phenyl group migrates in preference to the methyl.

$$\underset{\text{Acetophenone}}{C_6H_5.\ CO.\ CH_3} \longrightarrow \underset{\text{Phenyl acetate}}{C_6H_5-O-CO.\ CH_3}$$

It must be made in mind that the reaction is intramolecular and thus the migrating group retains its configuration as shown by the fact that optically active 3-phenyl-2-butanone affords optically active α-methylbenzyl acetate with the opposite sign of rotation.

$$\underset{\text{3-Phenyl-2-butanone}}{C_6H_5-\overset{\overset{CH_3}{|}}{C}H-COCH_3} \xrightarrow{RCO_3H} \underset{\alpha\text{-Methylbenzyl acetate}}{C_6H_5-\overset{\overset{CH_3}{|}}{C}H-O-CO.CH_3}$$

Also, it is probable that the loss of the carboxylate anion and rearrangement are concerted.

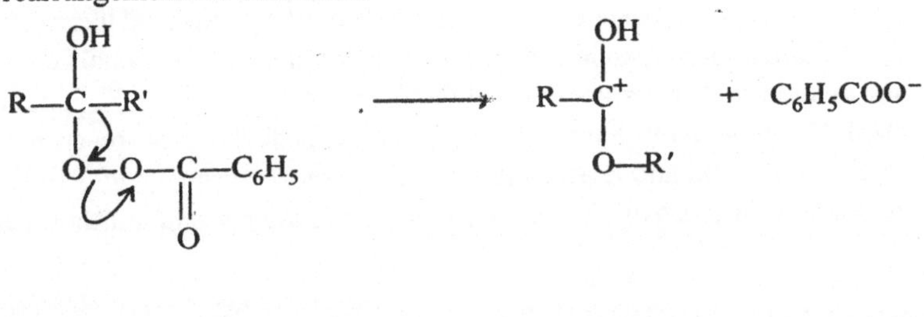

The electron-donating groups in the ketone accelerate the rate of reaction while the electron-withdrawing groupin the peracid accelerates the rate of reaction.

The above mechanism has been demonstrated by the observation that the labelled carbonyl oxygen atom of the ketone would become the carbonyl atom of the ester.

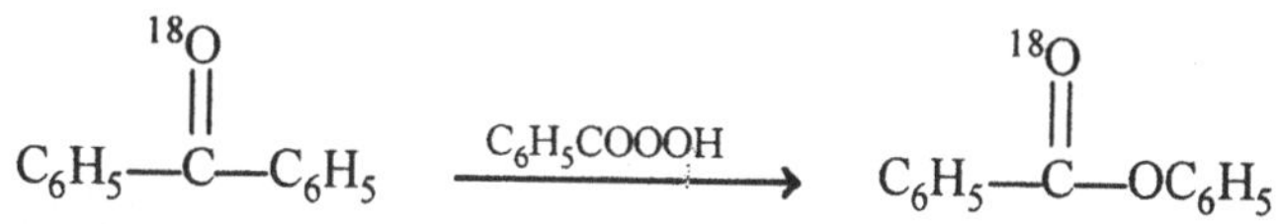

APPLICATIONS

(*a*) Baeyer-Villiger oxidation has been used in synthesising the following types of compounds

(i) Esters and acids :

$$R.CO.R' \xrightarrow{RCO_3H} R—O.COR' \xrightarrow{HOH} R'COOH + ROH$$

(*ii*) *Anhydrides* : These are formed when α-diketones or *o*-quinones are treated with peracides, *e.g.*

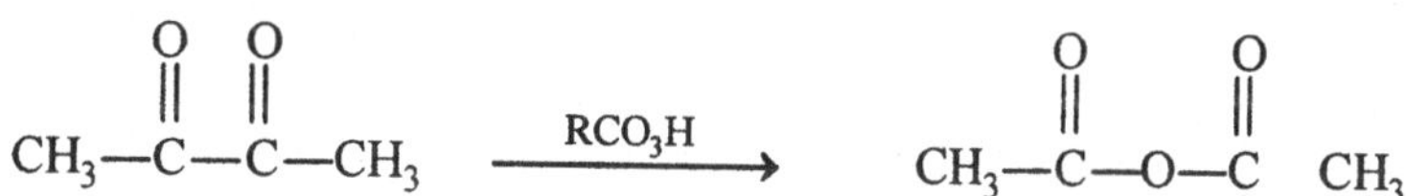

Diacetyl Acetic anhydride

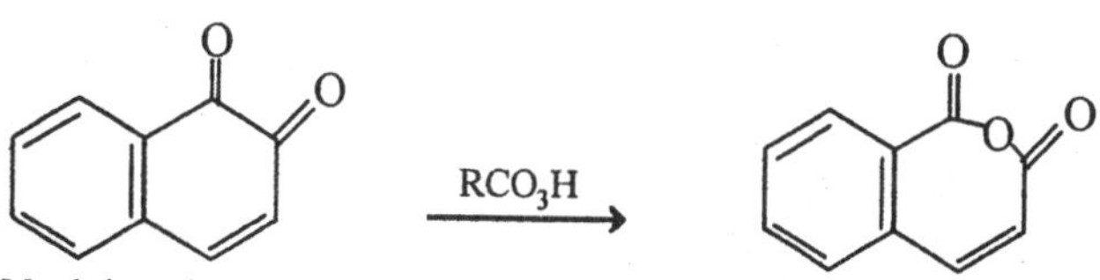

α-Naphthaquinone

(*iii*) *Lactones* : These are formed when cyclic ketones undergo ring expansion with peracids, *e.g.*, cyclohaxanone yields *o*-caprolactone.

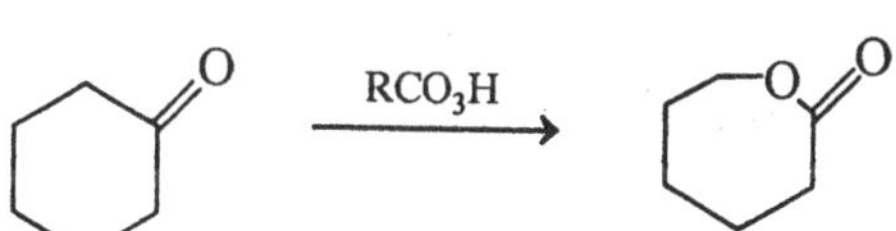

Cyclohexanone ε-Caprolactone

This method has been used to synthesise lactones having large rings (fourteen to eigteen membered rings) by the oxidation of the respective ketones.

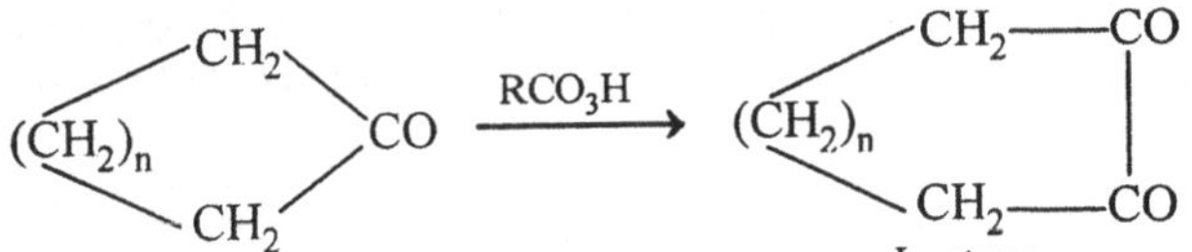

Lactone

$$\xrightarrow[C_2H_5OH]{H_2SO_4} HOCH_2\text{–}(CH_2)_n\text{–}CH_2 - COOC_2H_5$$

With some condensed cyclic ketones, two lactones in varying proportions are formed. For example, camphor gives two lactones (I) and (II).

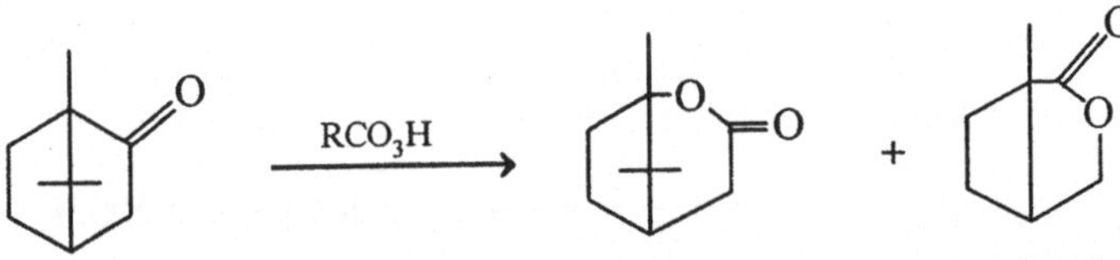

I (75%) II (75%)

Lactone (I) is the normal product formed by the migration of the tertiary bridgehead carbon while lactone (II) has been formed by the migration of the methylene group. The reason for the formation of two lactones in different proportions is steric factor.

(*b*) In order to elucidate the structures of various organic compounds, the Baeyer-Villiger oxidation is used ot degrade these compounds in which a carbon-carbon bond is broken and the ester formed can get easily hydrolysed to form two fragments for easier identification.

The reaction is not successful with aldehydes. Aliphatic aldehydes are oxidized to acids by the migration of the hydrogen.

$$R\text{—}\overset{\overset{O}{\|}}{C}\text{—}H \xrightarrow{R''CO_3H} R\text{—}\overset{\overset{O}{\|}}{C}\text{—}OH$$

A few aromatic aldehydes are converted to formates by the migration of the aryl group.

$$Ar\text{—}\overset{\overset{O}{\|}}{C}\text{—}H \xrightarrow{R''CO_3H} H\text{—}\overset{\overset{O}{\|}}{C}\text{—}OAr$$

Aryl formate

PROBLEM BASED ON B.V.OXIDATION

Problem-1

$$\underset{\displaystyle\mathrm{OH}}{\mathrm{CH_3\,.\,\underset{|}{CH}\,.\,CH_2\,.\,CH_3}} \xrightarrow{\mathrm{NBS}} \mathbf{A} \xrightarrow{\mathrm{SeO_2}} \mathbf{B}$$

$$\xrightarrow[\mathrm{H_2O_2}]{\mathrm{BF_3}} \mathbf{C} \xrightarrow[\mathrm{NaOAc}]{\text{Ø.CHO}} \mathbf{D} \xrightarrow[\Delta]{} \mathbf{E}$$

Identify **A, B, C, D, E.**

Ans.			
	A	Methylethyl ketone	$CH_3 \cdot CO \cdot CH_2 \cdot CH_3$
	B	Diacetyl or Dimethylglyoxal	$CH_3 \cdot CO \cdot CO \cdot CH_3$
	C	Acetic anhydride	$CH_3 \cdot CO \cdot O \cdot CO \cdot CH_3$
	D	Cinnamic acid	$Ph \cdot CH = CH \cdot CO_2H$
	E	Styrene	$Ph \cdot CH = CH_2$

11

BIRCH REDUCTION

A two electron reduction of aromatic ring by IA group element/ $NH_3(l)$ is known as Birch reduction. In this, reaction intermediate is radical anion.

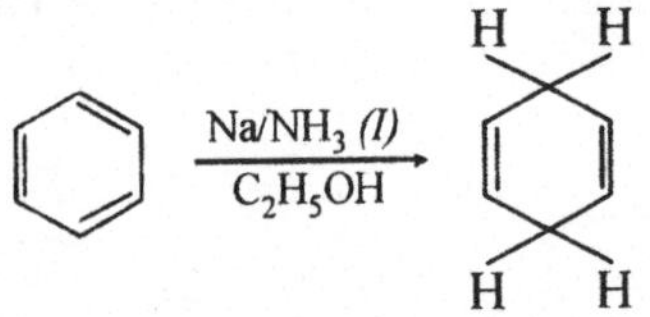

In this reaction addition of hydrogen takes place at position I and 4 to give an unconjugated diene.

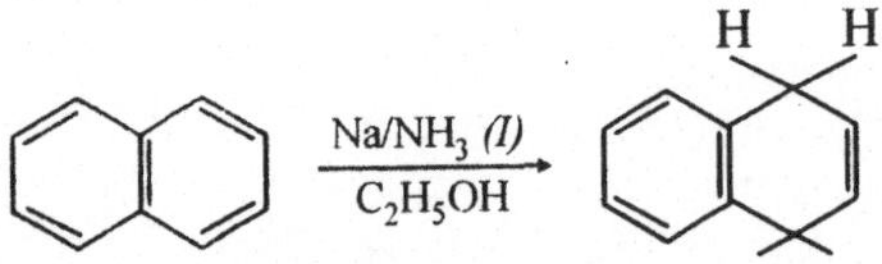

Liquid ammonia serves as solvent. Primary amines can also be used as solvent with advantage, because it allows higher temperature of reaction. (b.pt of ethyl amine is 19°C and b.p. of liquid ammonia is -33·0°C.)

MECHANISM

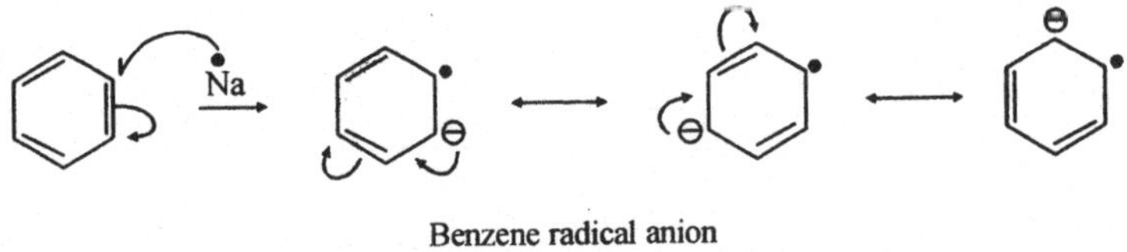

Benzene radical anion

The first electron transfer product a delocalised benzene radical anion.

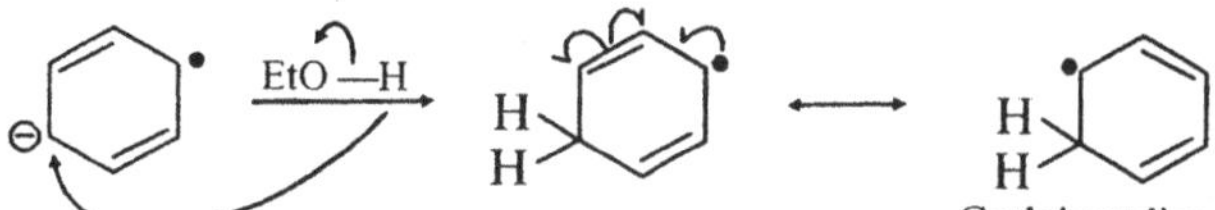

Cyclohexadienyl radical

Cyclohexadienyl radical is produced by proton of alcohol and it is also a delocalised species *i,e.* free radical.

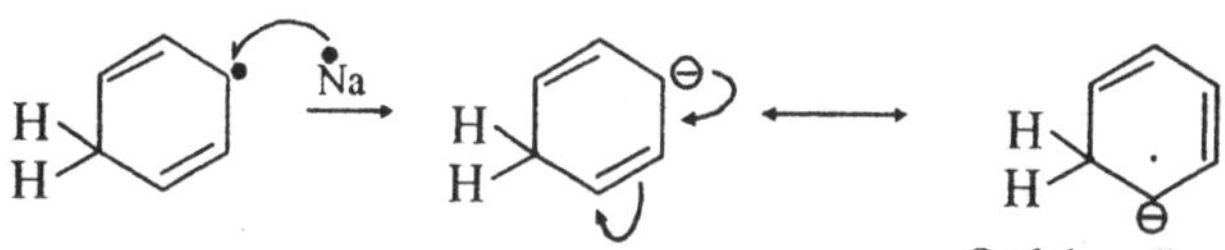

Cyclohexadienyl anion

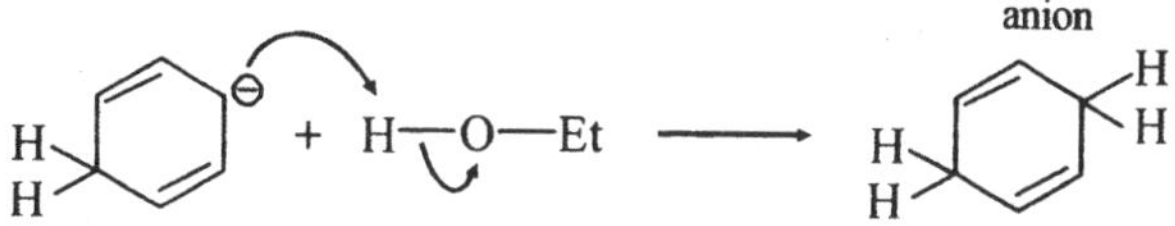

Formation of a 1, 4-cyclohexadiene is quite general in this type of reaction, but the reason for its formation in preference to the more stable conjugated 1, 3-cyclohexadiene is not understood.

The reson is that the two carbon atoms which are reduced on the aromatic ring go through carbanionic intermediates. Electron withdrawing substituents stabilise the carbanions whereas electron donating substituents destabilise them. Thus, reduction takes place on a carbon bearing an electron withdrawing substituent and not on a carbon bearing an electron releasing substituent. Substituents which are strongly electron releasing (*e.g.*, $-OCH_3$) deactivate the aromatic ring toward Birch reduction.

An electron-releasing group

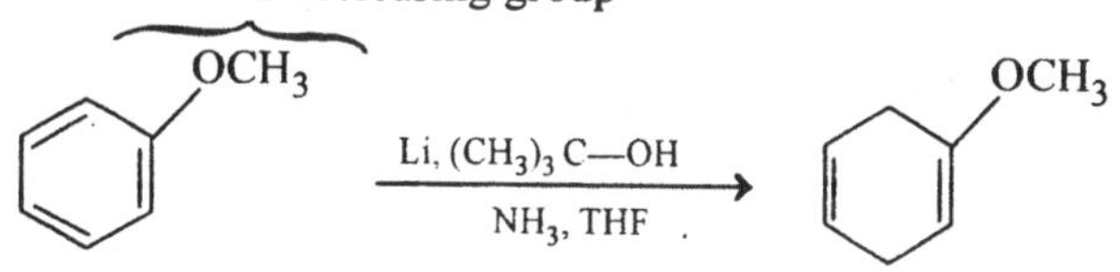

Methoxybenzene (anisole) | 1-Methoxy-1, 4-cyclohexadiene

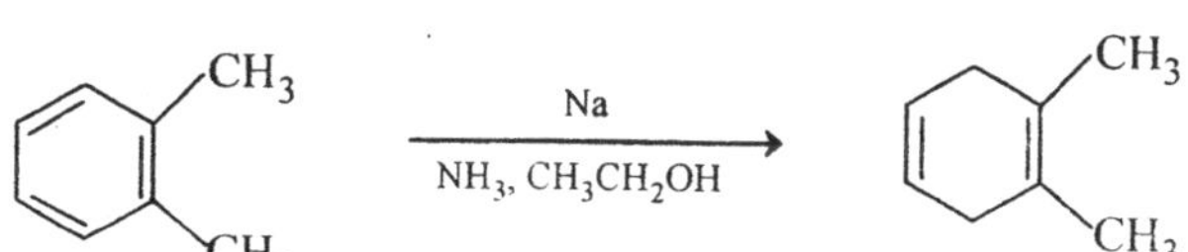

1, 2-Dimethylbenzene | 1, 2-Dimethyl-1, 4-cyclohexadiene

An electron-withdrawing group

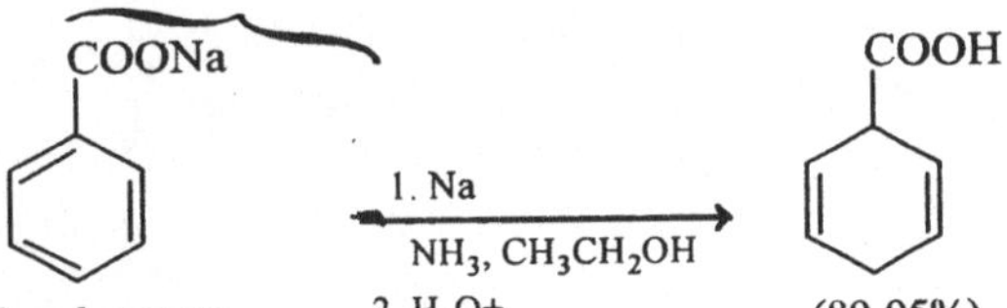

Sodium benzoate (89-95%)

With substituted benzene the electron-donating group remains on the unsaturated carbon while the electron-withdrawing group remains on the saturated carbon in the products.

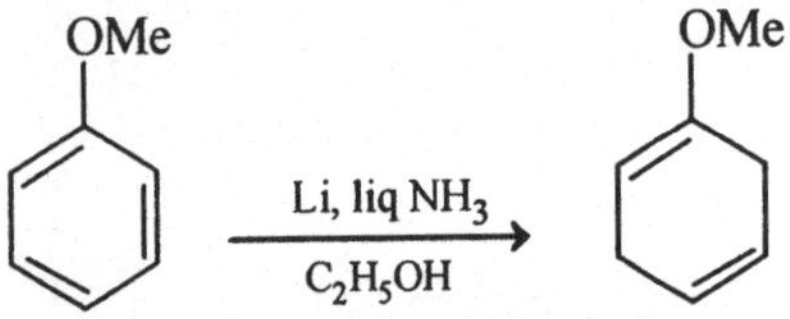

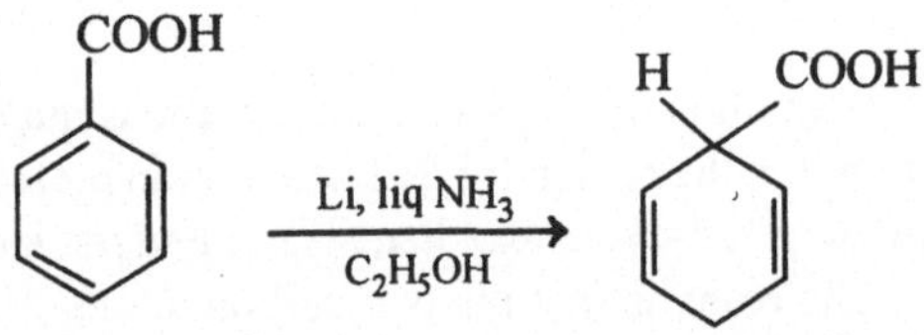

The reason is that calculations by molecular orbital method indicate that in the anion-radical

(*a*) the electron density has been the greatest at the ortho or meta positions with respect to the electron-releasing substituent in benzene and

(*b*) the electron density has been the greatest at the para position with respect to the electron withdrawing substituent in benzene.

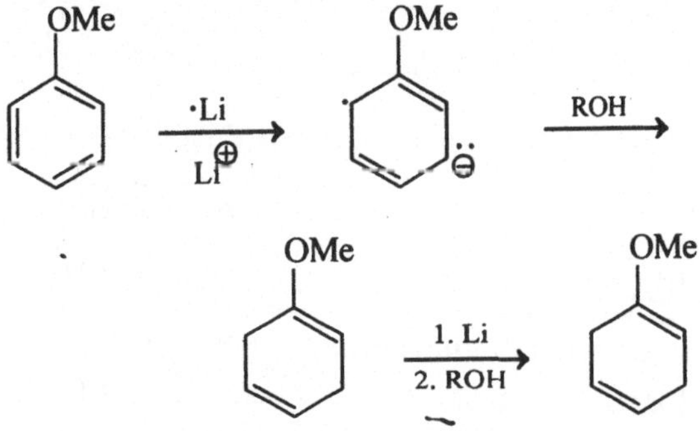

When a substituted aromatic compound is employed, a mixture of isomers may be obtained whose composition depends on the nature of the substituent (s).

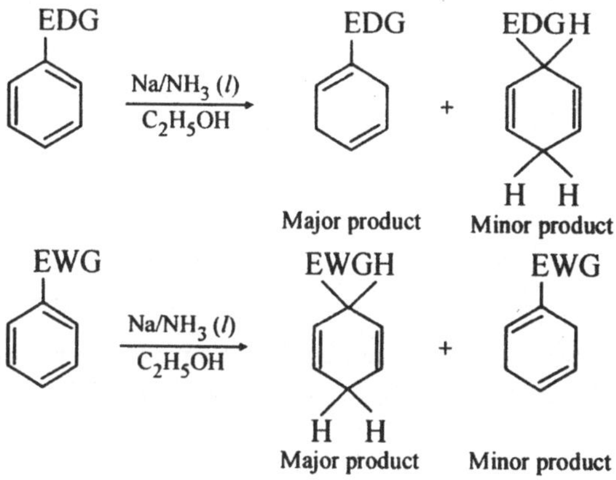

12

BLAISE REACTION

Condensation of cyanides with α-bromoester in presence of Zn is known as Blaise reaction. Product of the reaction is β-ketoester.

$$R-C\equiv N + Br-CH_2-COOC_2H_5 \xrightarrow{Zn}$$

$$R-\underset{}{\overset{\overset{N-ZnBr}{\|}}{C}}-CH_2-COOC_2H_5 \xrightarrow{HOH/H^{\oplus}} R-\overset{\overset{O}{\|}}{C}-CH_2-COOC_2H_5$$

$$C_6H_5-C\equiv N + Br-CH_2-COOC_2H_5 \xrightarrow[\text{(ii) } HOH/H^{\oplus}]{\text{(i) } Zn}$$

$$C_6H_5-\overset{\overset{O}{\|}}{C}-CH_2-COOC_2H_5$$

13

BOUVEALT-BLANCE REACTION

Reduction of ester by Na/C_2H_5OH into primary alcohol and other alcohol (*p*, Sec or Ter) is known as Bouvealt-Blance reaction.

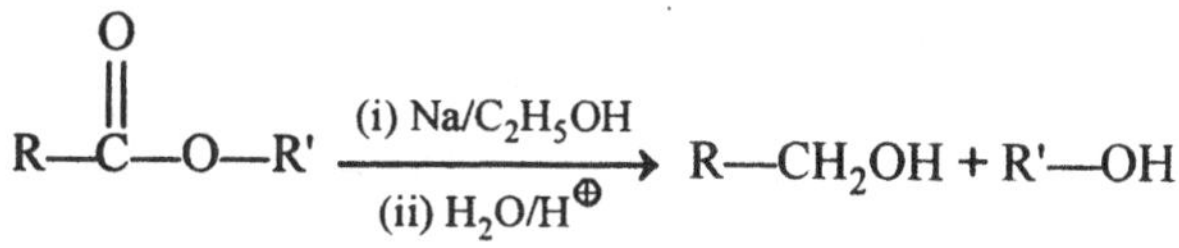

The same result is also obtained with $LiAlH_4$.

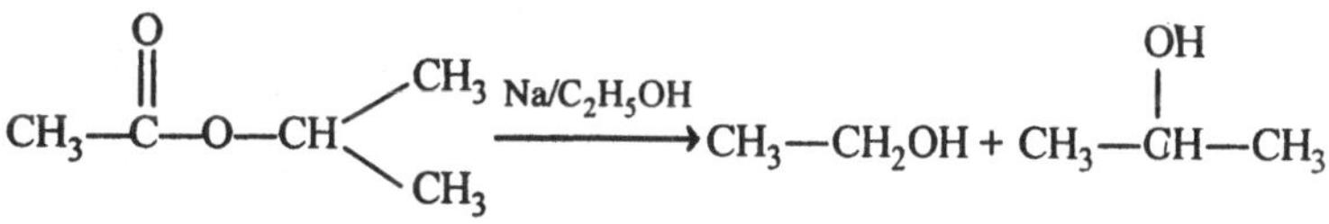

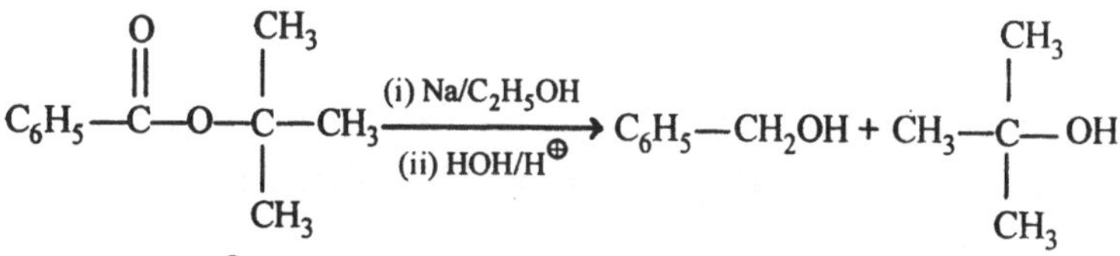

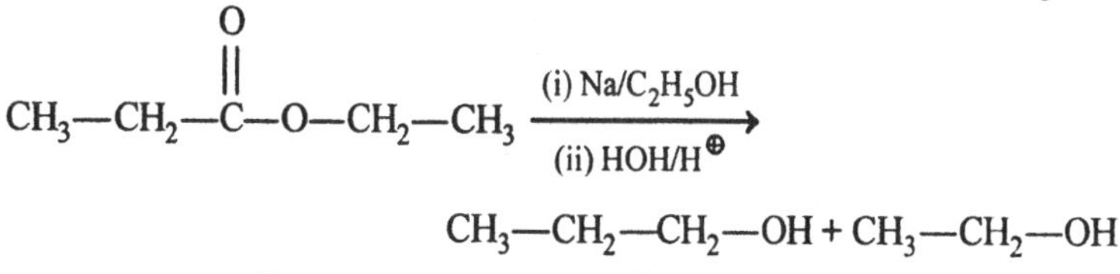

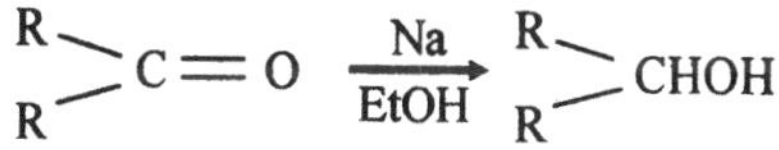

MECHANISM

The mechanism involves the electron transfer from the metal

$$Na - e \longrightarrow Na^{+}$$

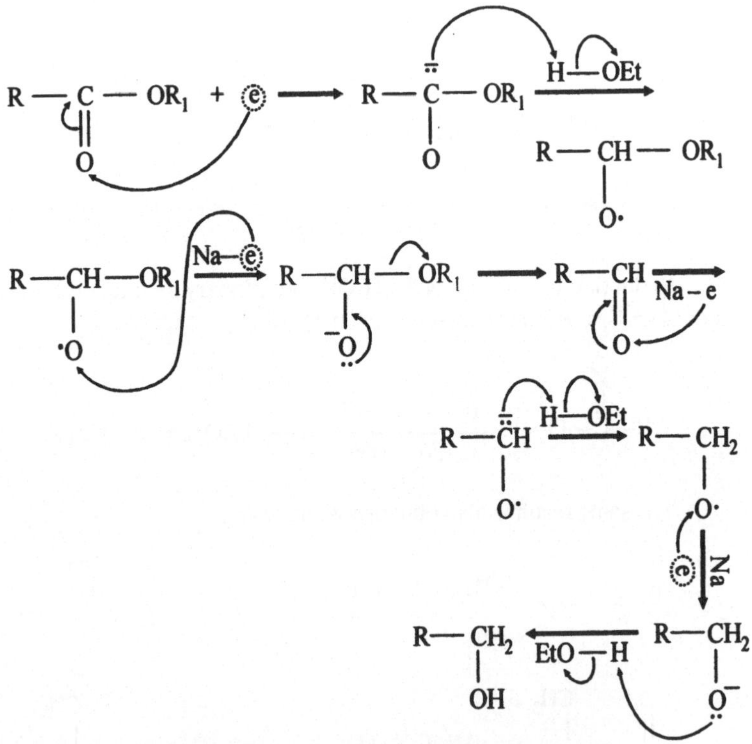

PROBLEM BASED ON BOUVEALT-BLANCE REACTION

Problem-1 $\underset{[A]}{C_{10}H_{18}O_2} \rightarrow \underset{[B]}{C_4H_{10}O} + \underset{[C]}{C_6H_{12}O}$

$$B \xrightarrow[\Delta]{PPA} C_4H_8 \xrightarrow{\text{Ozonolysis}} \text{Propionic Acid} + \text{Formic Acid}$$

$$C \xrightarrow[\Delta]{PPA} C_6H_{10} \xrightarrow{\text{Ozonolysis}} \text{Adipic acid}$$

Identify **A, B, C.**

Solution

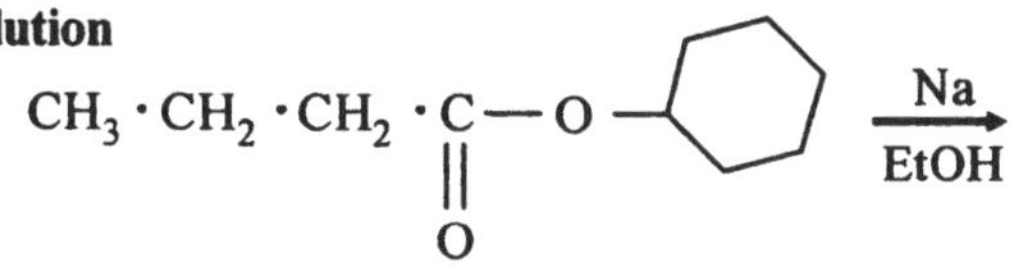

[A]

Cyclohexyl butyrate

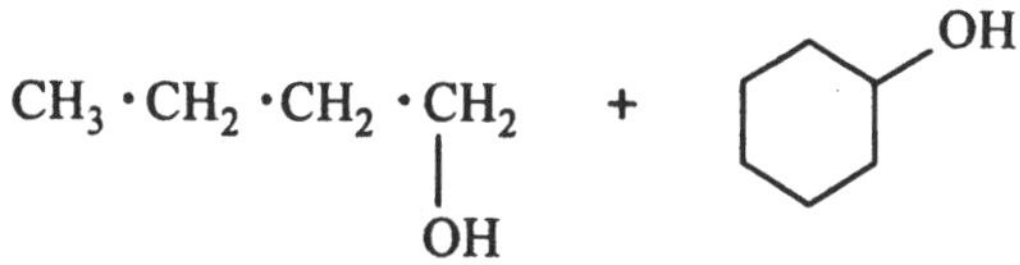

[B] [C]

Conclusion:	**A**	Cyclohexyl butyrate
	B	1-Butanol
	C	Cyclohexanol

Problem-2

$$\underset{[A]}{C_{13}H_{16}O_2} \xrightarrow[\text{EtOH}]{\text{Na}} \underset{[B]}{C_6H_6O} + \underset{[C]}{C_7H_{14}O}$$

B is acidic, forms sodioderivative and on heating with zinc dust produces benzene.

C on heating with conc. H_2SO_4 produces methylene cyclohexane.

Identify **A, B, & C.**

Solution

$$\left.\begin{array}{l} \mathbf{B} \longrightarrow \text{acidic} \\ \mathbf{B} \xrightarrow{\text{Zn}} \text{benzene} \end{array}\right\} \text{So, } \mathbf{B} \text{ is phenol}$$

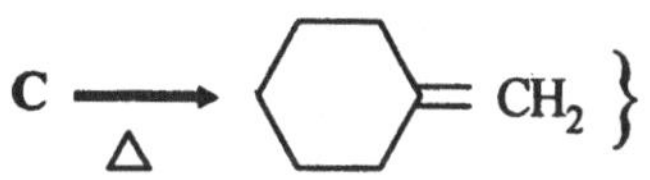

So, C sports

group

Ans.

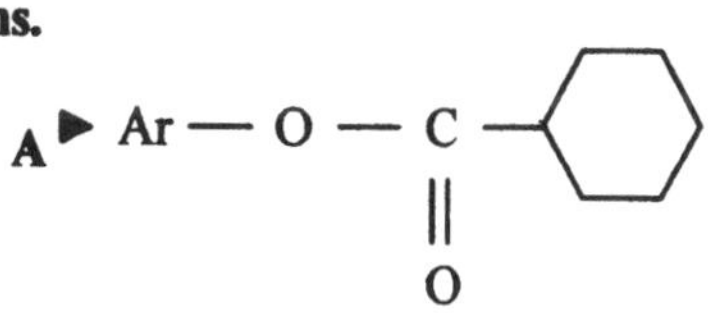

Phenyl cyclohexanoate

i.e. the compound A is Phenyl cyclohexanoate

B ▶ Ar — OH Phenol

C ▶ 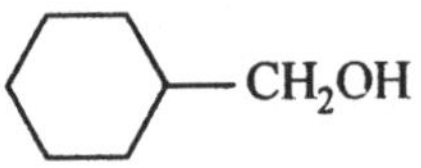 Cyclohexyl methanol

Problem-3

$$\underset{[A]}{C_4H_6O_2} \xrightarrow[\text{EtOH}]{\text{Na}} \underset{[B]}{C_4H_{10}O_2} \xrightarrow[\Delta]{Al_2O_3} \underset{[C]}{C_4H_6}$$

$$\underset{[C]}{C_4H_6} \xrightarrow{\text{Ozonolysis}} 2HCOOH + [\mathbf{D}]$$

C adds to two molar equivalents of bromine.
D has the following percentage composition:

C	26.66%
H	2.22%
O	71.12%

Indentity **A, B, C & D**

A ▶ H— C(=O) — CH_2 — CH_2 — C(=O) — H Butanedial

B ▶ CH_2(OH) — CH_2 — CH_2 — CH_2(OH) Butane-1, 4-diol

C ▶ CH_2 = CH — CH = CH_2 1,4-Butadiene

D ▶ COOH | COOH Oxalic acid

14

BRADSHER REACTION

Diaryl methane containing a carbonyl group in the ortho position undergoes cyclisation to an anthracene. Formation of anthracene takes place by 1, 4-dehydration reaction. This reaction is known as Bradsher reaction. The reaction takes place in the presence of acid, mainly polyphosphoric acid.

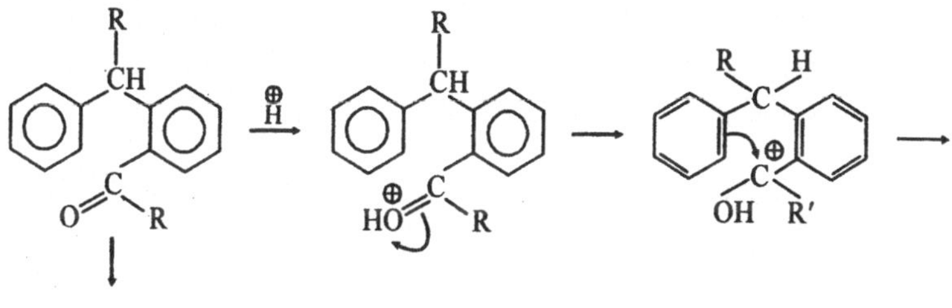

15

CANNIZZARO REACTION

The disproportionation (self redox) of aldehydes lacking α-hydrogen atoms[1],[2](viz. C_6H_5CHO HCHO R_3CCHO etc.) in presence of strong base to form salt of an acid and a primary alcohol is known as Cannizzaro reaction. The reaction best proceeds with the aromatic aldehydes.

$$\underset{\text{Benzaldehyde}}{2C_6H_5CHO} \xrightarrow{NaOH} \underset{\text{Benzyl alcohol}}{C_6H_5CH_2OH} + \underset{\text{Sod. benzoate}}{C_6H_5COONa}$$

$$\underset{\text{Formaldehyde}}{2HCHO} \xrightarrow{NaOH} \underset{\text{Methyl acohol}}{CH_3OH} + \underset{\text{Sod. formate}}{HCOONa}$$

In the above reaction note that one melecule of the aldehyde is oxidized to the salt of a carboxylic acid, while the second one is reduced to the corresponding alcohol. Such *self oxidation reduction (redox) reactions are called disproportionation or dismutation reaction.*

MECHANISM

The mechanism involves two distinct steps - ***Nucleophilic Attack*** and ***Hydride Ion Transfer.***

Step-[I] NUCLEOPHILIC ATTACK

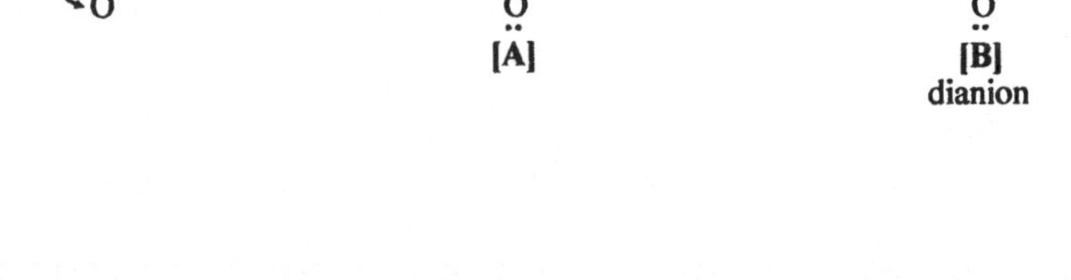

Step-[II] HYDRIDE ION TRANSFER

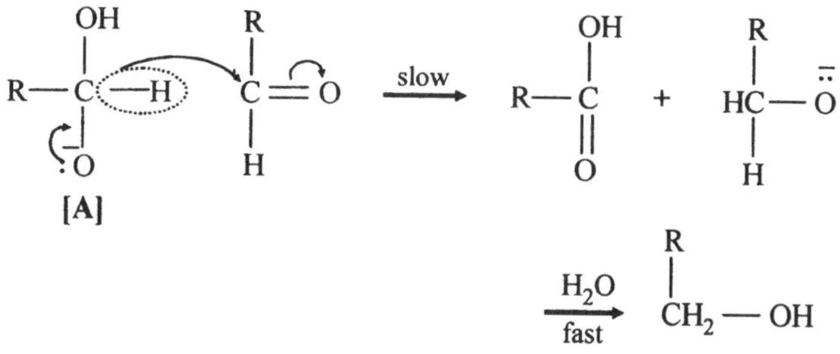

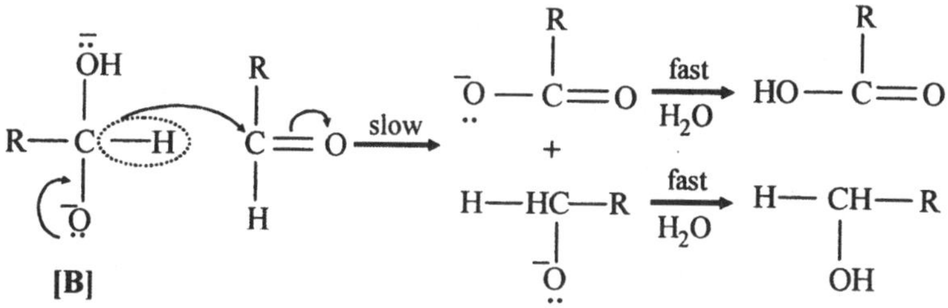

Note:

(I) All the aldehydes void of α-hydrogen atoms give rise to this reaction

(II) Benzaldehydes and other aromatic aldehydes having no α-hydrogen atoms undergo Cannizzaro reaction

EVIDENCE IN SUPPORT OF THE MECHANISM

(*i*) Under the usual conditions, the reaction follows third-order law (second order in aldehyde and first order in base) i.e.,

$$\text{Rate} \propto [\text{Ald}]^2[\text{OH}^-]$$

This reveals that the reaction occurring between the first-formed anion (from base and aldehyde) and another molecule of the aldehyde is the srate-determining step.

In the presence of a high concentration of base, the reaction follows fourth-order law;

$$\text{Rate} \propto [\text{Ald}]^2[\text{OH}^-]^2$$

However, the kinetics of Cannizzaro reactions of furfural and formaldehyde follows the fourth order law. This suggests the formation of a second anion from the first by the transfer of its proton to another hydroxide ion.

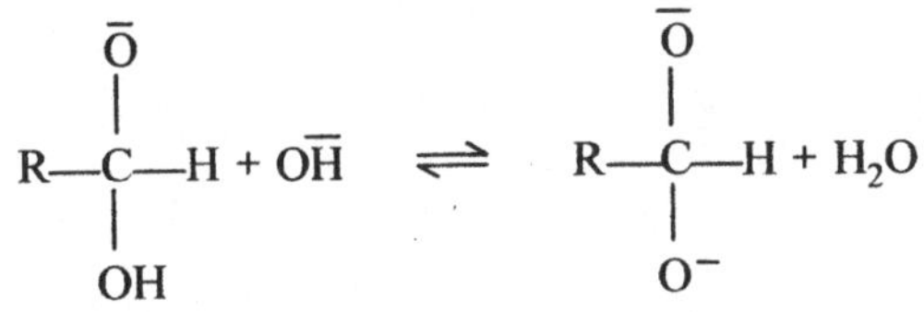

This doubly charged anion then furnishes a hydride ion to the carbonyl carbon of the aldehyde more readily.

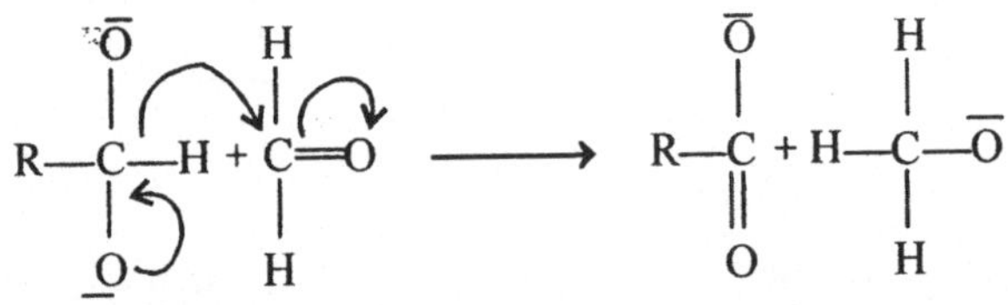

By keeping the above facts in view, the overall machanism for the Cannizzaro reaction of formaldehyde is as follows.

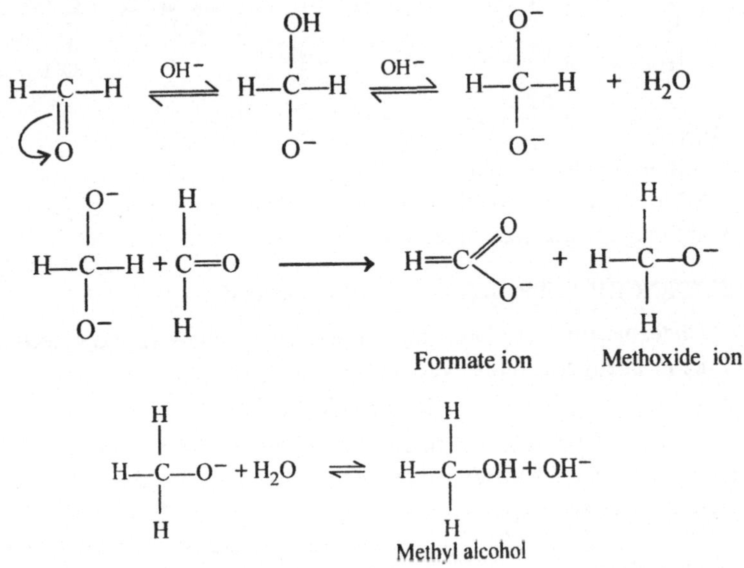

(*ii*) That the hydride ion is directly transferred from one molecule of the aldehyde to the other, and does not become free in solution has been proved by the observartion that the recovered alcohol does not contain deuterium when the reaction is performed in the presence of D_2O.

It can be seen that the reaction depends on the nucleophilic attack on the carbonyl carbon. Hence, factors which reduce the positive charge of the carbonyl carbon retard the reaction. In extreme cases the reaction may not take place. *e.g.*, *p*-dimethylamino benzaldehyde does not undergo Cannizzaro reaction.

Similarly, sterically hindered aldehydes fail to undergo the reaction.

INTRAMOLECULAR CANNIZZARO'S REACTION

There are certain carbonyl compounds such as 1, 2-dialdehydes and α-ketoaldehydes which may undergo internal Cannizzaro because they are capable of transferring a hydride ion. An example of these is the conversion of phenylglyoxal (II) to the salt of mandelic acid (III) in the presents of alkali.

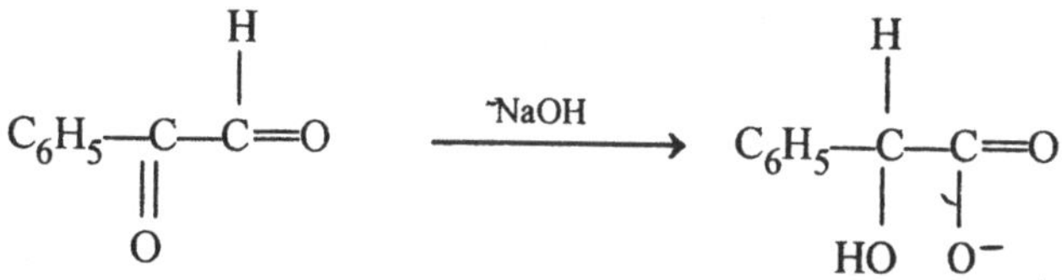

(II) (III)

This reaction has been found to be of second order, *i.e.*, first order in hydroxide and first order in phenylglyoxal. Also, no deuterium is incorporated in the product when this reaction is performed in D_2O. Keeping the above facts in view, the following machanism has been postulated :

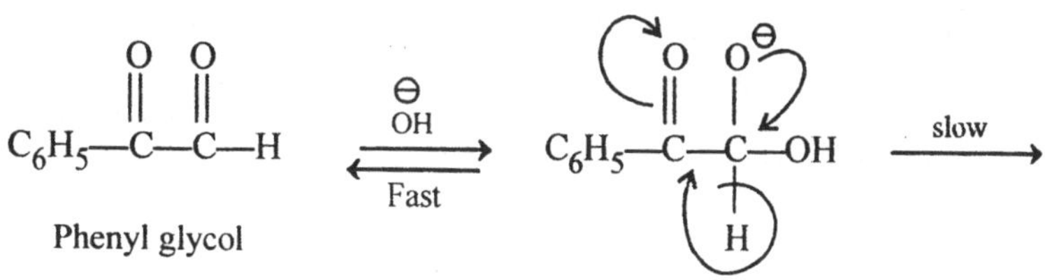

Phenyl glycol

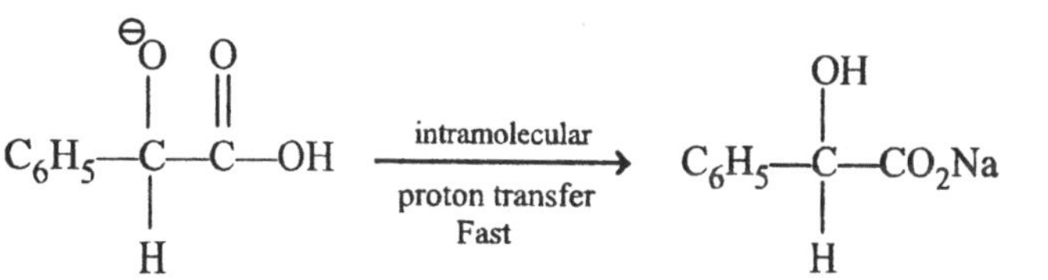

Sod. salt of mandelic acid

Another example of intramolecular Cannizzaro's reaction is as follows:

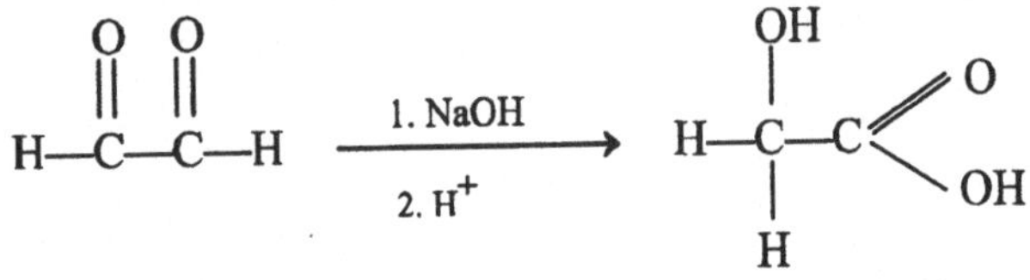

Glycol   Glycolic acid

EXTENSION AND APPLICATIONS

1. Crossed Cannizzaro reaction : The Cannizzaro reactions taking place between two different aldehydes is known as crossed Cannizzaro reaction. However, if one of the aldehydes is formaldehyde, the formate ion and the alcohol corresponding to the other aldehyde are formed exclusively. The exclusive formation of the formate ion is that formaldehyde is more reactive than other aldehydes towards nucleophiles and rapidly yields a high concentration of the donor anion.

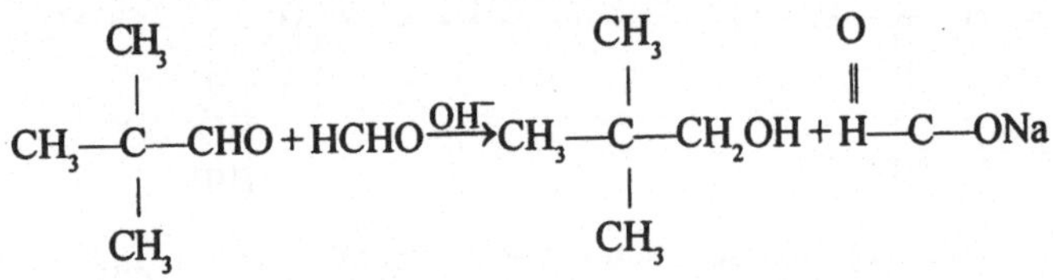

This reaction has been used to reduce aromatic aldehydes and hence for the prepartion of aromatic alcohols of the type $C_6H_5CH_2OH$.

However, if a mixture of HCHO and another aldehyde having α-hydrogen atom or atoms is warmed with alkali, formaldehyde first undergoes aldol condensation with another aldehyde, forming a hydroxyaldehyde lacking any α-hydrogen which then undergoes crossed Cannizzaro reaction with formaldehyde. An example is as follows :

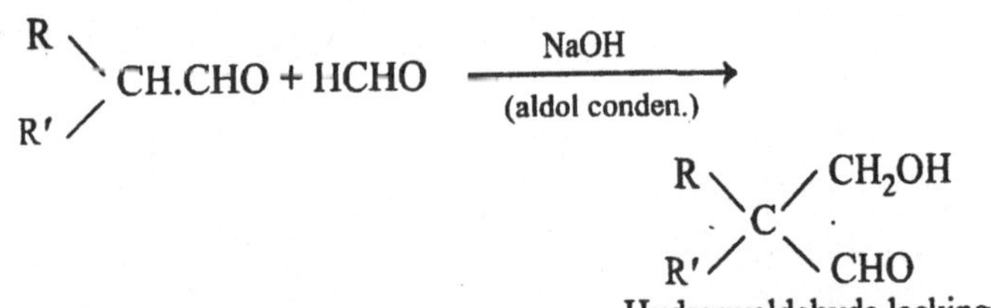

Hydroxyaldehyde lacking
α-hydrogen atom

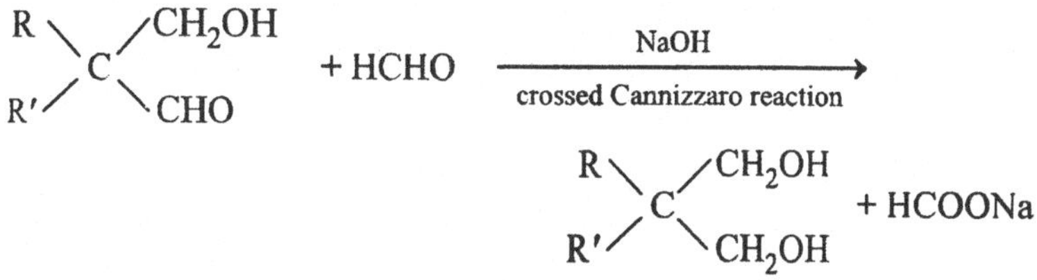

Thus, aldehydes containing one, two or three α-hydrogens react as given below.

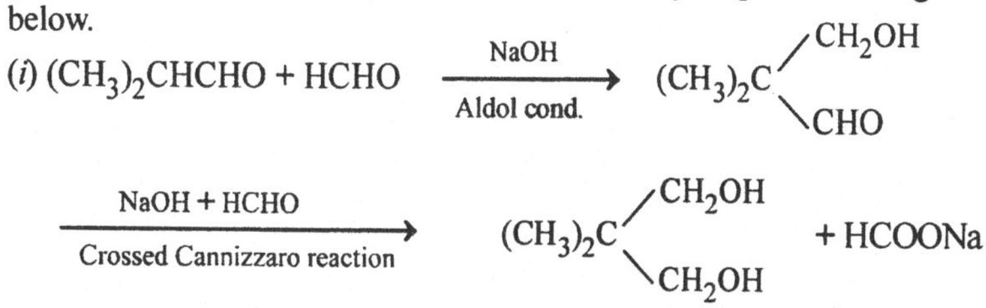

Substituted trimethylene glycol

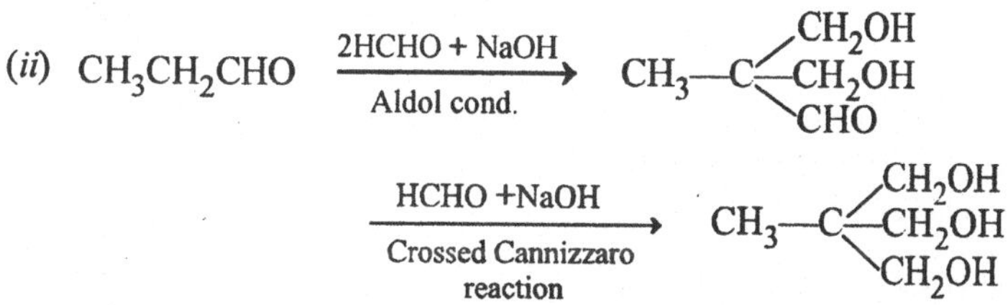

1, 1, 1-Trihydroxymethyl ethane

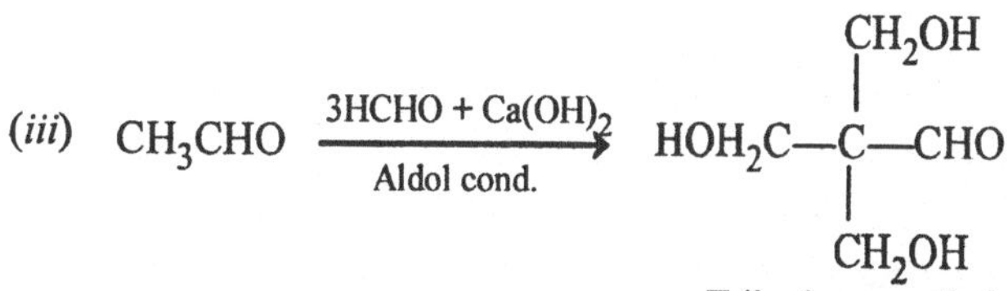

Trihydroxymethyl acetaldehyde

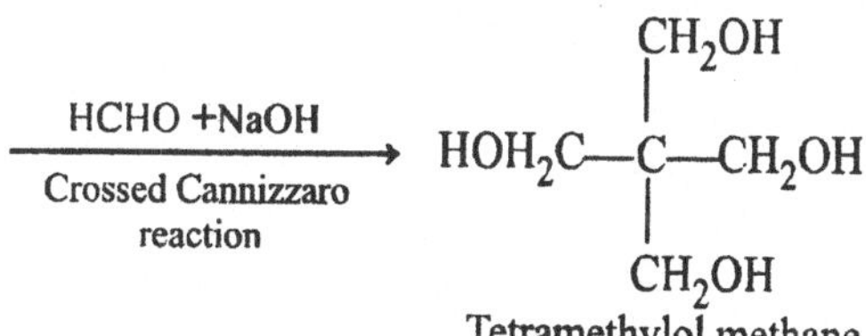

Tetramethylol methane (Pentaerythritol)

On the basis of this reaction, pentaerythritol (a tetrahydric alcohol) can be obtained from acetaldehyde and HCHO.

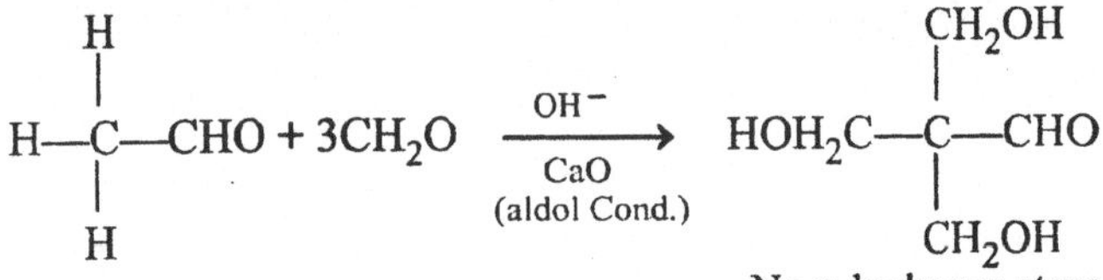

No α-hydrogen atom

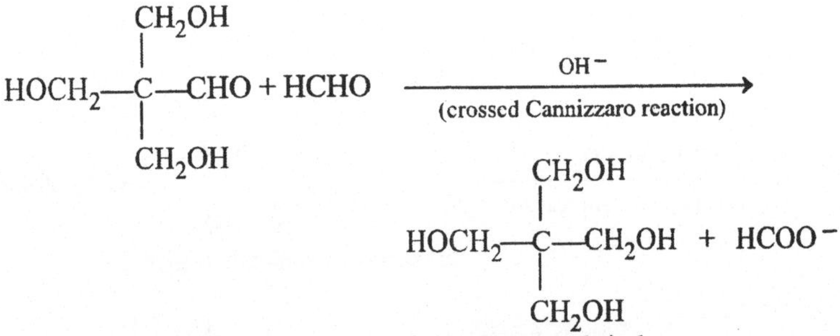

Pentaerythritol

Pentaerythritol when esterified with polybasic acids yields alkyal resin polymers. Also, its nitrate esters are used as explosives and medicinally as vasodilators.

$$\underset{\text{Pentaerythritol}}{C(CH_2OH)_4} + 3HNO_3 \xrightarrow{H_2SO_4} \underset{\text{Pentaerythritoltrinitrate (PETRIN)}}{HOCH_2.C(CH_2ONO_2)_3}$$

$$C(CH_2OH)_4 + 4HNO_3 \xrightarrow{H_2SO_4} \underset{\text{Pentaerythritoltetranitrate (PETN)}}{C(CH_2ONO_2)_4}$$

2. Heterocyclic compounds, which have aldehyde group, undergo Cannizzaro's reaction.

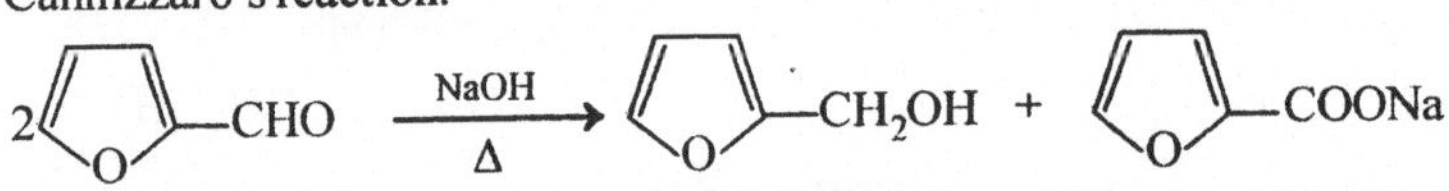

Furfural | Furfuryl alcohol | 2-Furoic acid (sod. salt)

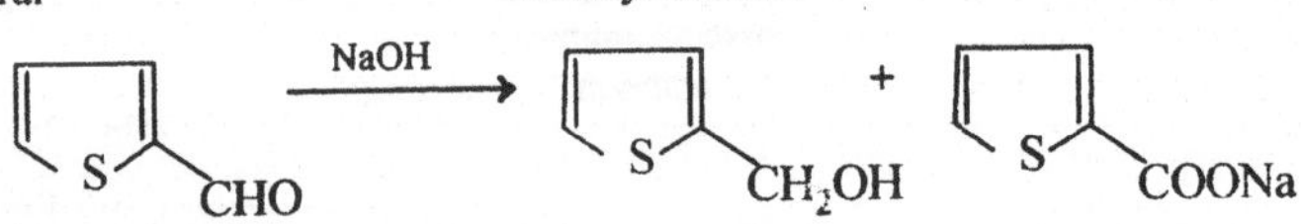

Thiophen-2-aldehyde

3. Glyoxal when warmed with alkali undergoes intramolecular Cannizzaro reaction to form glycollic acid.

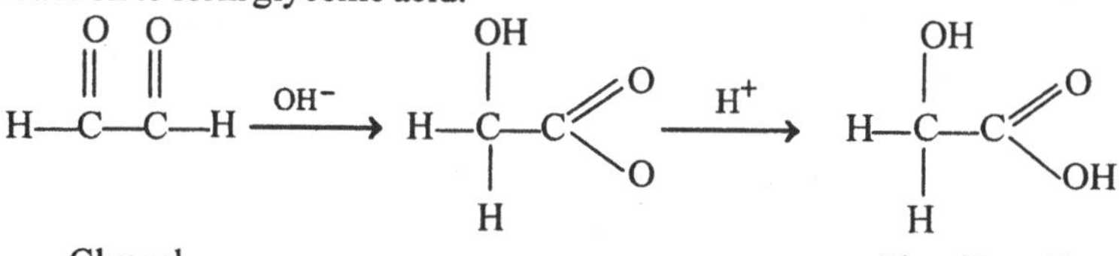

Glyoxal Glycollic acid

4. Cannizzaro reaction is used for the preparation of various alcohols and acids, *e,g.*,

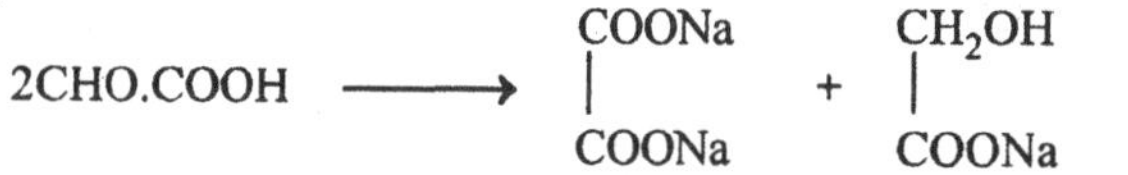

Glyoxalic acid Sod. oxalate Sod. glycollate

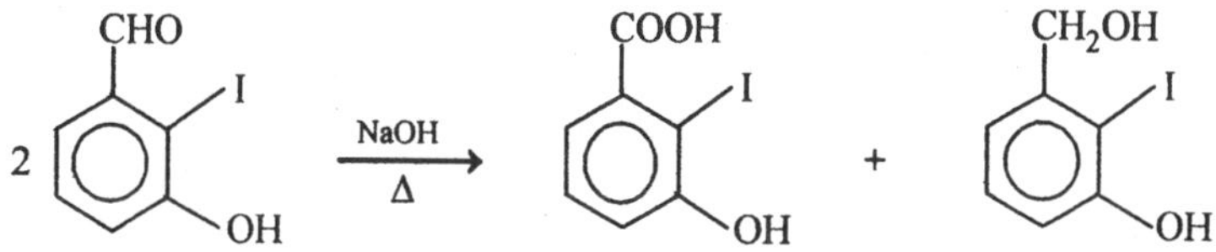

2-Iodo-3-hydroxy benzaldehyde 2-Iodo-3-hydroxy benzoic acid 2-Iodo-3-hydroxy benzyl alcohol

In some cases, Cannizzaro reaction fails due to steric hindrance. For example, di-*o*-substituted benzaldehyde does not undergo Cannizzaro reaction normally

5. Synthesis of ***para***-methoxy benzyl alcohol

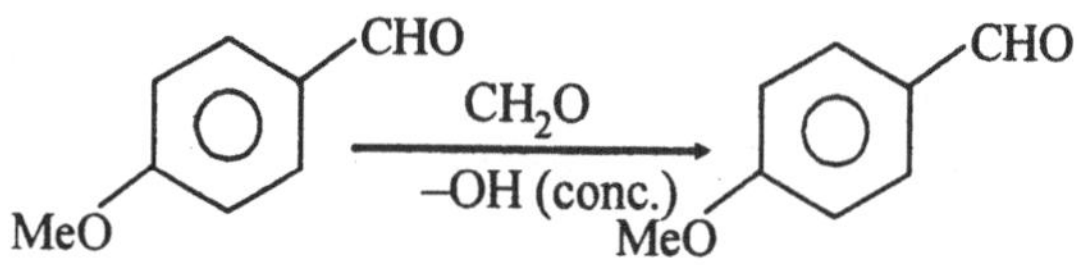

LIMITATION

In some cases, cannizzaro reaction fails due to steric hindrance. For example, di-o-substituted benzaldehyde does not undergo Cannizzaro reaction normally.

PROBLEMS

Problem-1

$$\underset{[A]}{C_8H_6O_2} \xrightarrow[\text{EtOH}]{CH_2O\,/\,\bar{O}H} \underset{[B]}{C_8H_{10}O_2} \xrightarrow{PBr_3} \underset{[C]}{C_6H_8Br_2}$$

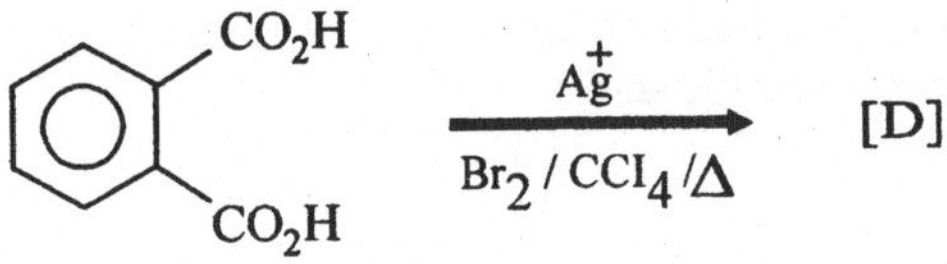

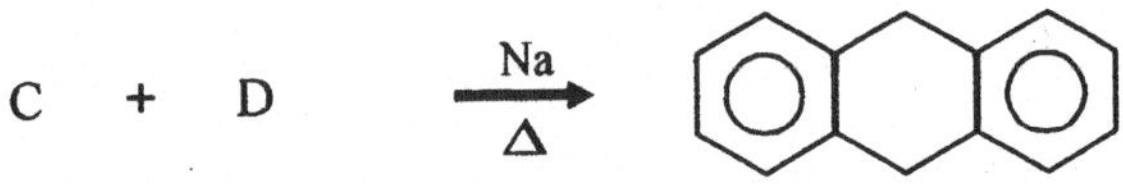

Identify **A, B, C & D**

D is definitely 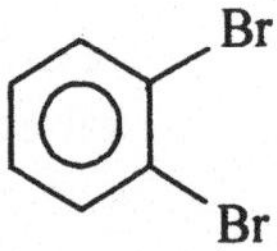

C and **D** upon fusion produce 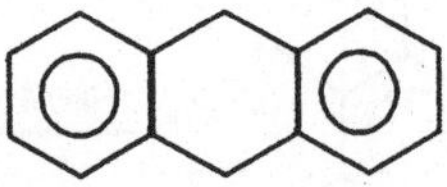

C is therefore 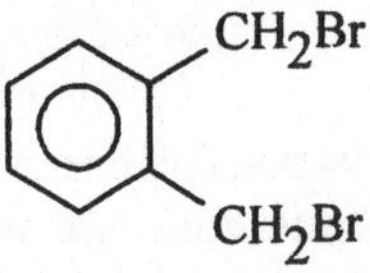

ANSWER

A ▶ Phthalaldehyde 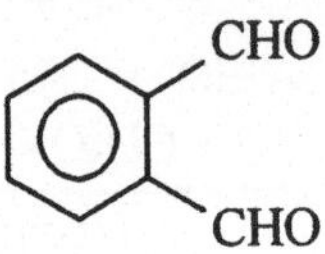

B ▶ Phenyl - 1,2 - dimethanol 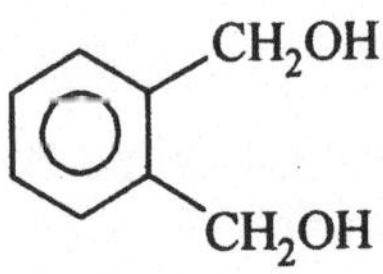

16

CARBYLAMINE REACTION

Alcoholic solution of *p*-amines (aliphatic and aromatic) react with chloroform in the presence of potassium hydroxide solution to yield isocyanide or carbylamine. This reaction is known as Carbylamine reaction.

The formation of isocyanides is readily noted by their characteristic nauseating odor. Due to this reason, this reaction has been used as a qualitative test for primary amines.

$$R{-}NH_2 \xrightarrow{CHCl_3/KOH/\Delta} R{-}\overset{\oplus}{N}{\equiv}\overset{\ominus}{C} + 3KCl + 3H_2O$$

MECHANISM:

Step I.

$$CHCl_3 \xrightarrow[\alpha\text{-elimination}]{\overset{\ominus}{O}H/\Delta} \underset{\text{Dichlorocarbene}}{\ddot{C}Cl_2}$$

Step II. 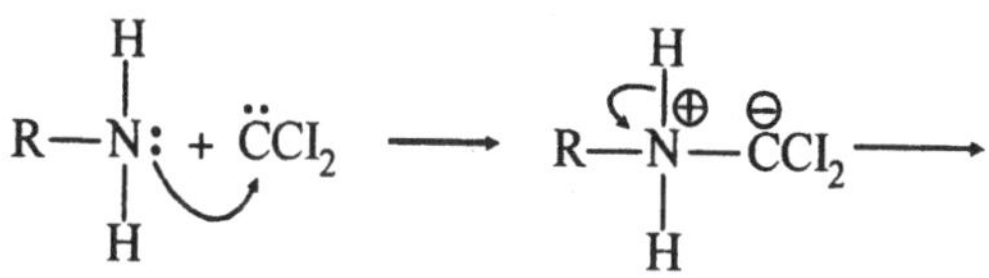

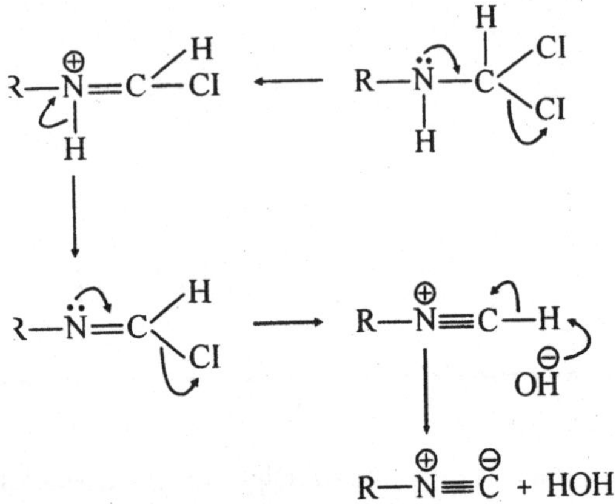

17

CHICHIBABIN REACTION

The amination of pyridine ring of pyridine and other heterocyclic compounds by amide ion is known as Chichibabin reaction. This reaction is aromatic nucleophilic substitution reaction which takes place at ortho position with respect to nitrogen. If *o*-positions are blocked, then the reaction takes place at the *p*-position.

Reaction takes place with $NaNH_2$ or KNH_2 in liquid ammonia at high temp. (100-200°C).

MECHANISM :

Step I. Nucleophilic attack of amide anion at α-position.

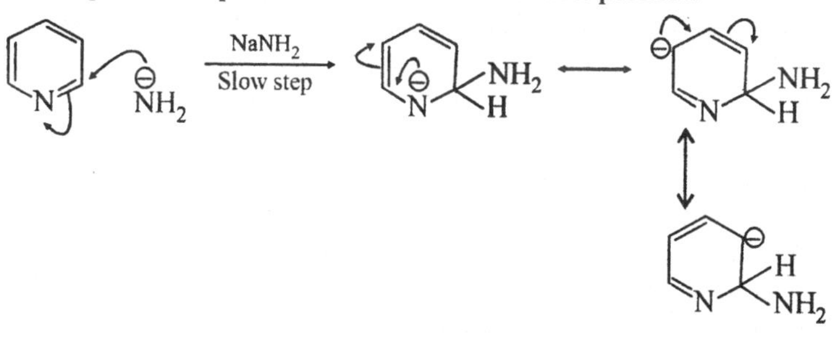

(Carbanion is resonance stabilised.)

Step II.

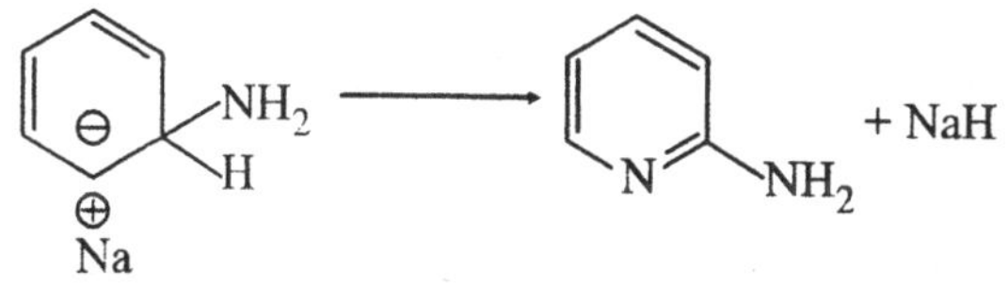

18

CHLOROMETHYLATION (or Blance Reaction)

Introduction of chloromethyl group by HCHO (as formaline or paraformaldehyde) and conc. HCl in presence of Lewis acid as catalyst in aromatic ring is known as chloromethylation.

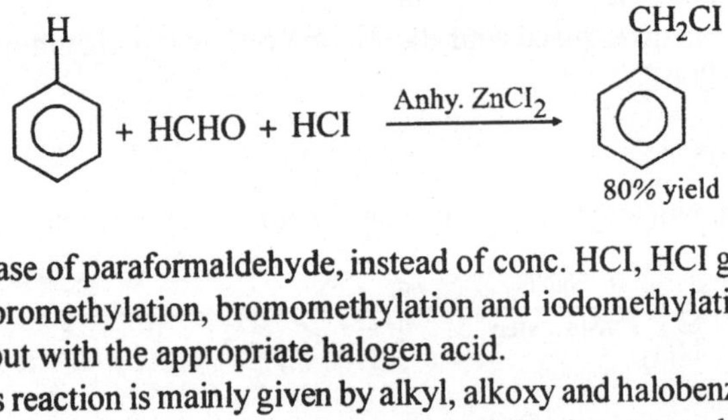

80% yield

In case of paraformaldehyde, instead of conc. HCl, HCl gas is used.

Fluoromethylation, bromomethylation and iodomethylation may be carried out with the appropriate halogen acid.

This reaction is mainly given by alkyl, alkoxy and halobenzenes. This reaction is also given by aromatic compounds having strong deactivating group such as nitrobenzene although *m*-dinitrobenzene and pyridine are inert.

This reaction is not significant when the compound has a strong activating group. For this reason chloromethylation is not a suitable procedure for phenols and aniline. These two compounds undergo polymerization.

MECHANISM:

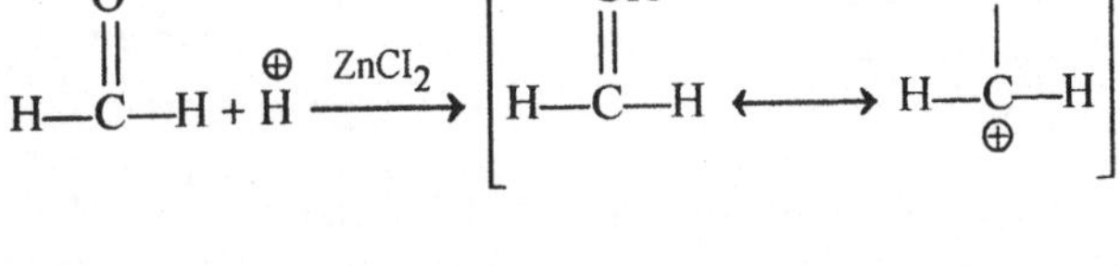

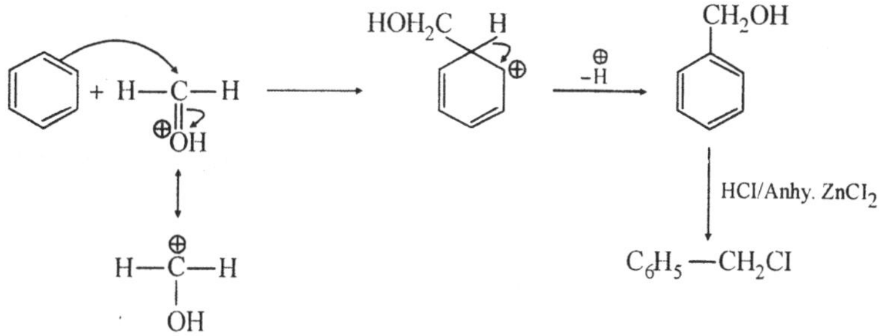

Since benzyl chloride is highly reactive for SN reaction, Cl group of the product can be replaced by different nucleophile.

$$C_6H_5-CH_2-Cl \xrightarrow{NaOH} C_6H_5-CH_2-OH$$

$$C_6H_5-CH_2-Cl \xrightarrow{KCN} C_6H_5-CN$$

Chloromethylation is an efficient process but it has a serious drawback. Small amount of the very carcinogenic compound, bis (chloromethyl) ether is formed in the reaction mixture so that the process has fallen out of favour.

$$2CH_2O \xrightarrow{HCl} Cl-CH_2-O-CH_2-Cl$$

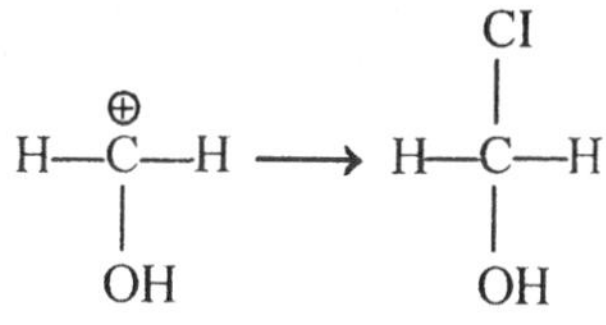

$$CH_2Cl-OH + HO-CH_2Cl \xrightarrow{\overset{\oplus}{H}} Cl-H_2C-\overset{+}{O}-CH_2-Cl$$

19

CLEMMENSEN REDUCTION

Reduction of carbonyl group to a methylene group by activated zinc and hydrochloric acid is known as Clemmensen reduction. Zinc is activated by treating it briefly with a small amount of $HgCl_2$(aq.)

$$C_6H_5-\overset{\overset{\displaystyle O}{\|}}{C}-CH_3 \xrightarrow[\text{(ii) } H_2O/H^{\oplus}]{\text{(i) } Zn/HgCl_2} C_6H_5-CH_2-CH_3$$

The reaction is useful, specially, when carbonyl compound contains phenolic or carboxylic groups which remain unaffected. This reaction is not useful for those compounds having acid sensitive groups and high molecular weight substrate. The reagent also reduces C=C bond in α, β-unsaturated ketones, acids and esters. Benzyl halides and alcohols are hydrogenated.

$$C_6H_5-\overset{\overset{\displaystyle O}{\|}}{C}-CH_2-CH_2-COOH \xrightarrow{\text{(i) } Zn\text{–}Hg\text{–}Cl_2} C_6H_5-CH_2-CH_2-CH_2-COOH$$

MECHANISM:

Step I.

$$R-\overset{\overset{\displaystyle \ddot{O}:}{\|}}{C}-R \underset{}{\overset{H^{\oplus}}{\rightleftharpoons}} R-\overset{\overset{\displaystyle \overset{\oplus}{O}-H}{\|}}{C}-R$$

Step II.

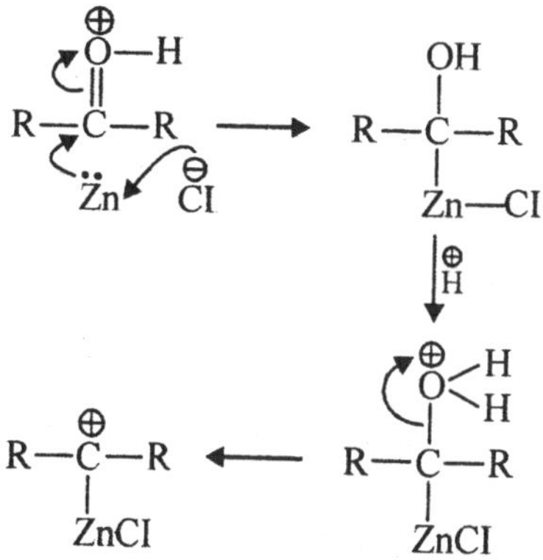

Step III.

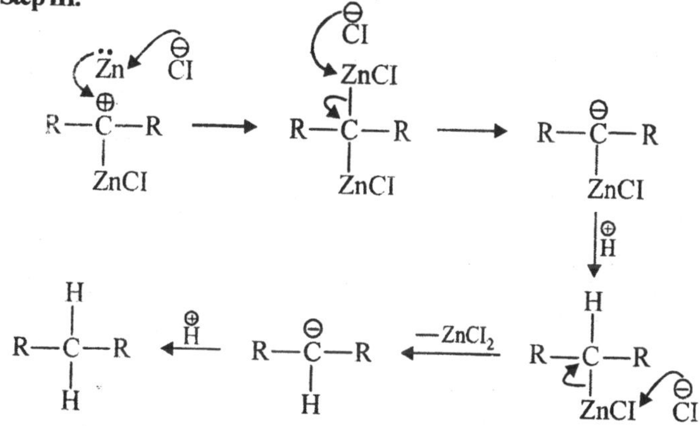

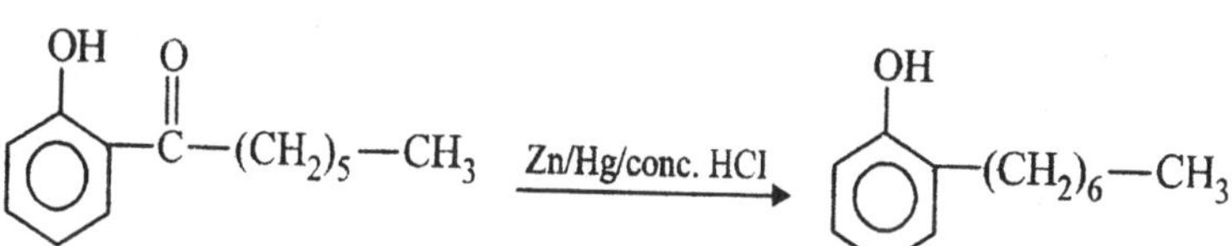

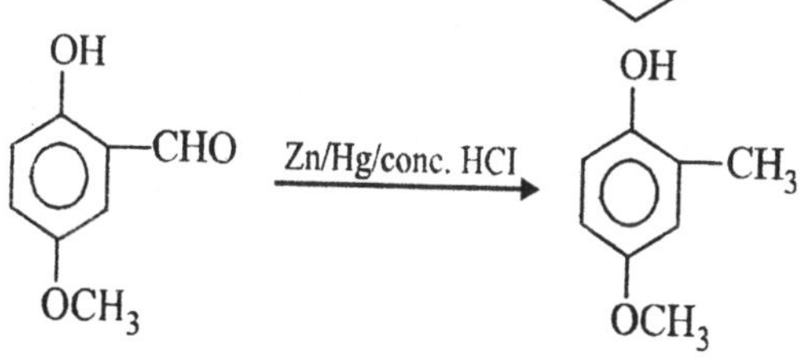

Certain cyclic 1, 3-diketones on Clemmensen reduction yield a fully reduced product along with a monoketone with ring contraction.

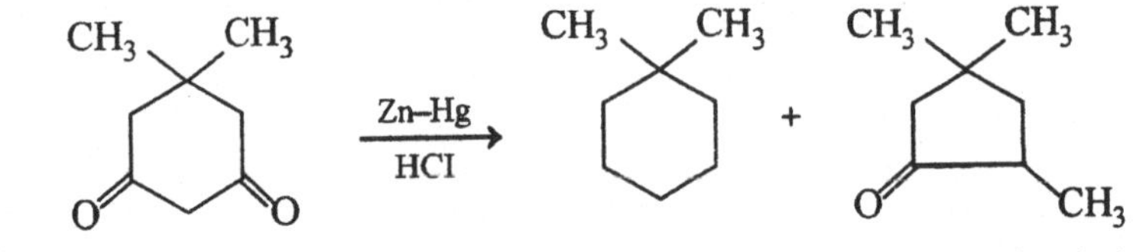

5, 5-Dimethyl cyclohexane-1, 3-dione   1, 1-Dimethyl cyclohexane   2, 4, 4-Trimethyl cyclopentanone

The latter probable is formed through the formation of a diradical with subsequent intramolecular C—C bond formation and pinacol-type rearrangement.

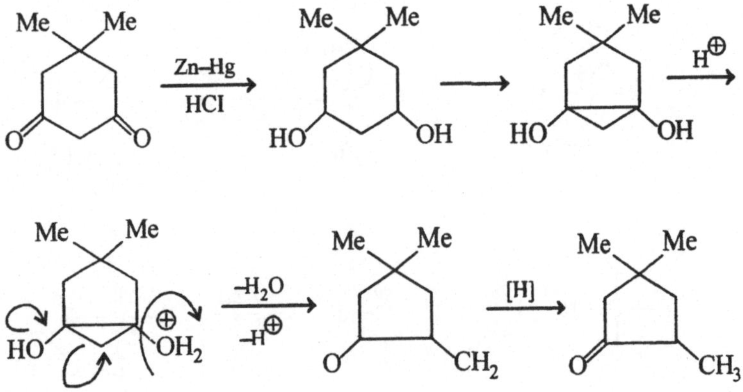

APPLICATIONS

1. **Coversion of Friedel-Craft's products by Clemmensen reductions** : The Clemmensen reduction converts products of Friedel-Crafts acylation to alkylbenzenes, including many of the alkyl benzenes that are impossible to make direct alkylation. For example, the Friedel-Crafts alkylation cannot be used to synthesize *n*-propyl benzene. Benzene reacts with *n*-propyl chloride and $AlCl_3$ to give isopropyl benzene, together with some diisopropylbenzene. Using the acylation, benzene reacts with propanoyl chloride and $AlCl_3$ to give ethyl phenyl ketone (propiphenone) which is reduced easily to *n*-propylbenzene by Clemmensen reduction.

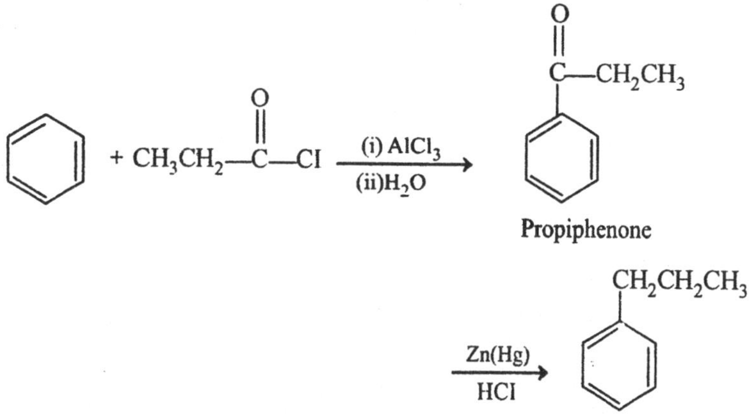

Propiphenone

n-Propylbenzene

2. Reduction of aliphatic and mixed aliphatic—aromatic carbonyl compounds,e,g.,

(a) $CH_3—(CH_2)_5—CHO \xrightarrow[HCl]{Zn–Hg} CH_3—(CH_2)_5—CH_3$

n-Heptaldehyde → *n*-Heptane

(b) $C_6H_5COC_3H_7 \xrightarrow[HCl]{Zn–Hg} C_6H_5CH_2C_3H_7$

n-Propylphenyl ketone → *n*-Butyl benzene

3. Reduction of keto acids, e,g.,

$C_6H_5COCH_2CH_2COOH \xrightarrow{Zn—Hg, HCl} C_6H_5CH_2CH_2CH_2COOH$

β-Benzoyl propionic acid → γ-Phenyl butanoic acid

α-And β-keto acids are generally not reduced

4. Redcuction of phenolic carbonyl compounds, e,g.,

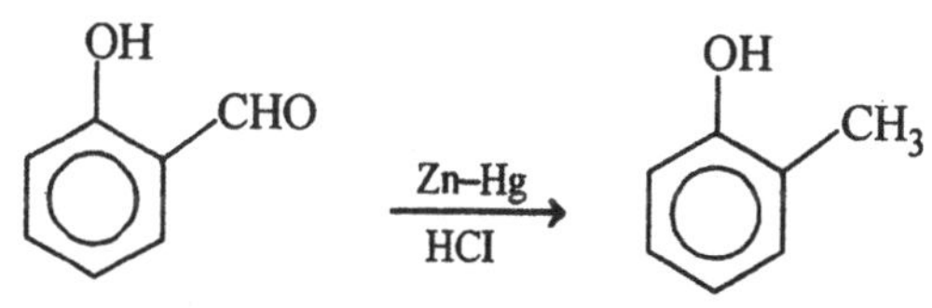

Salicylaldehyde → *o*-Cresol

5. Reduction of cylic ketones, e,g.,

(*a*)

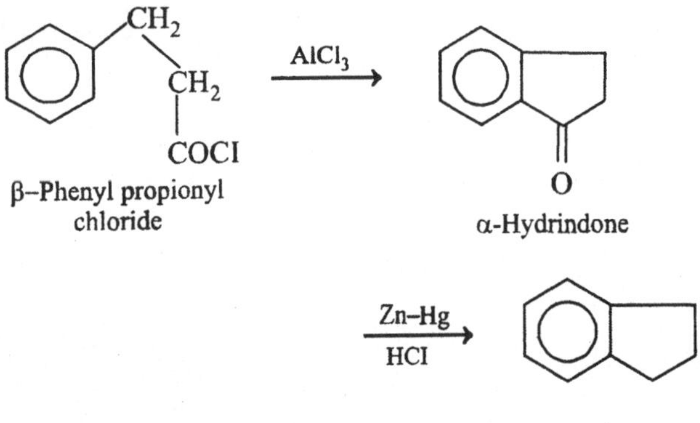

β–Phenyl propionyl chloride

α-Hydrindone

Indane

(b) In the synthesis of naphthalene, e,g.,

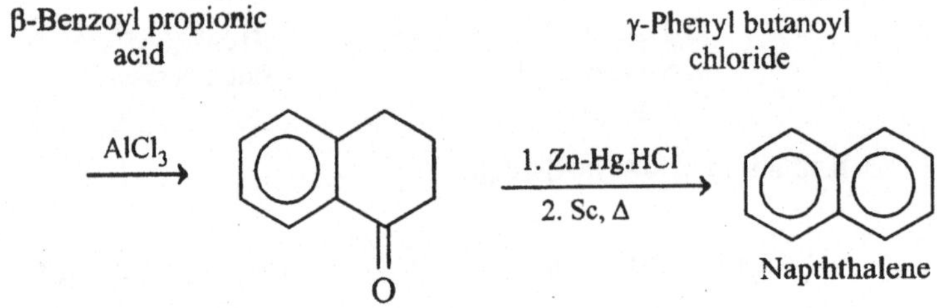

β-Benzoyl propionic acid

γ-Phenyl butanoyl chloride

α Tetralone

Napththalene

6. Reduction with ring expansion, e.g.,

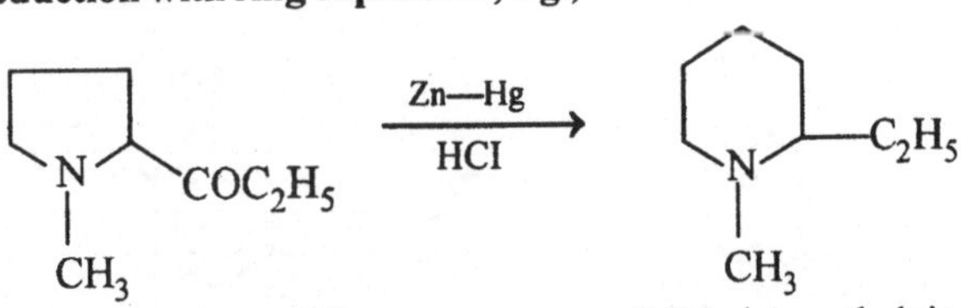

1-Methyl-2-propionyl pyrroidine

2-Ethyl-1-methylpiperidine

7. Reduction with ring contraction, e.g.,

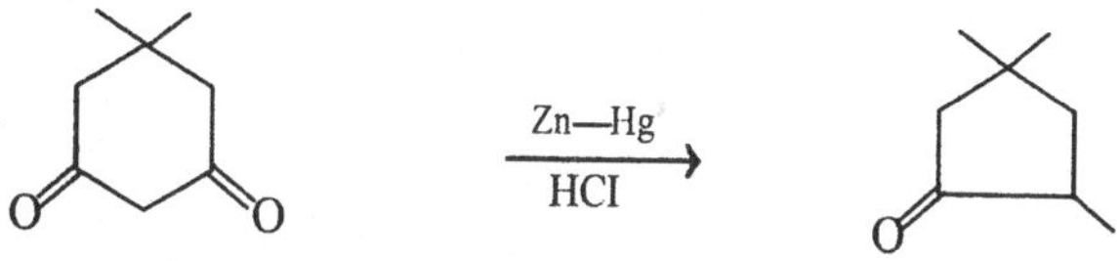

5, 5-Dimethyl cyclohexane-1, 3-dione 2, 4, 4-Trimethyl cyclopentanone

8. Synthesis of cycloparaffins, e.g.,

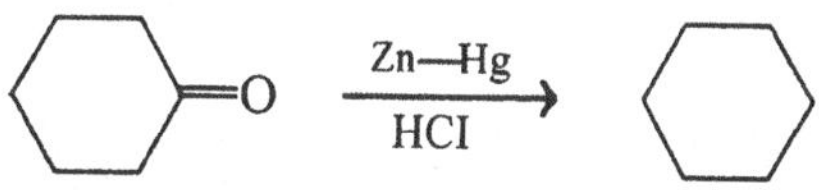

Cyclohexanone Cyclohexane

9. Zinc (dust) and hydrochloric acid has also been used in the reduction of pyrrole which gives both tetrahydro and 2, 5-dihydro derivatives.

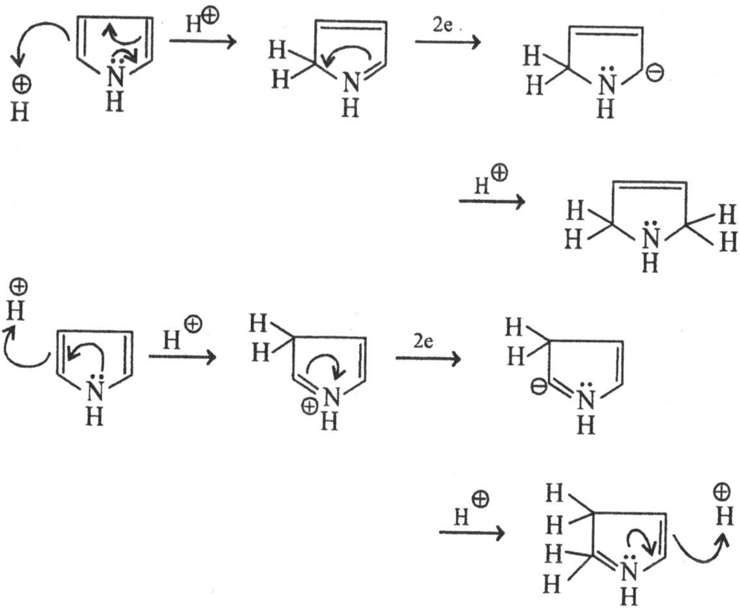

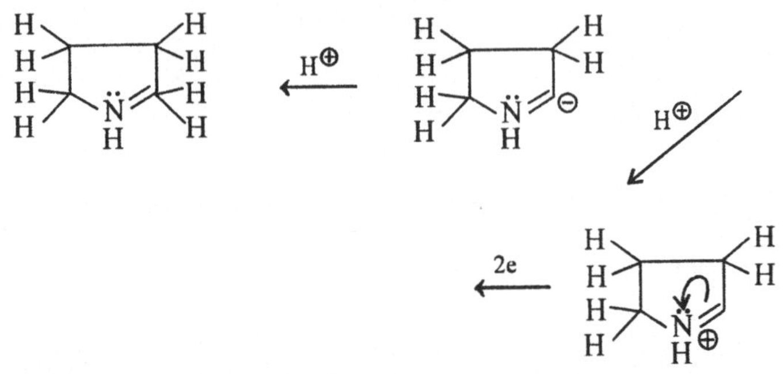

PROBLEMS BASED ON CLEMMENSEN REDUCTION

Problem-1.

$$\underset{[A]}{C_8H_8O_2} \xrightarrow{PCl_5} \underset{[B]}{C_8H_7OCl} \xrightarrow[EtOH]{KCN} \xrightarrow{H_3\overset{+}{O}} \underset{[C]}{C_9H_8O_3}$$

$$\xrightarrow[\text{HCl conc.}]{\text{Zn-Hg}} \underset{[D]}{C_8H_{10}O_2}$$

A is soluble in aqueous $NaHCO_3$ solution but insoluble in mineral acids.

D on heating produces ***ortho***-xylene. Identify **A, B, C, & D.**

Solution

$$\underset{[A]}{C_7H_7COOH} \xrightarrow{PCl_5} \underset{[B]}{C_7H_7CO.Cl} \xrightarrow{KCN} C_7H_7CO.CN \xrightarrow{H_3\overset{+}{O}}$$

$$\underset{[C]}{C_7H_7CO.CO_2H} \xrightarrow[\text{HCl conc.}]{\text{Zn-Hg}} \underset{[D]}{C_7H_7.CH_2.CO_2H}$$

CONCLUSION

A ▶ *ortho* - Methyl benzoic acid

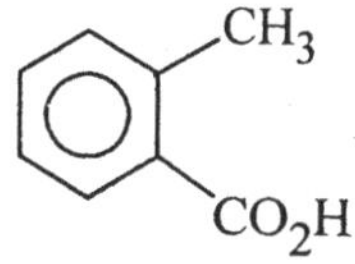

B ▶ *ortho* - Methyl benzoyl chloride

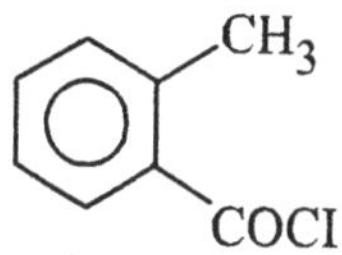

C ▶ *ortho* - Methyl carbobenzoxy acid

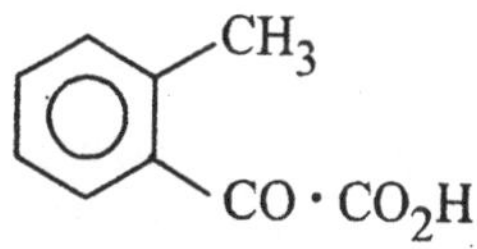

D ▶ *ortho* - Methyl phenyl acetic acid

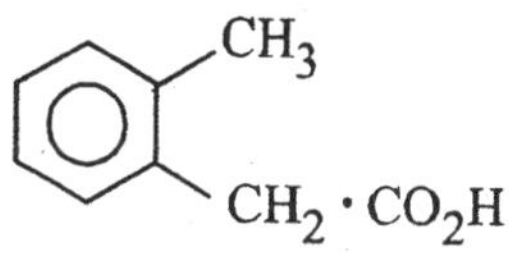

20

CLAISEN CONDENSATION

The condensation of an ester and an α hydrogen containing ester, ketone or nitrile to form a β-ketoester, ketone or nitrile respectively is known as Claisen condensation. The condensation is catalysed by sodium ethoxide, sodamide, triphenylmethyl sodium, etc. The most common and the simplest example is the condensation of ethylacetate in presence of ethoxide ion to form ethyl acetoacetate.

$$CH_3-\overset{\overset{\displaystyle O}{\|}}{C}-OC_2H_5 + CH_3-\overset{\overset{\displaystyle O}{\|}}{C}-OC_2H_5 \xrightarrow[\text{(ii) } H^{\oplus}]{\text{(i) } C_2H_5ONa/C_2H_5OH} \underset{\beta\text{-keto ester}}{CH_3-\overset{\overset{\displaystyle O}{\|}}{C}-CH_2-COOC_2H_5} + C_2H_5OH$$

MECHANISM :

Step I.

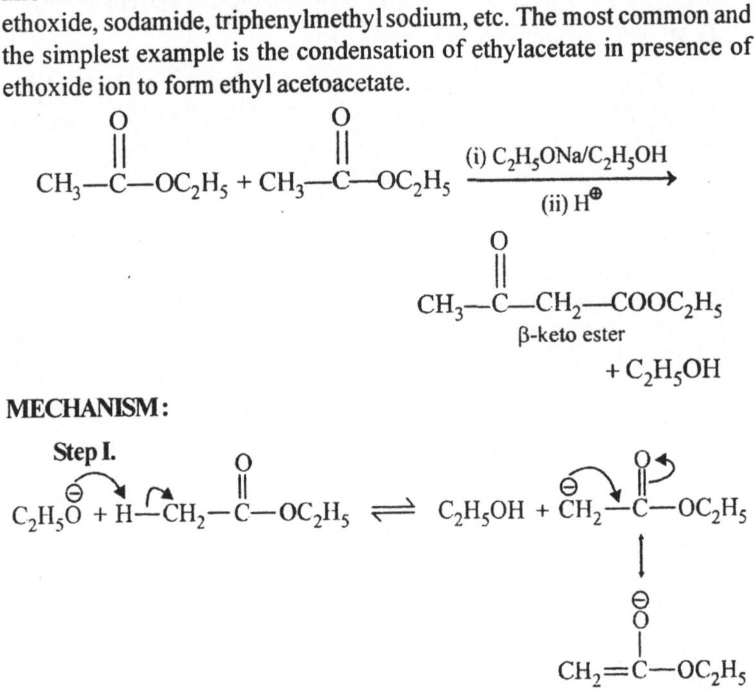

Enolate anion is satabilised by delocalisation

Step II.

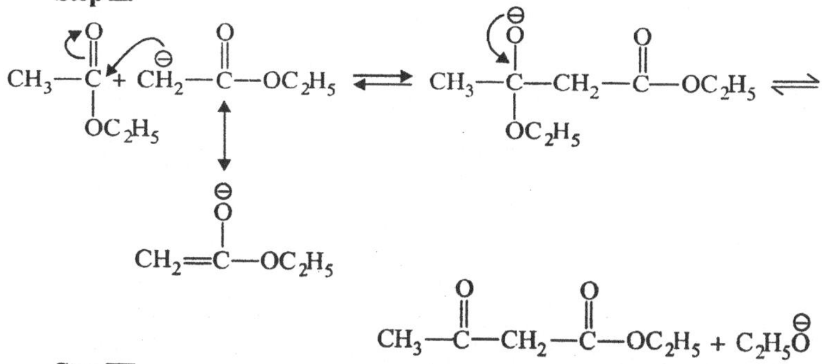

Step III.

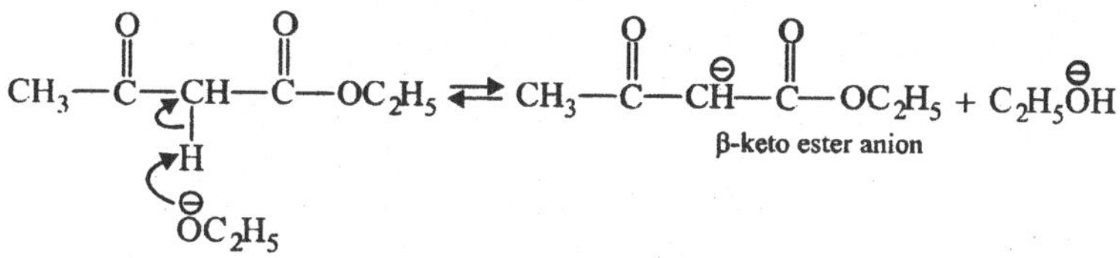

In this β-keto ester anion is stablished by delocalisation.

Step IV.

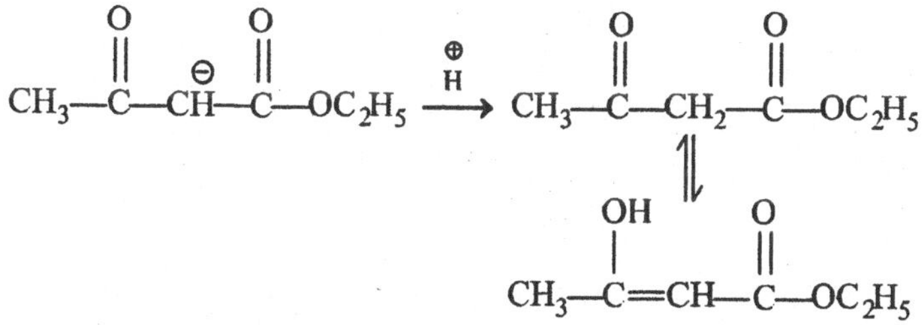

Cross-Claissen condensation : Cross Claissen condensation is generally carried out between two different ester, one having no α-hydrogen and other having at least one α-hydrogen.

In this case, ester having no α-hydrogen does not undergo enolisation and hence behaves as the substrate and ester having α hydrogen undergoes enolisation and hence behaves as the source of nucleophile.

The esters having no α-hydrogens which behave as substrate are

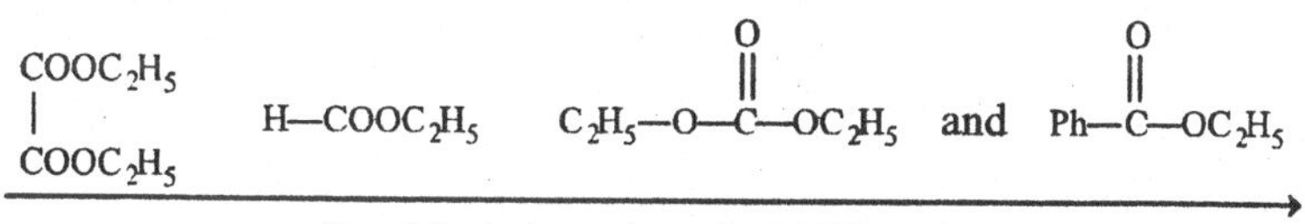

Reactivity in decreasing order for SN reaction

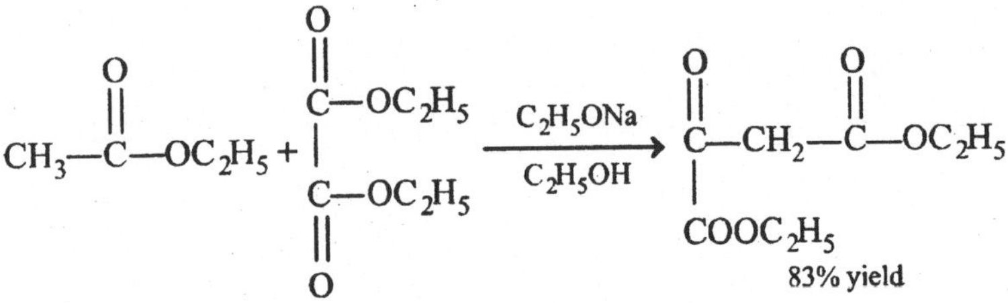
C2H5ONa
C2H5OH
83% yield

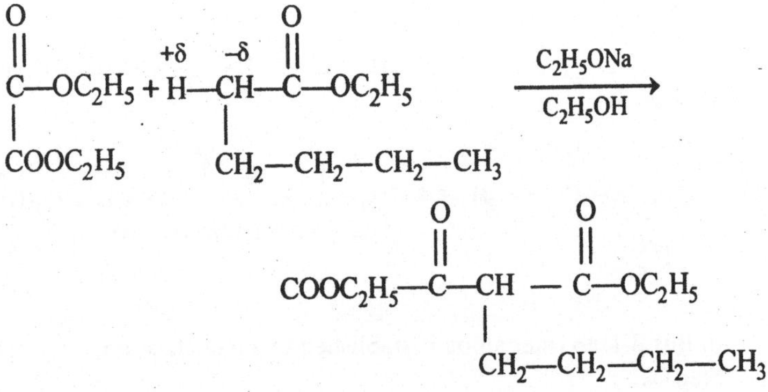
C2H5ONa
C2H5OH

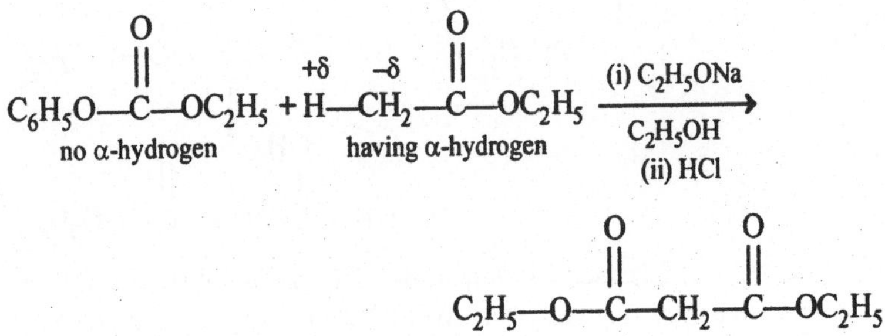
no α-hydrogen
having α-hydrogen
(i) C2H5ONa
C2H5OH
(ii) HCl

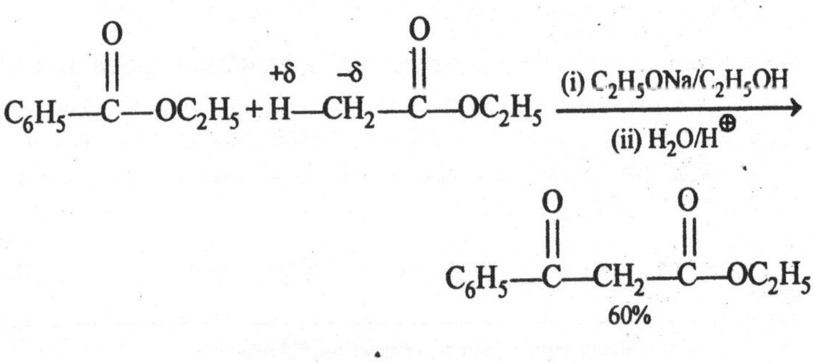
(i) C2H5ONa/C2H5OH
(ii) H2O/H⊕
60%

Cross-Claissen condensation also takes place between ester (generally having no α-hydrogen) and ketone having at least one α-hydrogen. In this case ketone behaves as a source of enolate (as nucleophile).

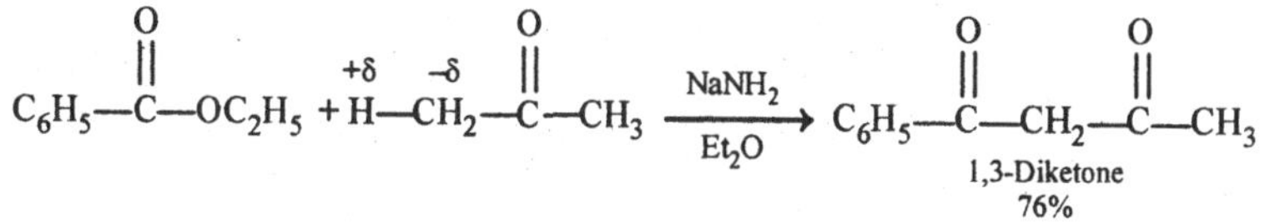

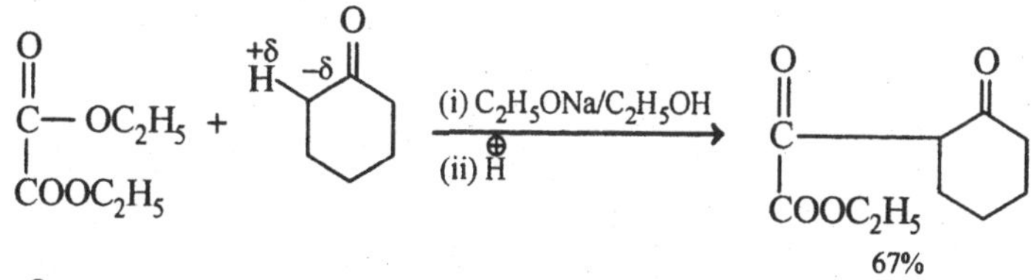

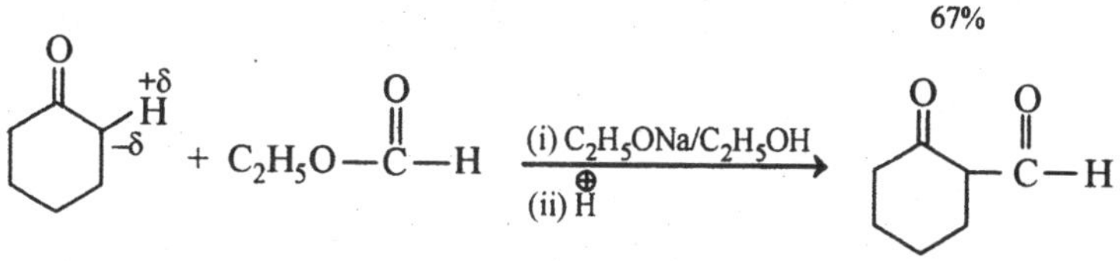

COMPARISON BETWEEN CLAISEN CONDENSATION AND ALDOL CONDENSATION

Although Claisen condensation is very similar to aldol condensation, yet the only difference is that in aldol condensation no group is lost whereas in Claisen condensation the group $C_2H_5O^-$ is lost because it is a good leaving group.

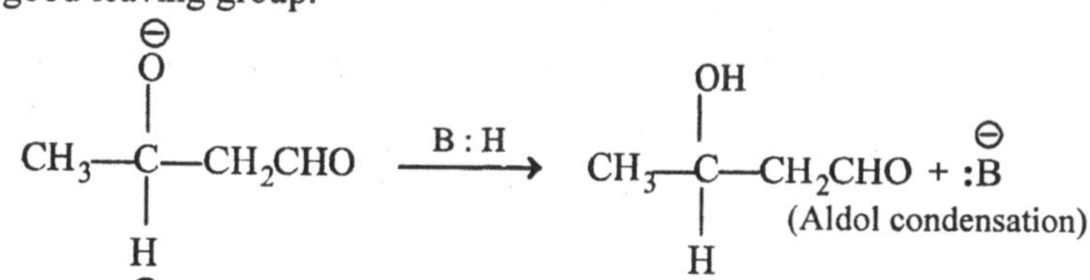

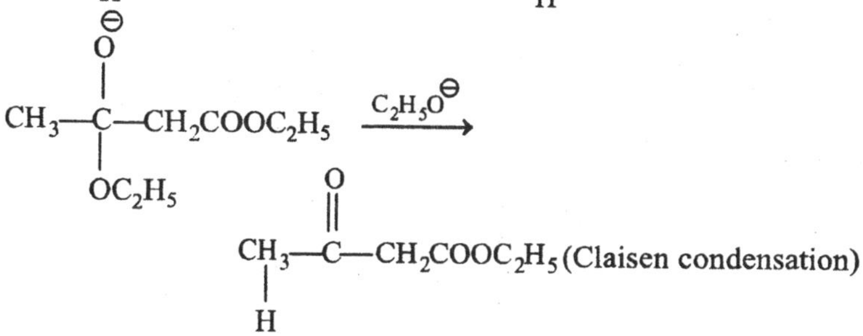

EXTENSION AND APPLICATIONS

Crossed Claisen Condensation

The Claisen condensation between two different esters, both having α-hydrogen, is called crossed Claisen condensation. However, this condensation is not of synthetic importance because a mixture of all the four possible ketoesters is produced *e.g.*, with ethyl acetate and ethyl propionate.

$$\textit{(i)}\quad CH_3CO\boxed{OC_2H_5 + H}CH_2COOC_2H_5 \xrightarrow{C_2H_5ONa} CH_3CO.CH_2.COOC_2H_5 + C_2H_5OH$$

$$\textit{(ii)}\quad CH_3CH_2CO\boxed{OC_2H_5 + H}\overset{\displaystyle CH_3}{\overset{|}{C}}H.COOC_2H_5 \xrightarrow{C_2H_5ONa} CH_3CH_2.CO.\overset{\displaystyle CH_3}{\overset{|}{C}}H.COOC_2H_5 + C_2H_5OH$$

$$\textit{(iii)}\quad CH_3CO\boxed{OC_2H_5 + H}\overset{\displaystyle CH_3}{\overset{|}{C}}H.COOC_2H_5 \xrightarrow{C_2H_5ONa} CH_3CO.\overset{\displaystyle CH_3}{\overset{|}{C}}H.COOC_2H_5 + C_2H_5OH$$

$$\textit{(iv)}\quad CH_3CH_2.CO\boxed{OC_2H_5 + H}CH_2.COOC_2H_5 \xrightarrow{C_2H_5ONa} CH_3CH_2.CO.CH_2COOC_2H_5 + C_2H_5OH$$

It is interesting to note that Claisen condensation can become of synthetic value, especially when one of the esters used is not capable of forming a carbanion, *i.e.*, it does not have α-hydrogen, *e.g.*, formate, oxalate, benzoate and ethyl carbonate.

(a) Crossed Claisen condensation with Formates

On the basis of the above reaction, ethyl formate finds use as a formylating reagent in the synthesis of various natural products.

$$\underset{\text{Ethyl formate}}{CH_3COOC_2H_5 + HCOOC_2H_5} \xrightarrow{C_2H_5ONa} \begin{array}{c} CHO \\ | \\ CH_2COOC_2H_5 \end{array}$$

$$\underset{\text{Ethyl phenylacetate}}{C_6H_5CH_2COOC_2H_5 + HCOOC_2H_5} \xrightarrow[2.\ H^{\oplus}]{1.\ C_2H_5ONa}$$

$$\underset{\text{Ethyl formylphenylacetate}}{\begin{array}{c} \quad\quad CHO \\ \quad\quad | \\ C_6H_5\text{—}CH\text{—}COOC_2H_5 \end{array}} + C_2H_5OH$$

Various natural products involving ethyl formate as a formylating reagent are synthesised as follows :

(*i*) *Synthesis of thiamine* :

$$\begin{array}{c} COOC_2H_5 \\ | \\ CH_2.CH_2OC_2H_5 \end{array} + HCOOC_2H_5 \xrightarrow{Na} \begin{array}{c} COOC_2H_5 \\ | \\ CH.CH_2OC_2H_5 \\ | \\ CHO \end{array}$$

(*ii*) *Synthesis of isoflavones, e.g.,* daidzein :

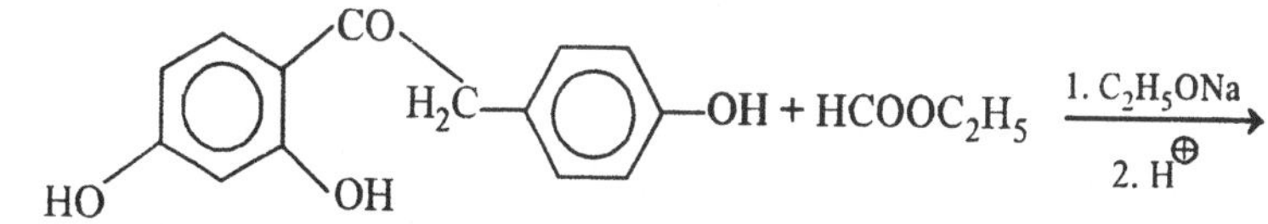

2-4-Dihydroxypheny-*p*-hydroxy benzyl ketone

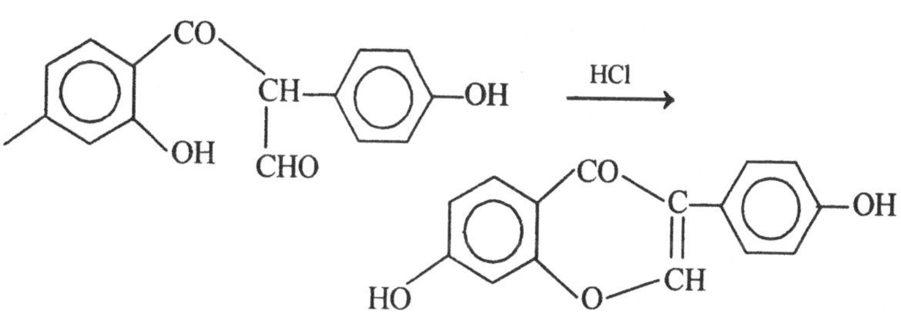

Daidzein(isoflavone)

(*iii*) *Synthesis of equilenine* :

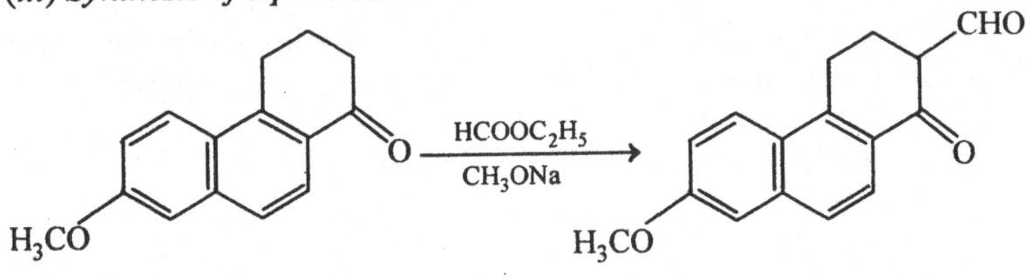

(*b*) Crossed Claisen condensation with oxalates

An example is as follow :

$$RCH_2COOC_2H_5 + \begin{matrix} COOC_2H_5 \\ | \\ COOC_2H_5 \end{matrix} \xrightarrow{C_2H_5ONa} RCH\begin{matrix} COOC_2H_5 \\ CO.COOC_2H_5 \end{matrix}$$

keto dibasic ester

The keto dibasic ester has been used in the synthesis of aryl substituted dibasic derivative and α-keto-acids.

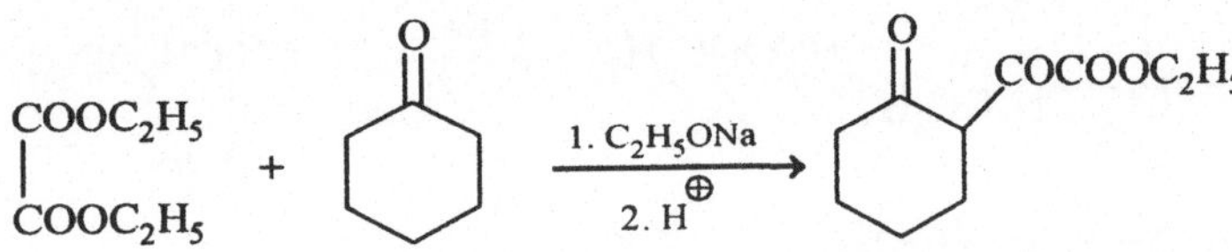

2-(Ethyl oxalyl)-cyclohexanone

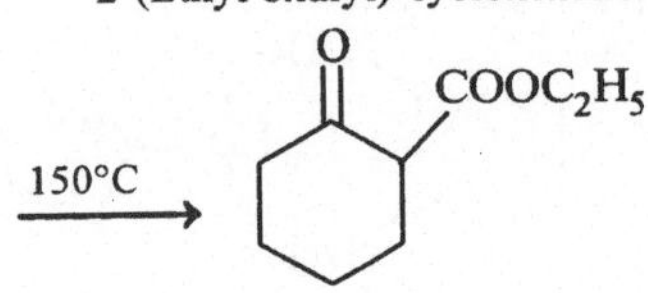

2-Carbethoxy cyclohexanone

$$\begin{matrix} COOC_2H_5 \\ | \\ COOC_2H_5 \end{matrix} + C_6H_5CH_2COOC_2H_5 \xrightarrow[2.\ H^{\oplus}]{1.\ C_2H_5ONa} \begin{matrix} C_6H_5{-}CHCOOC_2H_5 \\ | \\ COCOOC_2H_5 \end{matrix}$$

Diethyl-α-oxalyphenyl acetate

$$\xrightarrow{150°C} C_6H_5CH\begin{matrix} COOC_2H_5 \\ COOC_2H_5 \end{matrix}$$

Diethyl phenylmalonate

Phenyl-substituted malonic ester has important synthetic applications, *e.g.*, phenobarbitone may be prepared. Diethyl-α-oxalylphenylacetate on hydrolysis and heating gives α-keto acid.

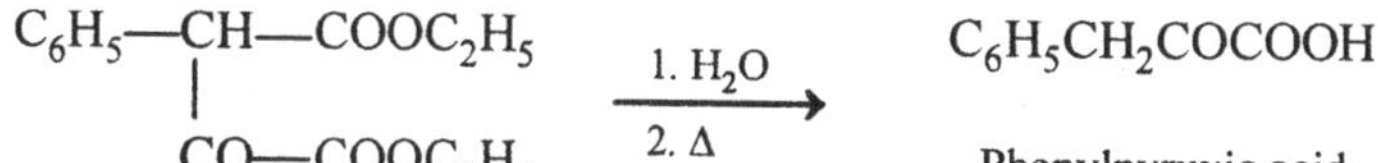

Phenylpyruvic acid

Also, the Claisen condensation using ethyloxalate has been used in the synthesis of menthone, camphoric acid, etc.

(i)

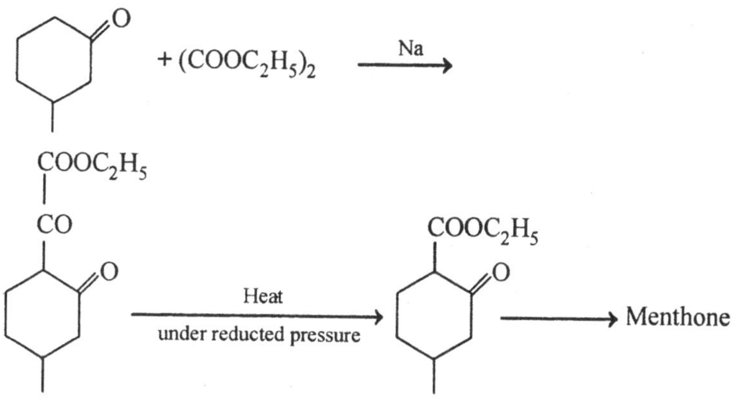

(ii)

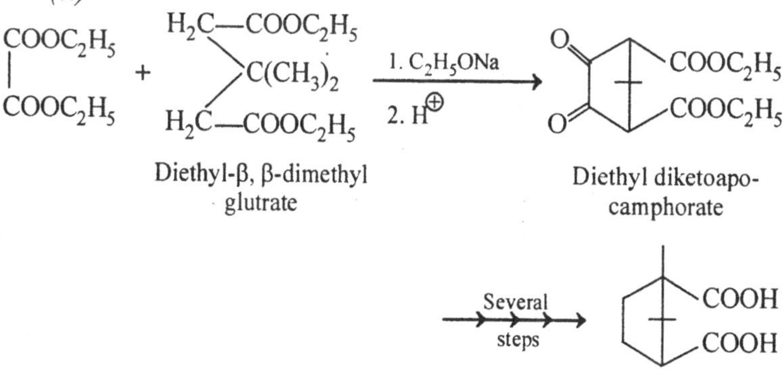

Diethyl-β, β-dimethyl glutrate

Diethyl diketoapo-camphorate

Camphoric acid

(c) Condensation with ethyl benzoate

$$C_6H_5COOC_2H_5 + CH_3COOC_2H_5 \xrightarrow[2.\ H^{\oplus}]{1.\ C_2H_5ONa} C_6H_5COCH_2COOC_2H_5 + C_2H_5OH$$

Ethyl benzote — Ethyl benzoylacetate

This has found applications in the synthesis of following compounds.

(*i*) *Synthesis of flavones, e.g., chrysin.*

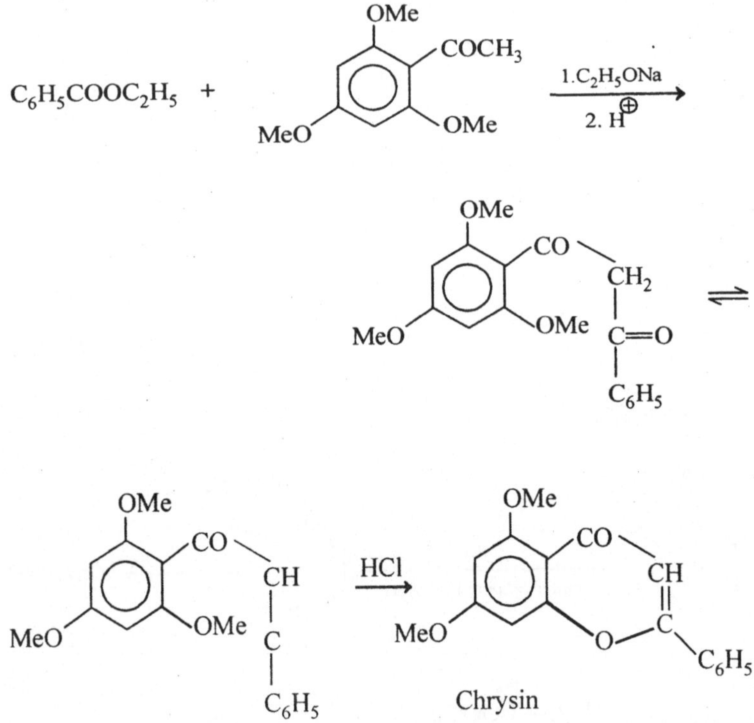

Chrysin

(*ii*) *Synthesis of nicotine from ethyl nicotinate.*

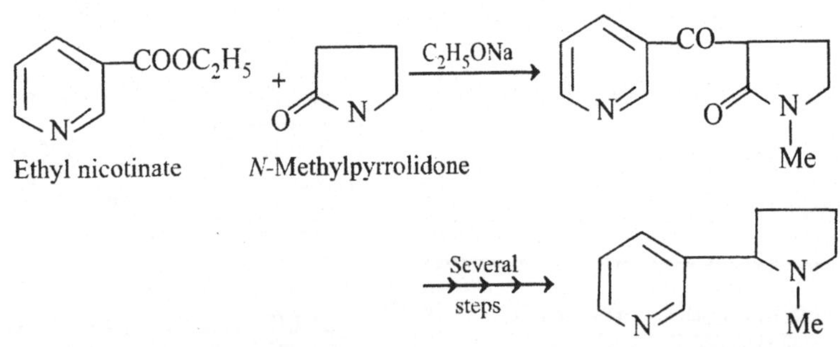

Ethyl nicotinate *N*-Methylpyrrolidone

Nicotine

(d) Condensation with diethyl carbonate

Ester condense with diethyl carbonate to form malonic ester derivatives which can be further converted into various compounds.

$$C_2H_5O-\overset{\overset{\displaystyle O}{\|}}{C}-OC_2H_5 + C_6H_5CH_2COOC_2H_5 \xrightarrow[2.\ H^{\oplus}]{1.\ C_2H_5ONa}$$

Diethyl carbonate

$$C_6H_5-\underset{\underset{\displaystyle COOC_2H_5}{|}}{CH}-COOC_2H_5 \xrightarrow[2.\ CH_3I]{1.\ C_2H_5ONa}$$

Diethyl phenylmalonate

$$C_6H_5-\overset{\overset{\displaystyle CH_3}{|}}{\underset{\underset{\displaystyle COOC_2H_5}{|}}{C}}-COOC_2H_5 \xrightarrow[2.\ \Delta]{1.\ H_2O} CH_3\underset{\underset{\displaystyle C_6H_5}{|}}{CH}-COOH$$

α – Phenylpropionic acid

(e) In the synthesis of quinine

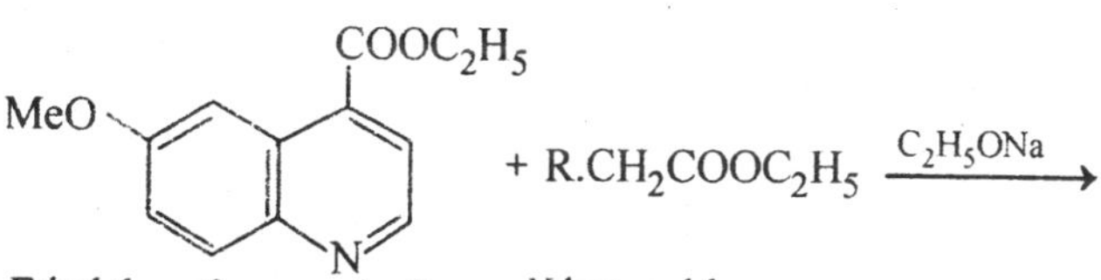

Ethyl 6-methoxy-quinoline -4-carboxylate   N-benzoyl-homomero-quinone

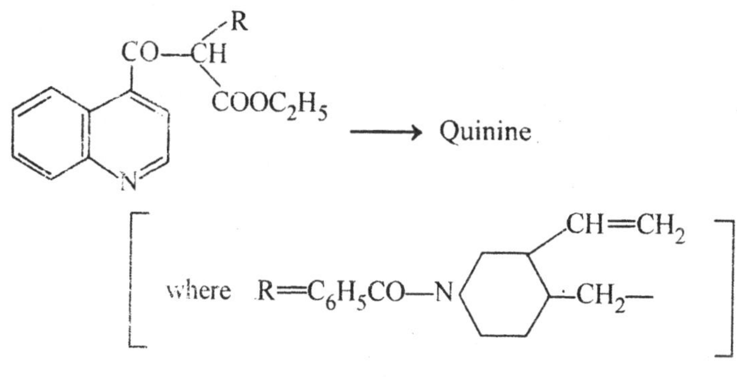

(f) Kostanecki's synthesis of flavones

An example is as follows

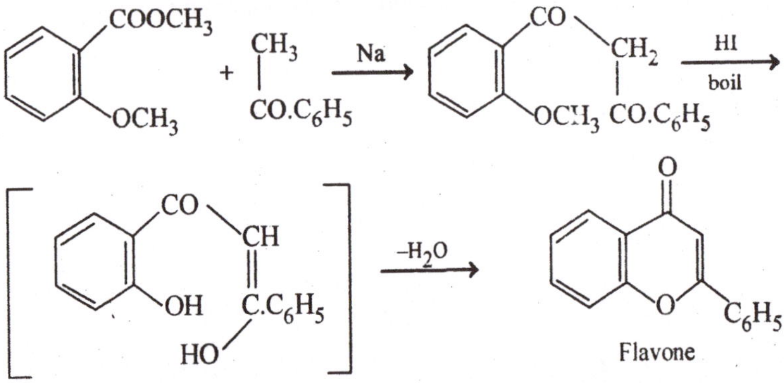

Flavone

(g) A very important product of Claisen condensation is the β-ketoester, ethyl acetoacetate (EAA). It can be easily alkylated. Ethyl acetoacetate and its alky derivatives under different conditions undergo ketonic and acidic hydrolysis to yield ketones and acids respectively.

$$CH_3COCH_2COOC_2H_5 \xrightarrow[2.\ RX]{1.\ C_2H_5ONa} CH_3CO\underset{\displaystyle R}{\overset{|}{C}}HCOOC_2H_5 + C_2H_5OH + NaX$$

$$RCH_2COCH_3 \xleftarrow[\text{(Ketonic hydrolysis)}]{\text{Dil Alc KOH,}} CH_3COCH(R)COOC_2H_5$$

$$\xrightarrow[\Delta\ \text{(Acidic hydrolysis)}]{\text{Conc. alc. KOH}} RCH_2COOH$$

This reaction has been used for the synthesis of a large variety of ketones, acids and many heterocyclic compounds. Intramolecular Claisen condensation in dibasic esters of six to eight carbons is known as Dieckmann reaction. The product is a cyclic ketone derivative.

21

CLAISEN REARRANGEMENT

INTRODUCTION

When a simple or substituted allyl ether of a phenol is heated at about 200°C, an isomerisation called Claisen rearrangement occurs in which the allyl group gets migrated from the ether oxygen atom to a carbon atom of the aromatic ring. The migrant allyl group goes preferentially to the ortho position or to the para position if both the *ortho* positions are blocked. Migration to a meta position has not been observed, *e.g.*,

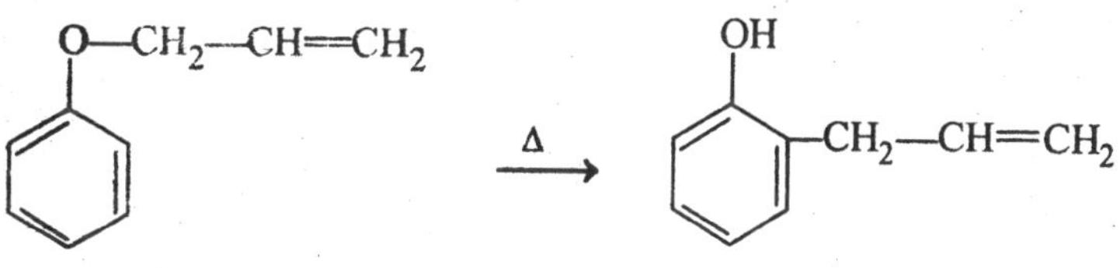

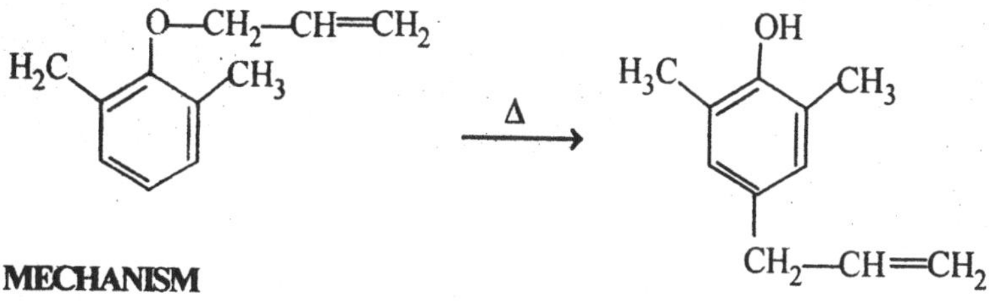

MECHANISM

The essential features of the Claisen's rearrangement are as follows :
(*i*) Claisen rearrangement does not need a catalyst and it also follows a first order kinetics. These facts support the idea that each individual act of rearrangement involves only one molecule.

(*ii*) If a mixture of two different ethers, in which the allylic groups are different, is heated together, no cross product is obtained. For example, a mixture of α-nephthylallyl ether and phenyl cinnamyl ether results in the formation of only 2-allyl-α-naphthol and *o*-α-phenylallyl phenol which are formed by the separate rearrangement of each.

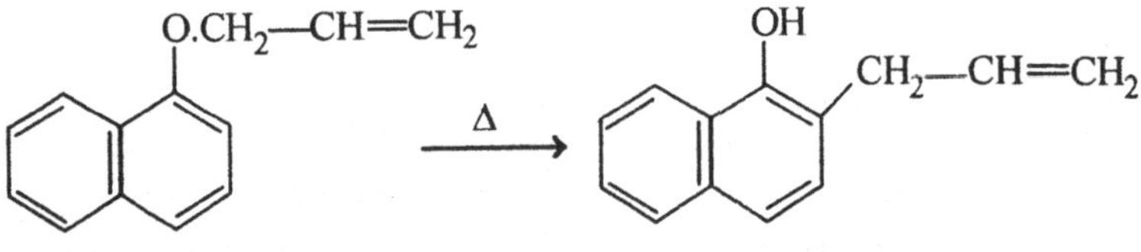

α-Naphthyl allyl ether 2-Allyl-α-naphthol

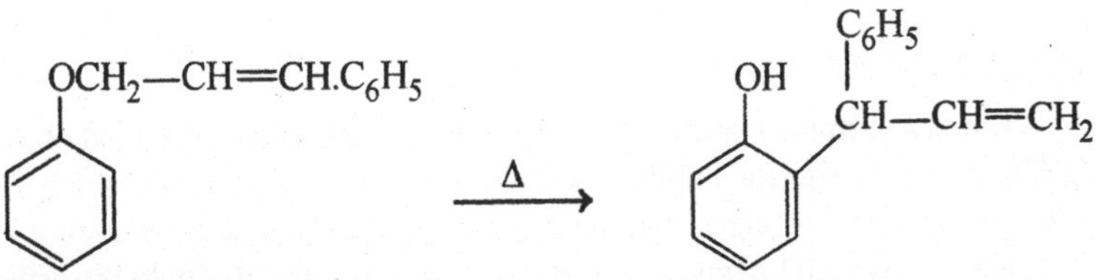

Phenyl cinnamyl ether α–Phenylallyl phenol

From the above experiment, it follows that the allyl group does not get separated during the migration, *i.e.*, the rearrangement is an *intramolecular one-step process.*

Mechanism of ortho-isomerisation. If all the above observations are considered together, it is found that the ortho rearrangement is a concerted process and it follows cyclic mechanism which proceeds through the formation of a cyclic six-membered transition state in which the rupturing of the oxygen-allyl bond is synchronous with the fomation of a carbon-carbon bond at an *ortho-position.* Thus, there occurs the formation of a cyclohexadienone intermediate (A) which by a prototropic change, is subsequently converted itself to the more stable aromatic compound, phenol.

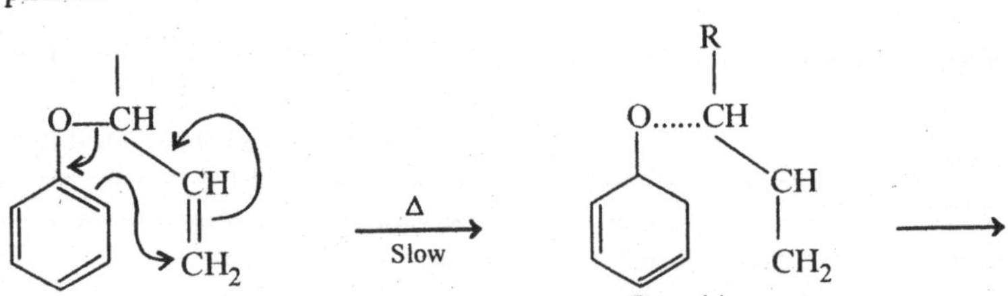

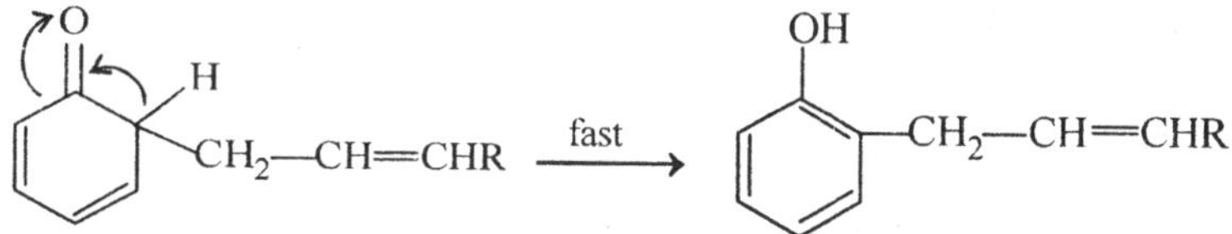

Dienone (I) Intermediate
(A)

Mechanism of para-isomerisation : If the *ortho*-positions are blocked, the very first step still involves the *ortho* migration, resulting in the fomation of *ortho*-substituted cyclohexadienone (II). As hydrogen is absent at the *ortho* position, this prevents aromatisation of (II) and this then undergoes further rearrangement involving migration of the allyl group, again through a cyclic six membered transition state yielding (III) which ultimately gets aromatised to the phenol.

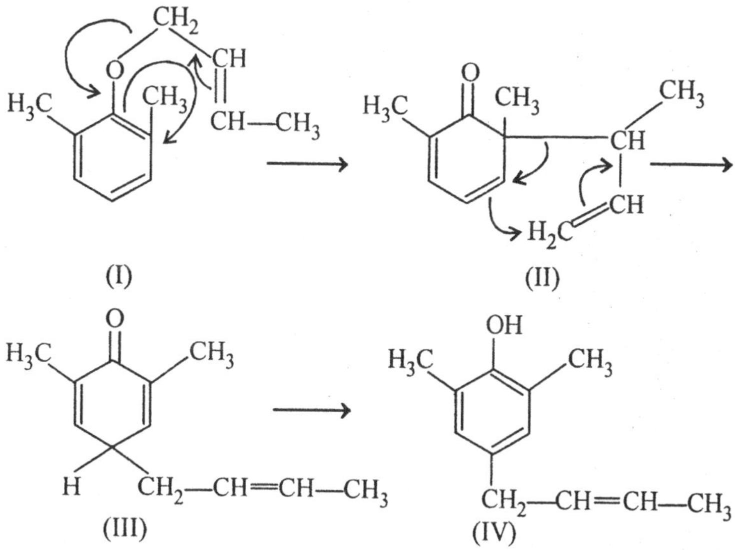

Hence the mechanism of *para*-rearrangement involves a double rearrangement and thus ultimately it is not rearranged during the reaction, *i.e.*, there is absence of inversion in the para Claisen rearrangement.

The formation of intermediate during para Claisen rearrangemnt has been proved bythe following facts :

(*i*) The intermediate (II) is trapped by means of a Diels-Alder reaction as its adduct with maleic anhydride.

(*ii*) The intermediate (II) has been synthesised independentlyand found to transform rapidly to (IV) on heating. The intermediate (II) is prepared by treating sodium 2, 6-dimethylphenate with $BrCH_2—CH=CH—CH_3$ in benzene at 15°C. The intermediate (II), when heated above 70°C, yields a mixture of (IV) and (I).

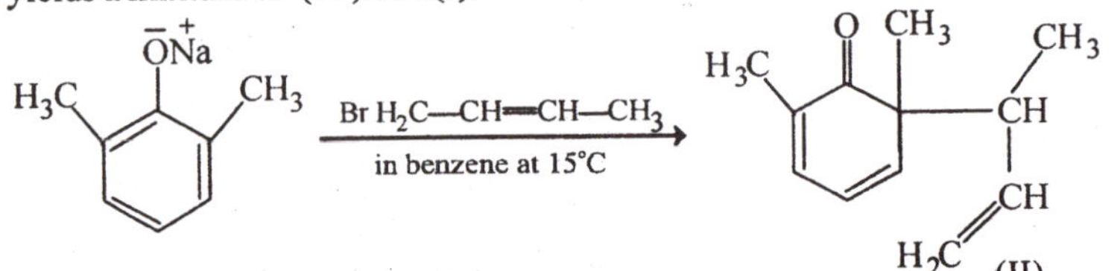

(II)

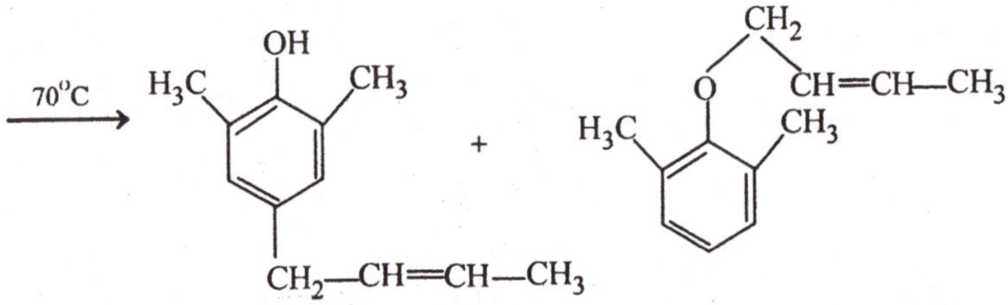

(IV)

(*iii*) When *ortho*-α-^{14}C-labelled allyl-2, 6-diallyl phenyl ether (V) is subjected to this rearrangement, the labelled carbon got divided equally between the *ortho* and *para* position in the rearrangement product (VII). This experiment proves that the intermediate is triallyl cyclohexadiene (VI) in which there occurs the migration of either the labelled or unlabelled allyl group to the *para* position with equal ease.

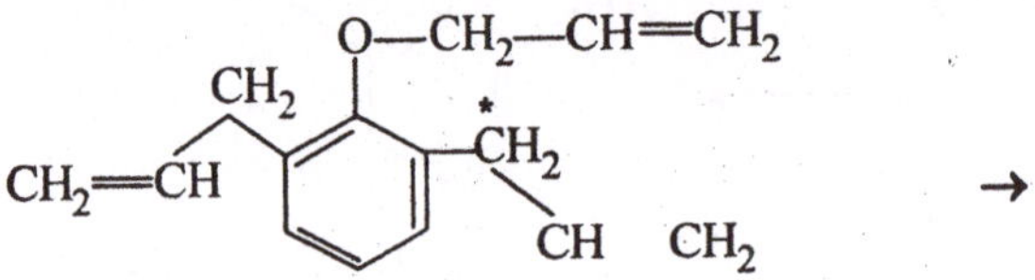

(V)

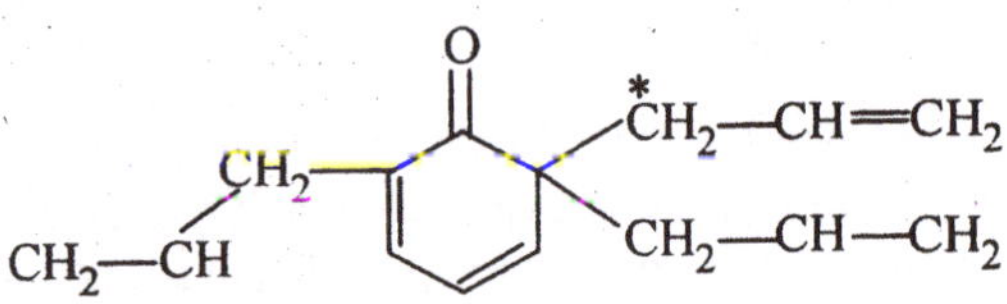

(VI)

↓

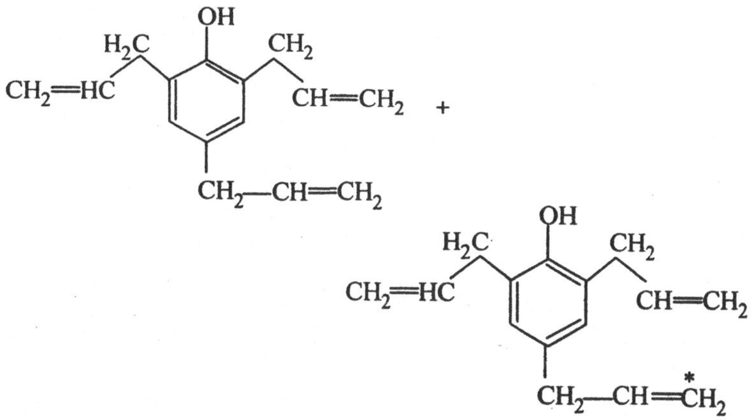

(VII)

(*iv*) When labelled allyl 2,6-dimethyl-4-allyl-phenol is heated, there occurs the distribution of ^{14}C between its original position and the terminal position of the O-allyl group. This experiment demonstrates reversibility of the cyclic mechanism operating in the Claisen rearrangement.

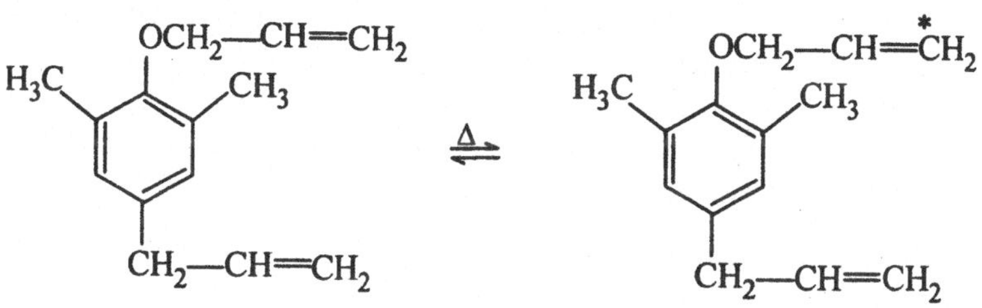

As Claisen rearrangement involves a cyclic mechanism, it does not greatly depend on the presence or absence of substituent groups on the ring. *Orho* isomerisation gets accompanied by an inversion of the position of substituent with respect to that of the starting compound. However, no such inversion takes place in para isomerisation because there occurs the double migration of the allyl group in each migration the allyl group gets inverted.

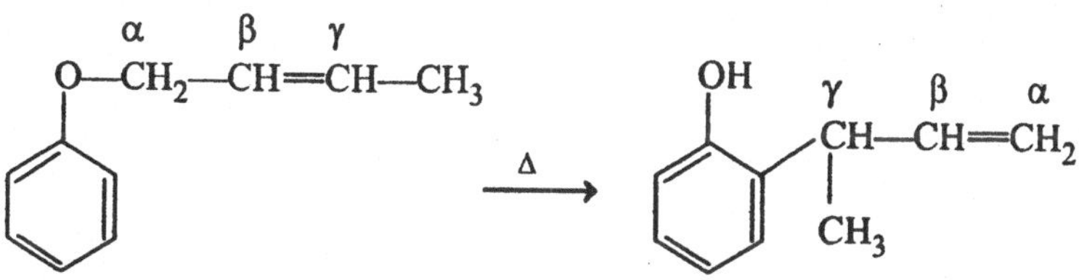

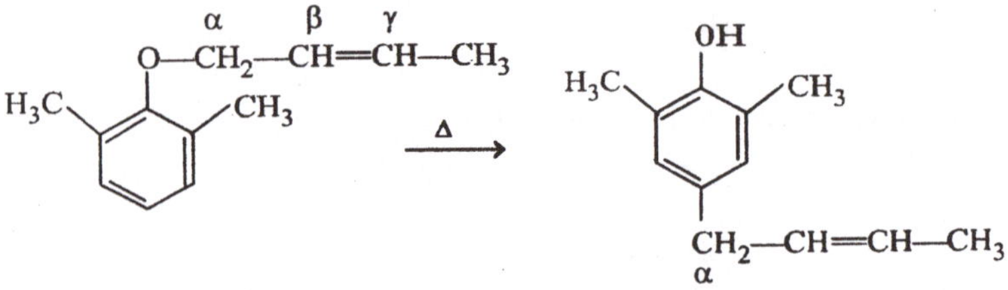

The fact that ortho isomerisation is accompanied by an inversion of the position of substituent with respect to that of the starting compund has been further supported by Schmidt who carried out the rearrangement of phenyl allyl ether labelled with ^{14}C in the γ-position and obtained the *o*-allyl phenol having ^{14}C in the α-position.

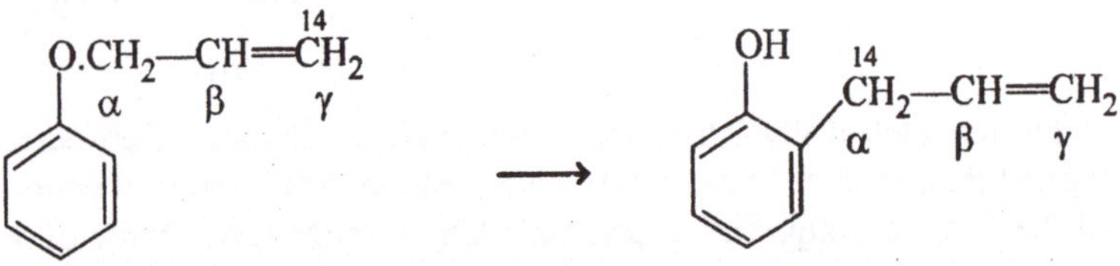

The fact that no inversion is observed in para isomerisation is proved by Schmidt who carried out rearrangement of 2, 6-dimethylallylether labelled with ^{14}C in the γ-position. This compound rearranges to 4-allyl-2, 6-dimethylphenol with the ^{14}C still in the position of the allyl group which occurs due to the double migration of the allys group; in each migration there occurs the inversion of allyl group.

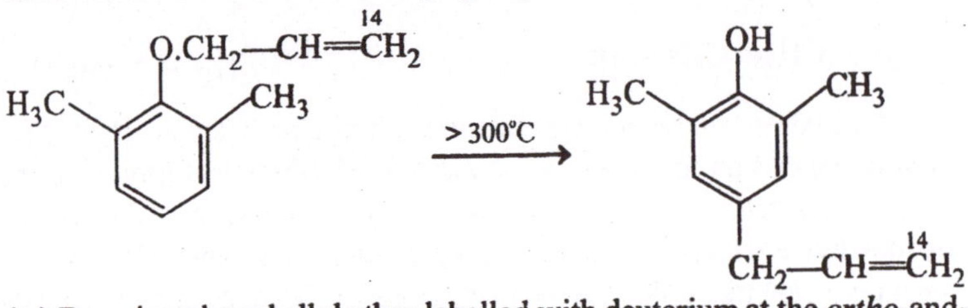

(*iv*) By using phenyl allyl ether labelled with deuterium at the *ortho*-and-*para*-position of the ring and observing the product, it was proved that there occurs the migration of hydrogen atom from the o-or *p*-position (as the case may be) of the benzene ring during rearrangement.

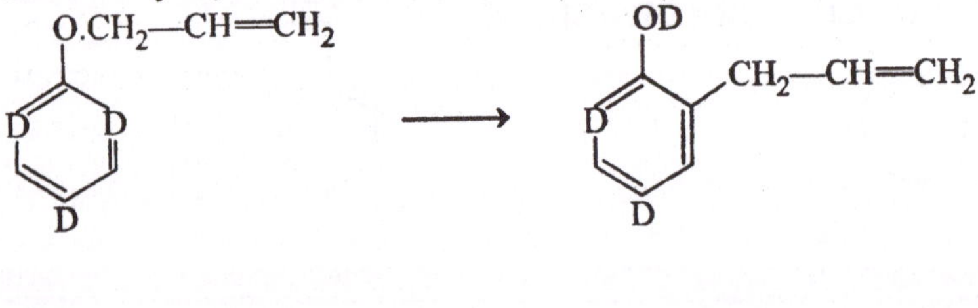

EXTENSION AND APPLICATION

Claisen rearrangement is extended to the following type of compounds :
(*a*) Claisen rearrangement is not only applicable to aromatic systems but also to allyl ethers of enols. It can be generalized in the following way :

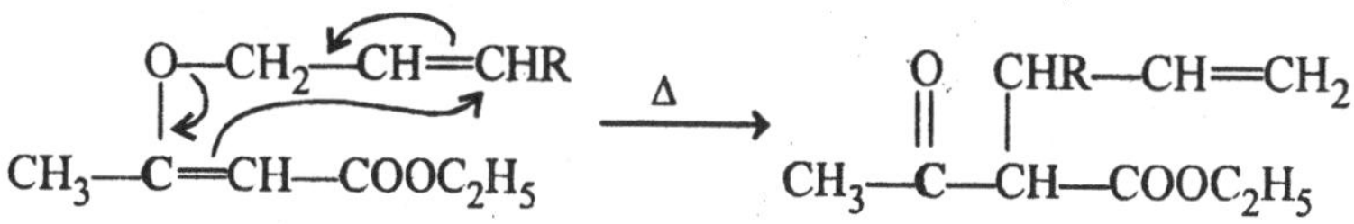

For example, *o*-allylacetoacetic ester also undergoes rearrangement on heating :

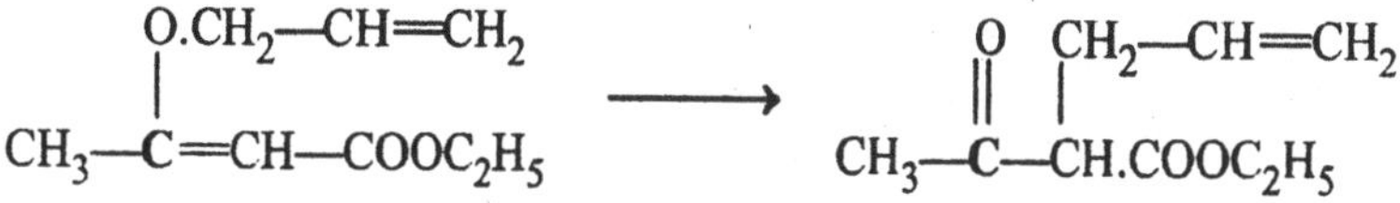

(*b*) *Rearrangement of allyl vinyl ethers, e.g.,*

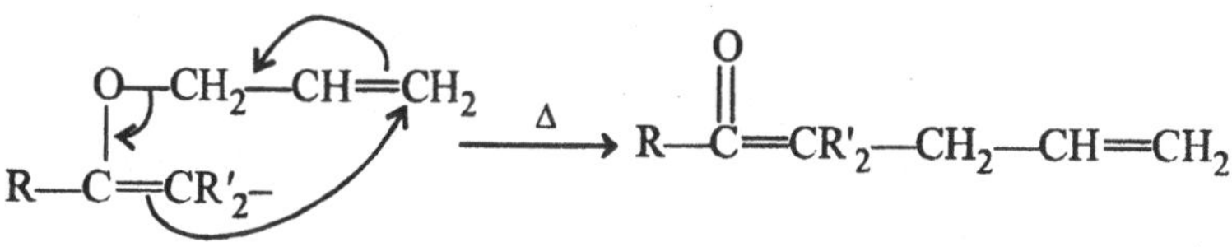

The product does not enolize (even when R or R′ = H) because ketones are more stable than enols.
(*c*) *Rearrangement of allyl esters, e.g.,*

(*d*) *Rearrangement of* 1 : 5-*heaxdienes* (Cope rearrangement) *e.g.,*

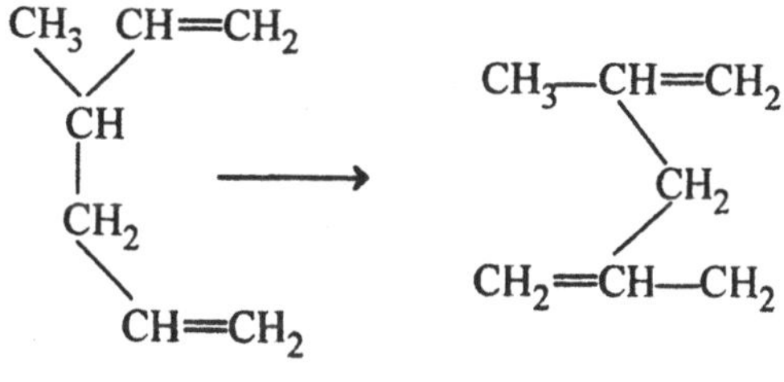

(*e*) *Rearrangement of allylamine oxides, e.g.,*

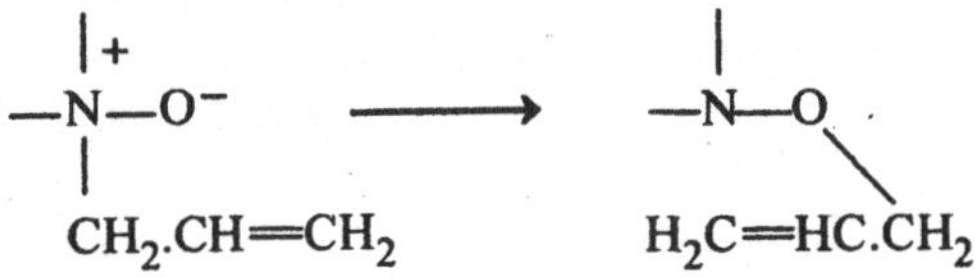

(*f*) *Oxygen atom of the allyl ether is replaced by sulphur.*

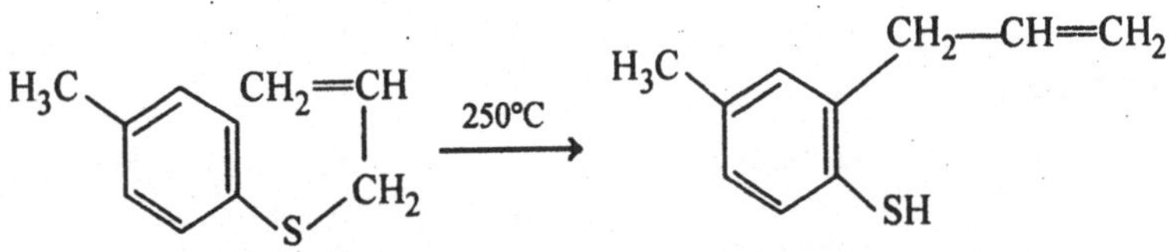

The Claisen rearrangement is frequently useful in synthesis, *e.g.*, guaiacol allyl ether yields 80—90% of *o*-eugenol.

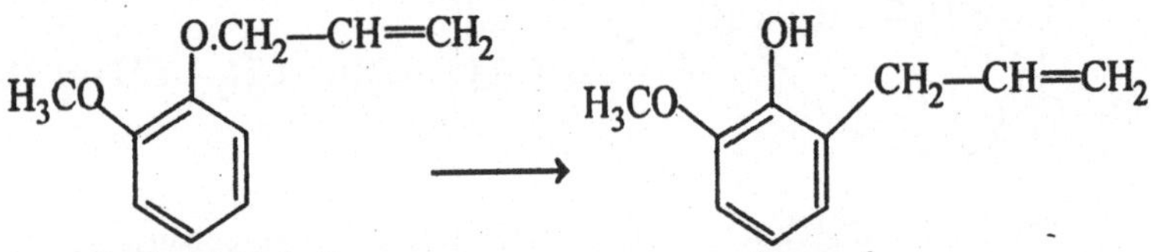

Guaiacol allyl ether *o*-Eugenol

(*g*) *Out-of-ring Claisen rearrangement* : Several example of out-of-ring Claisen rearrangements are known of the type given below.

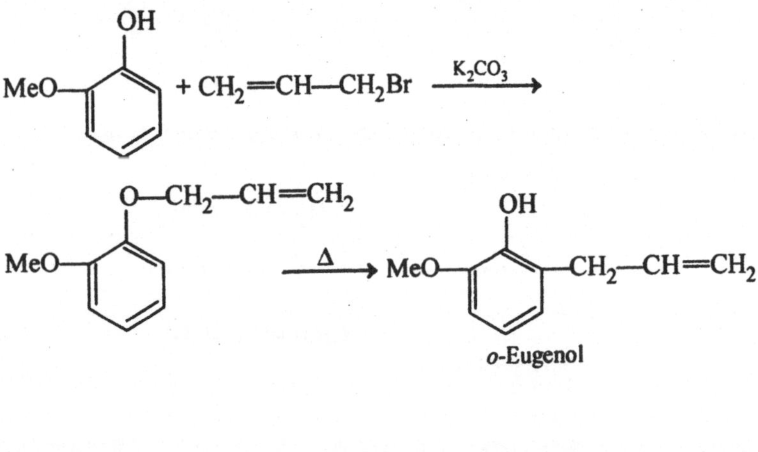

o-Eugenol

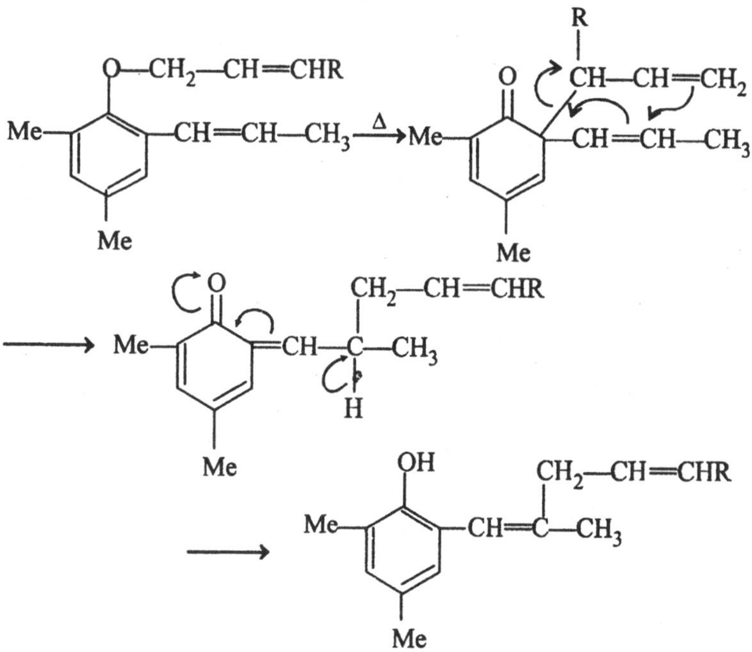

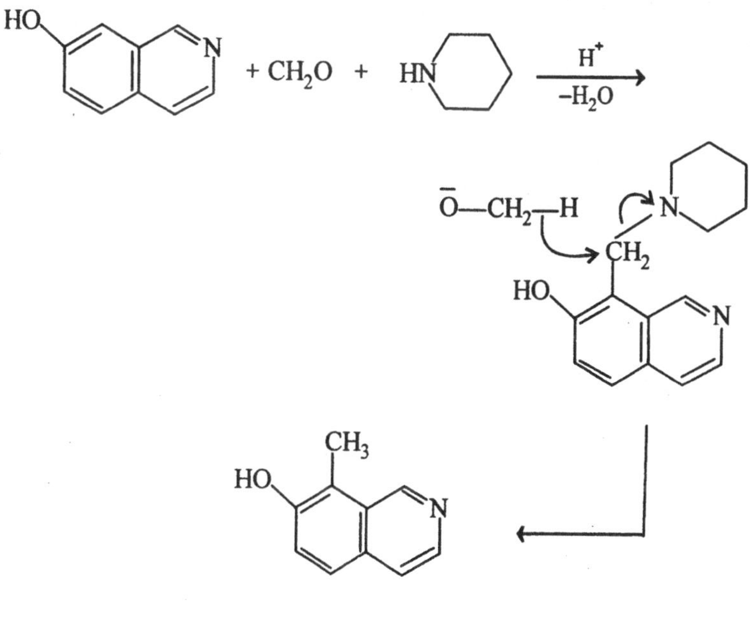

(*iv*) *Synthesis of gramine* : The iminium salt, being a strong electrophile, replaces the active hydrogens of indole, phenol, nitroalkane, etc., *e.g.*,

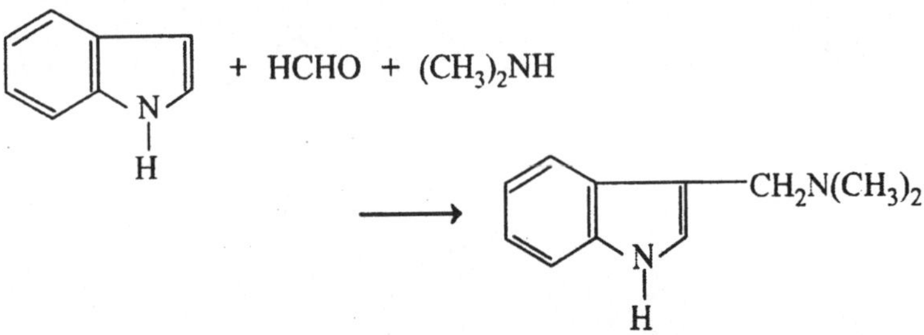

Gramine

The product may be quaternised and the amino group may be substituted by other groups.

22

CLAISSEN-SCHMIDT REACTION OR CLAISSEN-SCHMIDT CONDENSATION

(*i*) This reaction takes place between two molecules of carbonyl compounds; one should have at least two α-hydrogen. In this reaction best result is obtained when both carbonyl compounds are same or one has no α-hydrogen and other has two α-hydrogens.

(*ii*) These reactions are practical when bases such as sodium hydroxide are used and the reaction is carried out at high temp. (≈ 100°C)

(*iii*) The reaction is a two step reaction. First step is aldol condensation and the second step is dehydration of aldol.

$$CH_3—CHO + CH_3—CHO \xrightarrow{NaOH/100°C} \underset{\text{Crotonaldehyde}}{CH_3—CH{=}CH—CHO} + H_2O$$

$$C_6H_5—CHO + CH_3—CHO \xrightarrow[100°C]{\overset{\ominus}{OH}} \underset{\text{Cinnamaldehyde}}{C_6H_5—CH{=}CH—CHO} + H_2O$$

$$C_6H_5—CHO + CH_3—\overset{\overset{O}{\|}}{C}—C_6H_5 \xrightarrow{OH} \underset{\text{Benzalacetophenone}}{C_6H_5—CH{=}CH—\overset{\overset{O}{\|}}{C}—C_6H_5}$$

MECHANISM:

Step I.

$$\overset{\ominus}{O}H + H—CH_2—\overset{\overset{O}{\|}}{C}—CH_3 \rightleftharpoons HOH + \left[\overset{\ominus}{C}H_2—\overset{\overset{O}{\|}}{C}—CH_3 \longleftrightarrow CH_2{=}\overset{\overset{\overset{\ominus}{O}}{|}}{C}—CH_3\right]$$

Step II.

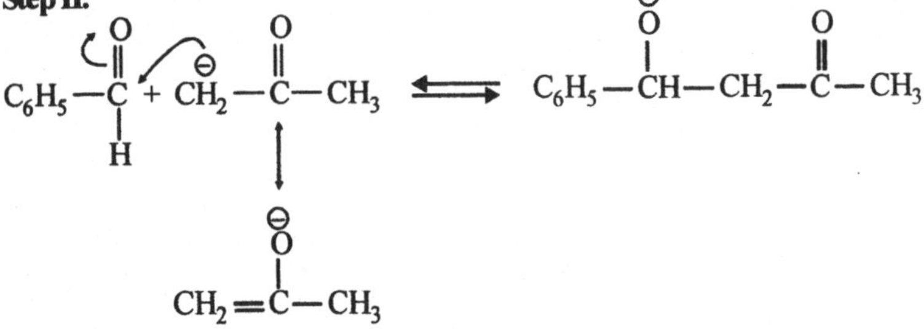

Step III.

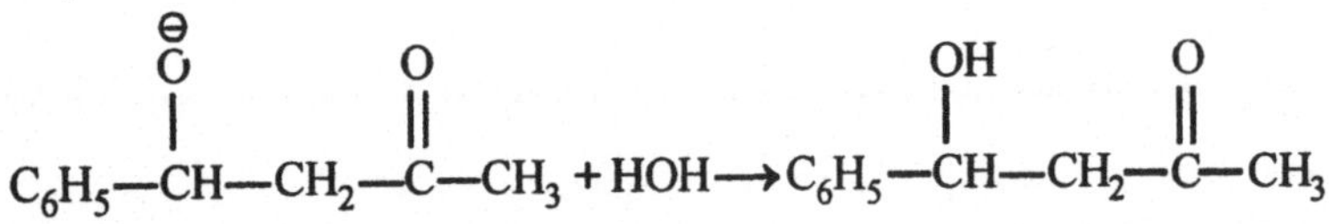

Step IV.

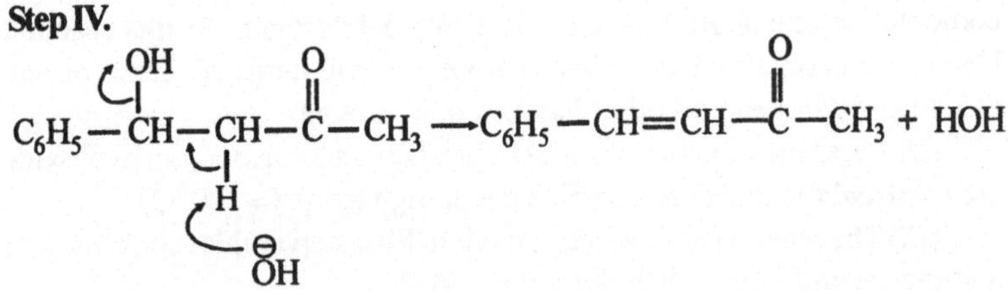

In the Claissen-Schmidt reaction, dehydration occurs readily because the double bond that forms is conugated, both with the carbonyl group and with the benzene ring. The conjugation system is thereby extended.

In short, product of the reaction can be written as follows :

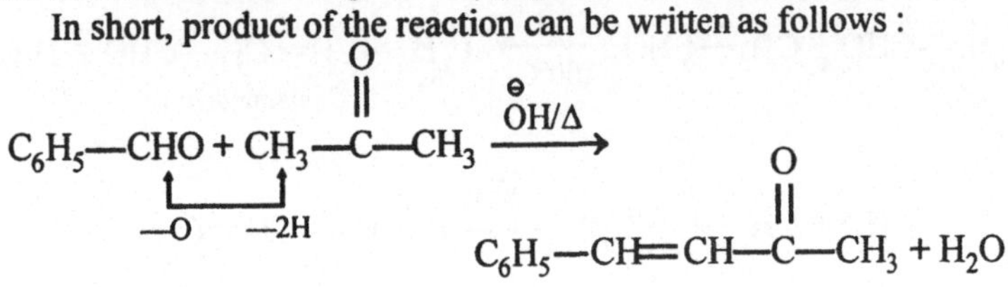

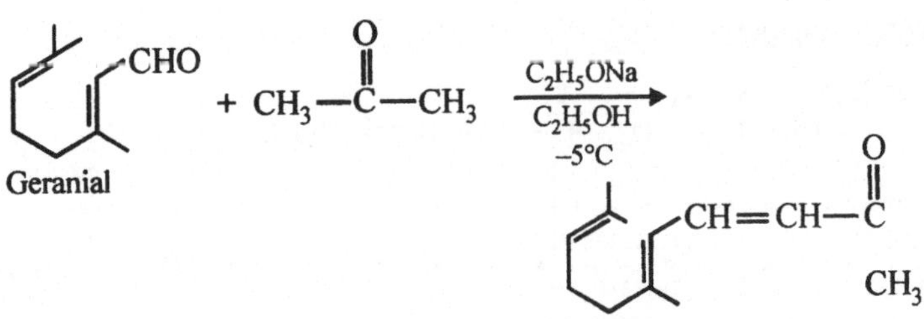

Geranial

Pseudo ionone 49%

Note 1 : Cross Claissen reaction between aromatic aldehyde and carbonyl compounds (aldehyde, ketone and also ester) having two α-hydrogens gives better result.

Note 2 : If reaction is carried out with aldehyde having two α-hydrogens then byproduct is also formed.

$$C_6H_5{-}CHO + CH_3{-}CHO \longrightarrow CH_3{-}CH{=}CH{-}CHO$$

$$C_6H_5{-}CHO + CH_3{-}CHO \longrightarrow C_6H_5{-}CH{=}CH{-}CHO$$

more resonance stabilised

Note 3 : If reaction is carried out with ketone having two α-hydrogens then ketone does not give self condensation reaction in basic medium because aldol gives retro aldol reaction in basic medium. Due to this reason, Aldol condensation of ketone is carried out in the presence of acid as catalyst.

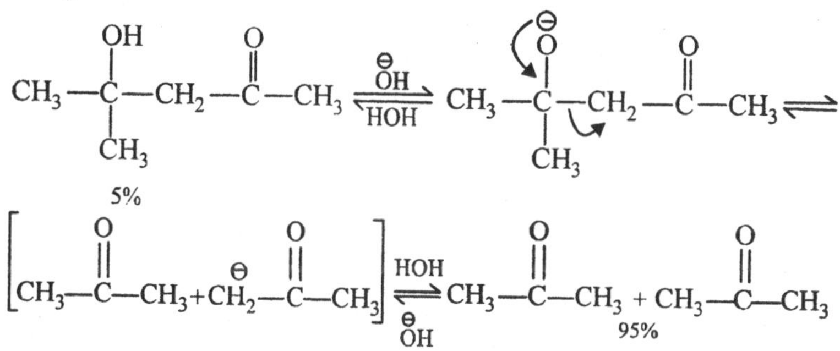

Note 4 : Claissen type of reaction is given by all types of compounds having at least two α-hydrogens with aromatic carbonyl compounds having no α-hydrogen and the reaction is catalysed by base on heating.

$$C_6H_5{-}CHO + CH_3{-}NO_2 \xrightarrow{\overset{\ominus}{O}H/\Delta} C_6H_5{-}CH{=}CH{-}NO_2 + HOH$$

APPLICATIONS

1. Claisen reaction has been used in synthesizing many α,β-unsaturated carbonyl compounds which find use in perfumery. Examples of such compounds are cinnamaldehyde, benzylidene, acetone, etc.

$$C_6H_5.CHO + H_3C.CHO \xrightarrow{NaOH} C_6H_5.CH{=}CH.CHO$$

Cinnamaldehyde

$$C_6H_5.CHO + H_3C.CO.CH_3 \xrightarrow{NaOH} C_6H_5CH{=}CH.COCH_3$$

Benzylideneacetone

Benzylidene acetone has been found to react with another molecule of benzaldehyde in the presence of ethanolic solution of NaOH, forming dibenzylidene acetone.

$$\underset{}{C_6H_5.CHO} + \underset{\text{Benzylideneacetone}}{H_3C.CO.CH{=}CH.C_6H_5} \xrightarrow[\text{KOH}]{\text{ethanolic}} \underset{\text{Dibenzylideneacetone}}{C_6H_5.CH{=}CH.CO.CH{=}CH.C_6H_5}$$

The reaction has also been used in the preparation of long-chain conjugated aldehydes, which are useful dye intermediates.

$$\underset{\text{Cinnamaldehyde}}{C_6H_5CH{=}CH{-}CHO} + \underset{\text{Crotonaldehyde}}{CH_3CH{=}CH{-}CHO} \xrightarrow[\text{+Alcohol}]{\text{Piperidine acetate}}$$

$$\underset{\text{7-Phenyl heptatrienal}}{C_6H_5{-}(CH{=}CH)_3{-}CHO} + \underset{\text{11-Phenylundecapentenal}}{C_6H_5{-}(CH{=}CH)_5{-}CHO}$$

2. By using the above reactions, benzaldehyde has been used to ascrertain the presence of certain groups such as —CH_2CO—and—CH_2COCH_2—in many natural products.

3. *Synthesis of natural products* : Claisen reaction has been used for synthesizing many natural products such as flavours, flavones, piperine, β-ionone, etc.

(*a*) *Kostanecki synthesis of flavonols, e.g.,* flavonol.

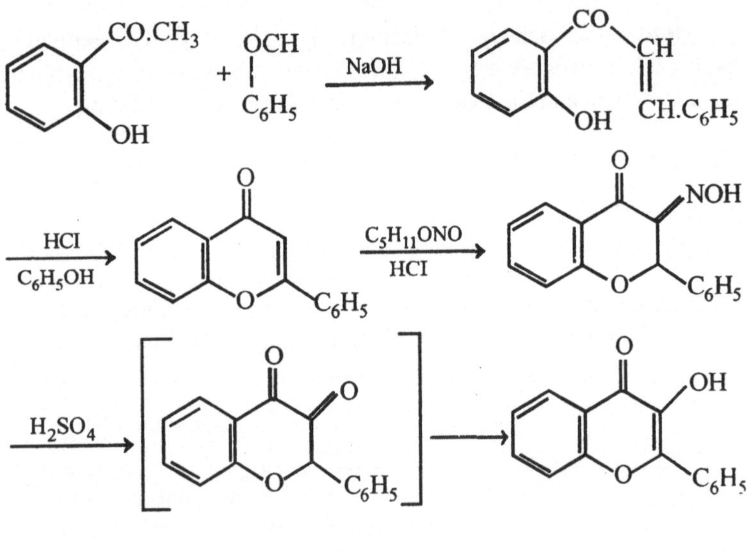

(*b*) *Kostanecki synthesis of flavones, e.g.,* flavone.

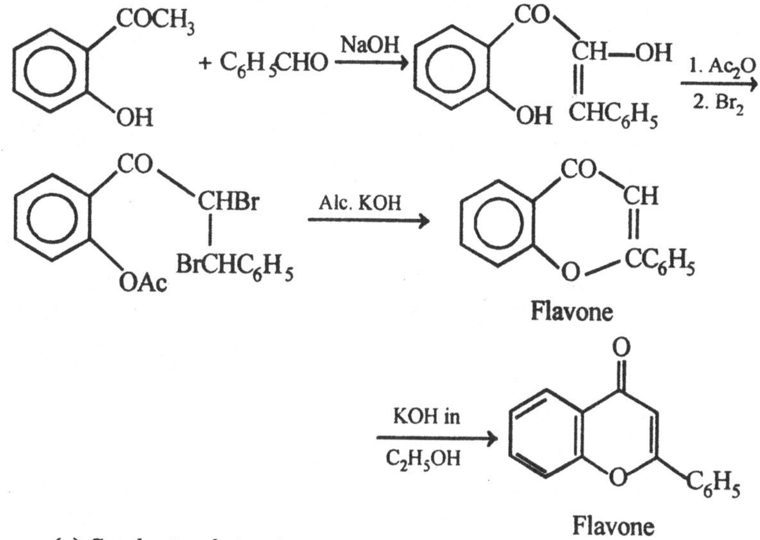

(*c*) *Synthesis of piperine* :

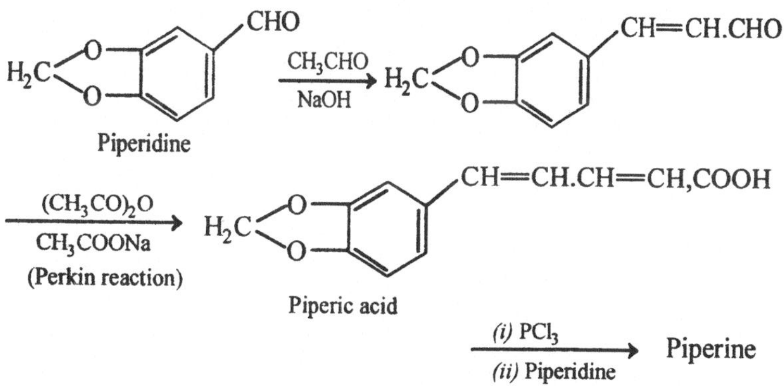

(*d*) *Synthesis of β-ionone* :

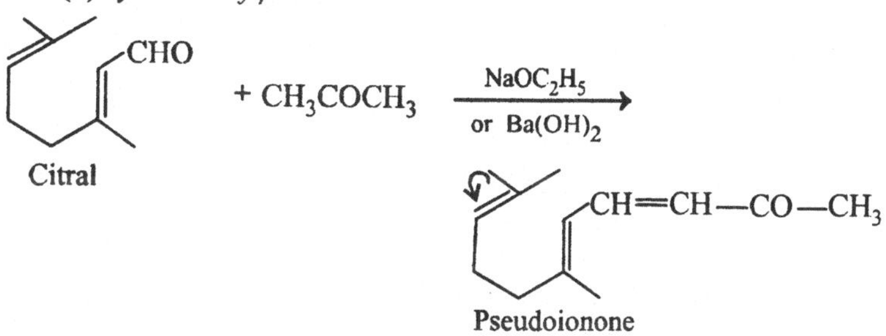

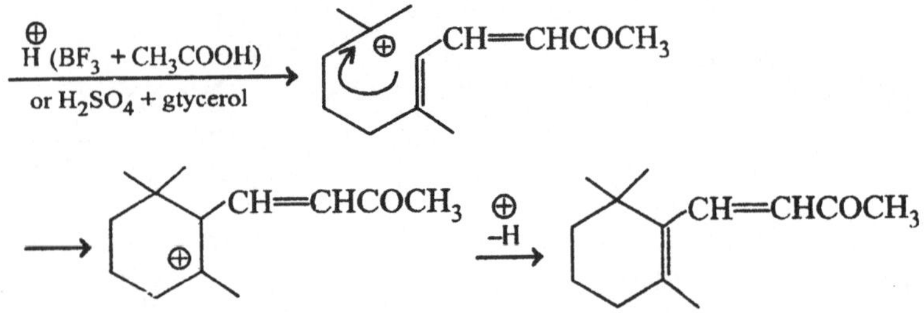

β-Ionone

(*e*) Cinnamic acid is technically obtained by the oxidaction of benzal acetone with sodium hypochlorite.

$$C_6H_5CHO + CH_3COCH_3 \xrightarrow[30^\circ C]{10\%\ NaOH} C_6H_5{-}CH{=}CH{-}CO{-}CH_3$$

$$\xrightarrow{NaOCl} C_6H_5CH{=}CHCOOH$$

(5) Test for the presence of —CH_2CO—Group

The reaction has been used to detect the presence of —CH_2CO— group in the natural products, for only if a compounds contains —CH_2CO— group, it will undergo the condensation.

$$C_6H_5{-}CH{=}O + H_2\underset{\substack{|\\H}}{C}{-}CO{-}CH_3 \xrightarrow{\overset{\ominus}{O}H}$$

$$C_6H_5CH{=}CHCOCH_3$$

23

CUMENE HYDROPEROXIDE REARRANGEMENT

It is the acid catalysed rearrangement of cumene hydroperoxide to phenol.

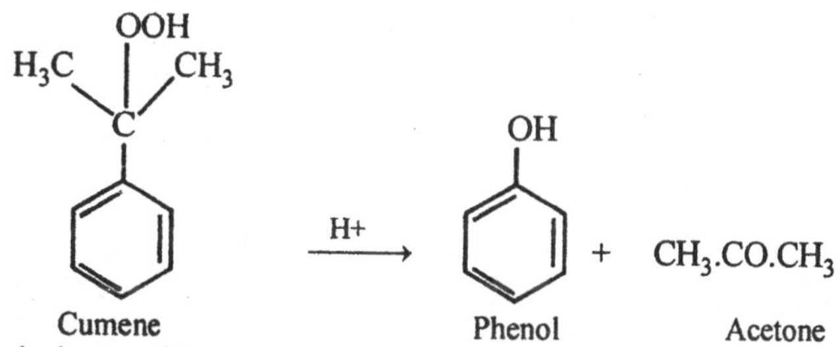

Cumene hydroperoxide — Phenol — Acetone

MECHANISM

The —OH group of hydroperoxide is first protonated in the acidic medium. Then the weakened O—O bond undergoes cleavage to form oxycation (I) along with the elimination of water.

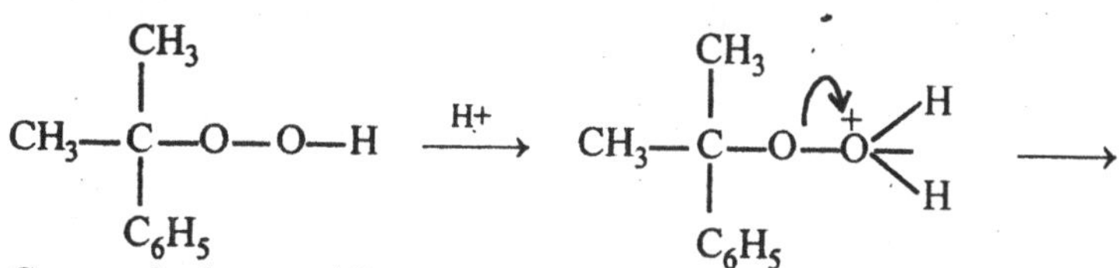

Cumene hydroperoxide

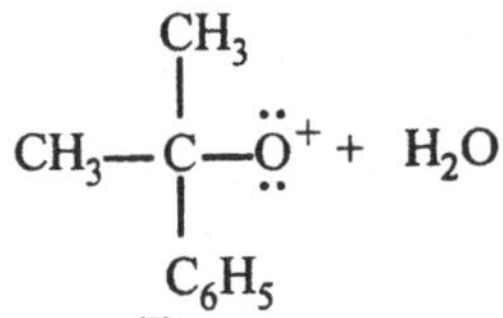

(I)

The electron-deficient oxygen atom in the cation (I) can attract any available nucleophile. But in aqueous acid solution there occurs a lack of any nucleophile which can form a stable product. Therefore, the electron-deficiency of the oxygen atom gets satisfied by the migration of the phenyl group with its pair of bonding electrons to the oxygen atom. This forms a more stable carbonium ion : this is the driving force for the migration. There occurs the migration of phenyl group rather than the methyl group as a result of its more electron donating power due to the presence of excess π electron cloud.

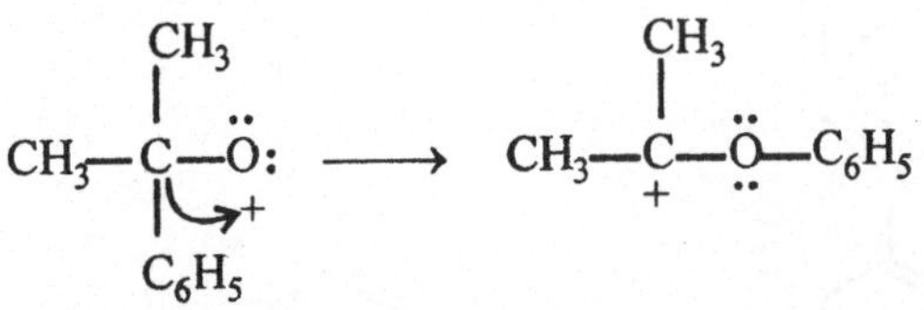

It may be mentioned here that the intermediate oxycation may not have a real existence but it is represented here only for convenience, because simultaneous elimination of water and migration of phenyl group is equally probable.

The carbonium ion accepts a molecule of water from the aqueous solution thereby forming hemiketal which gets converted into phenol and acetone.

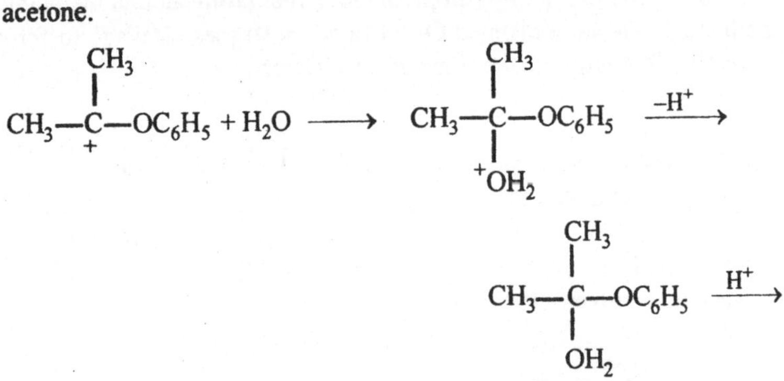

Hemiketal

APPLICATION

The reaction finds use as a modern commercial method for the synthesis of phenol, because cumene hydroperoxide is cheaply available from benzene and propene by using the following series of reactions.

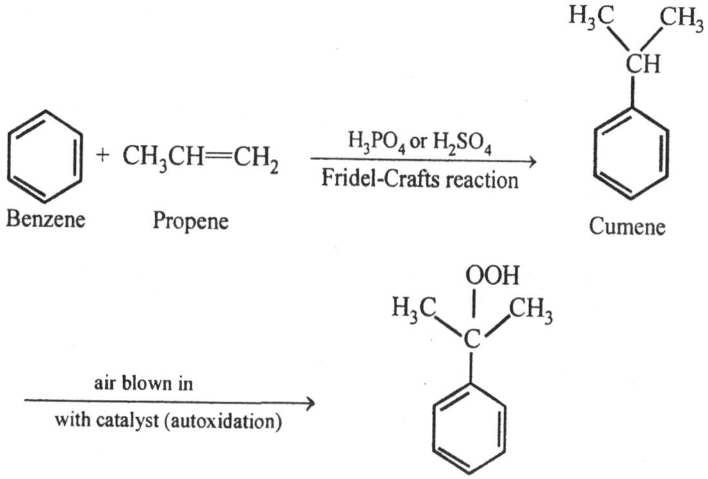

Cumene hydroperoxide

Acetone is a very useful by-product obtained during the reaction.

24

CURTIUS REARRANGEMENT

The importance of Curtius Rearrangement lies in its general application to acyl azides of all carboxylic acids—aliphatic, alicyclic, heterocyclic as well as aromatic — in their conversion to amines followed by successful degradation of carbon number by unity.

REACTION

It is the pyrolysis of acyl azides to isocyanates

$$R \cdot CO\ N_3 \xrightarrow[\Delta]{} R{-}N{=}C{=}O$$

The isocyanate readily undergoes hydrolysis to amine

$$R{-}N{=}C{=}O \xrightarrow{H_2O} R{-}NH_2 + CO_2$$

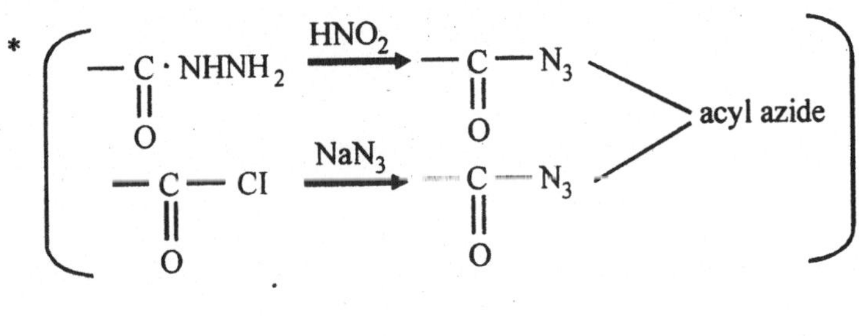

MECHANISM

The rearrangement is **intramolecular**

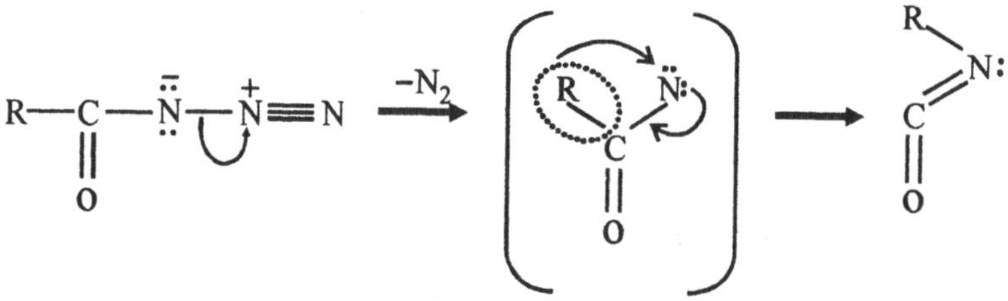

SUPPORT OF THE MECHANISM

The optical activity of the migrating R group is never lost confirming that R is never free during the course of the rearrangement. That is, the rearrangement is **intramolecular**.

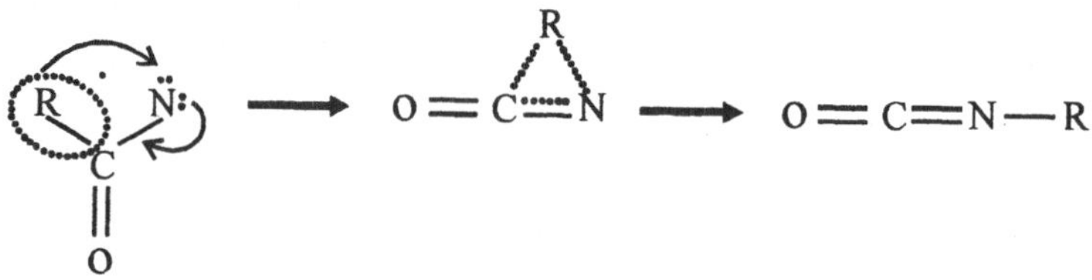

Note :

(I) This is a very general reaction applicable to all carboxylic acids—aliphatic, alicyclic, heterocyclic as well as aromatic.

(II) It gives good yield of isocyanate, for water is absent

(III) However, the reaction can be carried out in water or alcohol

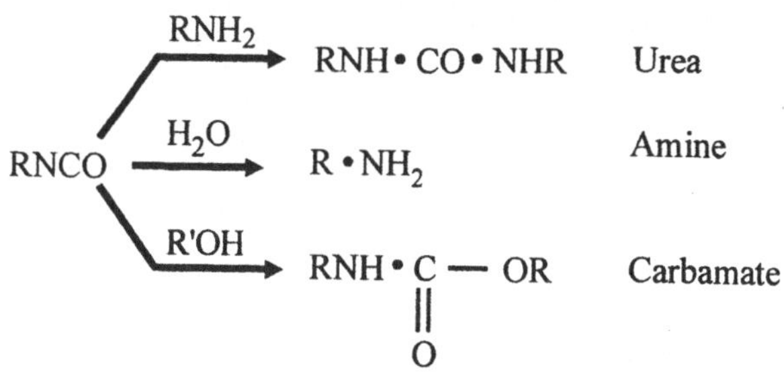

(IV) Lewis acids or protic acids catalyze the reaction but these are unnecessary for good results.

(V) The alkyl or aryl azide on heating gives an analogous reaction

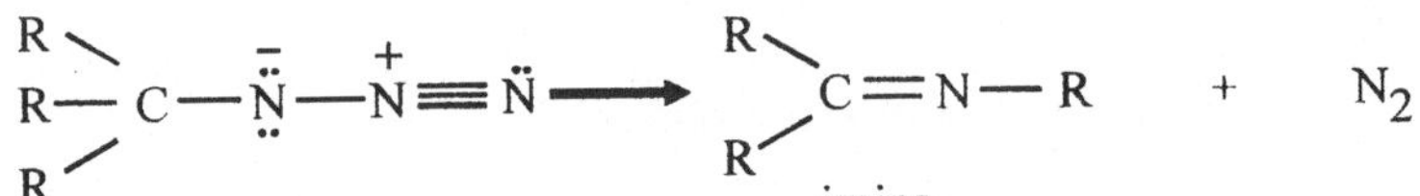

imine

R = H / Ar / Me, Et etc.

For cycloalkyl azides ring expansion occurs

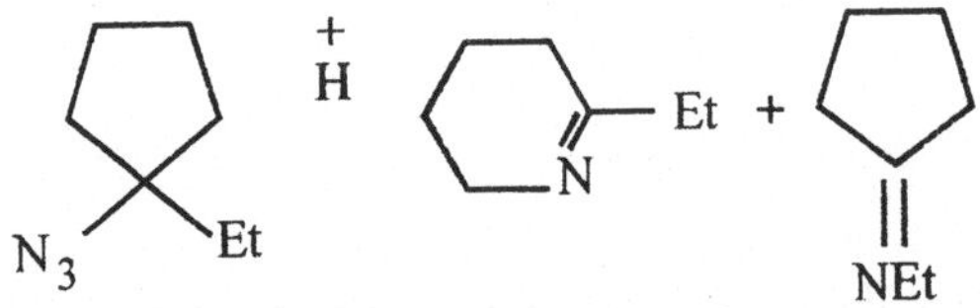

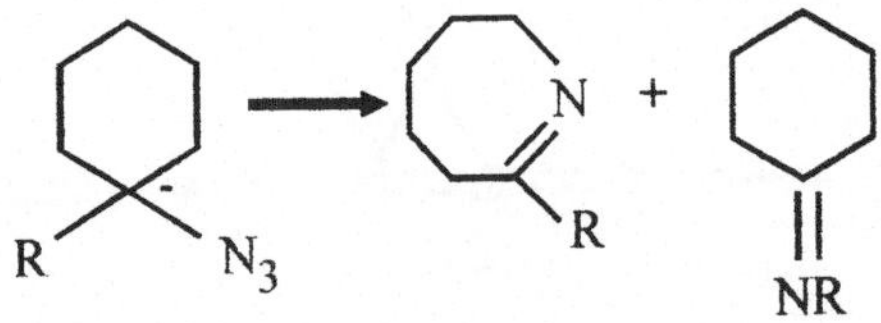

(~ 80%)

APPLICATION

(*a*) *Preparation of primary amines* : The Curtius reaction is of interest as it gives a primary amine which is free from secondary and tertiary amines, the impurities which are generally present with other methods of preparation.

$$C_6H_5CH_2COOC_2H_5 \xrightarrow[2.\ HNO_2]{1.\ NH_2NH_2} C_6H_5CH_2CON_3$$

Ethylphenyl acetate

$$\xrightarrow[2.\ H_2O]{1.\ \Delta,\ C_6H_6} C_6H_5CH_2NH_2$$

Benzylamine

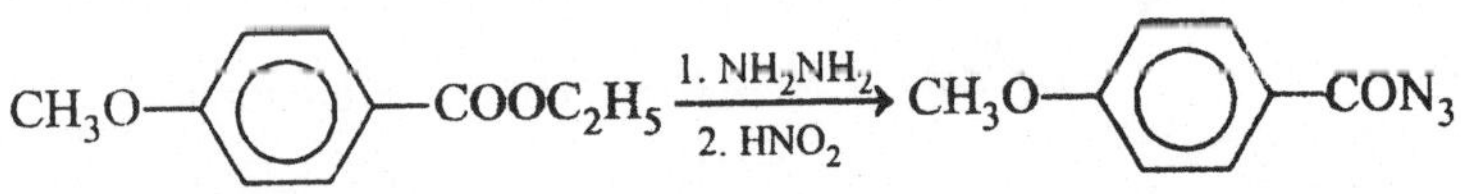

Ethyl anisate

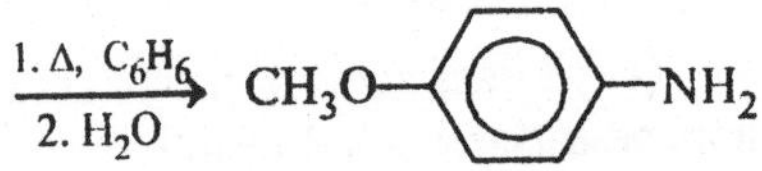

p-Anisidine

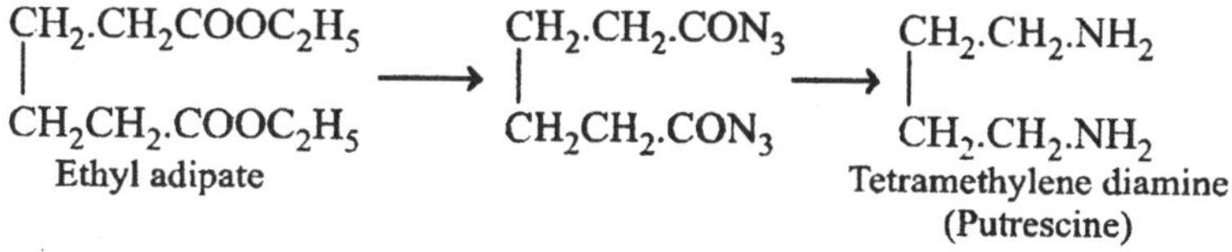

(*b*) *Preparation of α-amino acids* : Curtius reaction has been utilised in the preparation of the important α-amino acids such as glycine, alanine, phenylalanine and valine.

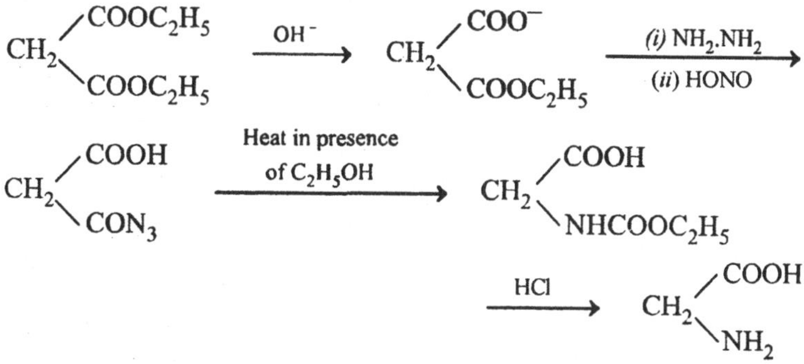

(*c*) *Preparation of urethanes* : Curtius reaction may be used for the preparation of N-substituted urethanes. The acid azide is refluxed in benzene solution and then an alcohol is added.

$$\underset{\text{Acid azide}}{R.CO.N_3} \xrightarrow{-N_2} R.NCO \xrightarrow{R'OH} \underset{\text{Urethane}}{R.NH.COOR'}$$

(*d*) *Preparation of aldehydes* : Like Hofmann reaction, Curtius reaction can be used for converting α, β-unsaturated acids and α-hydroxy acids into aldehydes, *e.g.*,

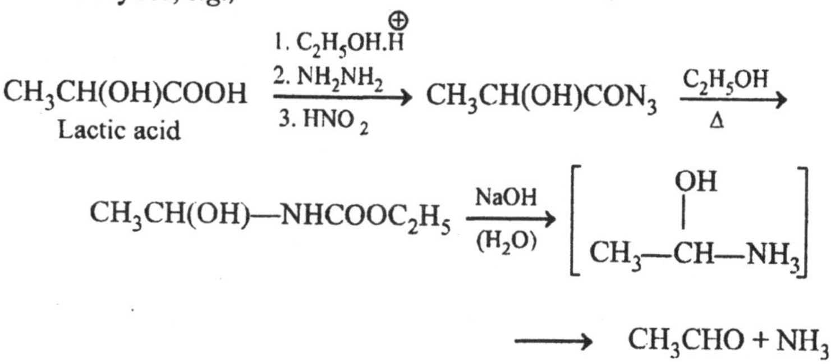

Similarly, benzylmalonic acid gives phenylacetaldehyde.

$$\underset{\text{Benzylmalonic ester}}{C_6H_5CH_2CH(COOR)_2} \xrightarrow{NH_2NH_2} C_6H_5CH_2CH(CONHNH_2)_2 \xrightarrow{HNO_2}$$

$$C_6H_5\,CH_2CH(CON_3)_2 \xrightarrow[2.\ NaOH]{1.\ C_2H_5OH} C_6H_5CH_2CH(NH_2)_2 \xrightarrow{HNO_2}$$

$$[C_6H_5CH_2CH(OH)_2] \xrightarrow{-H_2O} \underset{\text{Phenyl acetaldehyde}}{C_6H_5CH_2CHO}$$

Curtius reaction is also used for converting substituted malonic ester into aldehyde *via* acid azides.

$$\underset{\text{Diethyl benzylmalonate}}{C_6H_5CH_2.CH\begin{cases}COOC_2H_5\\COOC_2H_5\end{cases}} \xrightarrow[(ii)\ HONO]{(i)\ H_2N.NH_2} C_6H_5CH_2.CH\begin{cases}CON_3\\CON_3\end{cases}$$

$$\longrightarrow \underset{\text{Phenyl acetaldehyde}}{C_6H_5CH_2CHO}$$

(*e*) Curitus reaction may be used for *stepping down the acid series.*

$$RCH_2COOH \xrightarrow[H^{\oplus}]{C_2H_5OH} RCH_2COOC_2H_5 \xrightarrow[2.\ HNO_2]{1.\ NH_2NH_2} RCH_2CON_3$$

$$\xrightarrow[2.\ H_2O]{1.\ \Delta,\ C_6H_6} RCH_2NH_2 \xrightarrow{HNO_2} RCH_2OH \xrightarrow{Oxdn.} RCOOH$$

PROBLEMS BASED ON CURTIUS REARRANGEMENT

PROBLEM-I

$$\underset{[A]}{C_6H_{10}O} \xrightarrow[\overset{+}{H_3O}]{Na - Hg} \underset{[B]}{C_6H_{12}O} \xrightarrow[\Delta]{HBr} \underset{[C]}{C_6H_{11}Br} \xrightarrow[\substack{EtOH\\ \Delta}]{KCN} \underset{[D]}{C_7H_{11}N} \xrightarrow{\overset{+}{H_3O}}$$

$$\underset{[E]}{C_7H_{12}O_2} \xrightarrow{SOCl_2} \underset{[F]}{C_7H_{11}OCl} \xrightarrow{N_2H_4} \underset{[G]}{C_{11}H_{14}ON_2} \xrightarrow{HNO_2} \underset{[H]}{C_{11}H_{11}ON_3}$$

$$\xrightarrow{\Delta} \xrightarrow{H_2O.} \underset{[I]}{C_6H_{13}O} \xrightarrow[EtOH\ /\ \Delta]{CHCl_3\ /KOH}$$ 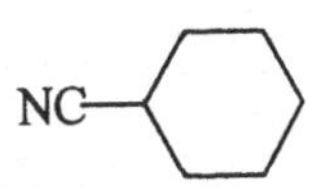

ANSWER

A ▶	Cyclohexanone;	**B ▶**	Cyclohexanol;
C ▶	Cyclohexylbromide;	**D ▶**	Cyclohexyl cyanide;
E ▶	Cyclohexanoic acid;	**F ▶**	Cyclohexanoyl chloride;
G ▶	Cyclohexanoyl hydrazide	**H ▶**	Cyclohexanoyl azide;
I ▶	Cyclohexyl amine		

PROBLEM-II

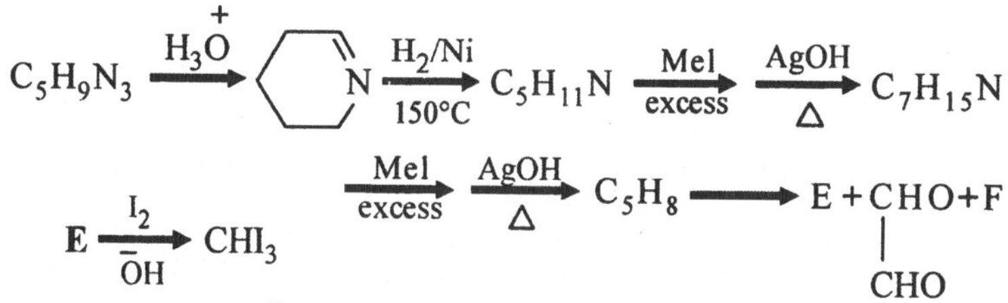

F + Ø.CHO $\xrightarrow{\bar{O}H}$ Ø.CH_2OH + HCO_2

ANSWER

A ▶	Cyclopentyl azide;	**B ▶**	Piperidine
C ▶	N,N-Dimethyl pent-4-ene;	**D ▶**	Pent-1, 4-dine
E ▶	Acetaldehyde	**F ▶**	Formaldehyde

PROBLEM-III

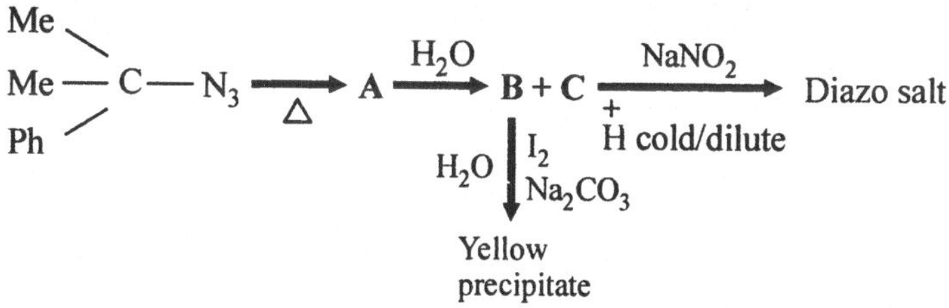

HINT

C forms a diazo salt. So it is an amine.

B responds to iodoform test. So it contains a CH_3 CO group.

ANSWER

A ▶	1, 1-Dimethyl-N-phenyl imine	**B ▶**	Acetone
C ▶	Aniline		

25

COPE REACTION OR COPE ELIMINATION

Tertiary amines are oxidised by hydrogen peroxide and per acids into amine oxides.

$$R_3\ddot{N} \xrightarrow[CF_3COOOH]{H_2O_2 \text{ or}} R_3\text{—}\overset{\oplus}{N}\text{—}\overset{\ominus}{O}$$

N-oxide

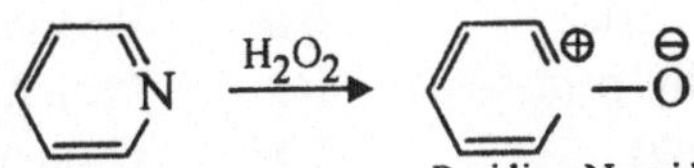

Pyridine-N-oxide

When an amine oxide of ter amines containing at least one β-hydrogen is heated at about 150°C, it decomposes to form an alkene and a derivative of hydroxylamine. This elimination is an example of thermal elimination known as cope elimination or cope reaction.

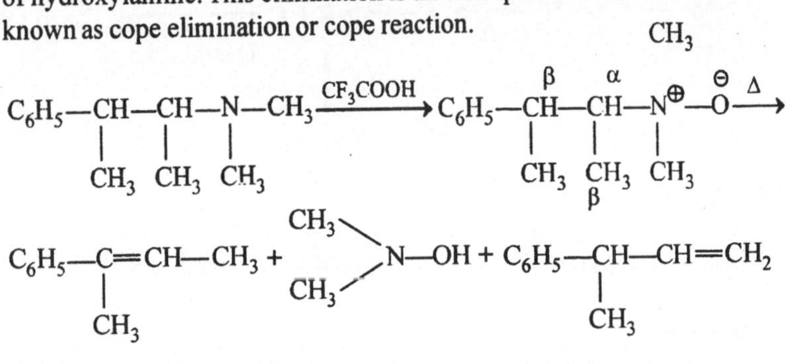

In this elimination product formation always takes place by the Hofmann rule.

26

COREY-HOUSE SYNTHESIS

Alkyl halides react with lithium in dry ether to form alkyl lithium.

$$R—X + 2Li \xrightarrow{\text{dry ether}} R—Li + LiX$$

This alkyllithium reacts with CuI to give Dialkyllithiumcuparate known as **Gilman reagent.**

$$2R—Li + CuI \longrightarrow R_2CuLi + LiI$$

This dialkyllithiumecuparate reacts with alkyl halide to give alkane and the reaction is known as Corey-House synthesis.

$$R_2CuLi + R'—X \longrightarrow R—R' + R—Cu + LiX$$

In this reaction R′—X may be methyl halide, *p*-alkyl halide or sec alkyl halide. The alkyl group of dialkyllithiumcuprate may be methyl, primary, secondary or tertiary.

In this reaction dialkyllithiumcuparate also reacts with aryl halide, alkyl halide and vinyl halide.

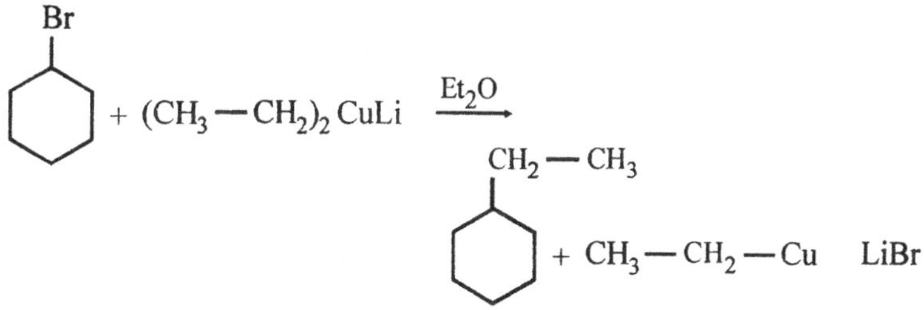

$$(CH_3—CH_2—CH_2)_2CuLi + CH_2{=}CH—CH_2—Br \xrightarrow{Et_2O} \underset{\text{1-Hexene}}{CH_3—CH_2—CH_2—CH_2—CH{=}CH_2}$$

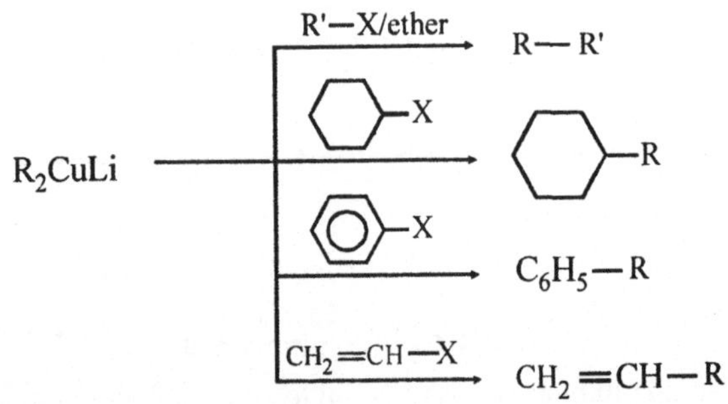

27

CRIEGEE REACTION

Oxidative cleavage of a 1, 2 diol by lead tertra acetate is known as Criegee reaction.

(1) 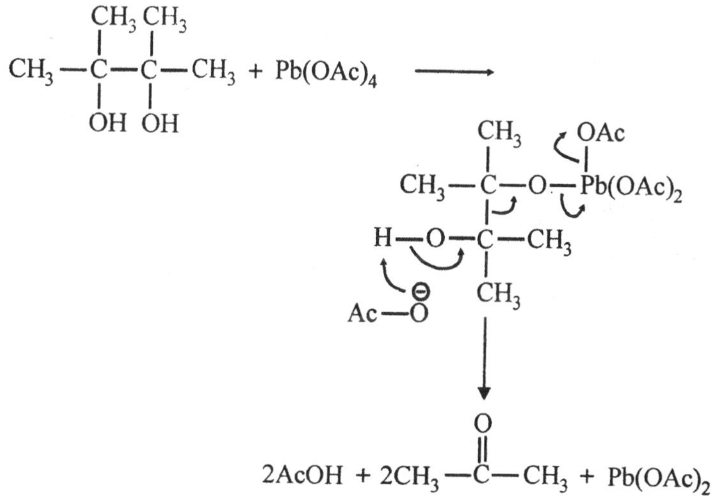

(2)

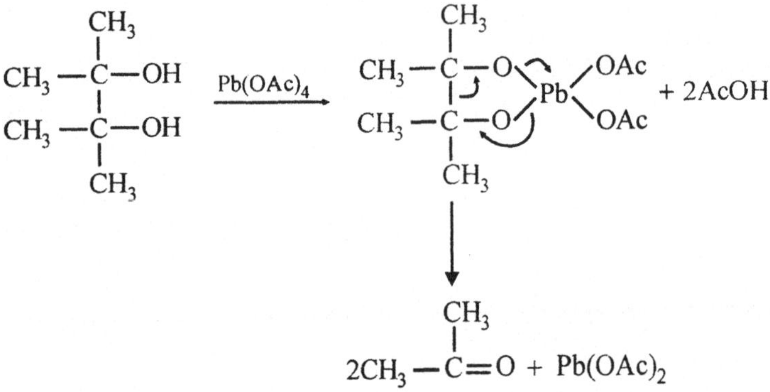

(3)

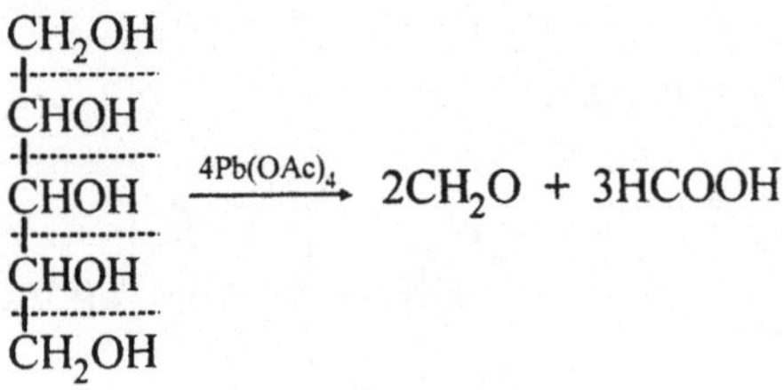

28

DARZENS REACTION

This reaction takes place between α-haloester having at least one α-hydrogen and a carbonyl compound in the presence of a strong base (NaOEt, NaH potassium or sodium ter butoxide) at 0–10°C.

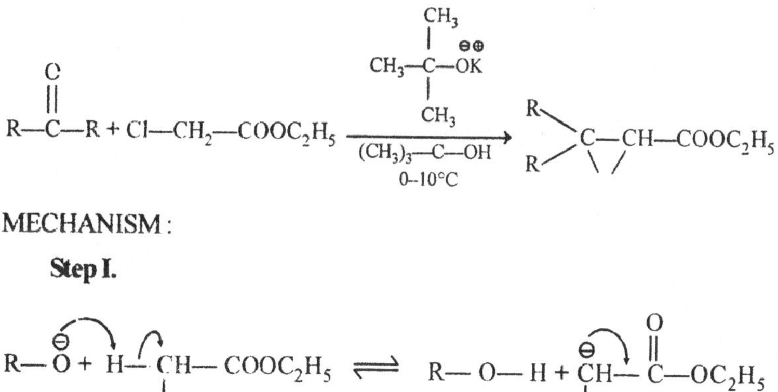

MECHANISM :

Step I.

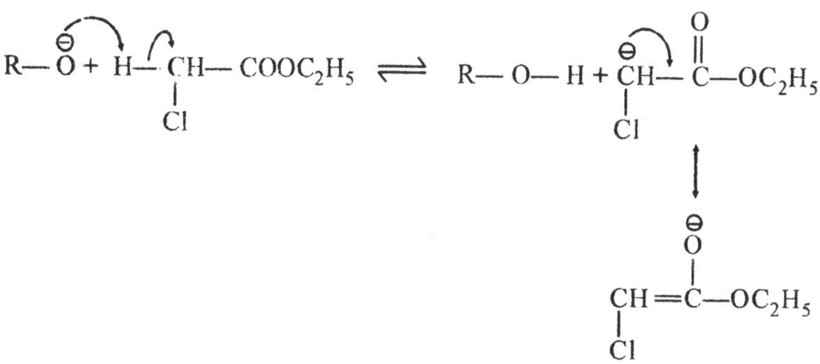

Enolate is delocalised

Step II. Nucleophilic addition of carbonyl compound with enolate ions followed by SN reaction.

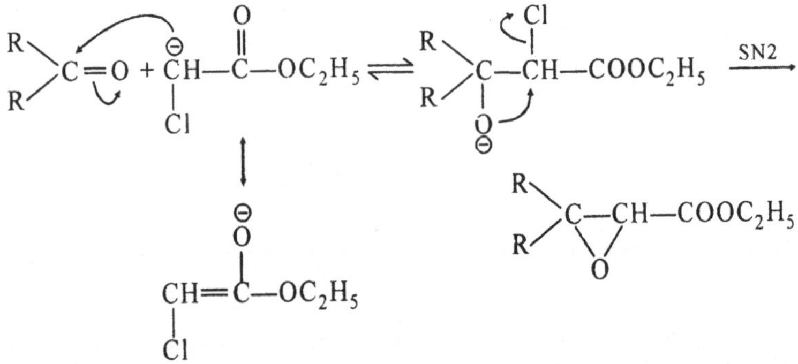

The presence of halogen on α-carbon of the ester assists in the formation of α-haloesterenolate.

Reaction is carried out in the presence of a strong base so that the intramolecular nucleophilic substitution should be SN2.

The synthetic utility of the reaction hinges on two successive steps, hydrolysis followed by decarboxylation, to give aldehydes or ketones that are difficult to prepare by other methods.

APPLICATIONS

Darzen condensation has been used in synthesising many organic compounds. Some of which are prepared as follows :

(*a*) *Synthesis of α-pinene* : Darzen reaction has been used in the synthesis of α-pinene [Ruzicka].

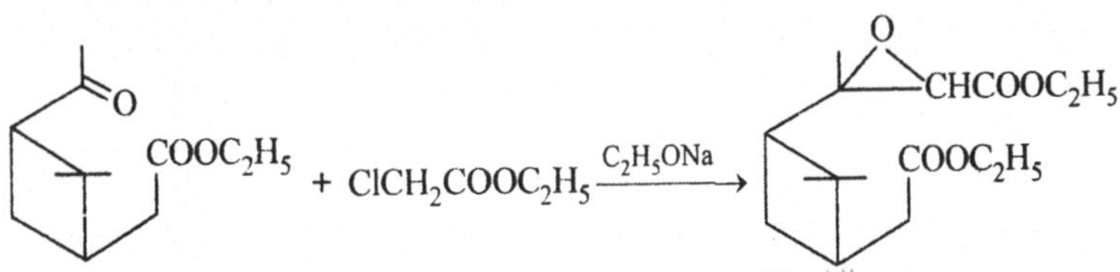

Etyl pinonate

Glycidic ester

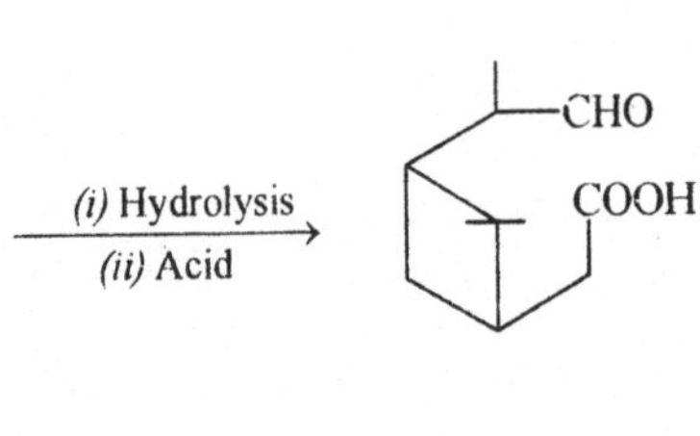

(*b*) *Synthesis of aldehydes and ketones* : Blanchard (1963) were successful in synthesising many aldehydes and ketones in quantitative yield by the pyrolysis of *t*-butyl glycidic esters.

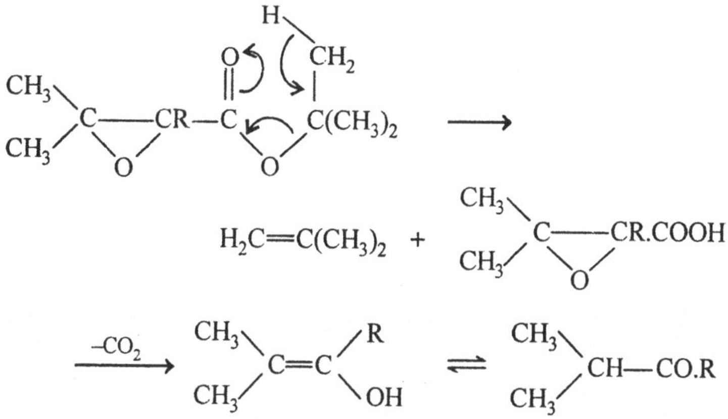

(*c*) *Synthesis of vitamin A_1* : Darzen reaction has been used in the synthesis of one of the reactions of vitamine A_1 synthesis [Isler; 1947].

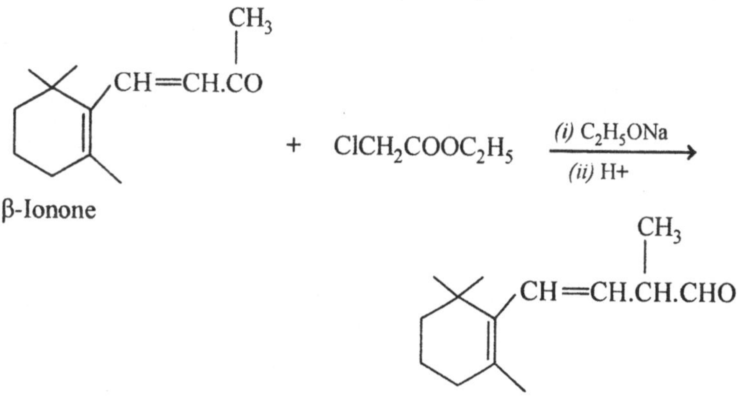

β-Ionone

29

DAKIN REACTION

The replacement of the aldehyde group of *o*-and *p*-hydroxy and *o*-amino-benzaldehyde (or ketone) by a hydroxyl group on reaction with alkaline hydrogen peroxide is called Dakin reaction. An example is as follows :

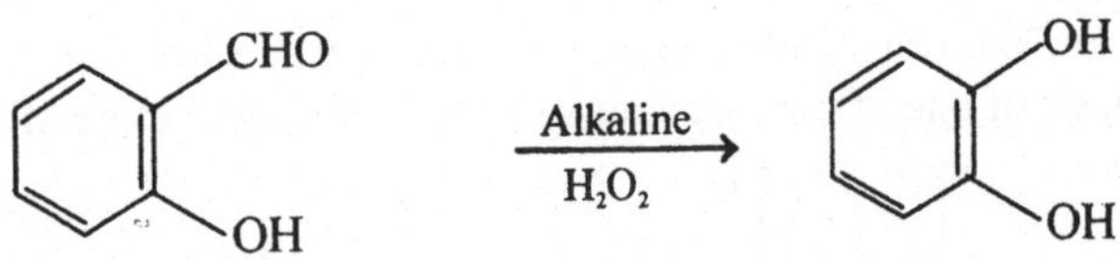

o-Hydroxybenzaldehyde Catechol

The Dakin reaction is limited to aromatic aldehydes and ketones that have a hydroxyl or amino group in the *ortho*-and *para*-position.

MECHANISM

This reaction involves a molecular rearrangement as in Baeyer-Villiger reaction. During the reaction, the formyl group will be removed as formate ion and then replaced by the hydroxyl group.

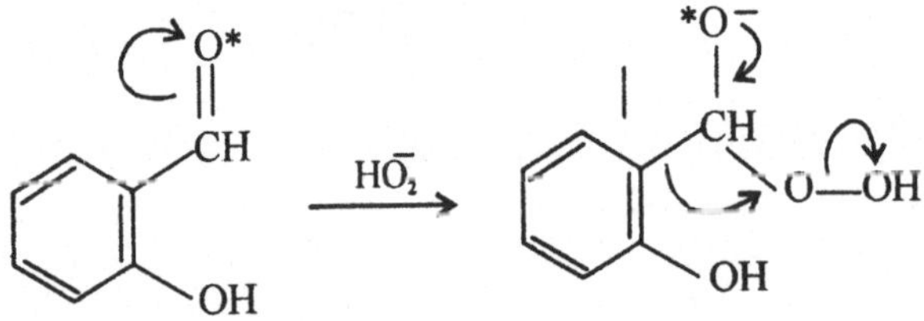

Salicyladehyde

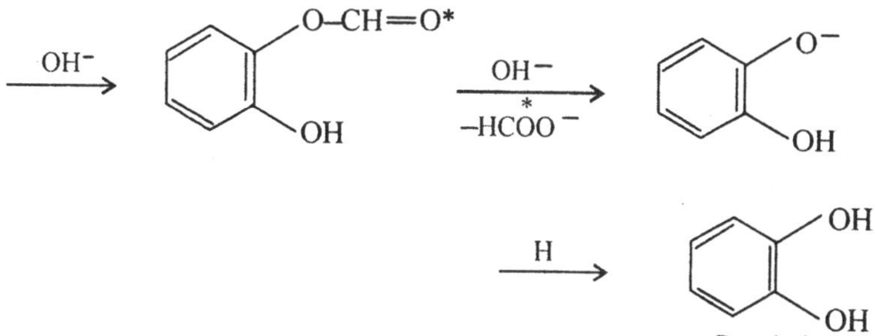

Catechol

The above mechanism has been supported by the tracer technique; all the labelled oxygen (O*) were found in formic acid.

APPLICATION

Daking reaction has been used in synthesising many polyhydric phenols from naturally occurring hydroxy aldehydes. Some examples are as follows :

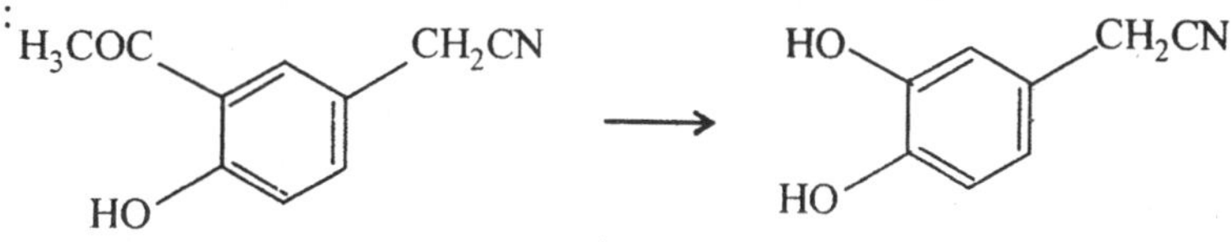

5-Cyanomethyl-2-hydroxyacetophenone → 3, 4-Dihydroxyphenylacetonitrile

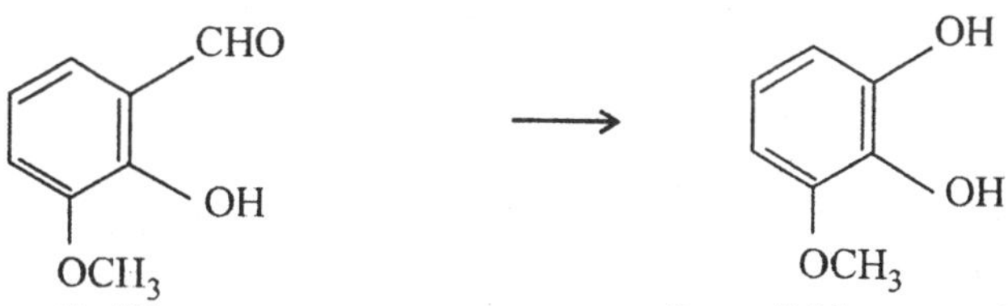

o-Vanillin → Pyrogallol l-monomethyl-ether (80%)

2-Hydroxy3, 4-dimethyl acetophenone → 3, 4-dimethylcatechol

Some other applications of Dakin reaction are as follows%

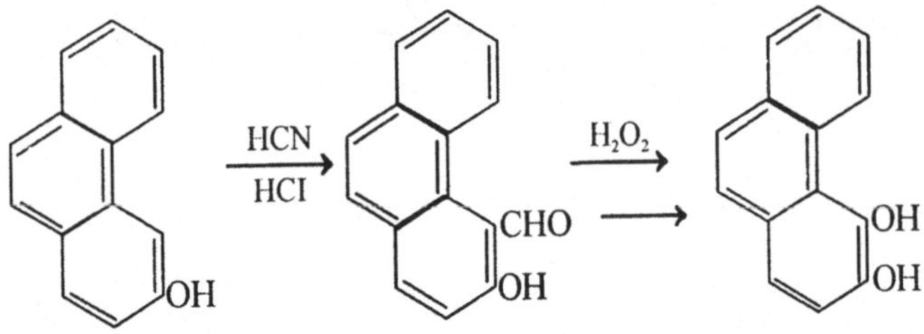

3-Phenanthrol

3, 4-Dihydroxy phenanthrene (morophol)

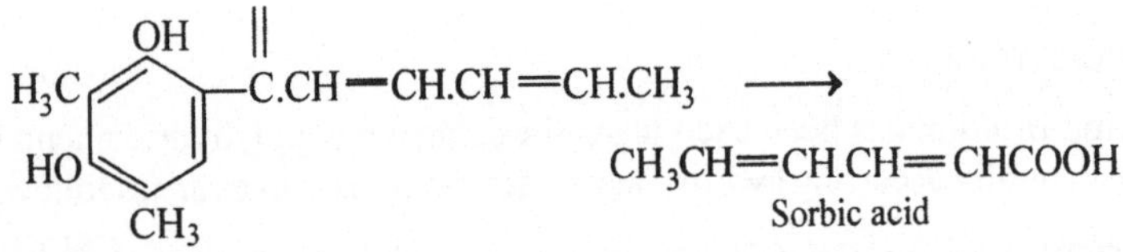

Sorbicillin

Sorbic acid

30

DAKIN -WEST REACTION

When α-amino acid is treated with an acid anhydride in the presence of pyridine, the carboxylic group is replaced by an acyl group and the amino group is acetylated. This reaction is known as Dakin-West reaction.

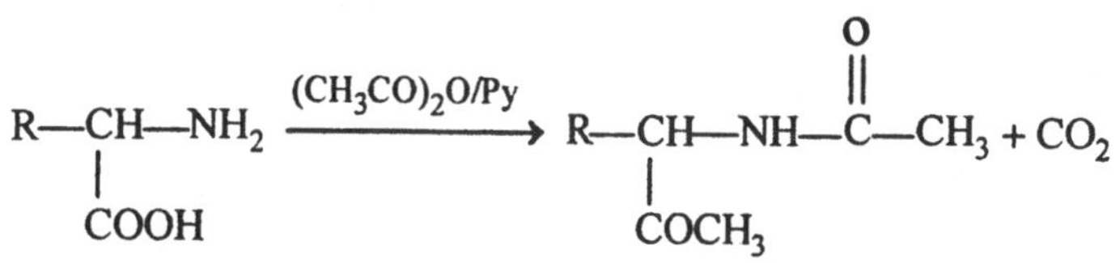

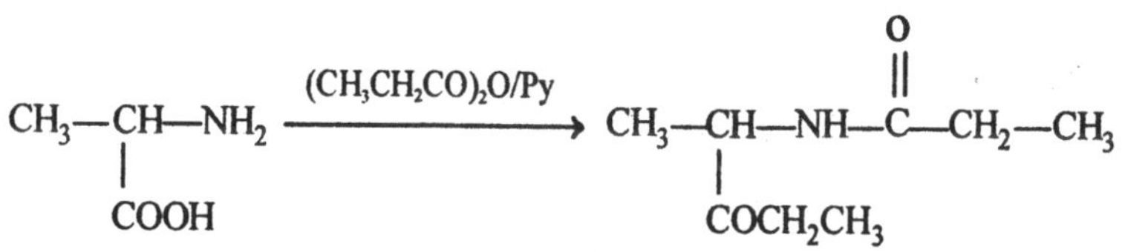

31

DUFF REACTION

Formylation reaction at o-position given by phenols and amines with $(CH_2)_6N_4$ followed by hydrolysis is known as Duff reaction.

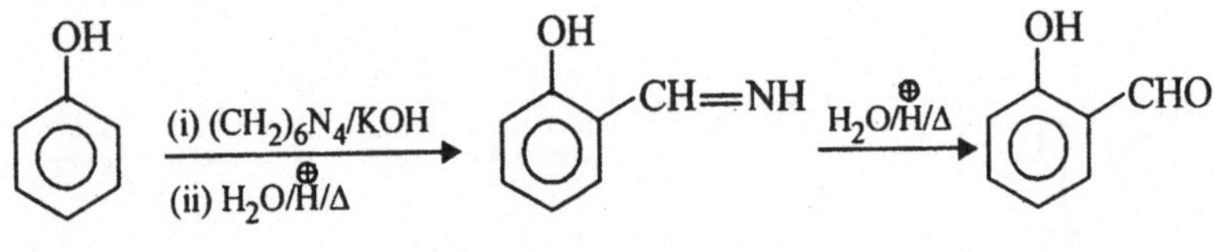

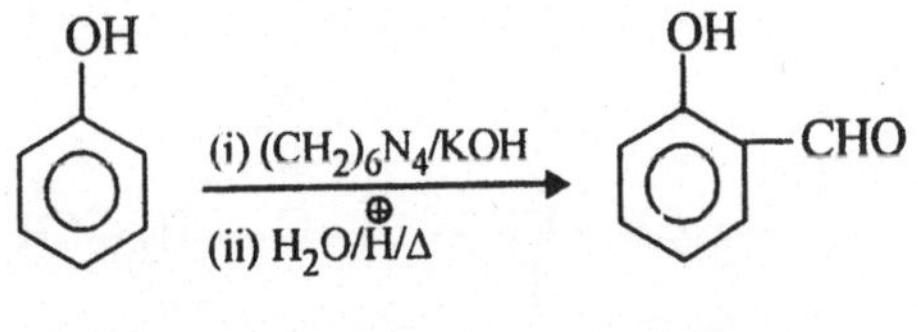

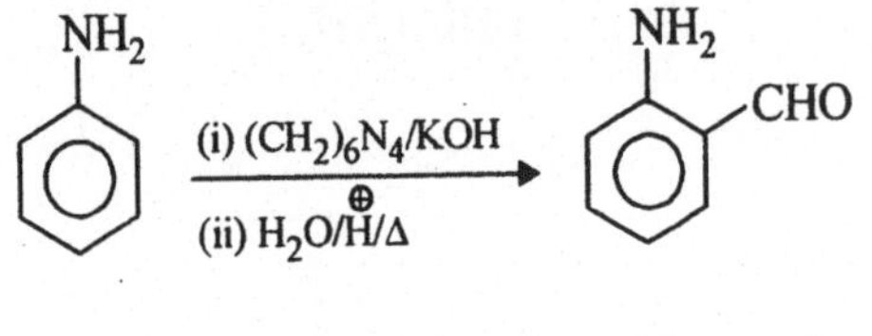

Alkyl benzene also gives this reaction when the reaction is carried out with $(CH_2)_6N_4/CF_3COOH$.

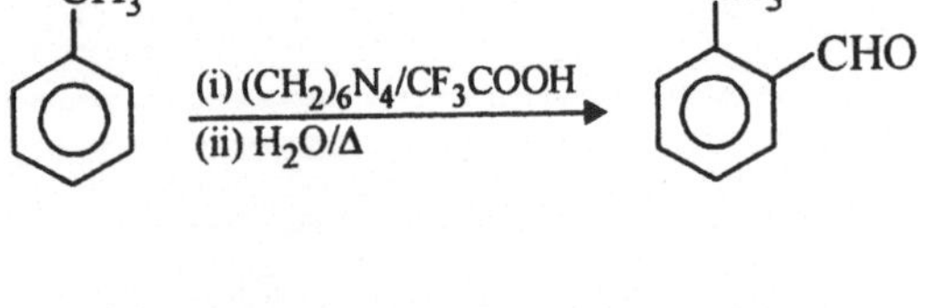

32

DIENONE-PHENOL REARRANGEMENT

When treated with acid, 4, 4-dialkyl cyclohexadienone is converted to phenol with migration of one of the alkyl groups to the adjacent carbon. This is called dienone-phenol rearrangement. The dienone is first of all dissolved in acetic anhydride and then treated with catalytic amount of sulphuric acid. The product on hydrolysis yields the phenol.

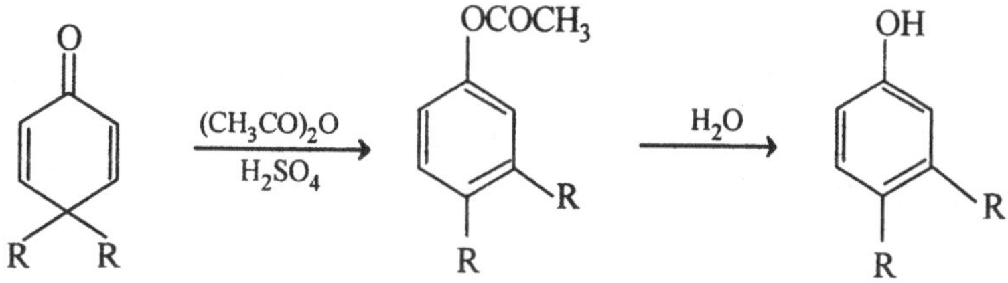

MECHANISM

When oxygen is protonated, a carbocation is generated which is stabilised by delocalization of the positive charge.

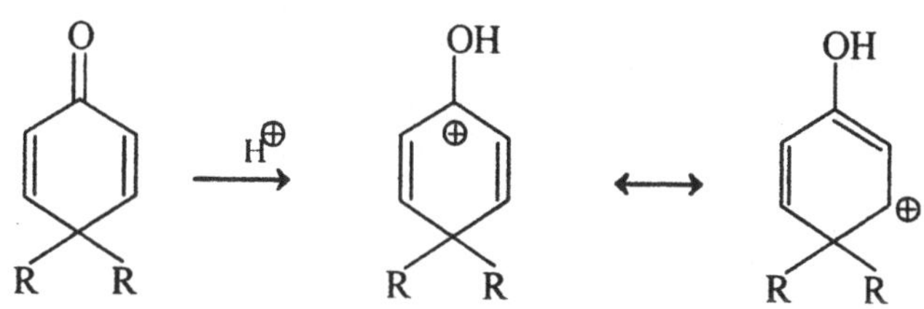

In one of the cononical structures, the positive charge lies on a carbon adjacent to a highly substituted carbon. Hence, a carbocation rearrangement takes place. Subsequent loss of a proton yields the 3, 4-disubstituted phenol.

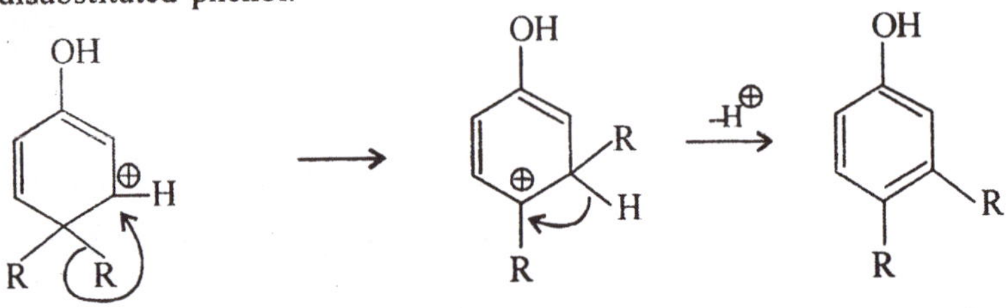

The ease of dienone-phenol rearrangement is because of the creation of a stable aromatic system. If one of the alkyl group forms a part of the cyslic system, either the alkyl group or the ring methylene group may migrate.

(*i*) *Alkyl migration*

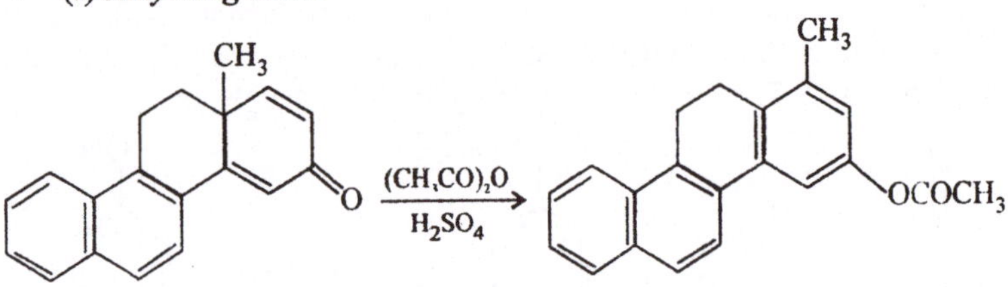

(*ii*) *Migration of ring methylene group*

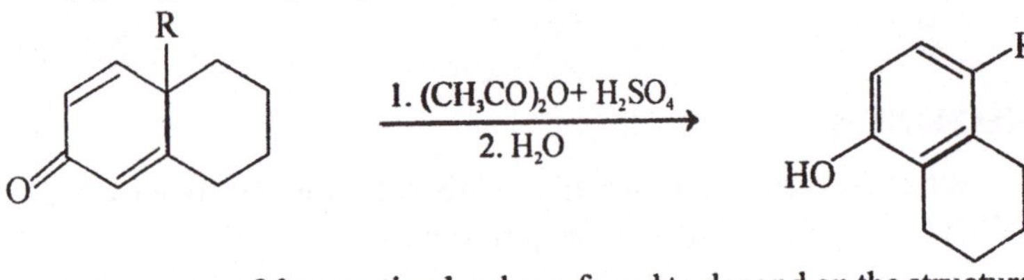

The course of the reaction has been found to depend on the structural or electronic factors and on the conditions of reaction.

A reverse rearrangement, *i.e.*, phenol-dieone rearrangement has been observed during the electrophilic substitution in phenols in some cases.

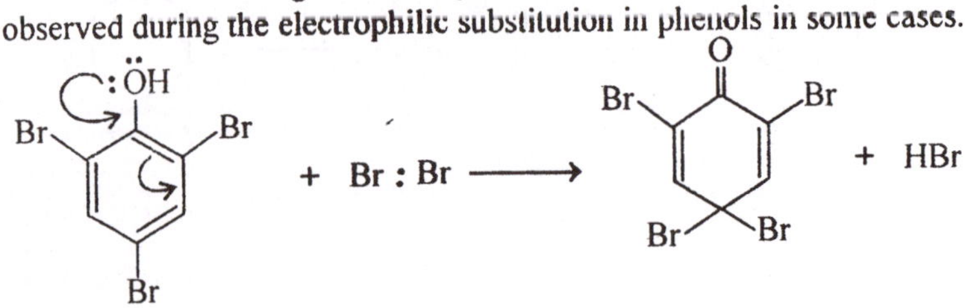

APPLICATION

The rearrangement finds many useful applications. A classical application is the rearrangement of santonin to desmotropo santonin.

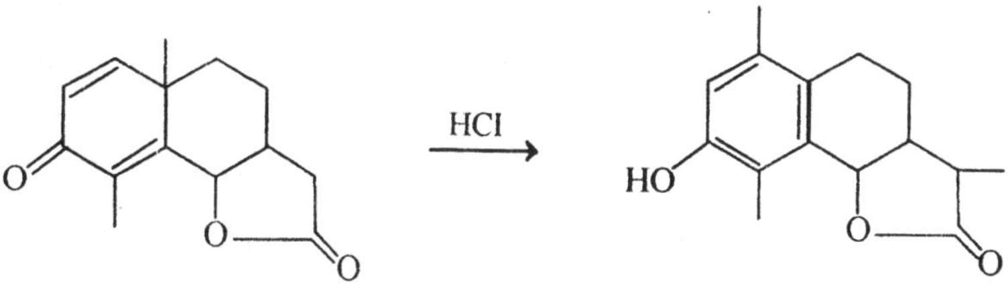

The anthrasteroid rearrangement is also closely related.

33

DIAZOTISATION (OR GRIESS) REACTION

Conversion of *p*-amino group into diazo group by HNO_2 or $NaNO_2$/conc. HCl at 0–5°C is known as diazotisation or Griess reaction. Although aliphatic diazonium compounds readily decompose but aromatic diazonium compounds are quite stable due to delocalisation.

$$Ar—NH_2 + NaNO_2 + 2HX \longrightarrow Ar—\overset{\oplus}{N}\equiv N\overset{\ominus}{X} + NaX + 2H_2O$$

MECHANISM:

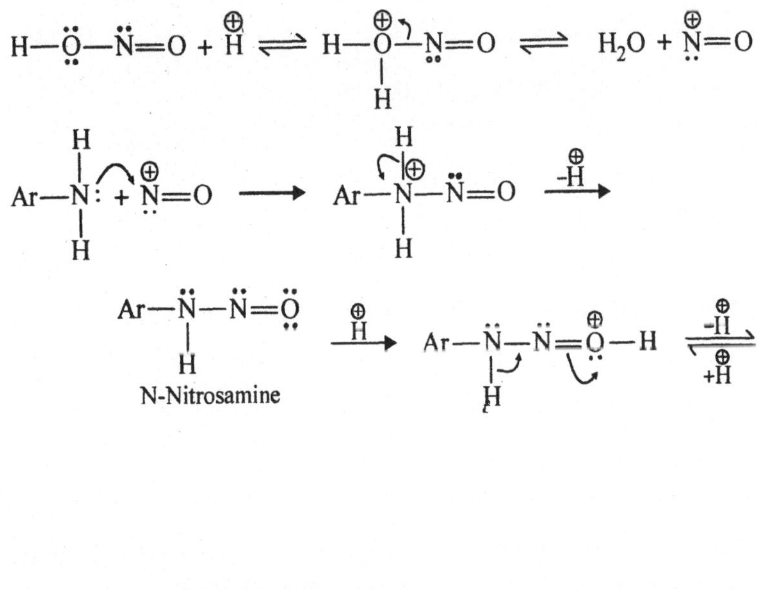

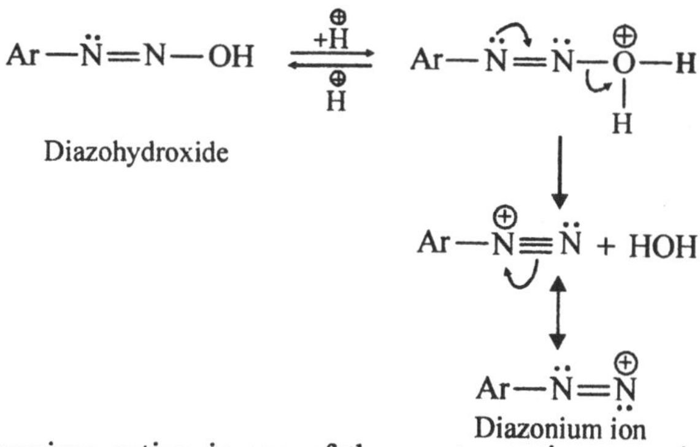

Benene diazonium cation is one of the most reactive organic compounds for aromatic nucleophilic substitution reaction.

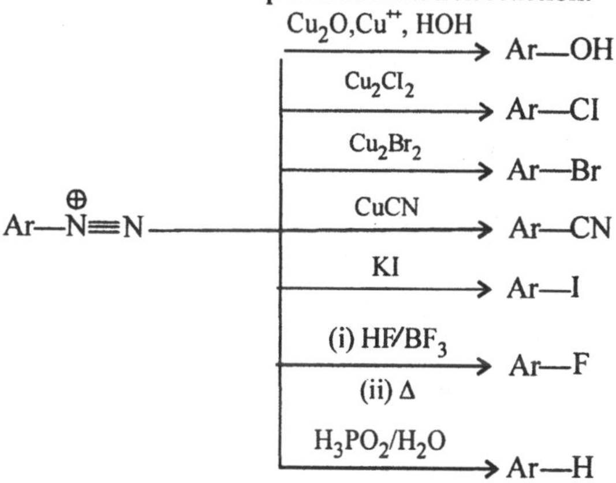

Arene diazonium cations act as weak electrophile. They react with highly reactive aromatic compounds (phenols and amines) to give aromatic electrophilic substitution reaction and this reaction is known as coupling reaction.

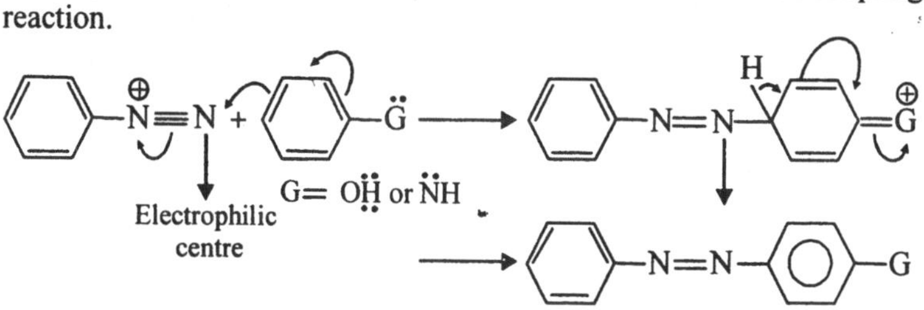

In case of phenol, coupling takes place in basic medium and in case of amine coupling takes place in acidic medium. (pH 5-7)

34

DIECKMANN CONDENSATION

Dieckmann condensation is an intramolecular Claissen condensation reaction given by diesters. The α-carbon atom having α-hydrogen and the ester group for the condensation come from the same molecule. In general, Dieckmann condensation is useful only for the preparation of five and six membered rings (*i.e.* diesters are mainly diesters of C_6 and C_7 dibasic acids).

The corresponding seven and eight membered ring compounds are formed in much lower yield and the rings between 9 and 12 members are in negligible yield.

Diesters of shorter-chain dibasic acids do not undergo the Dieckmann condensation because of the strain which would result in the small ring formed.

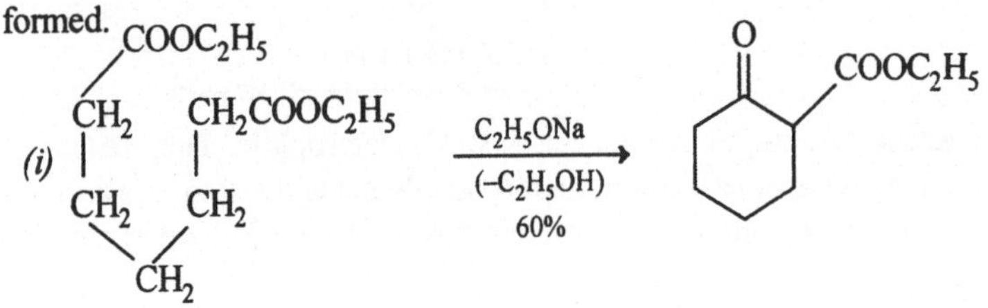

Ethyl pimelate — 2-Carbethoxycyclohexanone

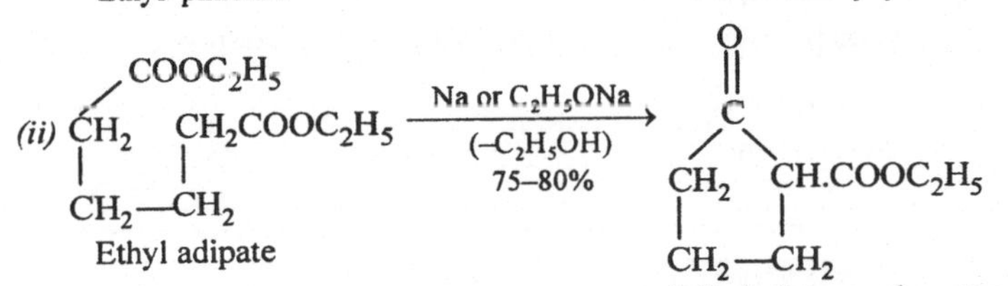

Ethyl adipate — 2-Carbethoxycyclopentanone

(*iii*) When ethylsuccinate is subjected to Dieckmann condensation, firstly an intermolecular condensation occurs which is followed by cyclistion to cyclohexanedione system.

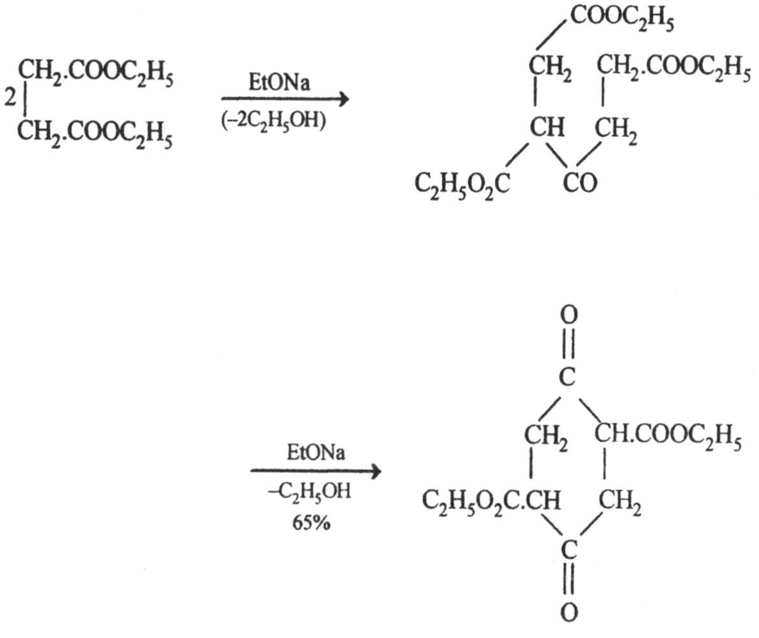

If esters of dibasic acids are having less than six carbon atoms, they do not undego dieckamann condensation owing to the large strain in the smaller rings formed. Also, if esters are having more than eight carbon atoms, they do not undergo Dieckmann reaction because the intramolecular reaction gets smaller with the increase in the chain length and intermolecular condensation gets facilitated.

MECHANISM

In Dieckman condensation, the carbanion attacks the carbonyl carbon of the second ester group in the same molecule. Expulsion of the ethoxide ion results in the formation of cyclic ketone derivative. The following examples of the Dieckmann condensation show that a 1, 6-diester gives a five-membered ring, and a 1, 7-diester gives a six-membered ring.

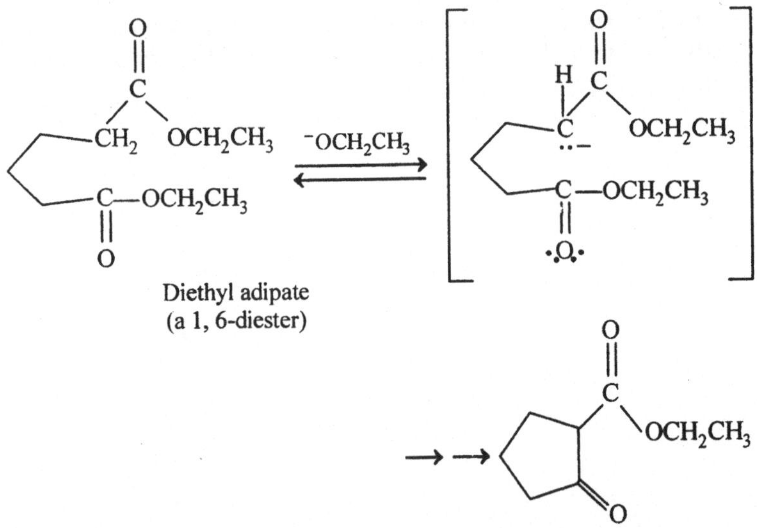

Diethyl adipate
(a 1, 6-diester)

Cyclic β-keto ester
(80%)

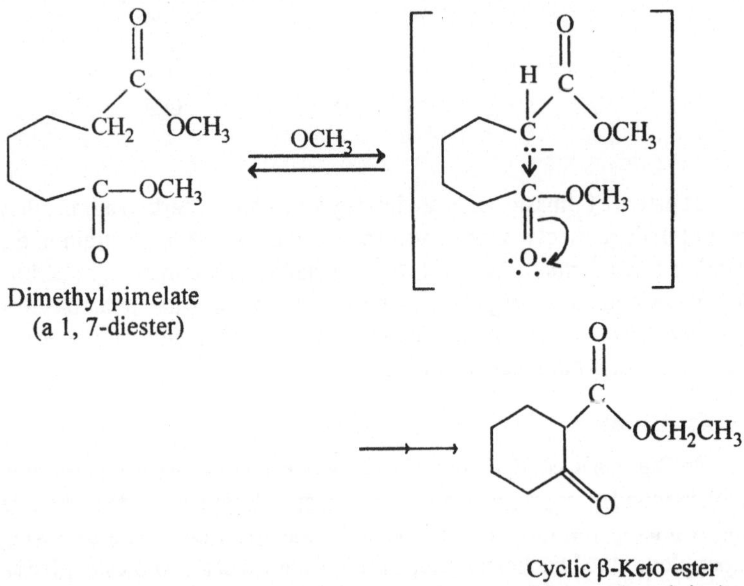

Dimethyl pimelate
(a 1, 7-diester)

Cyclic β-Keto ester
(six-membered ring)

The effect of experimental condition on the ring size has been seen in some compounds.

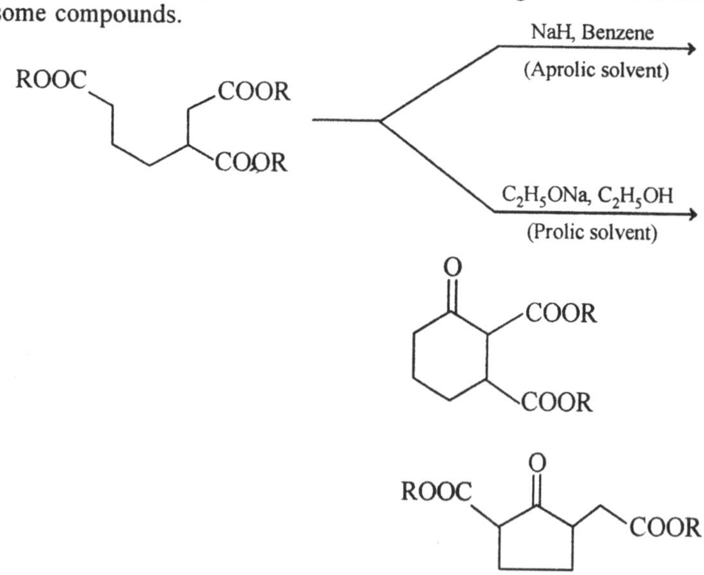

APPLICATION

Dieckmann condensation has provided an important route to the synthesis of various cyclopentane and cyclohexane derivatives.

The ester formed in the above reactions can be converted to ketone, in the following way :

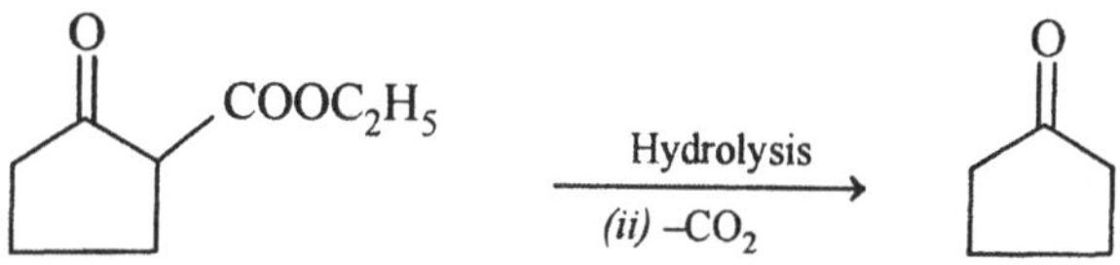

Thus, Dieckmann reaction has been employed to introduce five and six membered rings during the synthesis of many natural product, *e.g.*, α-pinene, epiandrosterone, oesterone, etc. The general steps are given as follows

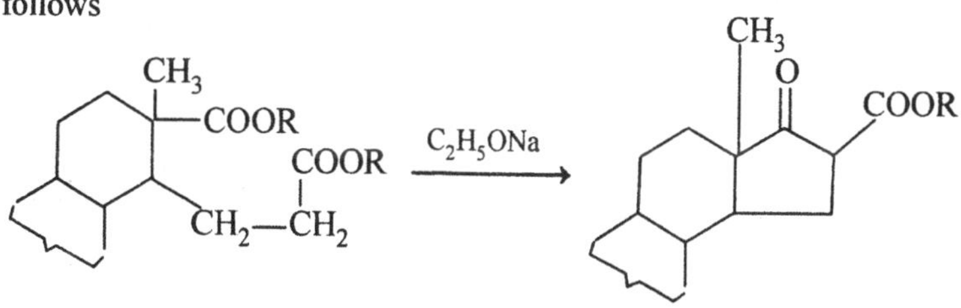

:

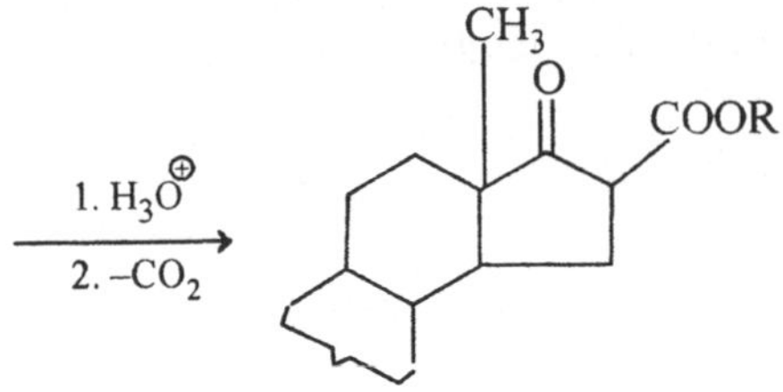

An interesting example of Dieckmann condensation is the formation of very useful *bicyclic diketone.*

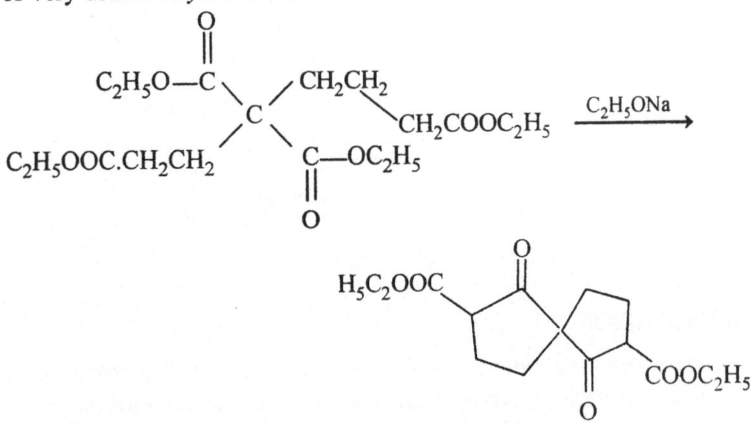

The diketone is required for various stereo-chemical and spectral studies.

2. Preparation of heterocyclic keto esters

(*a*) Piperidone derivative

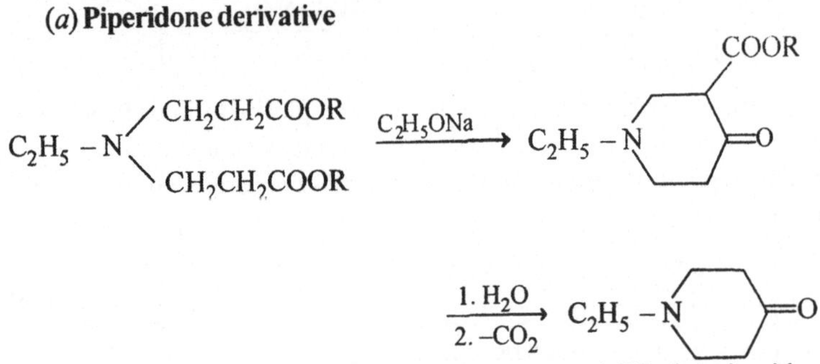

1-Ethyl-4-piperidone

(*b*) **Thiophene drivatives :** Depending on the conditions of the reaction, two isomers may be obtained.

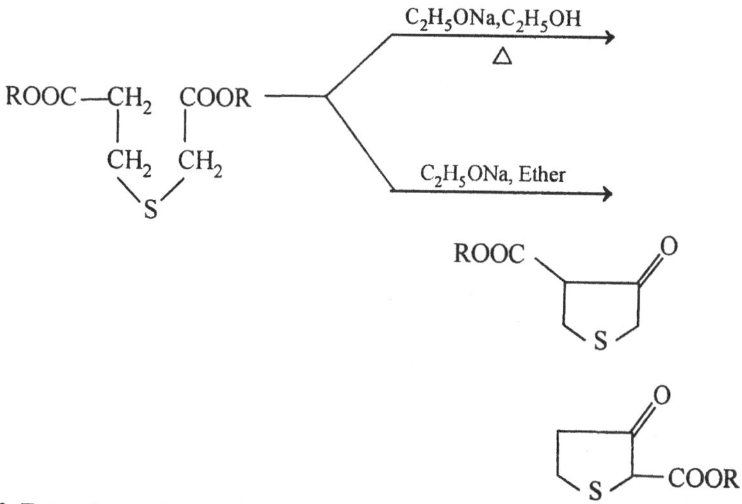

3. Extension of Dieckmann reaction

An interesting variation of Dieckmann reaction is the condensation of oxalic ester with the esters of dibasic acids, *e.g.*,

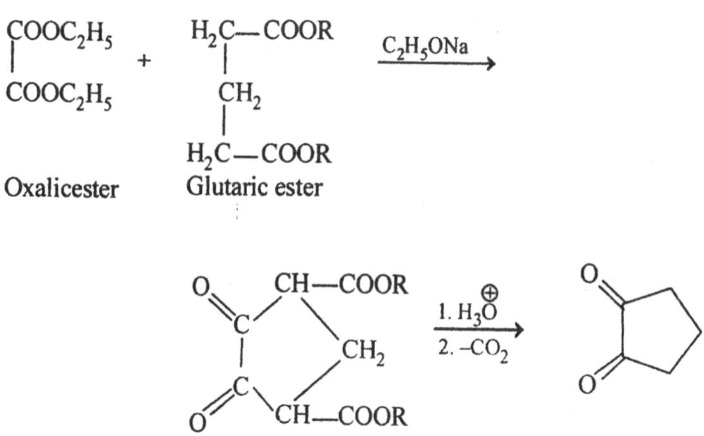

Cyclpentane-1, 2-dione

Kampa applied this extension in the synthesis of camphoric acid and thus helped in ascertaining the structure of camphor.

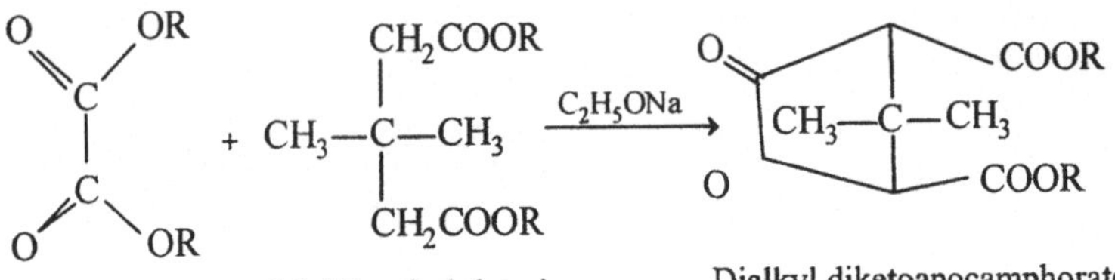

Oxalic acid ester

3,3-Dimethylglutaric acid ester

Dialkyl diketoapocamphorate

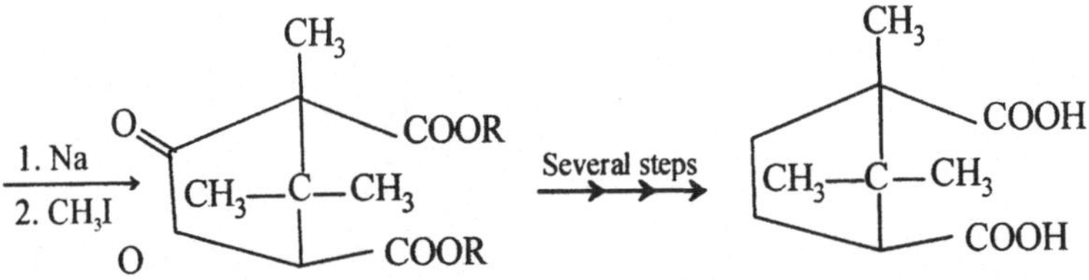

Dialkyl diketocamphorate

Camphoric acid

(*c*) Ziegler applied Dieckmann reaction on dinitriles to prepare large-size rings using lithium ethyl anilide and high dilution technique.

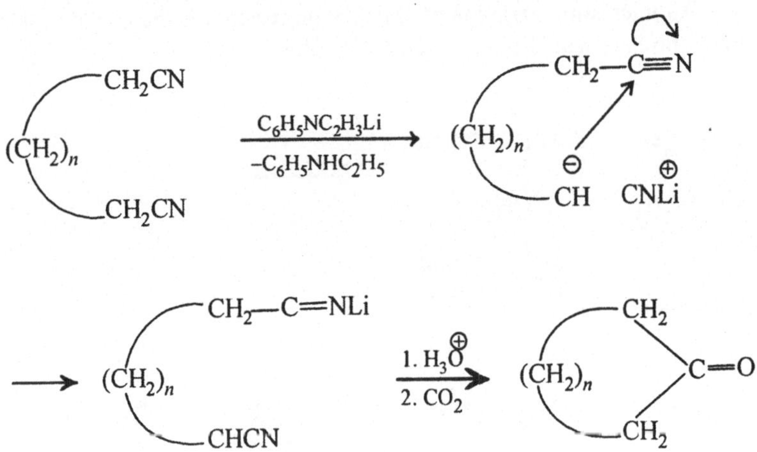

35

DIEL'S-ALDER REACTION

1, 4-addition reaction of conjugated diene with alkenes and alkynes is known as Diel's-Alder reaction. Conjugated diene of the reaction is known as diene and the other component is as the dienophile. This reaction is an example of both pericyclic reaction (a reaction that occurs in one step by a cyclic flow of electrones) and a cycloaddition (an addition reaction that forms a ring).

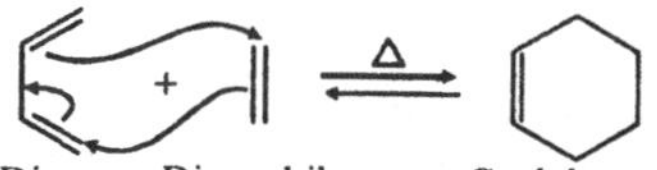

Diene Dienophile Cyclohexene adduct 30% yield

In this reaction two new σ bonds are formed at the expense of two π bonds of the diene and dienophile. the adduct contains a new six membered ring with a double bond. Since σ bonds are usually stronger than π bonds, formation of adduct is usually favoured energetically but most of the Diel's-Alder reactions are reversible.

Factor favouring the Diel-Alder reaction : Diel's-Alder reaction is favoured by the presence of electron withdrawing groups in the dieophile and by electron releasing groups in the diene or presence of electron donating groups in the dienophile and electron withdrawing groups in the diene, without reducing the yield of the adduct.

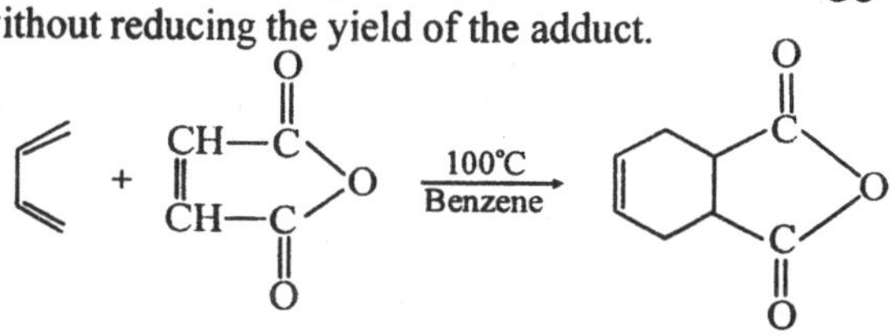

Adduct 100% yield

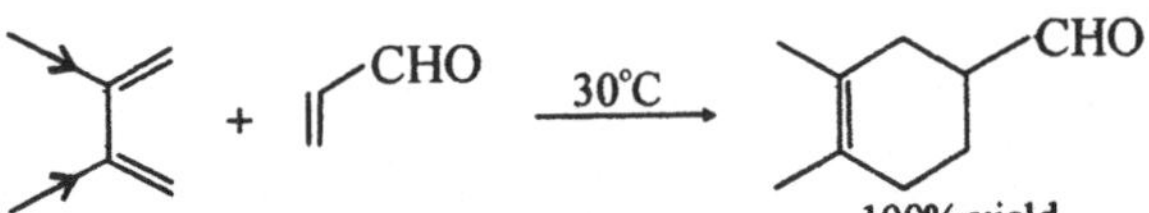

100% yield

Dienes with electron-withdrawing group have been found to react readily with dienophile containing electron-releasing groups High temperature and high pressure also enhance the rate of Diel's-Alder reaction. Some Diel's-Alder reactions take place at ambident temperature in the presence of Lewis acid catalyst.

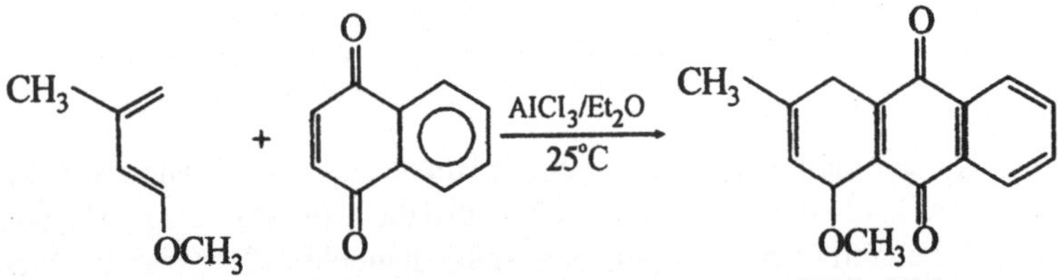

80% yield

STEREOCHEMISTRY OF THE DIEL'S'-ALDER REACTION

(*i*) This reaction is highly stereospecific. The reaction is syn addition and configuration of the dienophile is retained in the product.

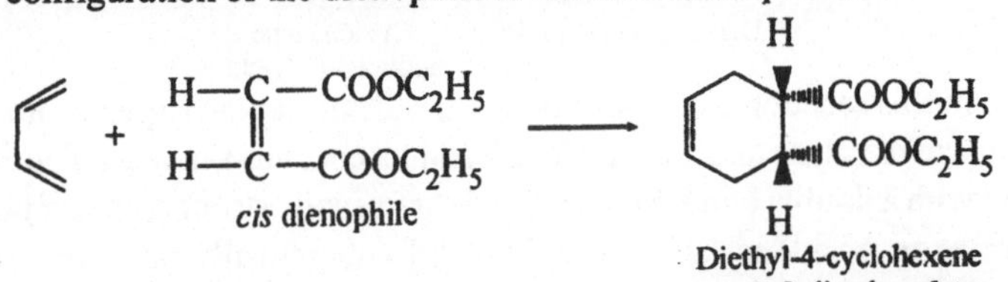

cis dienophile

Diethyl-4-cyclohexene *cis*-1. 2-dicarboxylate (Adduct has *cis* geometry)

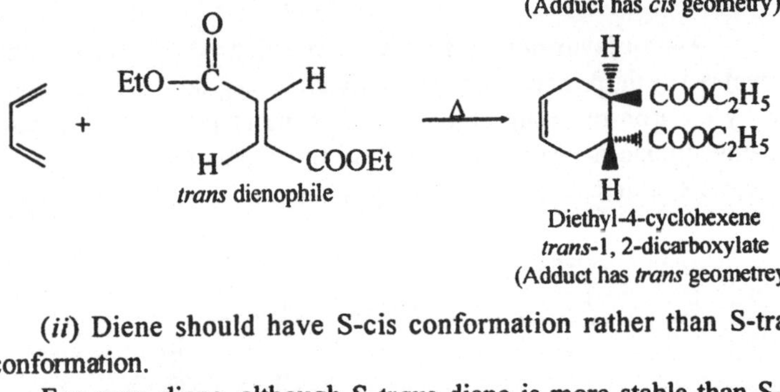

trans dienophile

Diethyl-4-cyclohexene *trans*-1, 2-dicarboxylate (Adduct has *trans* geometrey)

(*ii*) Diene should have S-cis conformation rather than S-trans conformation.

For open diene, although S-*trans* diene is more stable than S-*cis* diene by = 3 Kcal/mole, but there is equilibrium between S-*cis* and S-*trans*.

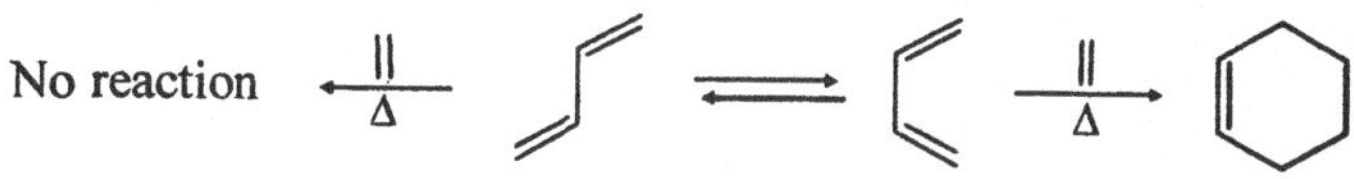

In open chain diene, the whole S-*trans* converts into S-*cis* due to re-establishment of equilibrium. Eventually, all the diene can be converted into adduct even though at any moment there is very little of the active S-*cis* form present.

When there is no possibility of any *cis* form, there can be no Diel's-Alder reaction. In cyclic system, S-*trans* does not convert into S-*cis* and hence S-*trans* does not give Diel's-Alder addition.

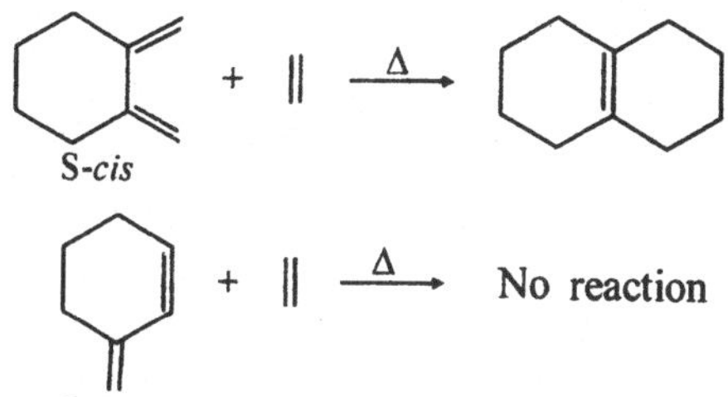

The Diel's-Alder reaction will produce a bicyclic system if diene is already cyclic.

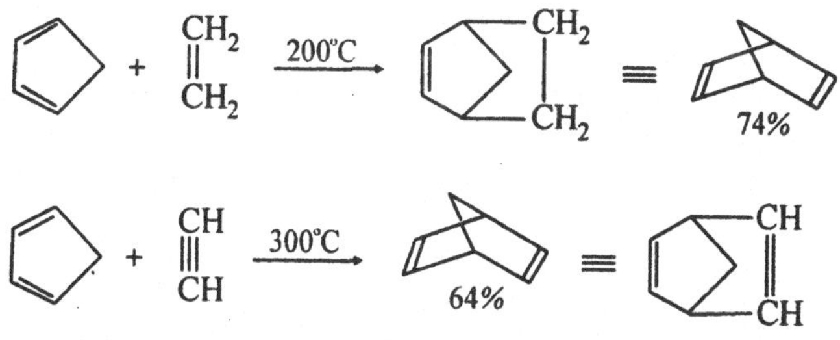

In such cases when a substituted alkene reacts with a cyclic diene, there are two possible stereoisomeric products.

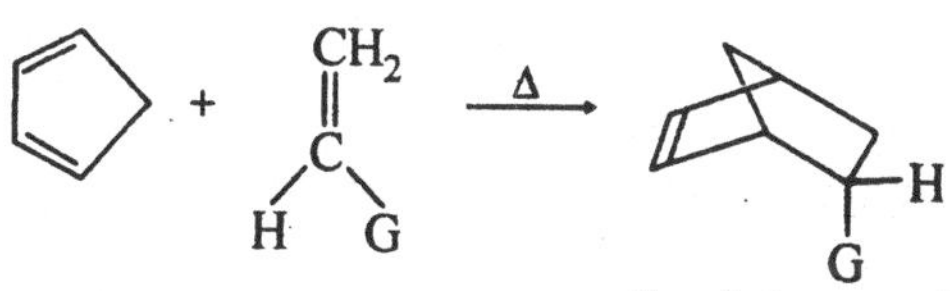

G pointing towards the double bond called endo product

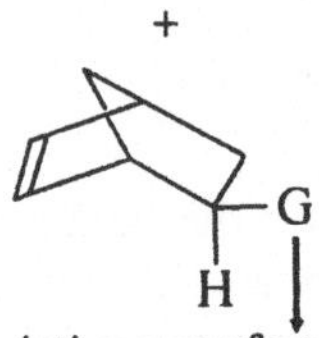

G pointing away from the double bond, called exo product

The exo isomers are usually more stable, but the endo compounds are kinetically favoured in this reaction.

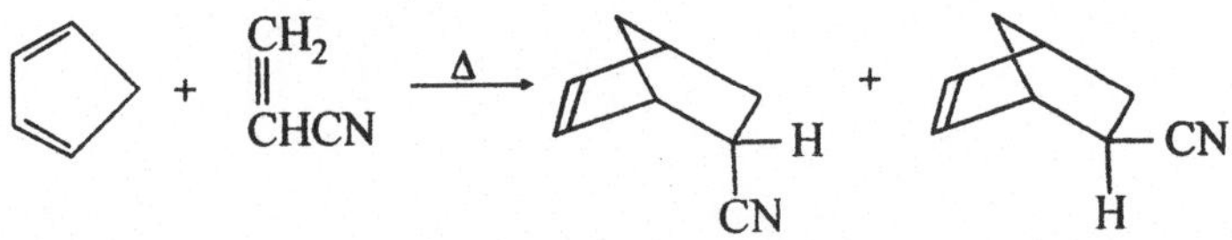

60% Endo 40% exo

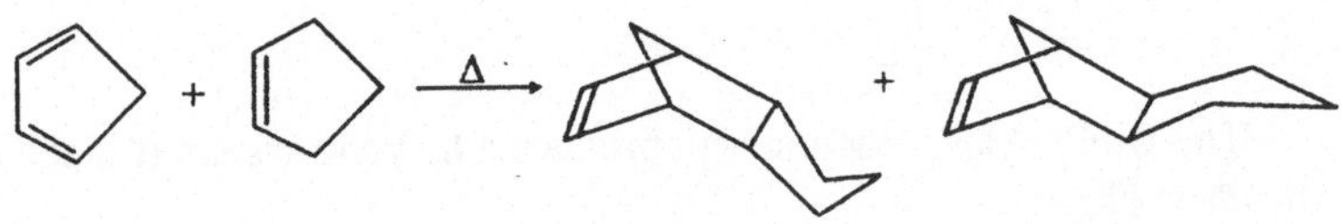

Endo product 98% Exo product 2%

In case of heterocyclic compound, furan behaves as diene because it is the least aromatic among pyrrole, thiophene and benzene. These three compounds do not lose their aromaticity but furan can.

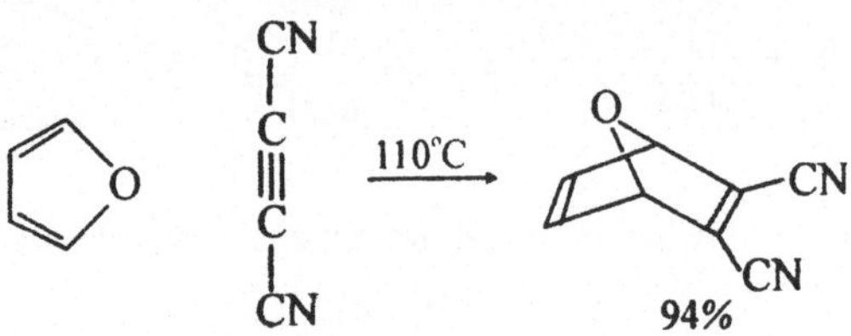

APPLICATIONS

Various applications of Diels-Alder reactions are as follows :

(*a*) *Detection of the conjugated double bonds* : We know that the maleic anhydride (dienophile) adds on to the conjugated double bonds forming a crystalline adduct. Thus, it is possible to use this as a positive test for the detection of conjugated system in the usaturated compounds. In actual practice, Diels-Alder reaction is used for detecting the presence of conjugated system of double bonds in various natural product, *e.g.*, b-carotene, vitamin A, abietic acid zinziberene, ergosterol, etc.

(*b*) *Determination of the structures of substituted homoannular cyclohexadienes* : A carbon-carbon triple bond may also particiate in the Diels-Alder reaction.

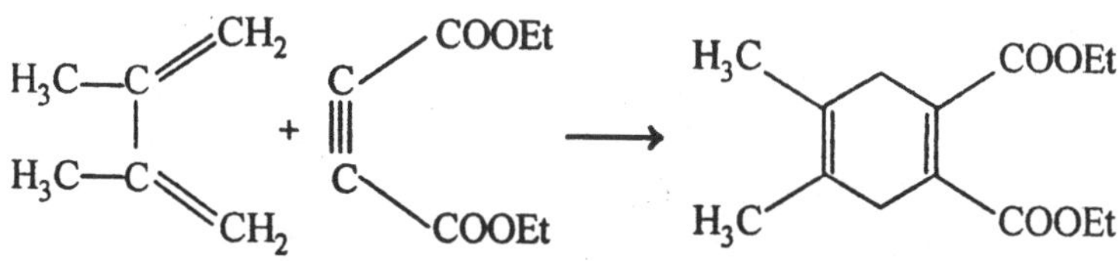

This reaction is used in determining the structures of substituted homoannular cyclohexadienes. For instance, it is possible to determine the structures of the following isomers methylcyclohexadiens.

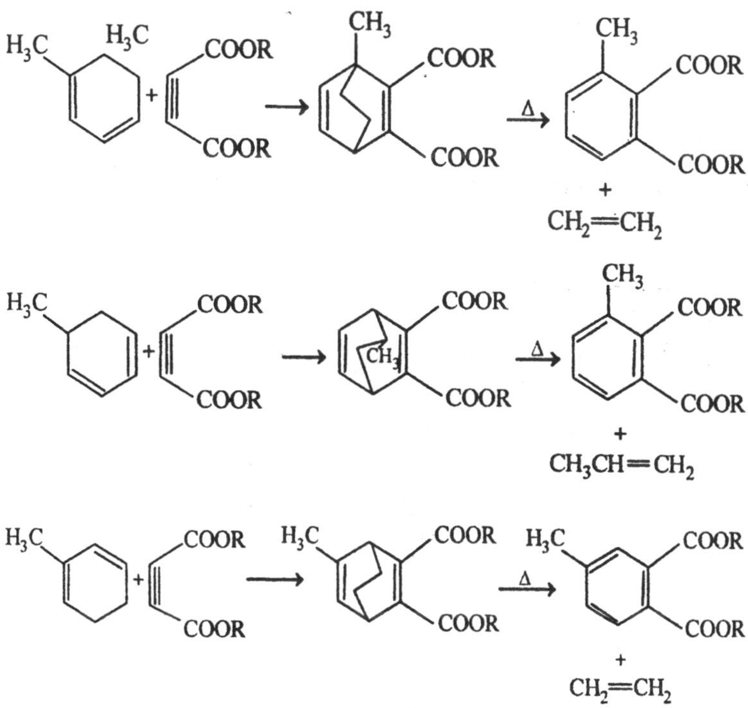

If we could be able to identify the phthalic esters through hydrolysis and pyrolysis to the corresponding phthalic anhydride, then we could determine the structures of the homoannular cyclohexadienes very conveniently.

(*c*) *Separation of anthracene and phenanthrene* : As anthracene but not phenanthrene is able to form and adduct with maleic anhydride, Diels-Alder reaction is used for separating anthracene from phenanthrene.

$$\text{Anth. + Phen.} \xrightarrow[\text{anthydride}]{\text{Maleic}} \text{Adduct of anth.} \xrightarrow[\text{vaccum}]{\text{Heat in}} \text{Anthracene}$$

(*d*) *Determination of configuration* : During Diels-Alder reaction, the configuration of the dienophile gets retained in the cyclic product. Also, it is easier to ascertain the configuration of the cyclic adduct. Thus, Diels-Alder reaction has been used for determining to configuration of various dienophiles, *e.g.*, maleic and fumaric acids; *cis*-and *trans*-cinnamic acids.

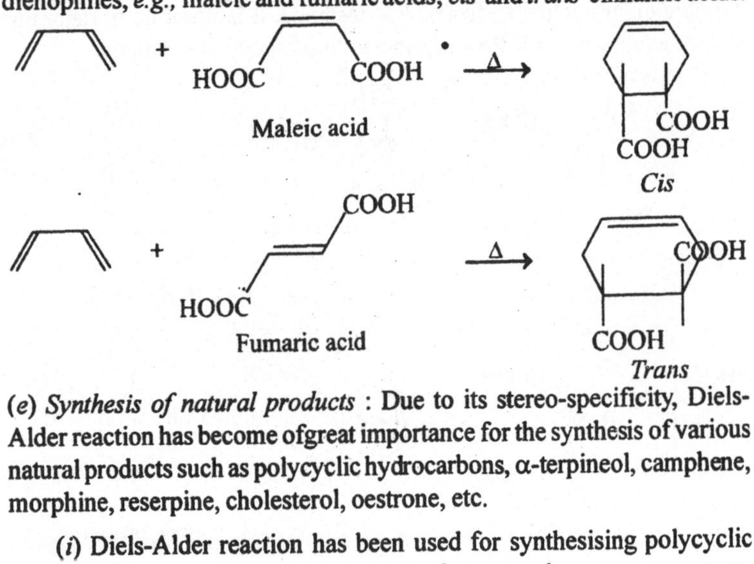

Maleic acid — *Cis*

Fumaric acid — *Trans*

(*e*) *Synthesis of natural products* : Due to its stereo-specificity, Diels-Alder reaction has become ofgreat importance for the synthesis of various natural products such as polycyclic hydrocarbons, α-terpineol, camphene, morphine, reserpine, cholesterol, oestrone, etc.

(*i*) Diels-Alder reaction has been used for synthesising polycyclic hydrocarbons, *e.g.*, anthracene, pentacene, hexacene, hexacene, coronene, etc.

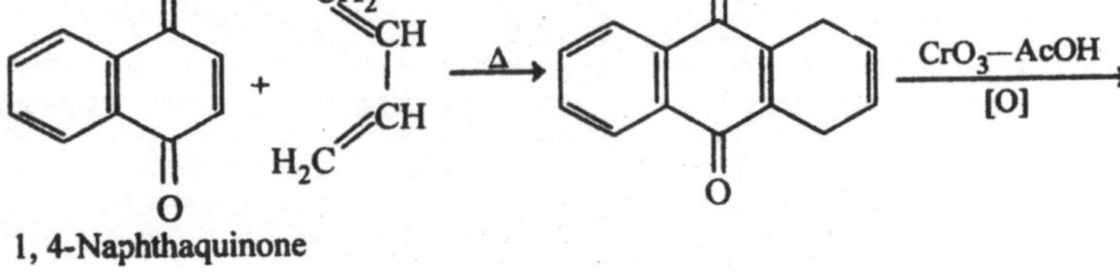

1, 4-Naphthaquinone

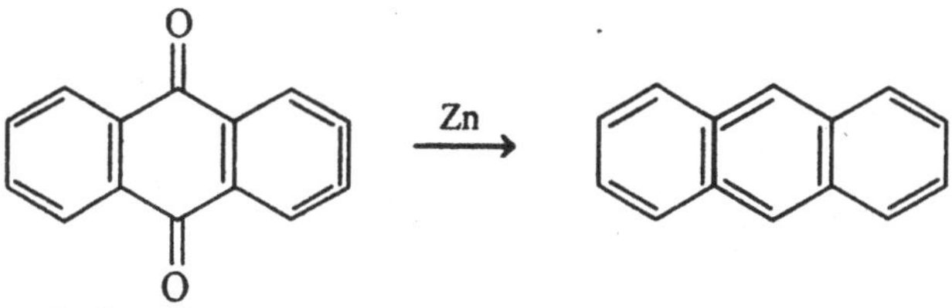

Anthraquinone Anthracene

It is also used for the synthesis of quinones

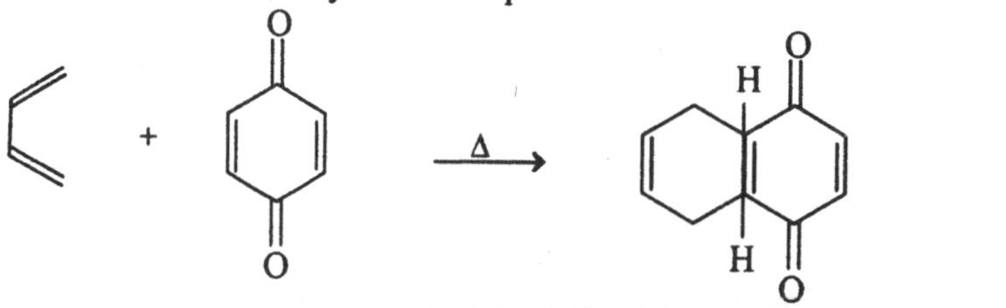

5, 8, 9, 10-Tetrahydro-1, 4-napththaquinone

(*ii*) Diels-Alder reaction has been used for synthesising α-terpineol.

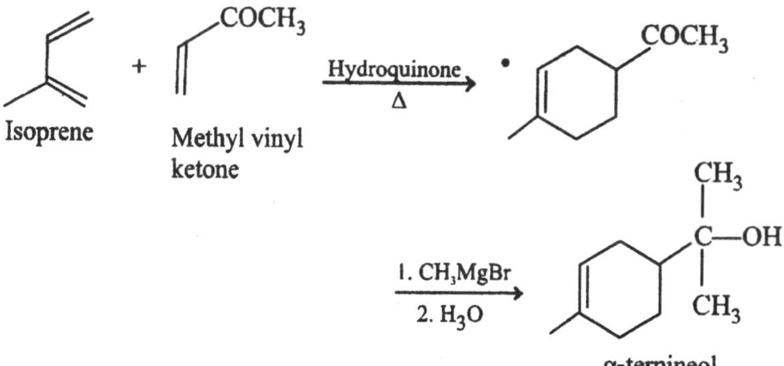

α-terpineol

(*iii*) Diels-Alder reaction has been used for synthesising camphene.

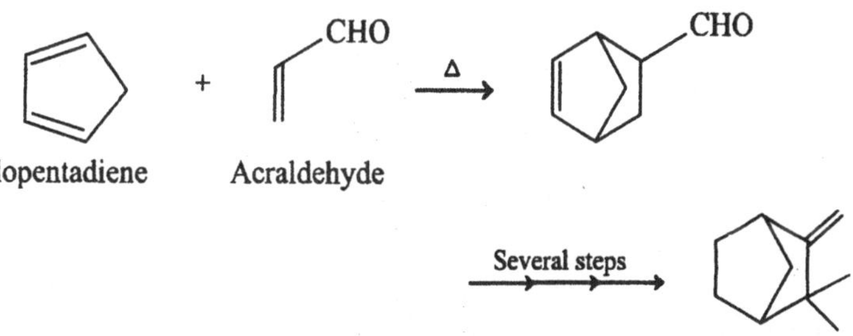

Camphene

(*iv*) *Synthesis of reserpine.* Diels-Alder reaction produces a particular stereochemical arrangement (as given below), which is required for the synthesis of reserpine, by the reeaction between *trans*-vinyl acrylic acid (diene) and *p*-benzoquinone (dienophile).

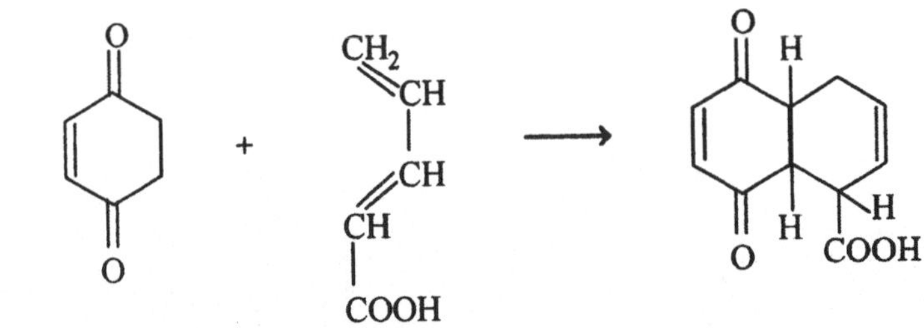

(*f*) *Woodward synthesis of cholesterol* : Diels Alder reaction brings about the reaction between the starting materials, *viz.*, 2-methoxy-5-methyl-benzoquinone and butadiene, for the cholesterol. The two reactants form a product with a *cis*-ring junction which gets converted into more stable *trans*-compound readily.

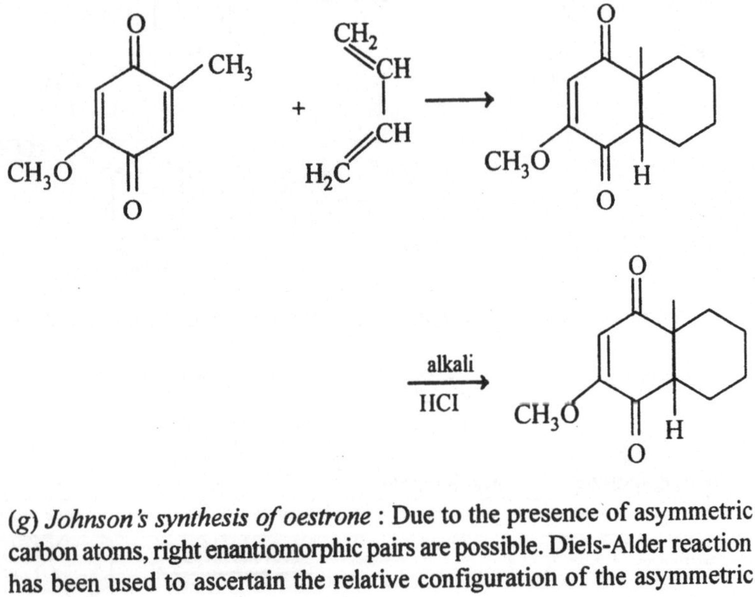

(*g*) *Johnson's synthesis of oestrone* : Due to the presence of asymmetric carbon atoms, right enantiomorphic pairs are possible. Diels-Alder reaction has been used to ascertain the relative configuration of the asymmetric carbon atoms present in oestrone.

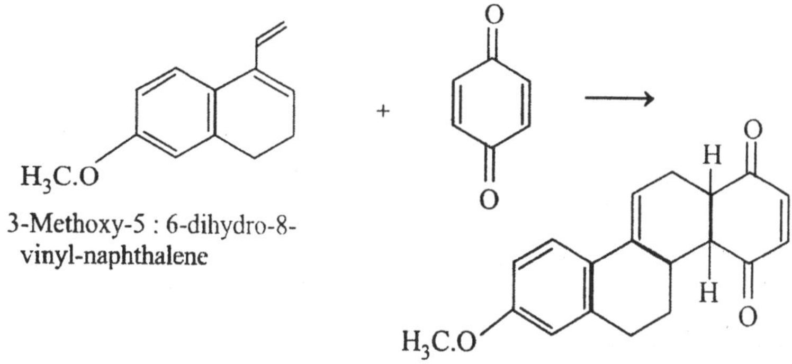

3-Methoxy-5 : 6-dihydro-8-vinyl-naphthalene

(*h*) *In the synthesis of cantharidine* : The Diels-Alder adduct from furan and acetylene dicarboxylate has been the starting products for a complicatied synthesis of natural cantharidine

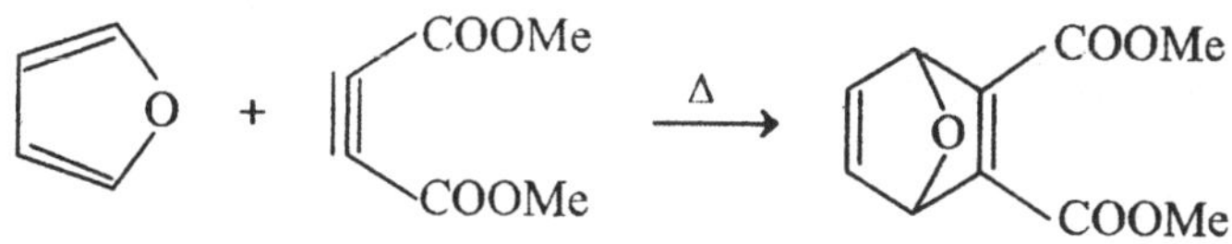

Furan Acetylene dicarboxylate

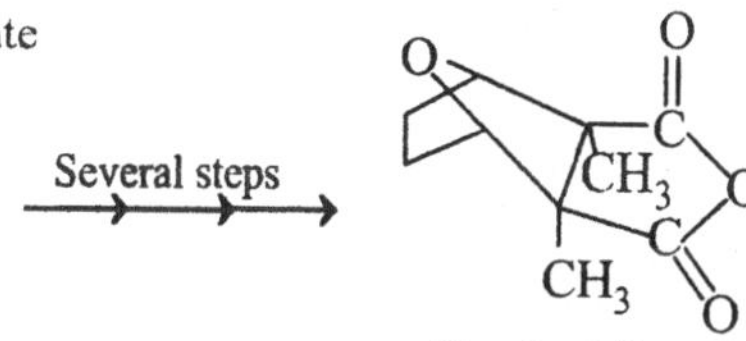

Cantharidine

The Diels-Alder adduct obtained from furan and dimethylmaleic anhydride on hydrogenation yields cantharidine which is not same as the natural product.

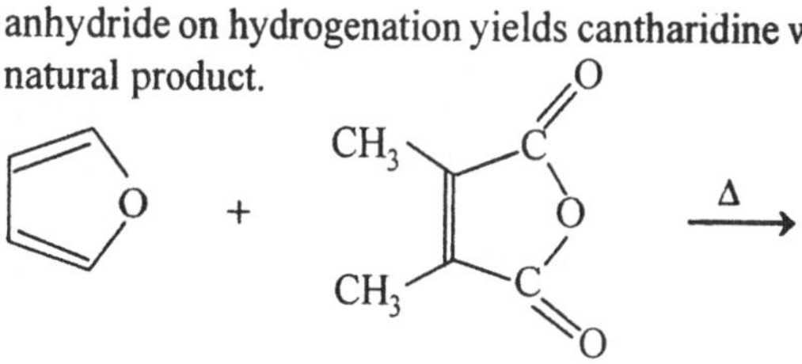

Dimethyl maleic anhydride

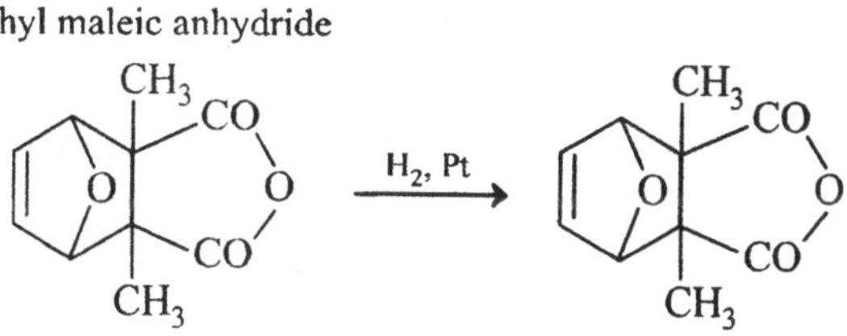

Cantharidine

36

ELBS PERSULPHATE OXIDATION

Elbs observed that when a phenol is oxidised by means of potassium persulphate a new hydroxyl group is introduced in the *para*-position to the hydroxyl group already present. Thus phenol on oxidation with potassium persulphate gives hydroquinone.

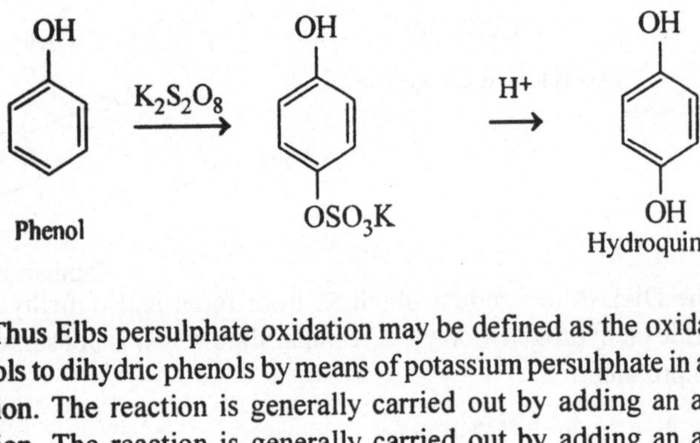

Phenol

Hydroquinone

Thus Elbs persulphate oxidation may be defined as the oxidation of phenols to dihydric phenols by means of potassium persulphate in alkaline solution. The reaction is generally carried out by adding an aqueous solution. The reaction is generally carried out by adding an aqueous solution of potassium persulphate to a cold alkaline solution of the phenol followed by acidification. Oxidation always takes place in the *para*-position provided it is free, in case it is bloked some oxidation occurs in the *ortho*-position.

MECHANISM

The mechanism postulated by Berman *et al.* (1963) is as follows :

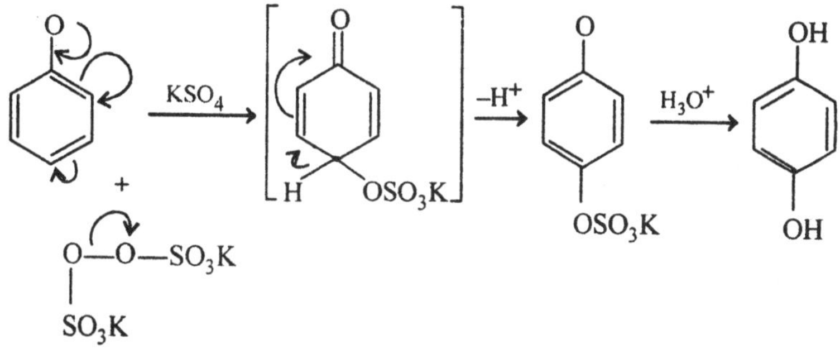

APPLICATION

Elbs persulphate oxidation has been used for preparing substituted phenols, *e.g.*,naphthols, coummarins, flavones, and *N*-heterocyclic compounds. Yields are not encouraging but the pure products are obtained which can be readily isolated.

(*i*) *Preparation of gentisic acid*

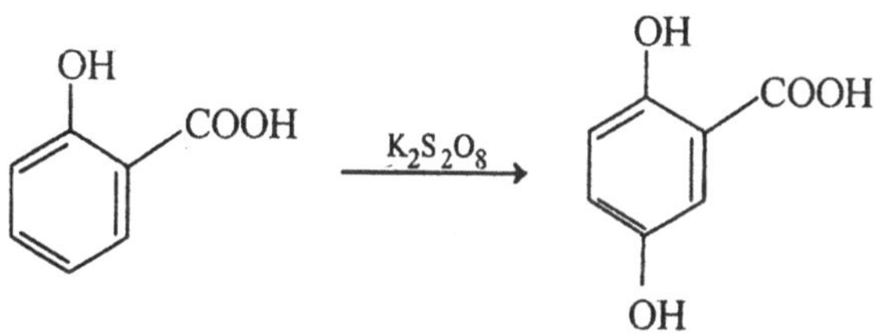

Salicylic acid Gentistic acid

(*ii*) *Preparation of vinylhyaroquinone*

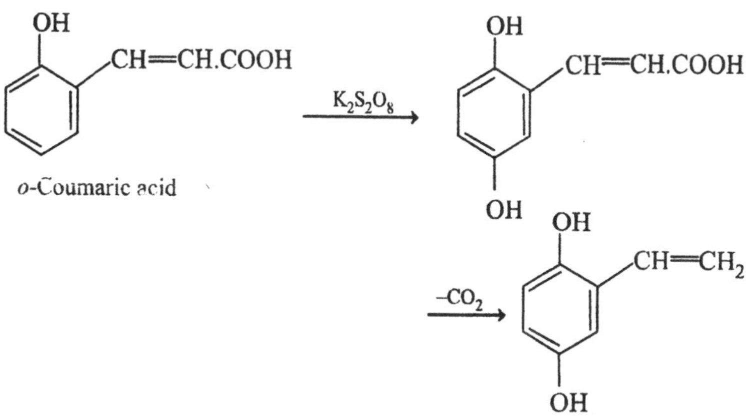

o-Coumaric acid

Vinylhydroquinone

(*iii*) *Preparation of coumarin derivatives*

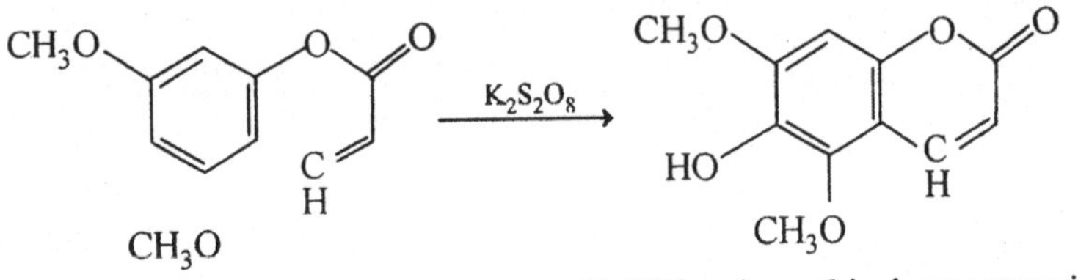

5, 7-Dimethoxycoumarin 5, 7-Dimethoxy-6-hydroxycoumarin

(*iv*) *Preparation of 2, 5-dihydroxypyridine*

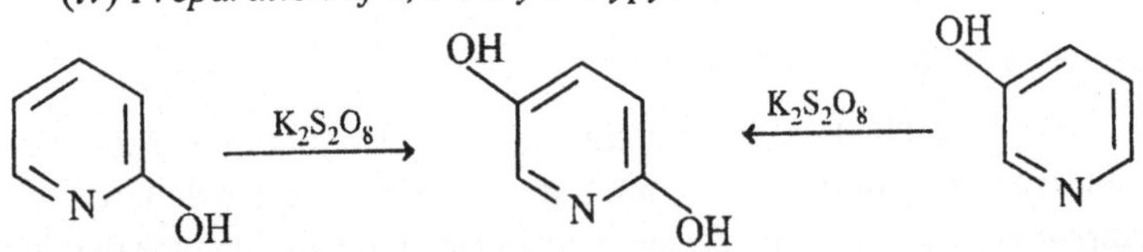

2-Hydroxy pyridine 2, 5-Dihydroxy-pyridine 3-Hydroxy pyridine

37

ELBS REACTION

The conversion of a diaryl ketone having a methyl or methylene group ortho to the carbonyl function is known as the Elbs reaction. For instance, when *o*-methyl benzophenone is heated at 450°C, anthracene is formed

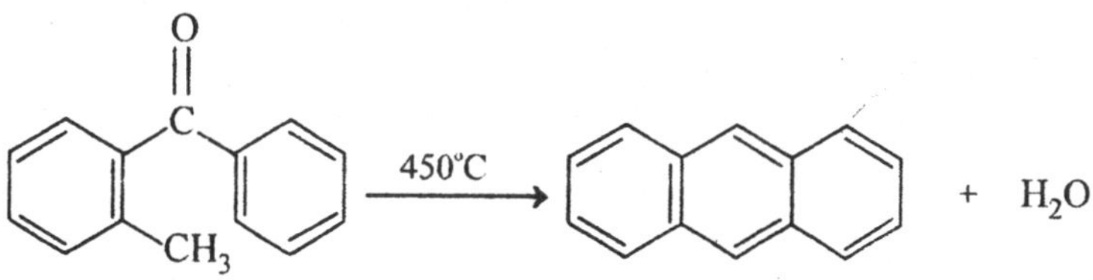

o-Methylbenzophenone Anthracene

38

ENE REACTION

Ene reaction involves the thermal reaction of alkene having an allylic hydrogen (known as ene) with a compound having multiple bond (X=Y or X≡Y, called enophile).

During the reaction transfer of allylic hydrogen (1, 5 migration of hydrogen), shift of allylic double bond and bonding between two unsaturated termini (X and Y) take place to form 1 : 1 adduct.

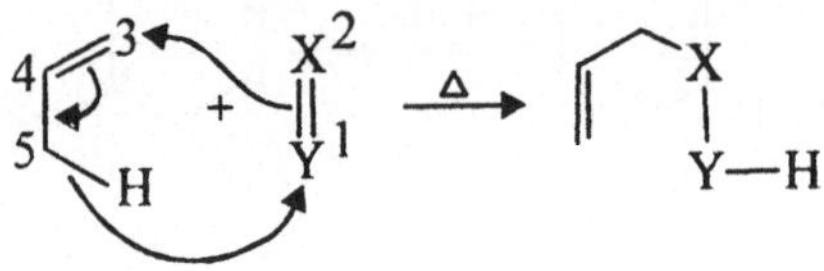

X=Y	X≡Y
C=C	C≡C
C=O	C≡N
C=S	
C=N	
N=O	
N=N	
O=O	

39

ESCHWEILER-CLARK REACTION

The conversion of primary or secondary amines to ter amines by reductive amination with a mixture of HCHO and conc. HCOOH is known as Eschweiler-Clark reaction.

i)

$$R_2NH \xrightarrow[\text{(ii) HCOOH}]{\text{(i) } CH_2O} R_2N{-}CH_3 \quad \text{Ter amine}$$

ii)

$$R{-}NH_2 \xrightarrow[\text{(ii) HCOOH}]{\text{(i) } CH_2O} R{-}N(CH_3)_2$$

iii)

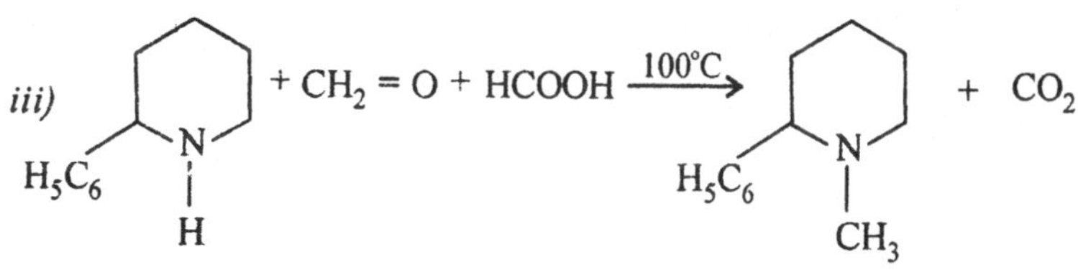

(94%)

iv)

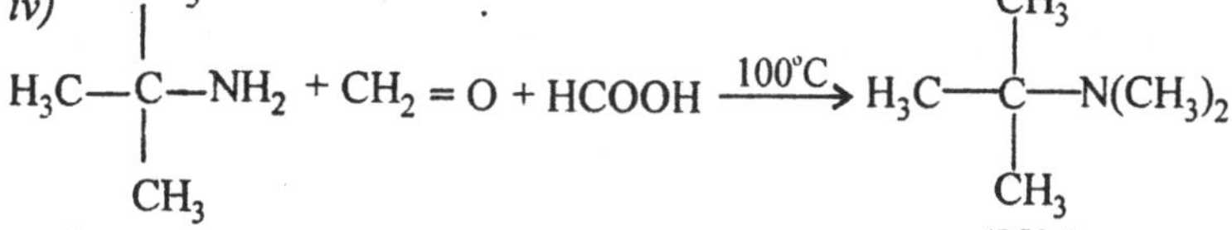

(95%)

t-Butyldimethylamine.

The reaction proceeds in excellent yield

MECHANISM:

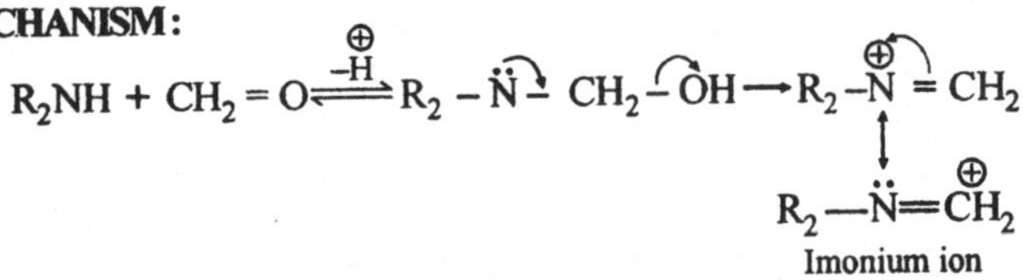

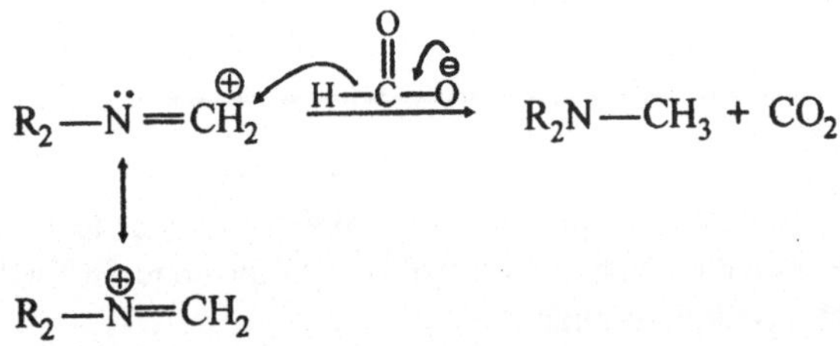

Formic acid behaves as a reducing agent in this reaction by transfering a hydride ion to the electron dificient C and is itself oxidised to CO_2.

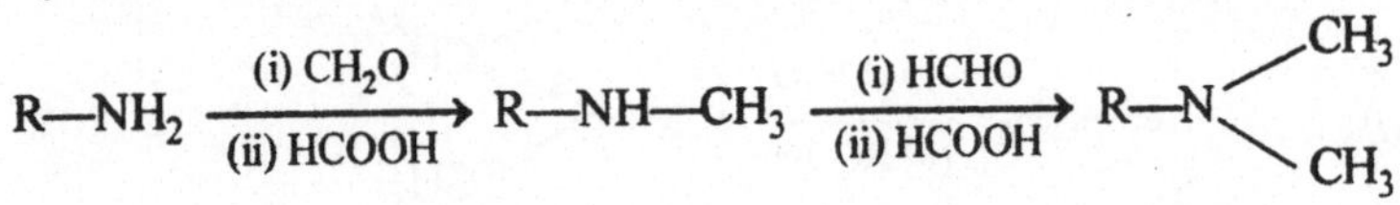

An earlier version of this reaction, called the Leukart reaction, gives lower yields, but is more general. It can be used to prepare primary, secondary or tertiary amines.

40

ETARD OXIDATION

Etard Oxidation

In this reaction, the alkyl side chain attached to the aromatic ring is specifically oxidised into aldehyde or ketone by chromyl chloride.

This reaction is carried out by treating the alkyl benzene with chromyl chloride (CrO_2Cl_2) in carbon tetrachloride solution when the complex is precipitated which is decomposed with water to get an aldehyde or ketone. This reaction is exemplified by toulene which is oxidised to benzaldehyde with chromyl chloride.

$$C_6H_5CH_3 + 2\,CrO_2Cl_2 \xrightarrow{CCl_4} \underset{\text{Complex}}{C_6H_5\text{—}CH(OCrCl_2OH)_2}$$

$$\xrightarrow{H_2O} C_6H_5CHO + H_2CrO_4 + CrCl_2 + 2HCl$$

However, stronger oxidising agents such as $KMnO_4$, $K_2Cr_2O_7$, etc., oxidise the side chain into carboxylic acid.

Side-Chains bigger than CH_3 are oxidised by CrO_2Cl_2 at the end carbon atom of the side chain.

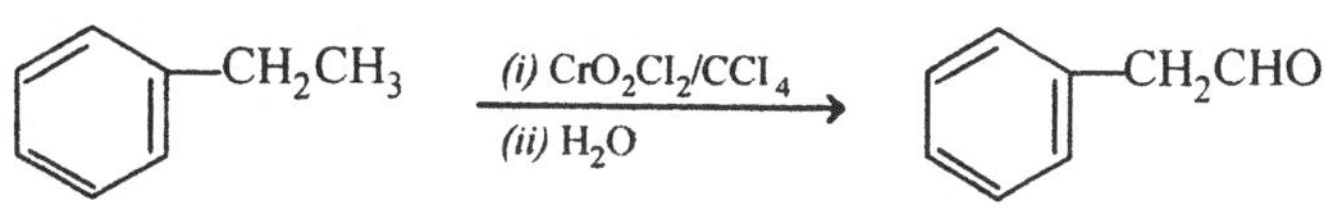

Ethylbenzene Phenylacetaldehyde

The oxidation of *n*-propyl benzene yields a mixture of propiophenone and benzyl methyl ketone. The latter compound is obtained through the involvement of a rearrangement.

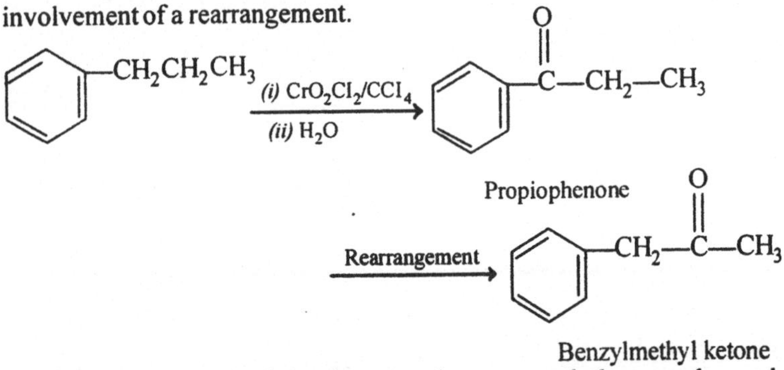

If benzene nucleus is having more than one methyl group, then only one is oxidised, *e.g.*, *m*-xylene yields *m*-tolualdehyde (80%)

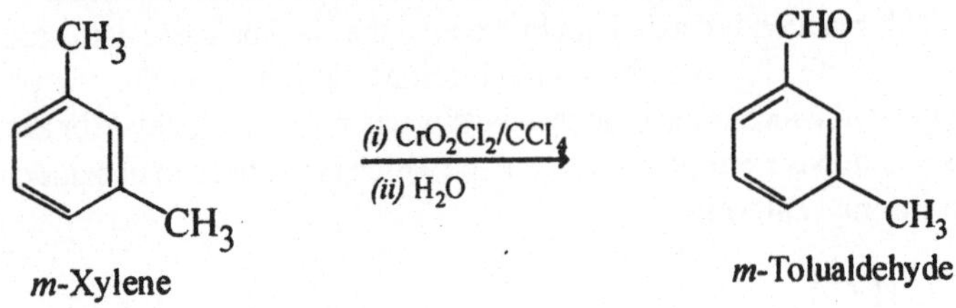

MECHANISM:

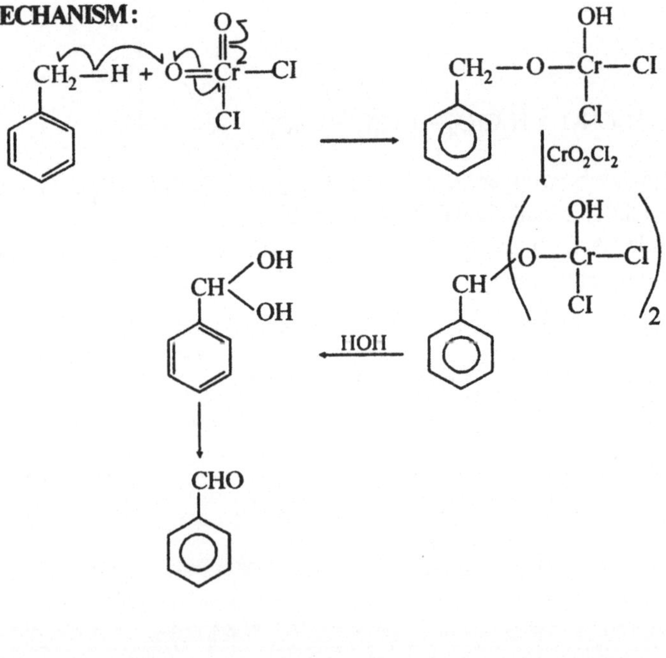

41

FAVORSKII REARRANGEMENT

The conversion of α-haloketones to esters with rearranged carbon skeleton by treating with alkoxide ions is known as Favorskii rearrangement. Alkali hydroxides or amines in place of metal alkoxides yield acids or amides respectively. Cyclic α-halo ketone yield esters with ring contraction.

$$C_6H_5CH_2COCH_2Cl \xrightarrow{R\overset{\ominus}{O}} RO{-}\overset{\overset{\displaystyle O}{\|}}{C}{-}CH_2CH_2C_6H_5;$$

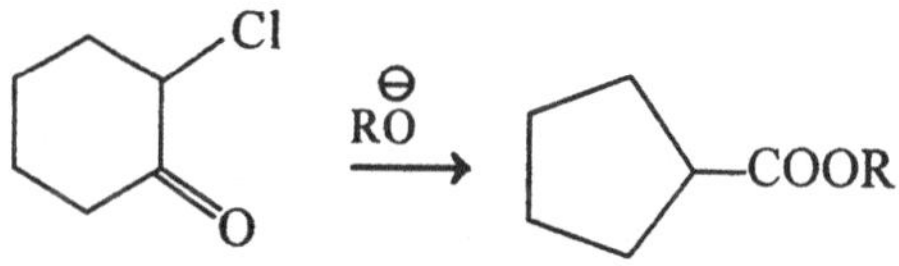

MECHANISM

The mechanism of this rearrangement is not understood and still has been the subject of much investigation. Both the isomeric ketones, (I) and (II) yield β-phenylpropionic acid on treatment with hydroxide ions.

$$\underset{\text{(I)}}{C_6H_5{-}CH_2{-}\overset{\overset{\displaystyle O}{\|}}{C}{-}CH_2Cl} \xrightarrow{\overset{\ominus}{O}H} C_6H_5CH_2CH_2COOH \xleftarrow{\overset{\ominus}{O}H} \underset{\text{(II)}}{C_6H_5CH(Cl){-}\overset{\overset{\displaystyle O}{\|}}{C}{-}CH_3}$$

From this observation it follows that the chlorine is not being directly replaced by the incoming group from the other side of the carbonyl group, as otherwise (II) would yield $C_6H_5CH(CH_3)COOH$ which is not obtained. Further, it was observed that a cyclic ketone (2-chlorocyclohexanone) with labelled carbon bearing the chlorine atom, on treatment with alkoxide ion gave a product in which equal amounts of ^{14}C were present at the α-carbon and the β-carbon. This reveals a symmetrical cyclopropanone intermediate which opens up with equal ease on either side of the carbonyl group.

On the basis of the above observations, the following mechanism has been postulated : The base abstracts an α-hydrogen to form a carbanion. Intramolecular nucleophilic attack on the carbon bearing the chlorine is able to displace the chlorine atom with the formation of a transient symmetirical cyclopropanone ring. Subsequent attack of the alkoxide ion on the carbonyl carbon opens the ring with equal ease on either side of the carbonyl carbon so that the product has 50% of ^{14}C at the α-position and 50% at the β-position.

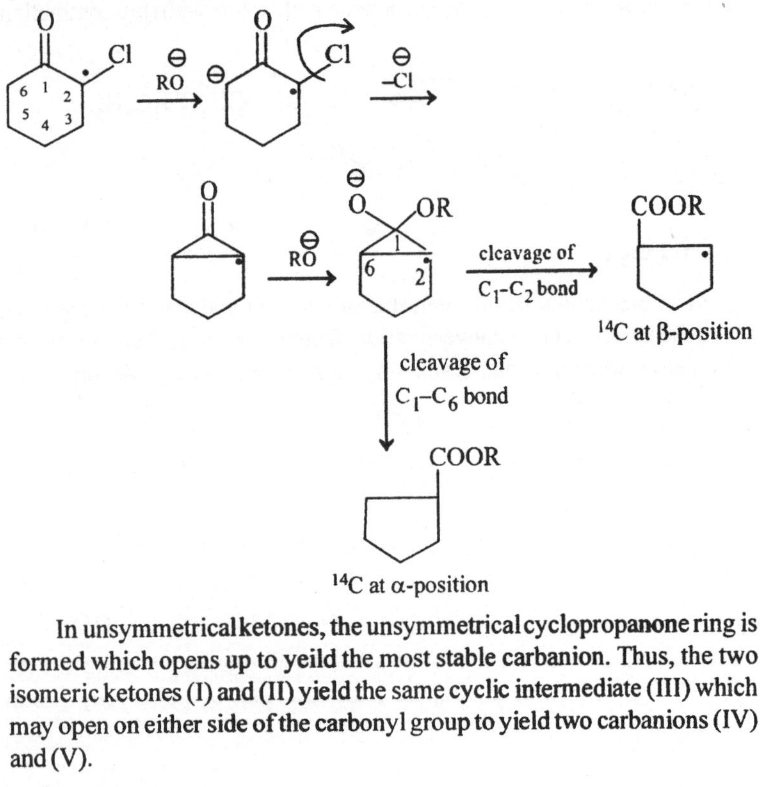

In unsymmetrical ketones, the unsymmetrical cyclopropanone ring is formed which opens up to yeild the most stable carbanion. Thus, the two isomeric ketones (I) and (II) yield the same cyclic intermediate (III) which may open on either side of the carbonyl group to yield two carbanions (IV) and (V).

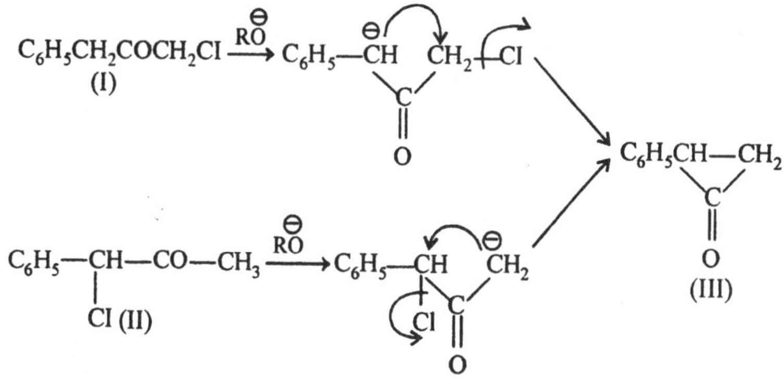

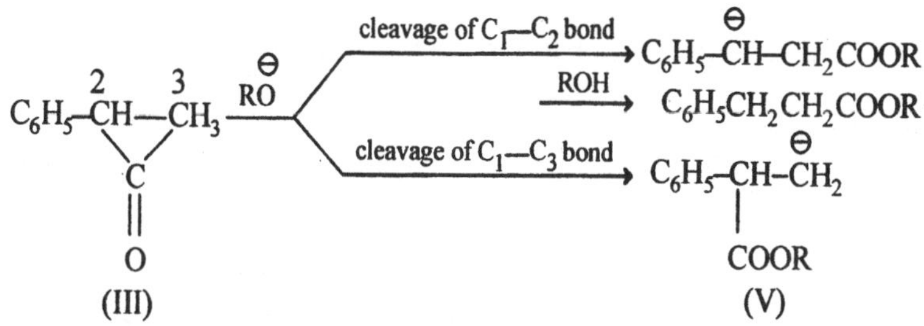

The carbanion (IV) which is reasonance-stabilized is preferentially formed so that the product is $C_6H_5CH_2CH_2COOR$. Hence, both (I) and (II) yield the same product.

Although the cyclopropanone intermediate could not be isolated, it has been trapped to yield a Diels-Alder type of adduct with furan in some cases.

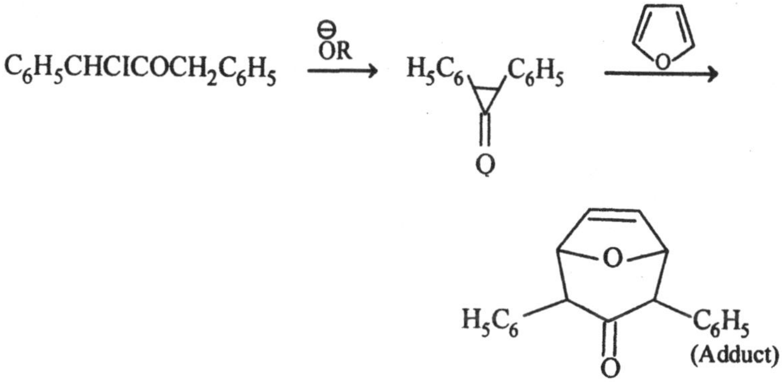

Ketones not possessing α-hydrogen also undergo similar rearrangement in some cases. The mechanism for this is in part, similar to benzilic acid rearrangement and is known as semibenzilic mechanism.

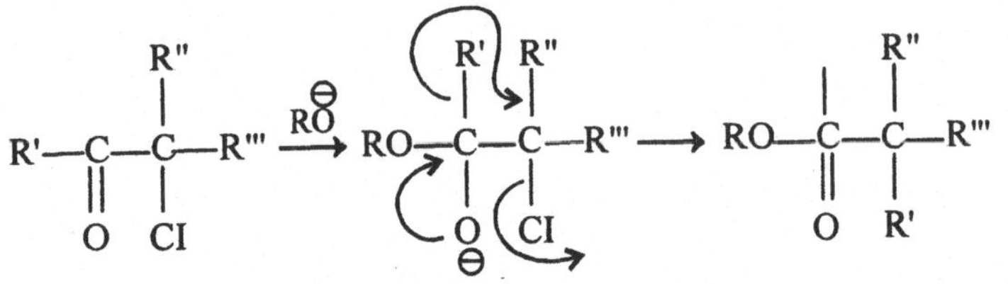

Even if there is a suitably placed α-hydrogen, the rearrangements may follow semibenzilic pattern in some compounds.

Hence, 2-bromocyclobutanone undergoes Favorskii rearrangement if it is treated with water as the base. When treated with D_2O, no deuterium is incorportated in the ring.

Cyclopropanone intermediate mechanism postulates incorporation of deuterium in the ring.

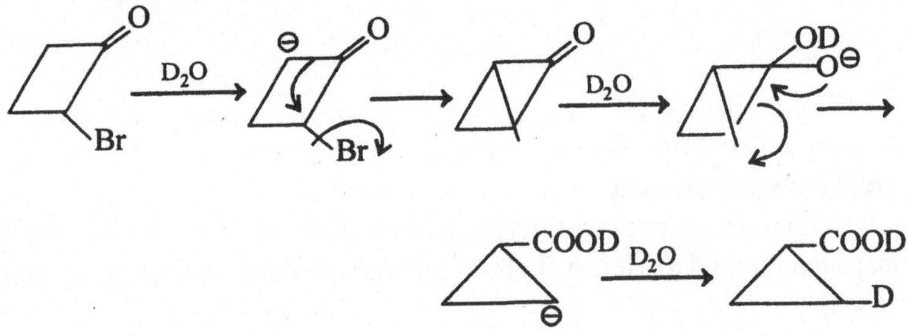

However, semi-benzillic mechanism is able to explain the formaion of the actual product without deuterium in the ring.

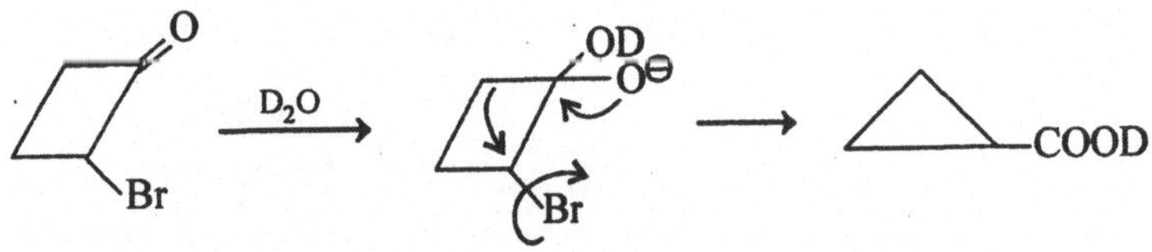

The strain in the bicyclobutanone ring is able to restrict the operation of cyclopropanone mechanism. Favorskii rearrangement of α, α (gem) and αα′ (vic) dihaloketones gives rise to α, β-unsaturated ester.

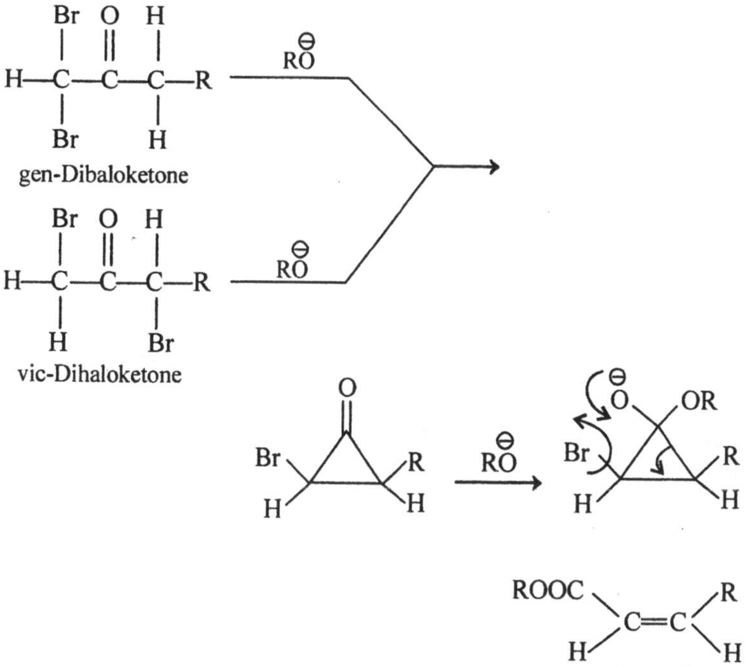
gen-Dibaloketone
vic-Dihaloketone

42

FISCHER-INDOLE SYNTHESIS

This is the most general method for the synthesis of indoles and *consists in heating the phenylhydrazone or substituted phenylhydrazone of an aldehyde, ketone or keto-acid with an acidic catalyst* (*e.g.* H_2SO_4, $ZnCl_2$, polyphosphoric acid or BF_3).

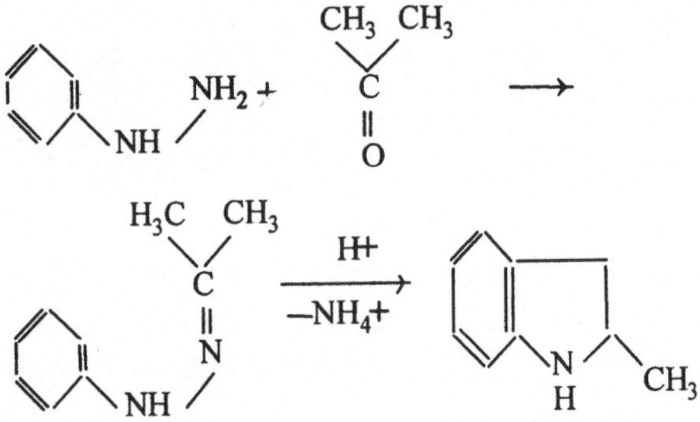

Acetone phenythydrazone 2-Methyl indole

MECHANISM :

Various mechanisms have been put forward to explain this reaction, but the most accepted one is that of Robinson (1918).

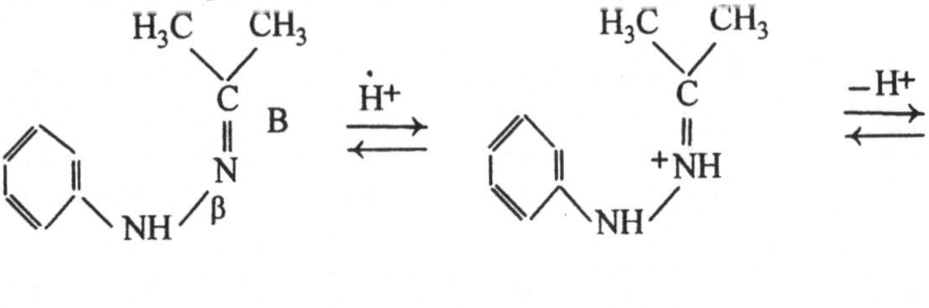

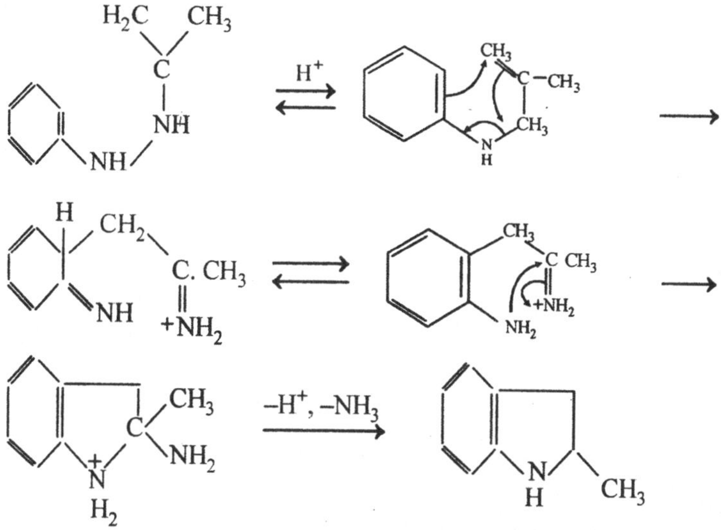

2-Methyl indole

It was shown by the use of isotopic N(N^{15}) that the α-nitrogen atom of the hydrazone is eliminatied as ammonia during the reaction.

It must be noted that the mechanism of the reaction is analogous to the *o*-benzidine rearrangement.

SYNTHETIC USES

Fischer-indole synthesis has been used in the synthesis of many heterocyclic compounds. Some of theses are described as follows :

1. *Synthesis of indole-3-acetic acid* : Indole-3-acetic acid is synthesised from phenylhydrazone of succinaldehydic acid.

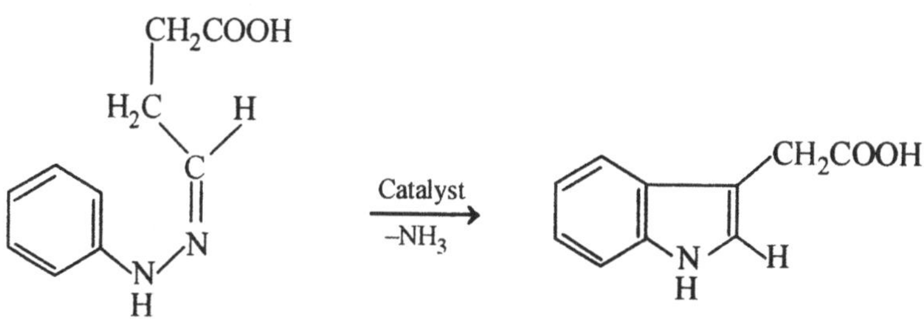

Succinaldehydic acid phenylhdrazone

Indole-3-acetic acid

2. *Synthesis of 3-methyloxindole* : 3-Methyloxidole is synthesised from β-propionylphenylhydrazone.

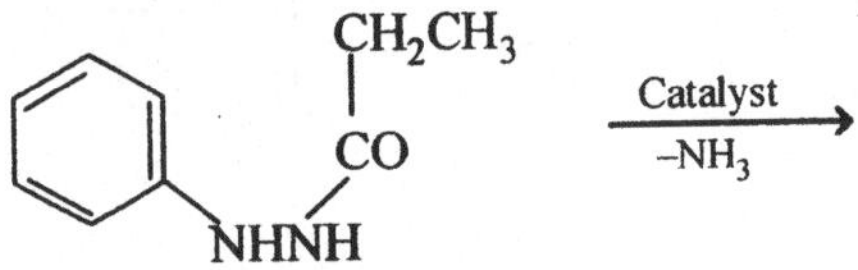

Propionylphenylhydrazone

3. *Synthesis of tryptamine* : In a variation of the reaction, Fischer used phenylhydrazine and the acetal of γ-aminobutyraldehyde for the preparation of tryptamine.

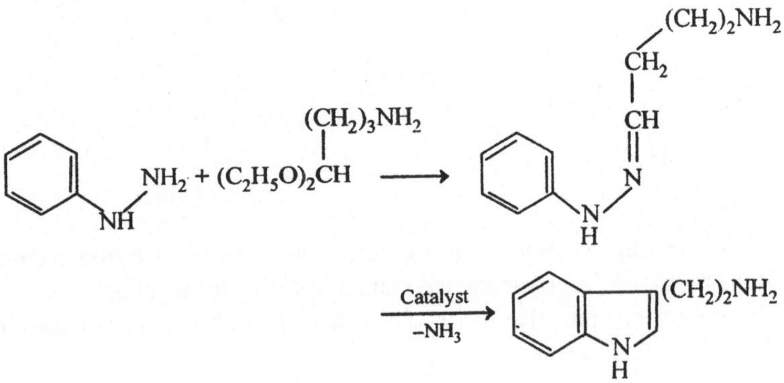

Tryptamine

4. *Synthesis of carbazoles* : Fischer i
ndole reaction is used for synthesising carbazoles (Borsche's' modification) from phenylhydrazine and cyclohexanone.

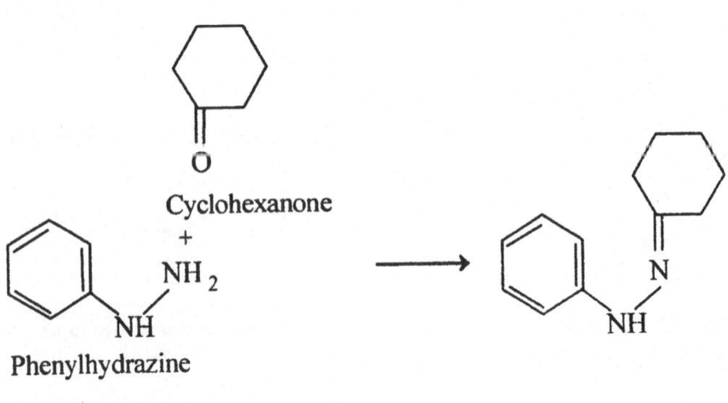

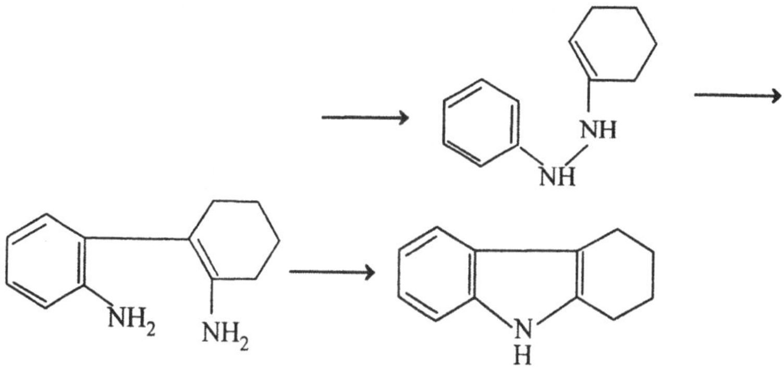

Tetrahydro carbazole

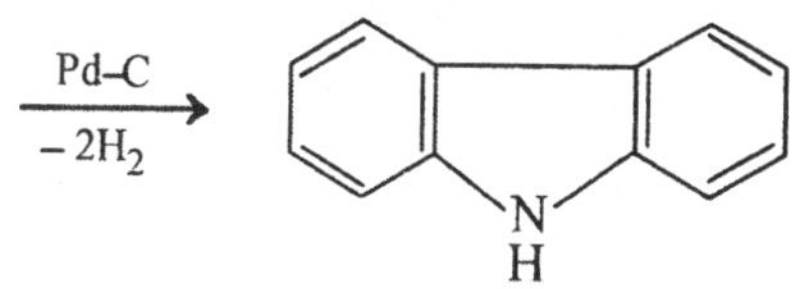

Carbazole

LIMITATIONS

(*i*) Aryhydrazones of acetaldehyde do not undergo this reaction. Thus, it is not possible to prepare indole directly from phenylhydrazone of acetaldehyde. However, it is possible to prepare indole indirectly by this synthesis using pyruvic acid (a keto acid) in place of acetaldehyde and decarboxylating the resulting indole-2-carboxylic acid thermally.

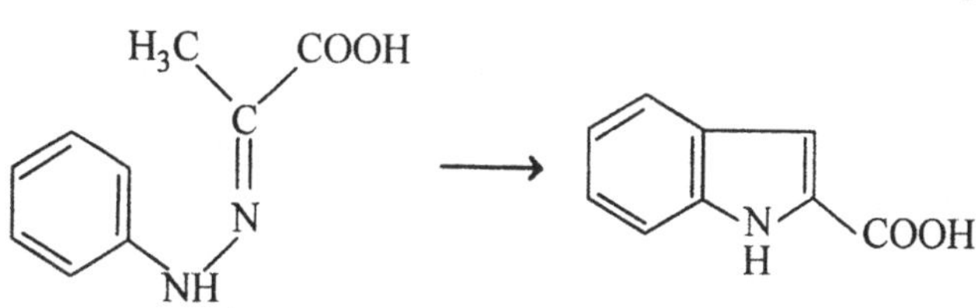

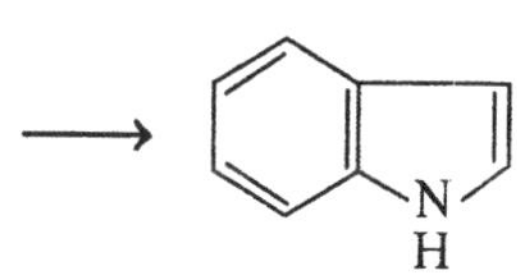

Indole

(*ii*) Arylhydrazones of β-*ketonesters* gernerally give pyrazolones and not indoles *e.g.*,

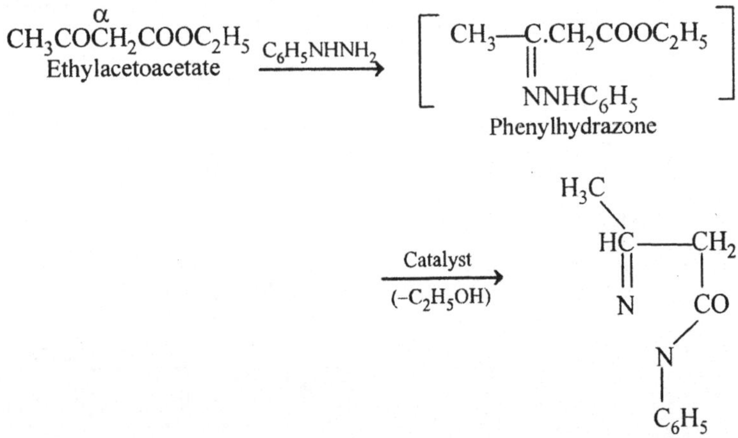

3-Methyl- 1-phenyl pyrazolone

43

FINKELSTEIN REACTION

This reaction is an example of halide exchange reaction.

This reaction is used to prepare iodine from bromide and chloride in the presence of NaI/acetone. This reaction is an example of SN2 reaction; hence it is given by *p*-halides, sec and ter alkyl halide also gives this reaction with excess of NaI/CS_2/$ZnCl_2$.

$$R{-}X \xrightarrow{\text{NaI/Acetone}} R{-}I + NaX \text{ (as ppt.)}$$

$$\begin{array}{c} CH_3 \\ | \\ CH_3{-}C{-}Br \\ | \\ CH_3 \end{array} \xrightarrow{\text{NaI}/CS_2/\text{Anhy. }ZnCl_2} \begin{array}{c} CH_3 \\ | \\ CH_3{-}C{-}I \\ | \\ CH_3 \end{array}$$

Vinyl halide also gives this reaction with KI/$NiBr_2$/Zn.

$$CH_2{=}CH{-}Cl \xrightarrow{KI/NiBr_2/Zn} CH_2{=}CH{-}I$$

When alkyl halide is treated with HF, AgF, KF, HgF_2, HF/SbF_3, then X is replaced by F.

$$C_6H_5{-}CH_2{-}X \xrightarrow{HF} C_6H_5{-}CH_2{-}F$$

$$CH_2{=}CH{-}CH_2{-}X \xrightarrow{AgF} CH_2{=}CH{-}CH_2{-}F$$

44

FRIEDAL -CRAFT ALKYLATION

Alkylation of aromatic compounds with aliphatic compounds like halides, alcohols, esters, ether, alkenes, aldehydes and ketones in presence of Lewis acid or proton acid such as $FeCl_3SnCl_4$, BF_3, $ZnCl_2$, $SbCl_5$, HF, H_2SO_4, H_3PO_4 etc. as catalyst is known as Friedal-Craft alkylation. This reaction is an example of aromatic electrophilic substitution reaction. If alkylating agent is halide, alcohol, ester or ether then catalyst is Lewis acid, but when alkylating agent is alkene, aldehyde and ketone then catalyst is proton acid.

Halides are the reagents of most frequent choice in alkylation and aluminium halides is usually employed as the catalyst.

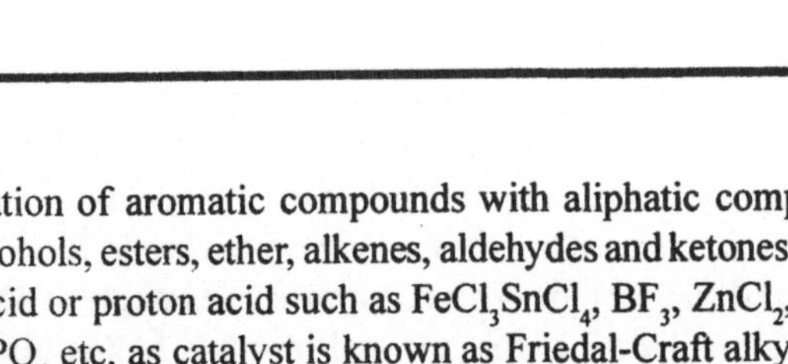

When R—X is primary halide then this forms ion-pair with Lewis acid and the carbocation part of the ion pair gives ArSE reaction with arenes.

MECHANISM:

$$R{-}CH_2{-}X + AlX_3 \longrightarrow \underset{\text{Ion pair}}{R{-}\overset{+\delta}{C}H_2{-}{-}{-}X\overset{\delta-}{A}lX_3} \rightleftharpoons [R{-}\overset{\oplus}{C}H_2\overset{\ominus}{A}lX_4]$$

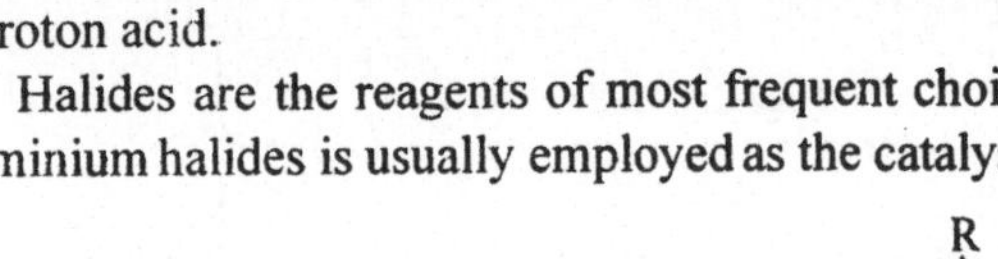

σ complex

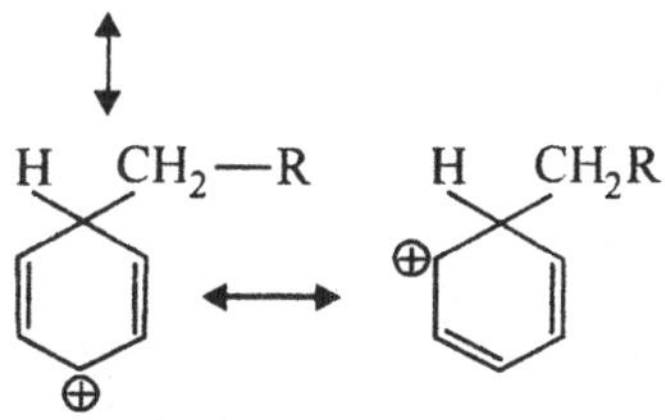

σ complex is delocalised

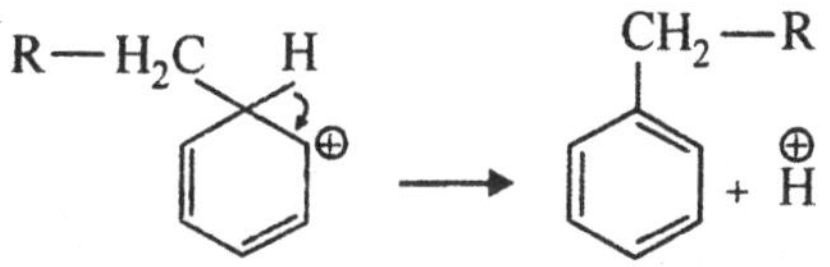

$$\overset{\ominus}{AlX_4} + \overset{\oplus}{H} \longrightarrow HX + AlX_3$$

If alkyl halide is sec and ter, then reaction starts with the formation of carbocation and this carbocation acts as the electrophile.

MECHANISM:

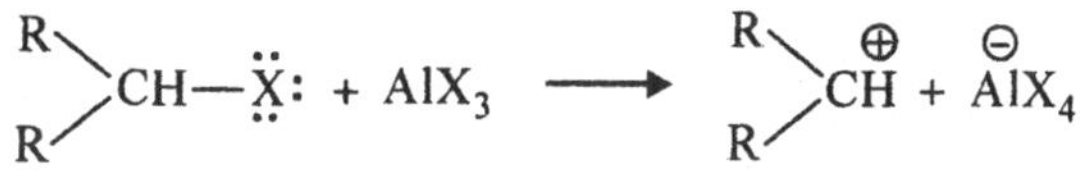

Step II.

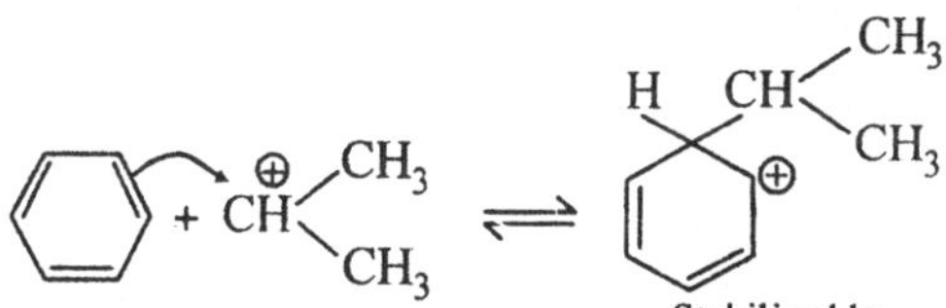

Stabilised by resonance

Step III.

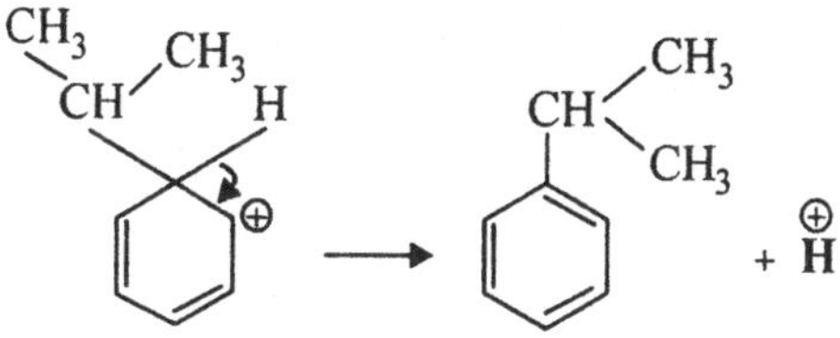

$$\overset{\oplus}{H} + \overset{\ominus}{AlX_4} \longrightarrow HX + AlX_3$$

Some examples :

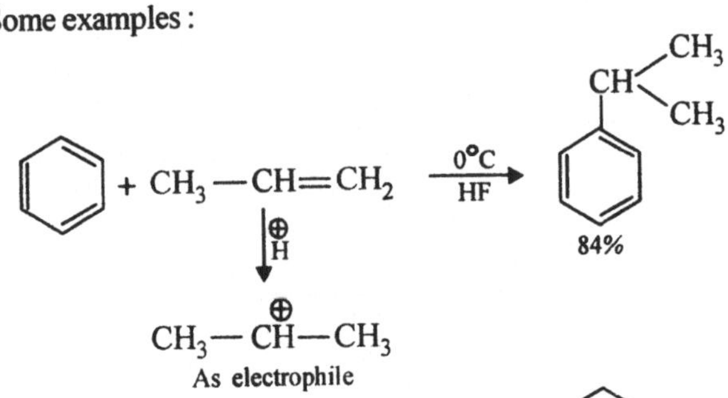

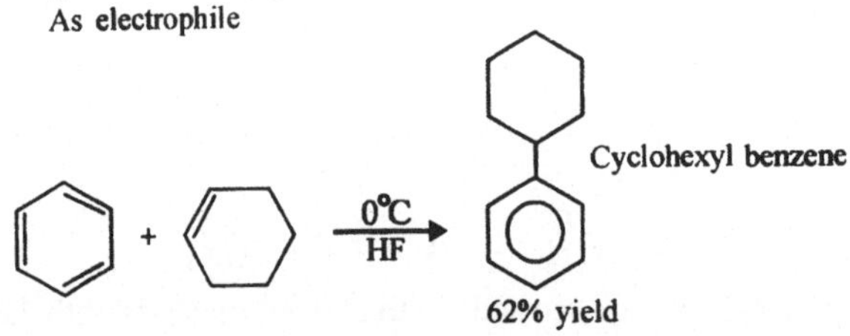

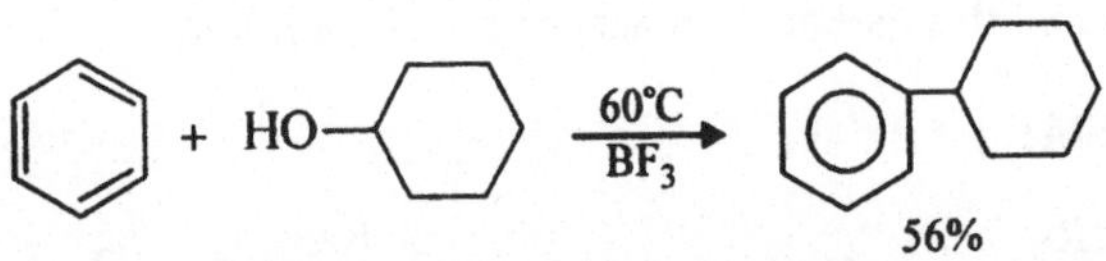

Benzyl halides are very reactive for this reaction but vinyl halides and aryl halides do not give this reaction. The use of dihalides and polyhalides leads to successive alkylations. For example :

(i) $C_6H_5—H + CH_2Cl_2 \xrightarrow{\text{Anhy. AlCl}_3/\Delta} C_6H_5—CH_2—Cl \xrightarrow[\text{Anhy. AlCl}_3]{C_6H_5—H}$

$$C_6H_5—CH_2—C_6H_5$$

(ii) $2C_6H_5—H + Br—CH_2—CH_2—Br \xrightarrow{\text{Anhy. AlBr}_3/\Delta}$

$$C_6H_5—CH_2—CH_2—C_6H_5$$

(iii) $3C_6H_5—H + CCl_4 \xrightarrow{\text{Anhy. AlCl}_3/\Delta} (C_6H_5)_3—C—Cl$

Aldehyde, ketone and ester mainly give intramolecular Friedal-Craft alkylation in the presence of proton acid.

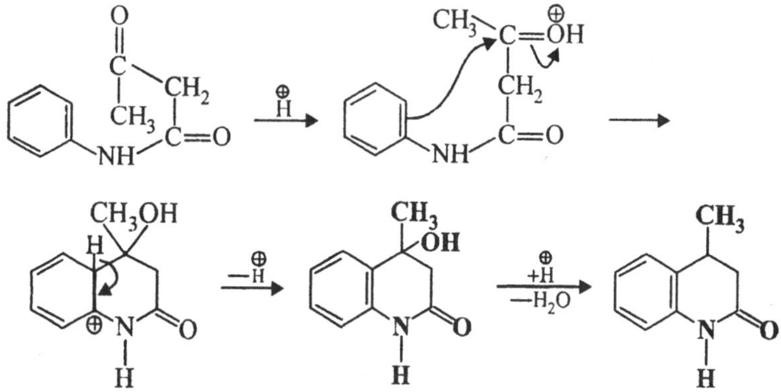

APPLICATIONS

(*a*) **In the synthesis of homologues of benzene**

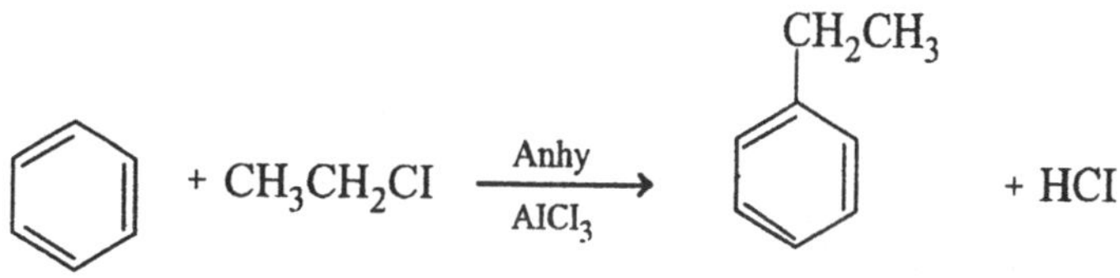

Ethyl benzene

$$C_6H_5—CH_3 + CH_3Cl \xrightarrow[AlCl_3]{Anhy} C_6H_5(CH_3)_2 + HCl$$

Toluene Xylenes

(*b*) **In the preparation of aromatic compounds containing more than one aromatic ring :** If instead of alkyl halide, a polyhalogenated alkane is used the reaction proceeds in a similar manner, thereby forming compounds having more thạn one aromatic ring, *e.g.*

(*i*)

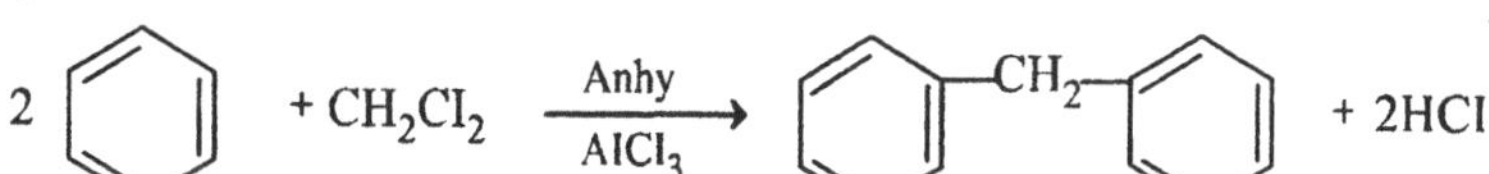

Benzene Methylene chloride Diphenylmethane

(*ii*)

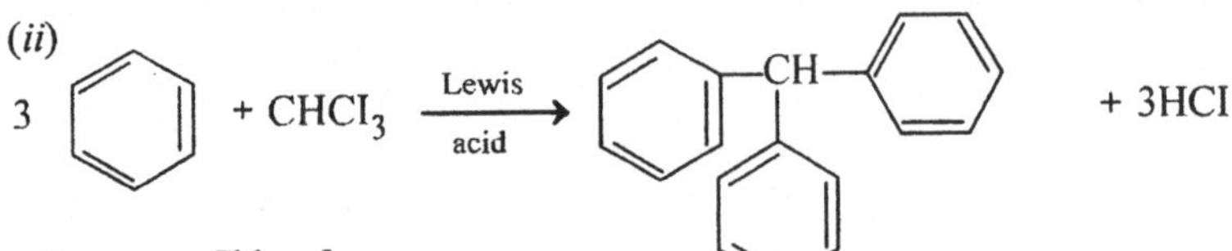

Benzene Chlorofom

Triphenylmethane

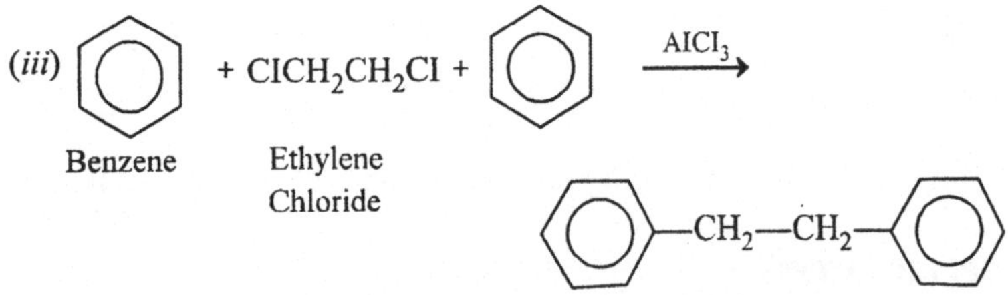

Dibenzyl

(*c*) **In the preparation of polynuclear hydrocarbons :**

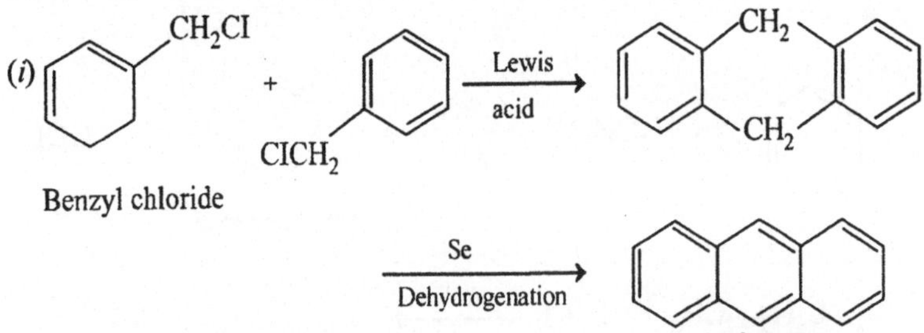

Anthracene

(*ii*)

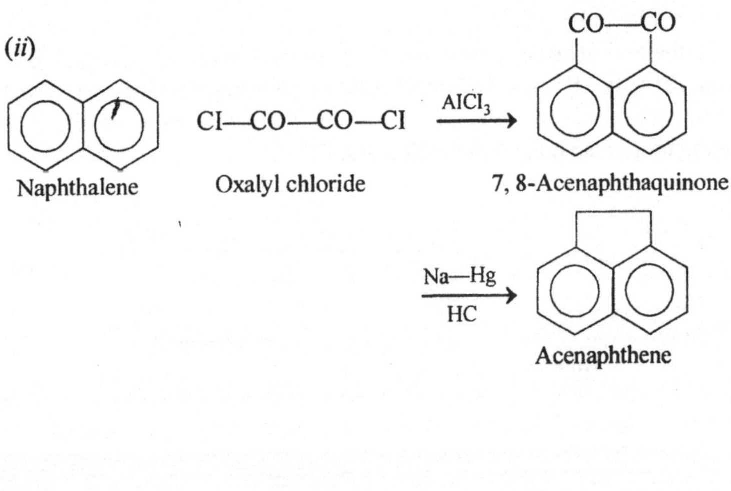

Acenaphthene

(*iii*) 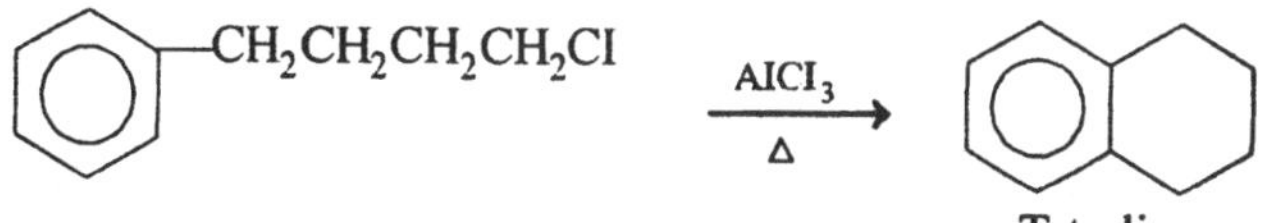

Tetralin

(*iv*) 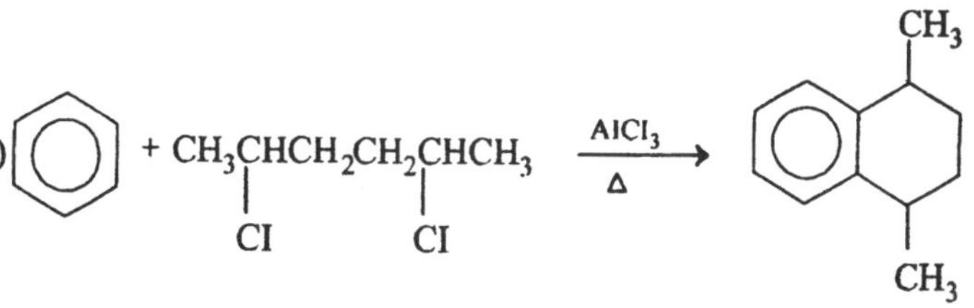

1, 4-Dimethyl tetralin

DRAWBACKS OF FRIEDEL-CRAFTS AKYLATION

The following are some of the drawbacks of alkylation reaction, that limit its synthetic importance :

(*i*) **Polyalkylation :** Polyalkylation often occurs, *i.e.*, the reaction may not stop after monosubstitution, *e.g.*, benzene on alkylation with methyl chloride gives a mixture of mono, di, tri, tetramethyl benzenes.

(*ii*) **Rearrangement :** It is difficult to introduce *n*-propyl, *n*-butyl or higher normal chain groups, because such groups undergo rearrangement to form *secondary or tertiary* groups, *e.g.*,

(*iii*) Alkylating agents such as aryl halides and vinyl halides, which do not form carbonium ion readily, connot be used.

(*iv*) *Effect on the reactivity of the substrates* : Aromatic compounds having electron withdrawing groups (negative groups) such as—NO_2, —CH, etc. or groups such as —NH_2, —NHR and —NH_2 on the aromatic ring cannot be alkylated. On the other hand, the activating groups such as OH, OR, NH_2, etc react with the catalyst to retard the reaction. Napthalene and heterocyclic compounds are very reactive and react with the catalyst. Naphthalene yields poor yield and heterocyclic compounds are not alkylated.

(*v*) *Isomerisation* : In the presence of excess catalyst and at high temperature, isomerization and disproportionation occur.

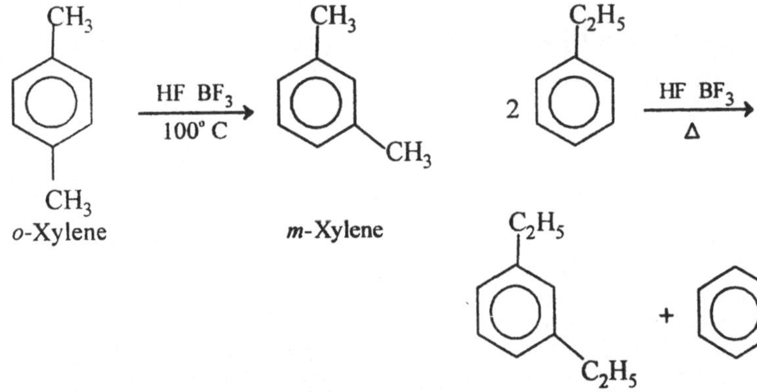

FRIEDAL-CRAFT ACYLATION

The $R-\overset{O}{\overset{\|}{C}}$ group is called an acyl group and a reaction whereby an acyl group is introduced into a compound called acylation reaction. Acylation of aromatic compounds takes place is the presence of excess amount of Lewis acid because the product of the reaction is ketone which forms complexes with Lewis acid.

$$C_6H_5-H + R-\overset{O}{\overset{\|}{C}}-Cl \xrightarrow[\Delta]{\text{Anhy. } AlCl_3/C_6H_6} C_6H_5-\overset{O}{\overset{\|}{C}}-R + HCl$$

97% yield if $R = CH_3$

$$C_6H_5-H + C_6H_5-\overset{O}{\overset{\|}{C}}-Cl \xrightarrow[\Delta]{\text{Anhy. } AlCl_3/C_6H_6} C_6H_5-\overset{O}{\overset{\|}{C}}-C_6H_5 + HCl$$

Friedal-Craft acylation also takes place with acid anhydride.

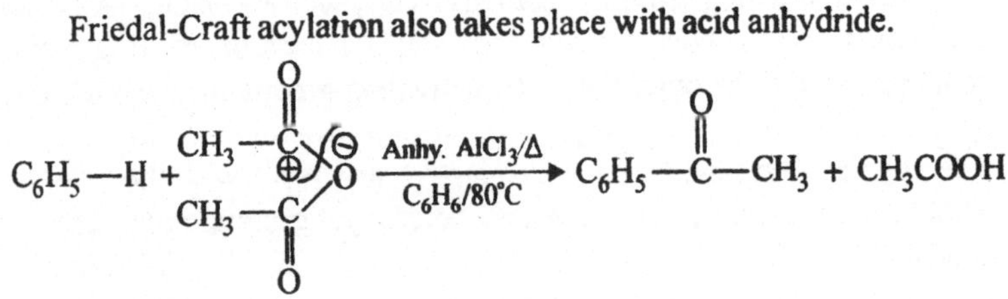

MECHANISM :

Step I.

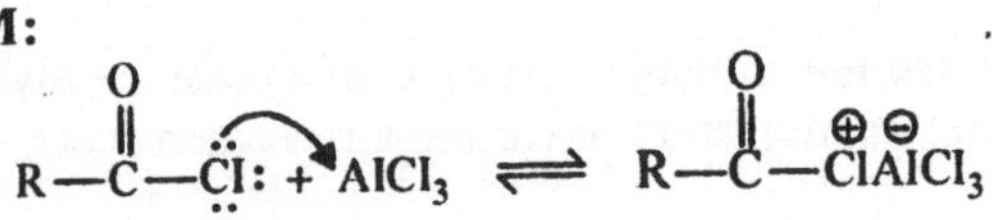

Step II.

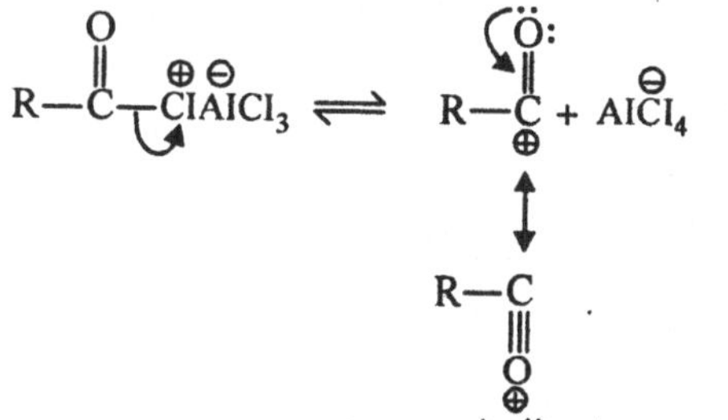

Acylium ion
stabilised by delocalisation

Step III.

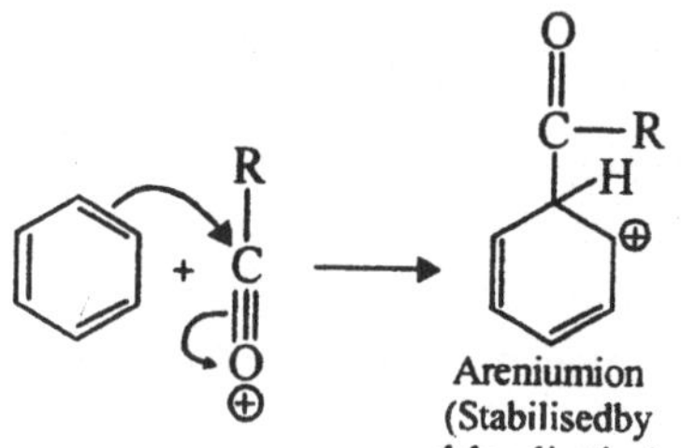

Areniumion
(Stabilisedby
delocalisation)

Step IV.

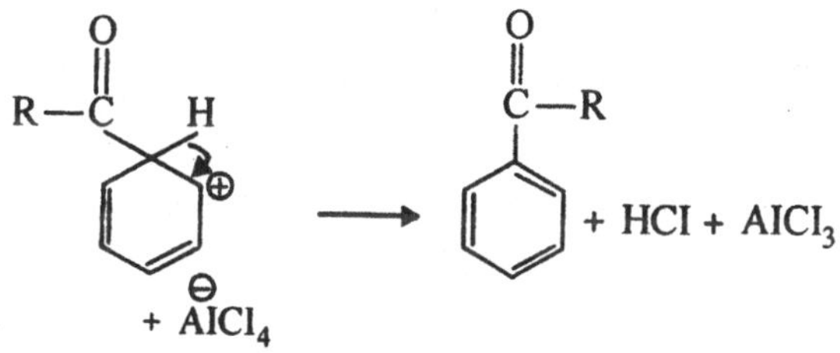

Step V.

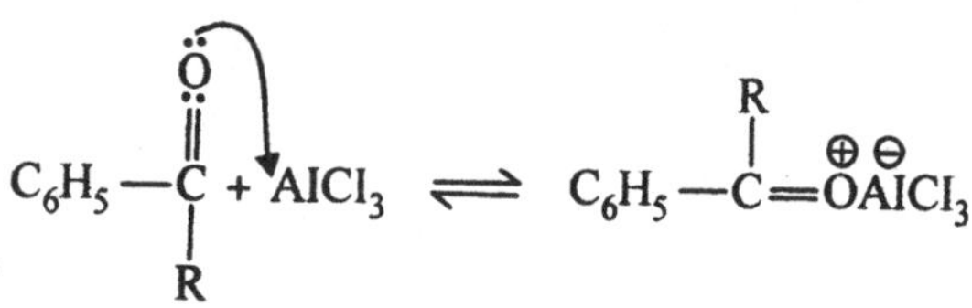

Step VI. Step VI.

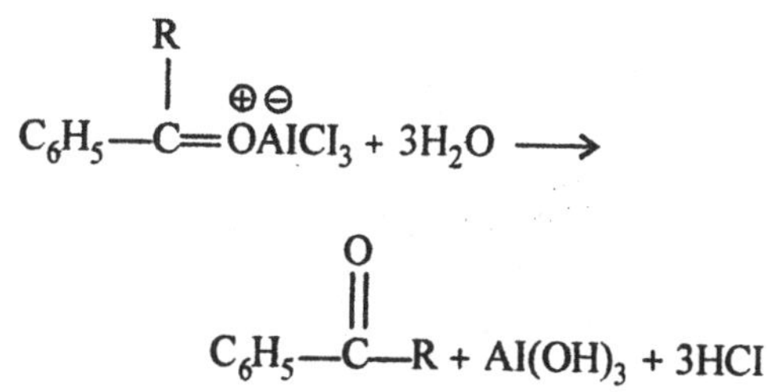

LIMITATIONS OF FRIEDAL-CRAFT REACTIONS

(*i*) **Reactivity of aromatic compound :** Aromatic compound whose reactivity is comparable with or greater than that of benzene gives this reaction. Compounds having strong deactivation group does not give this reaction. Compounds having —NH_2, —NHR, —NR_2 and —OH groups also do not give this reaction.

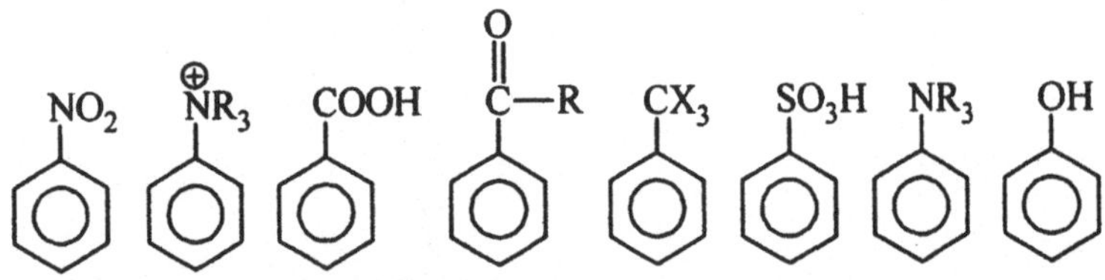

These compounds either do not give Friedal-Craft reaction or the yield is very-very poor.

Althrough NH_2, NHR and NR_2 are activation groups, but these group react with $AlCl_3$ to form complex which has a strong deactivating group.

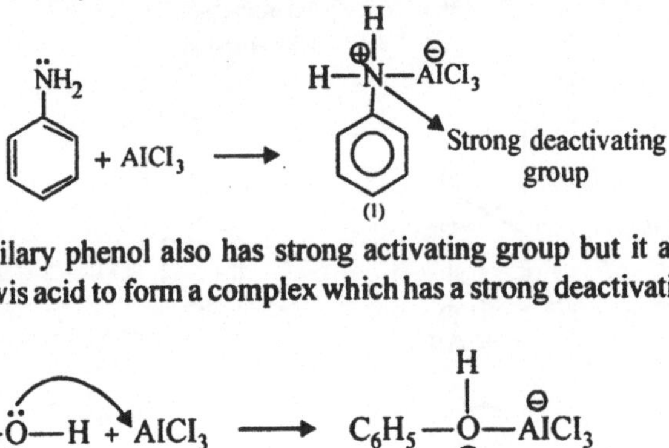

Similary phenol also has strong activating group but it also reacts with Lewis acid to form a complex which has a strong deactivating group.

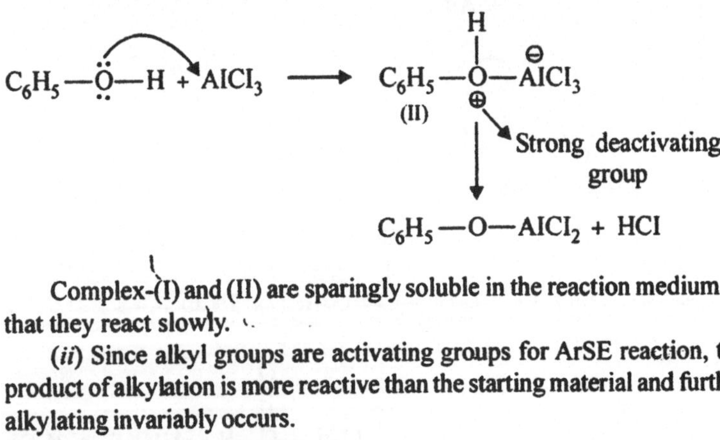

Complex-(I) and (II) are sparingly soluble in the reaction medium so that they react slowly.

(*ii*) Since alkyl groups are activating groups for ArSE reaction, the product of alkylation is more reactive than the starting material and further alkylating invariably occurs.

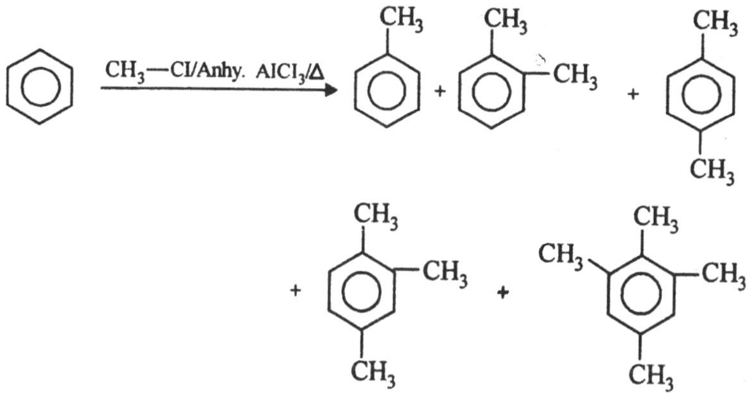

Polyacylation in Friedal-Craft reaction is not possible because acyl group is a deactivating group and only monosubstituted product is formed.

(*iii*) The carbocation formed from alkyl halide, alkene or alcohol can rearrange to a more stable carbocation, it usually does so, and the major product is obtained from the most stable carbocation.

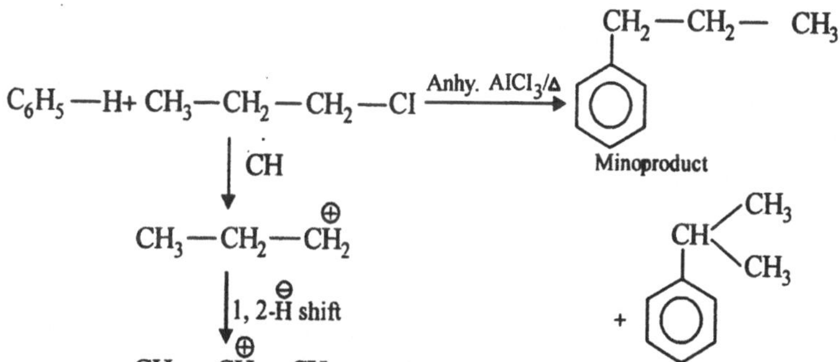

(*iv*) Alkylation reaction is a reversible reaction, so that reaction is thermodynamically controlled. Thermodynamically controlled reaction is carried out at high temp. Meta alkylated derivatives are theremodynamically most stable; hence at higher temp. *m*-derivatives are formed.

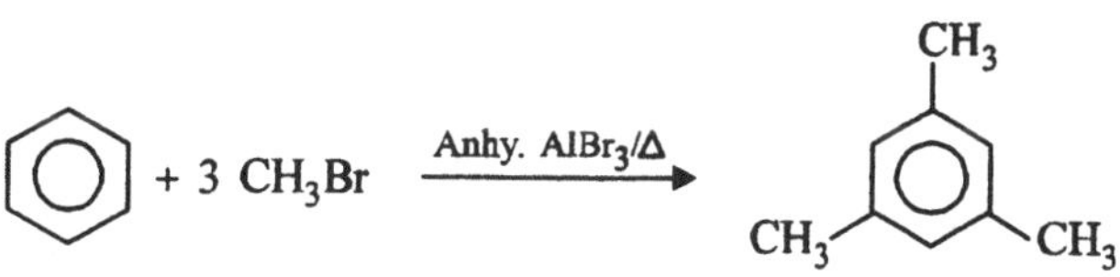

Tertiary alkyl groups are most readily introduced during alkylation; so they are most readily removed because the reaction is a reversible reaction.

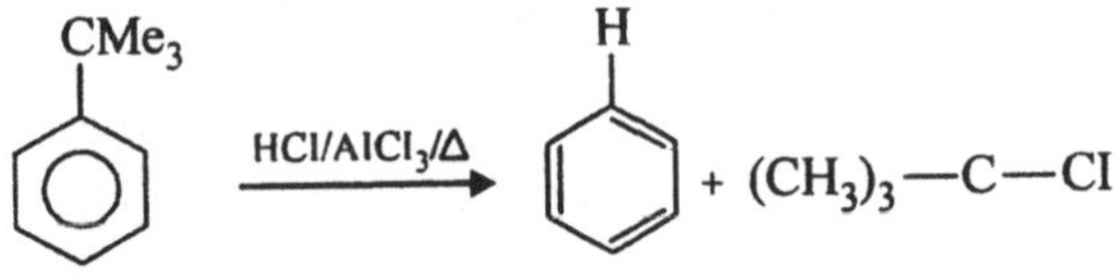

Application of this reaction is very important.

Suppose we want to convert benzene into *o*-methyl acetophenone. This conversion can be carried out as follows :

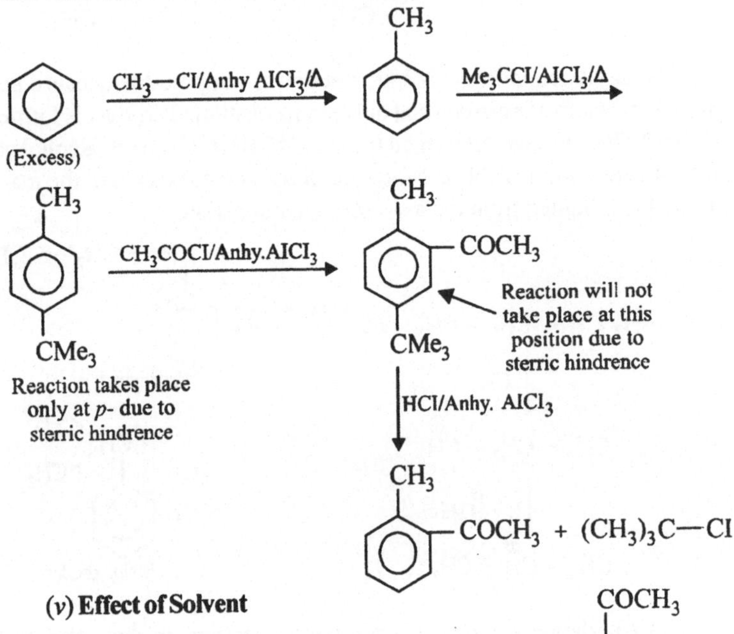

(*v*) Effect of Solvent

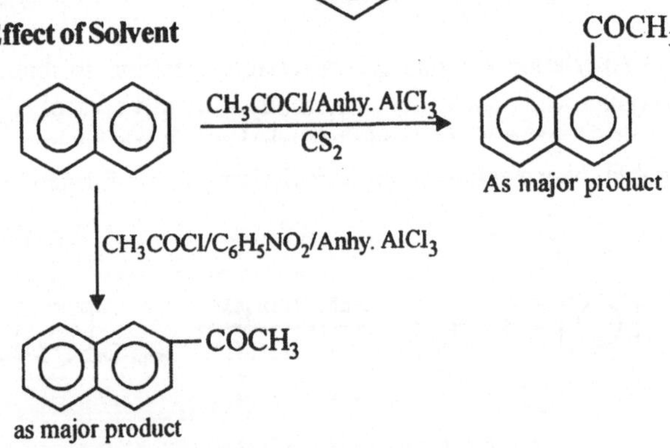

APPLICATION

1. It is mainly used for preparing different aromatic ketone.

(*i*) 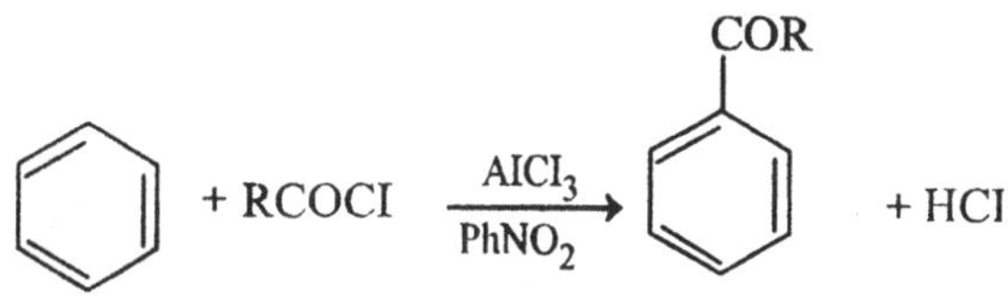

(*ii*) 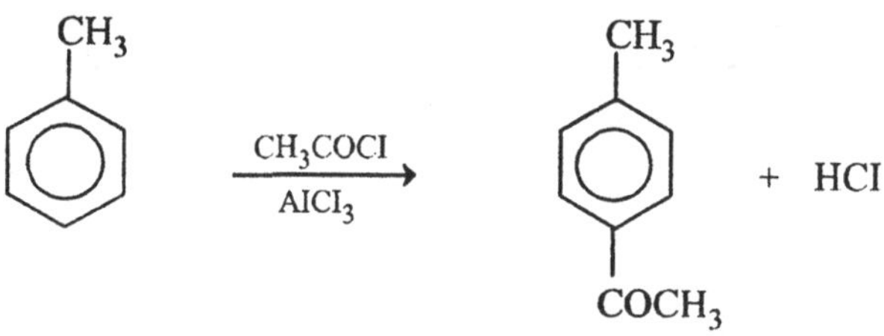

p-Tolyl methyl ketone

(*iii*) 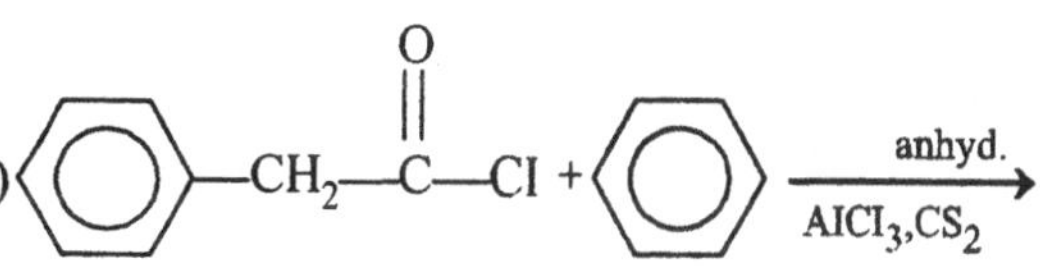

Phenyl acetyl chloride

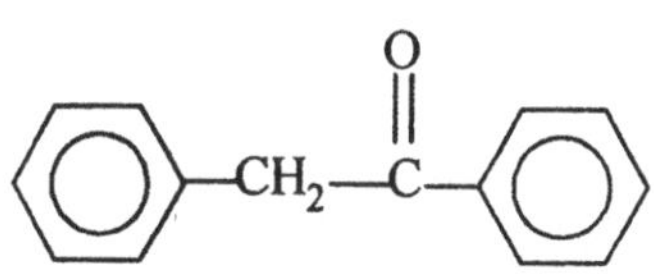

Benzyl phenyl ketone

(*iv*) 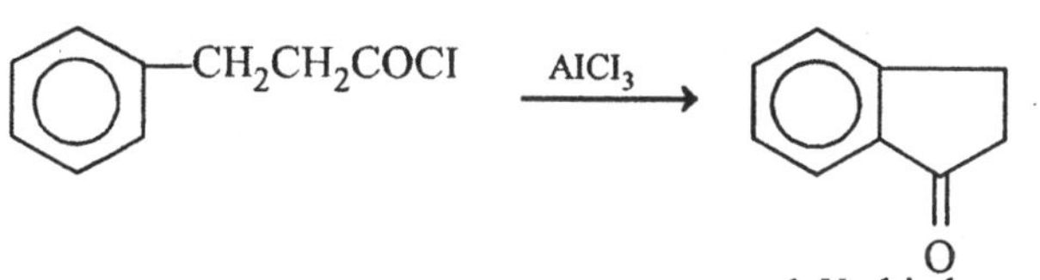

1-Hydrindone

2. The acylated product can be converted into alkyl benzenes, by Clemeasen reduction. The introduction of straight chain alkyl groups in aromatic ring by Friedel-Crafts alkylation is difficult because of rearrangement.

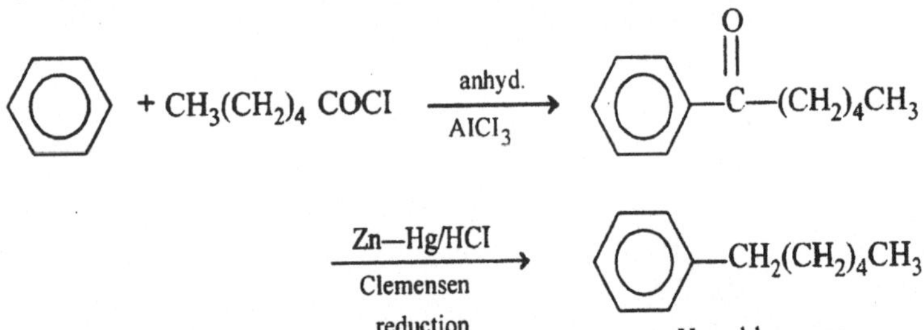

n-Hexyl benzene

3. The products of acyliation can be sometimes cyclized under mild conditions. This can be used in the preparation of higher carbocyclic rings (Haworths Synthesis).

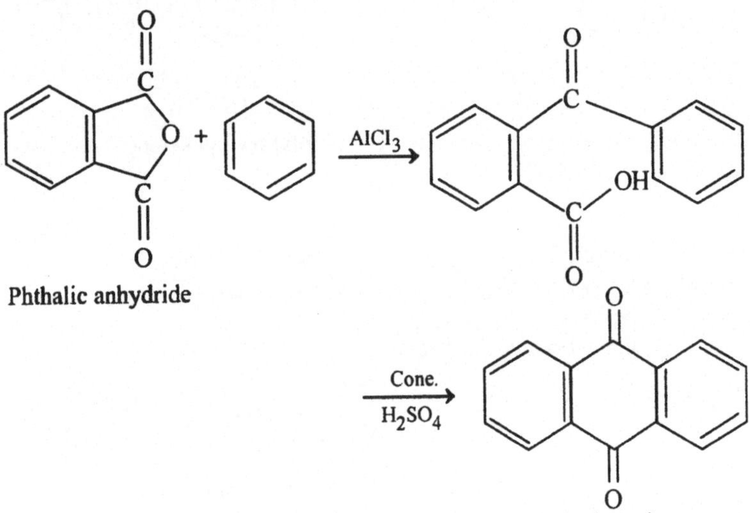

Phthalic anhydride

Anthraquinone

4. The acyl derivatives of heterocyclic compounds can be prepared by using other catalysts like stannic chloride, phosphoric acid etc.

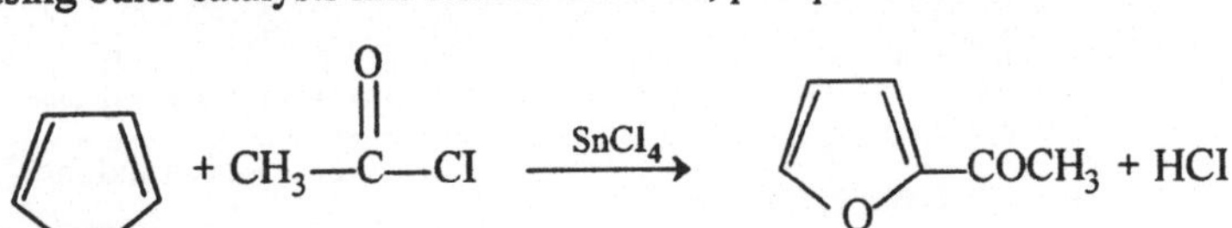

Furan

2-Acetyl furan

5. 2-Naphthyl methyl ketone can be prepared as follows :

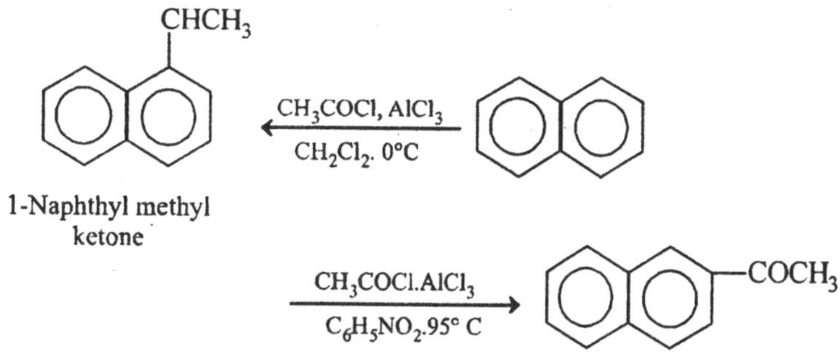

1-Naphthyl methyl ketone

2-Naphthyl methyl ketone

6. Some polynuclear hydrocarbons (cyclisations) can be prepared as follows :

(*i*)

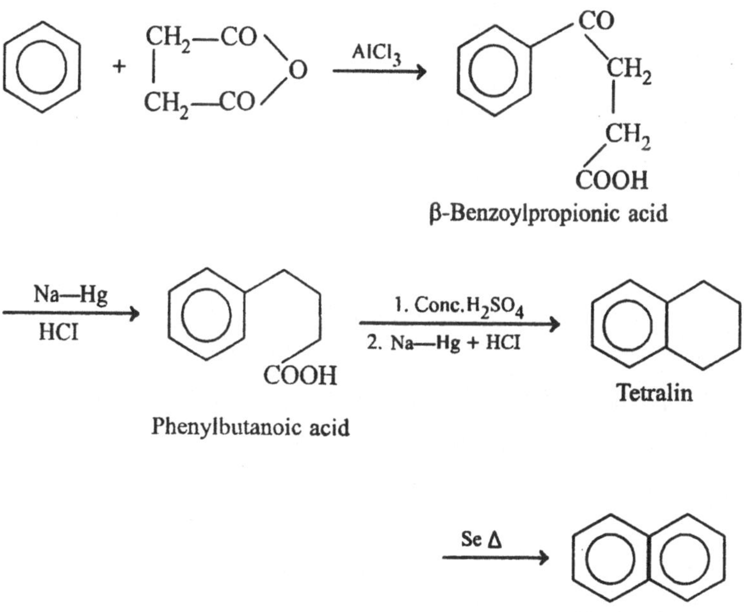

β-Benzoylpropionic acid

Phenylbutanoic acid

Tetralin

Naphthalene

(*ii*)

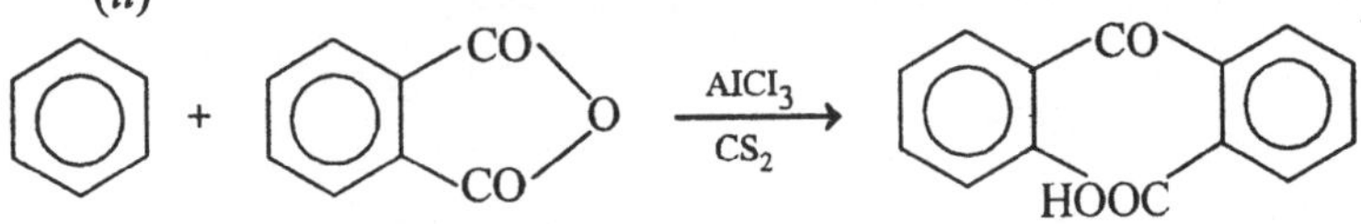

Phthalic anhydride | *o*-Benzoylbenzoic acid

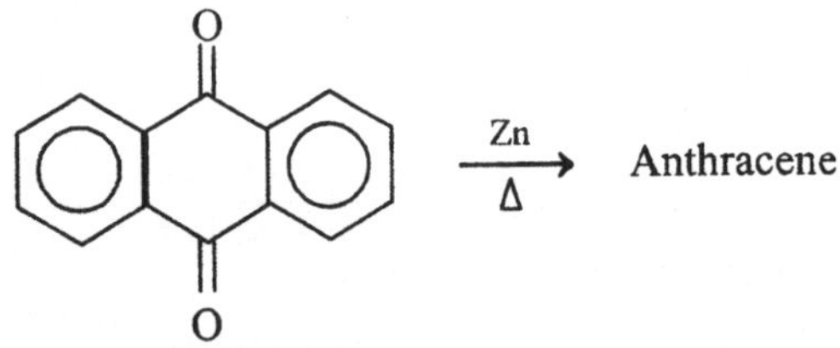

Anthraquinone

DIFFERENCES BETWEEN ACYLATION AND ALKYLATION

Acylation differs from alkylation in the following ways :

(*i*) Acylation unlike alkylation is performed in solvents such as nitrobenzene, carbon disulphide., petroleum ether, methylene chloride etc.

(*ii*) Acylation requires more catalyst than alkylation for complexing with the carbonyl oxygen of the reagent.

(*iii*) Since acyl group is deactivating unlike alkyl group, a monosubstituted product is always obtained.

(*iv*) In acylation rearrangement in the substituent is not observed.

This fact has been utilized for preparing long chain substituted arenes without rearrangement.

$$C_6H_6 + CH_3CH_2COCl \xrightarrow{AlCl_3} C_6H_5COCH_2CH_3 \xrightarrow[HCl]{Na\ Hg} C_6H_5CH_2CH_2CH_3$$

n-Propyl benzene

ADVANTAGES OF ACYLATION OVER ALKYLATION REACTIONS

(*i*) Since the product in acylation reaction (mono-acylated product) is less reactive than the starting material, polysubstitution as observed in alkylation, does not occur in acylation. This is because introduction of an acyl group into the aromatic ring, deactivates the ring.

(*ii*) The mono-alkyl derivatives can be obtained in a better yield by Friedel-Crafts acylation followed by reduction of carbonyl group to methylene group, than by direct alkylation.141

45

FRIES REARRANGEMENT

Phenolic esters, when heated with anhydrous aluminium chloride, undergo a rearrangement in which the acyl group migrates to the aromatic ring to give principally *ortho*-phenolic ketones. This arrangement is called Fries rearrangement. If the ortho position is not vacant or the reaction is carried out at low temperature, the *para*-product predominates. Electron-withdrawing grups in the substrate retard the reaction as in the case of Friedel-Crafts reaction. In general , low temperatute favours the *para*-product and high temperture favours the ortho-product. Solvent variation and catalyst concentration also affect the *ortho* and *para* ratio.

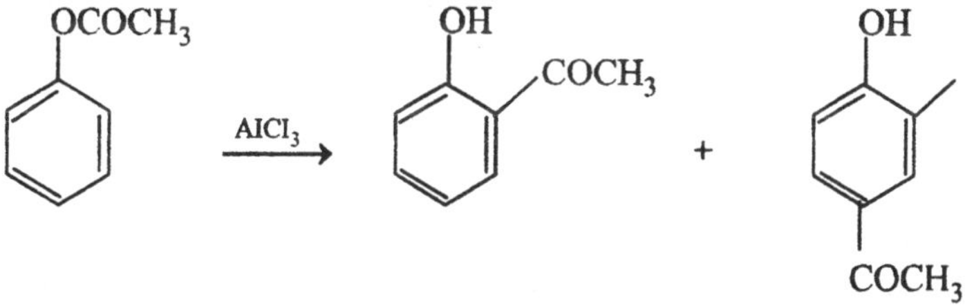

Phenyl acetate | *o*-Hydroxy acetophenone | *p*-Hydroxy acetophenone

The Fries rearrangement can also be carried out photochemically by exposure to UV light.

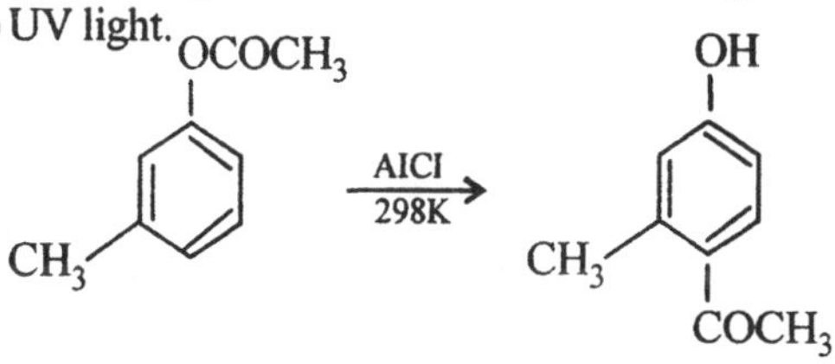

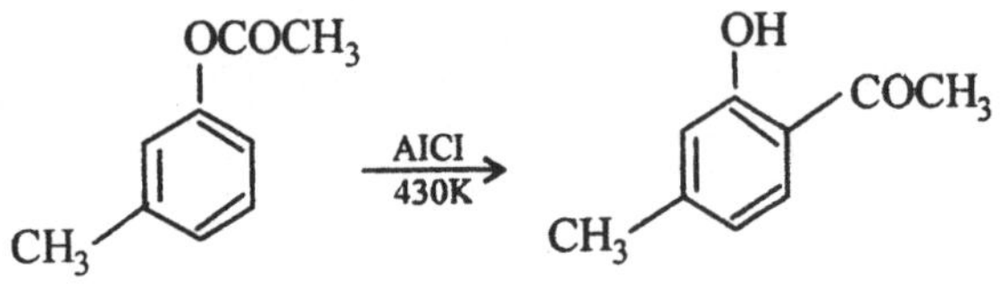

Para-isomer can be converted to ortho-isomer by heating with aluminium chloride.

MECHANISM

Several mechanisms have been postulated to explain the Fries rearrangement but without any success. Sufficient evidence is available for the intermolecular as well as intramolecular mechanism. Thus, the actual mechanism is regarded as a combination of these two mechanisms occurring simultaneously.

INTERMOLECULAR MECHANISM

It may be regarded as Friedel-Crafts acylation in which the acylium ion. (MeCO) is supplied by the substrate, *i.e.*, it is a self-acylation.

At first, $AlCl_3$ complexes with the oxygen of the phenoxy group from which the acylium ion is produced. The acylium ion then attacks the benzene ring as in the case of Friedel-Crafts reaction.

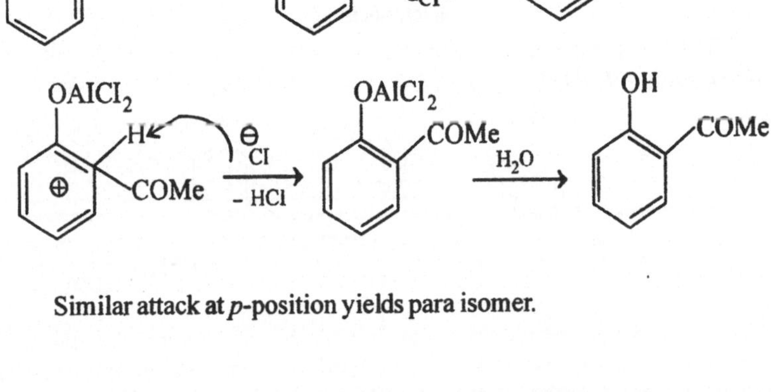

Similar attack at *p*-position yields para isomer.

INTRAMOLECULAR MECHANISM

According to this mechanism, aluminium withdraws electron from the acyl-oxygen bond and facilitates the initial heterolysis. The resulting acyl cation then attacks the ring of the anion at the available ortho position (tripositive aluminium stabilises *ortho* product by chelation). Finally the complex upon hydrolysis yield phenolic ketone.

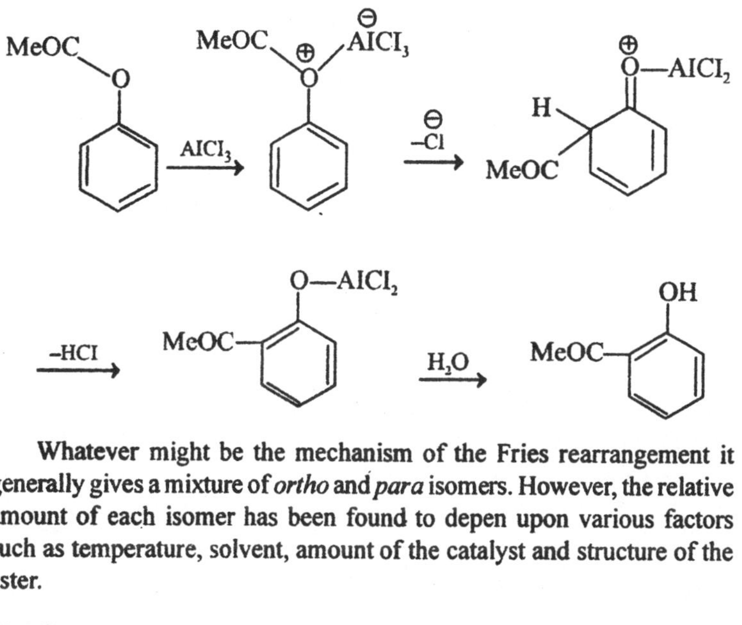

Whatever might be the mechanism of the Fries rearrangement it generally gives a mixture of *ortho* and *para* isomers. However, the relative amount of each isomer has been found to depen upon various factors such as temperature, solvent, amount of the catalyst and structure of the ester.

(*i*) *Effect of temperature* : Generally low temperatutes (60°C or less) favour the formation of the *p*-isomer whereas the higher temperatures (above 60°C) favour the *o*-isomer. This follows from the following example :

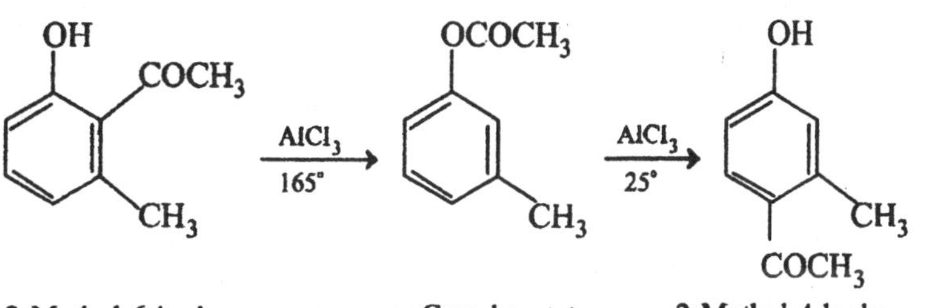

2-Methyl-6-hydroxy acetophenone (9%) | *m*-Cresyl acetate | 2-Methyl-4-hydroxy-acetophenone (80%)

The reason for this is that the *para* derivative is formed faster (kinetic control) whereas the ortho derivative is thermodynamicallymore stable because it possesses a chelate system

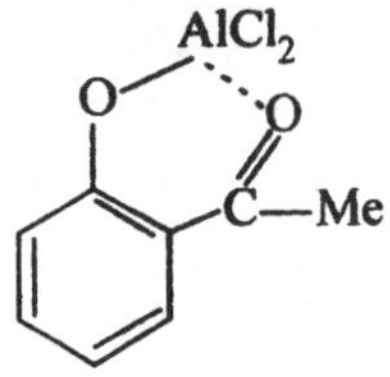

One the other hand, if the *p*-position is blocked as in *p*-tolyl acetate, a high yield of the *ortho* isomer is obtained.

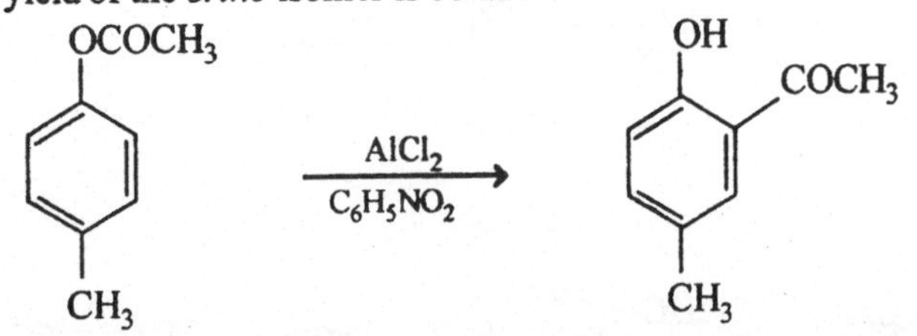

(*ii*) *Effect of solvent* : Generally Fries rearrangement does not need any solvent. However, it is sometimes observed that the use of some solvent reduces the temperature at which the rate of reaction becomes appreciable, *e.g.*,

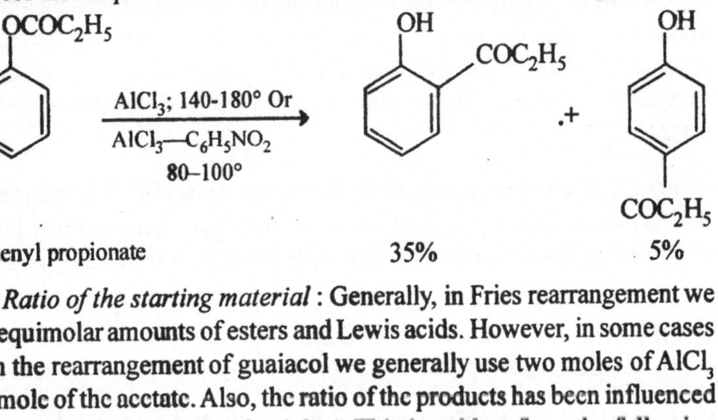

Phenyl propionate 35% 5%

(*iii*) *Ratio of the starting material* : Generally, in Fries rearrangement we use equimolar amounts of esters and Lewis acids. However, in some cases as in the rearrangement of guaiacol we generally use two moles of $AlCl_3$ per mole of the acetate. Also, the ratio of the products has been influenced by the ratio of the ester and catalyst. This is evident from the following example :

Ratio of catalyst ($AlCl_3$)to ester (phenyl acetate)	Ratio of *o*-to *p*-hydroxy-acetophenones
0.5	2.93
1.0	2.14
2.0	10.1

The Fries rearrangement can also be be carried out photochemically by exposure to UV light (no catalyst is needed). Here it proceeds by intramolcular free radical mechanism in which the attacking moiety is RCOO.

(*iv*) *Structure of esters* : An important role is played by the nature and position of the substituent in the aromatic nucleus of the ether. Generally, the presence of a *m*-directing group tends to retard the reaction. From the following examples in which methyl group is present in all the three different positions, it is possible to know the effect of the position of the group in the aromatic nucleus to the product.

o-Methyl phenylacetate — Major product

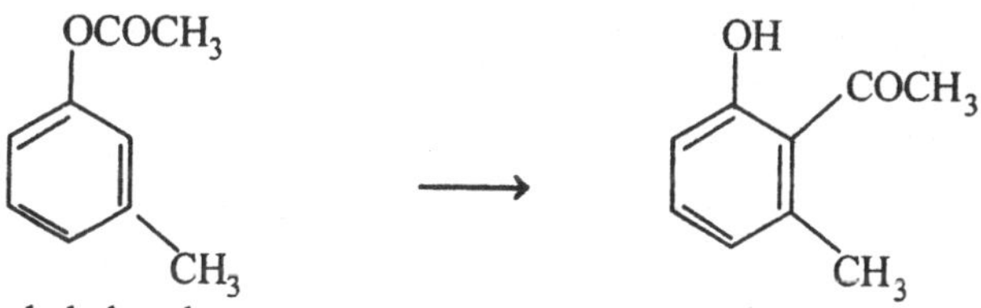

m-Methyl phenylacetate — Major product

p-Methyl phenylacetate — Exclusive product

APPLICATIONS

(*i*) This reaction is used for the preparation of aromatic hydroxyketones because direct acylation of phenols gives very poor yield. The effects of this rearrangement on substrates bearing methyl groups at ortho, meta and para positions reveals that if the *para* position is occupied, the acyl group substitutes exclusively to the *ortho* position.

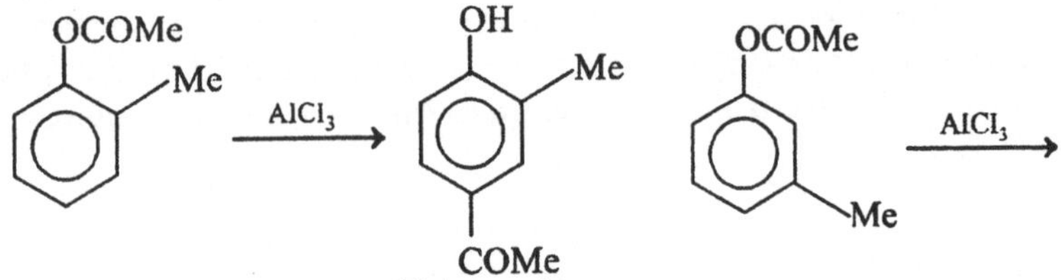

Major product

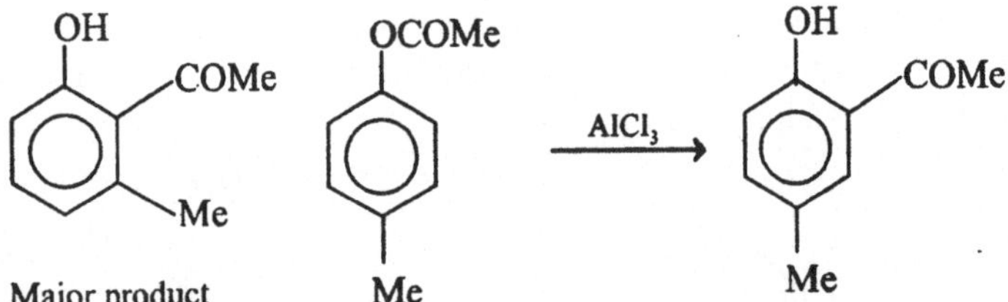

Major product

100%

It is possible to control the proportions of *o*-and *p*-isomers by controlling the conditions of reactions, *viz*., temperature, solvent and the amount of catlalyst. The reaction is useful because it tields boththe isomers which can be easily separated by steam distillation. The *ortho*-isomer having lower boiling point because of H-bonding distills over first, followed by the *para*-isomer.

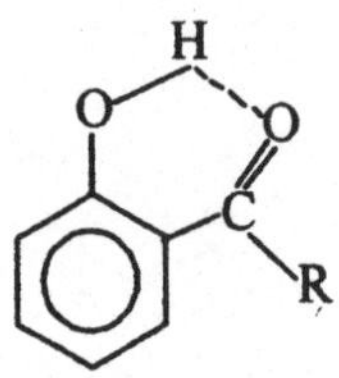

(*ii*) Esters of catechol yield acyl catechols.

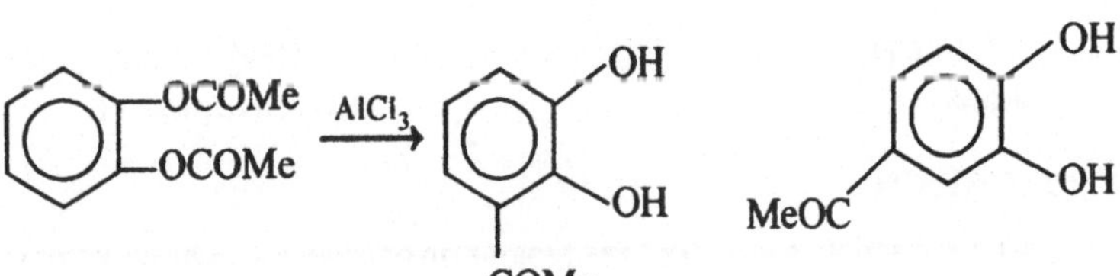

2, 3-Dihydroxy acetophenone(minor)

3, 4-Dihydroxy-acetophenone(major)

(*iii*) Diphenyl esters also undergo Fries rearrangement.

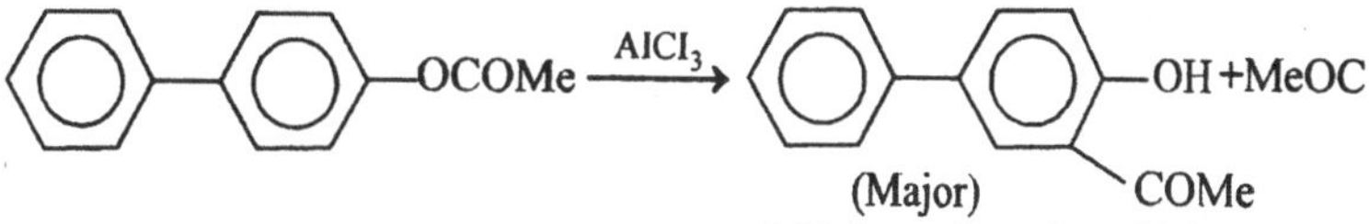

(Major)
4-Hydroxy-3-methoxybiphenyl

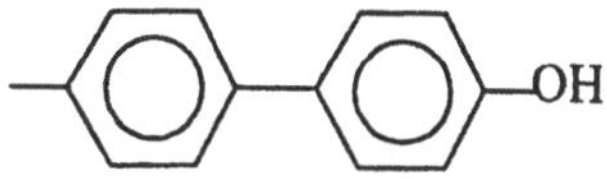

(Minor)
4-Hydroxy-4 -methoxybiphenyl

(*iv*) The main application is in the synthesis of (±) adrenaline which is a heart stimulant.

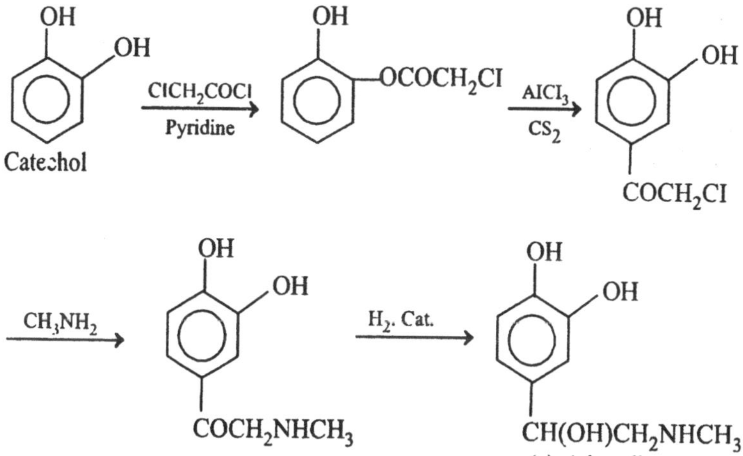

( ) Adrenaline

PHOTO-FRIES REACTION

One can carry out Fries rearrangement in the presence of ultraviolet light.

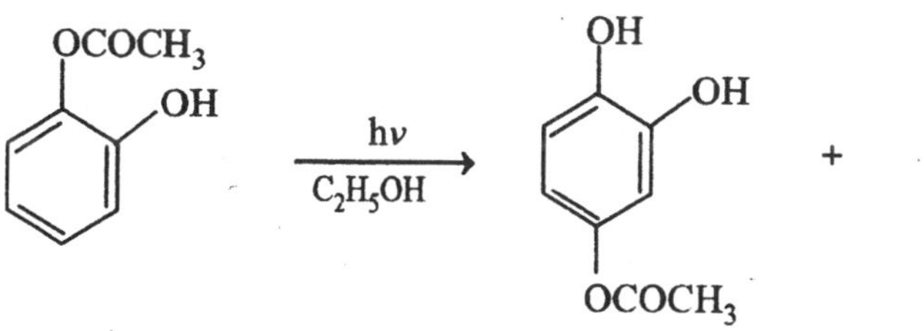

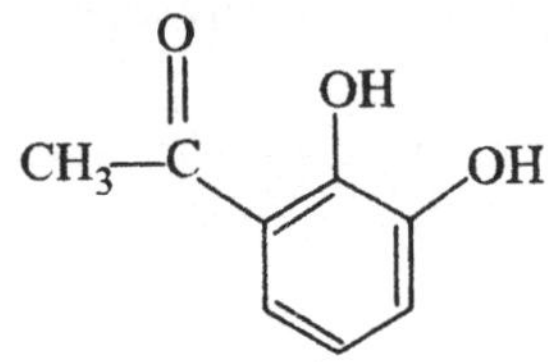

MECHANISM

When a solution of (I) having phenol is irradiated, the irradiated solution does not show that presence of cross product (2-hydroxy acetophenone). This fact reveals that the photo-Fries reaction is an *intramolecular* process. The reaction is assumeed to take place by the ester followed by its dissociation into radical within a solvent cage. Rearrangement product are formed by the recombination of the radicals before diffusion from the solvent cage.

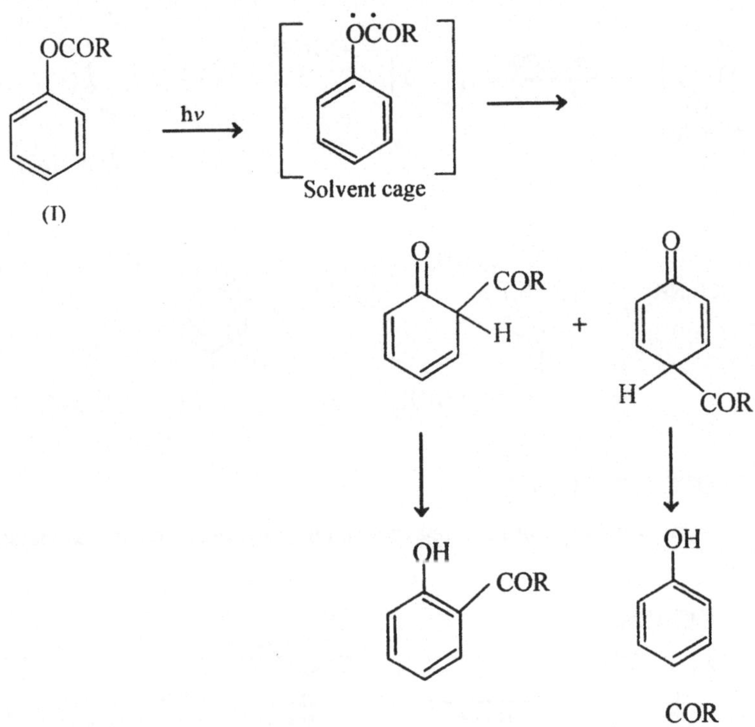

46

GATTERMANN REACTION

One of the successful means in formylating aromatic ring is afforded by GATTERMANN REACTION.

REACTION

Formylation effected by HCN & HCl mixture in presence of suitable catalyst is called GATTERMANN

The reaction is carried out by passing a mixture of HCN and HCl into the solution of the aromatic compound having aluminum chloride as the suspended catalyst

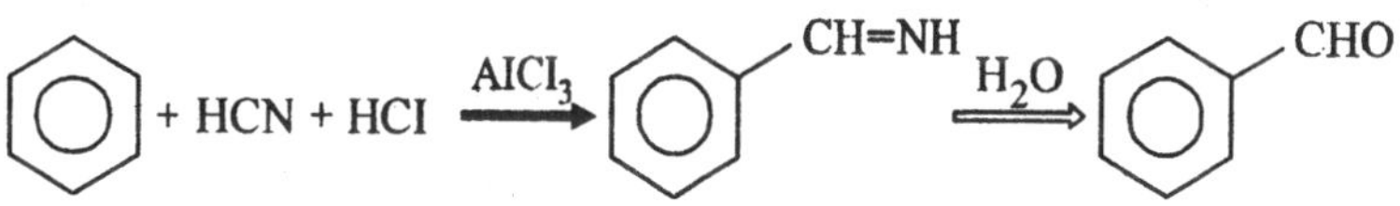

The toxic HCN can be substituted by Zn $(CN)_2$.

MECHANISM

The reaction mechanism is under investigation. A widely accepted mechanism being.

(I) *Formation of Imidoformyl chloride*

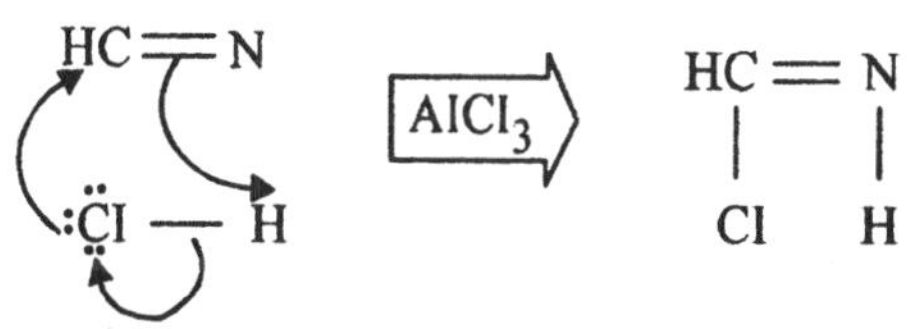

(II) *Nucleophilic Attack by* $CH(Cl){=}NH$

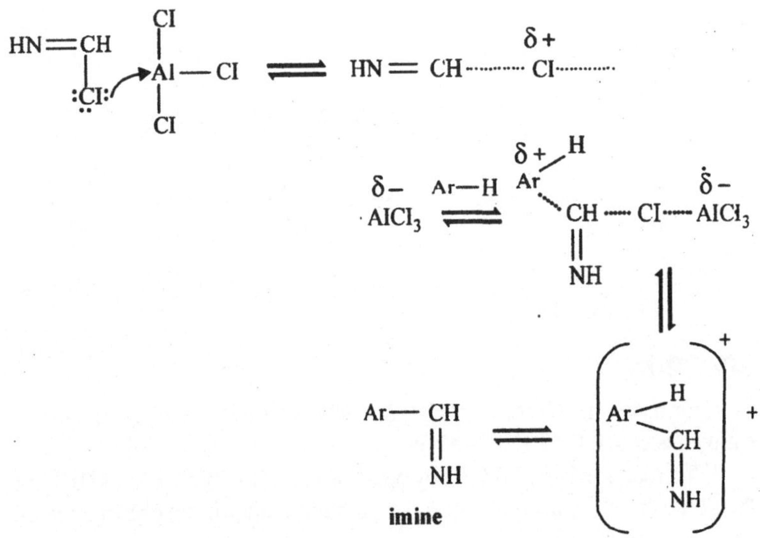

(III) *Hydrolysis of Imine*

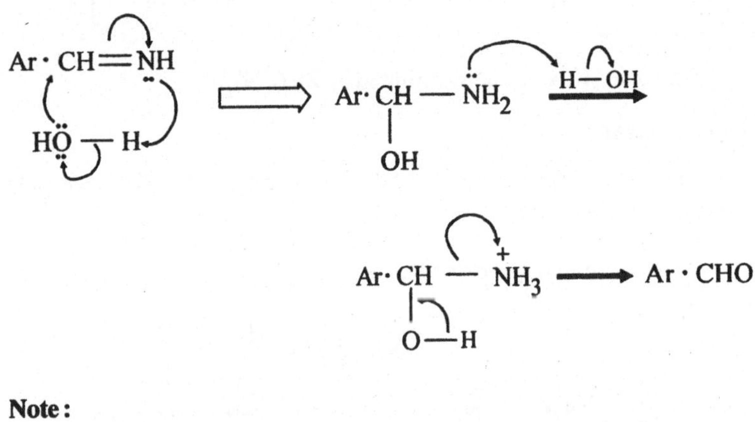

Note :

(I) The reaction is applicable to benzene, phenols and their ethers

(II) Certain heterocyclic compounds can also be formylated by this way

(III) However, this reaction cannot be applied to aromatic amines

APPLICATION

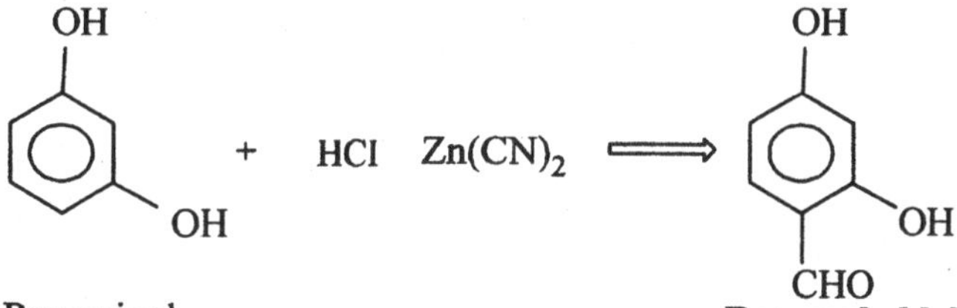

Resorcinol

Resorcyl aldehyde

YIELD 95%

PROBEMS

Problem 1.

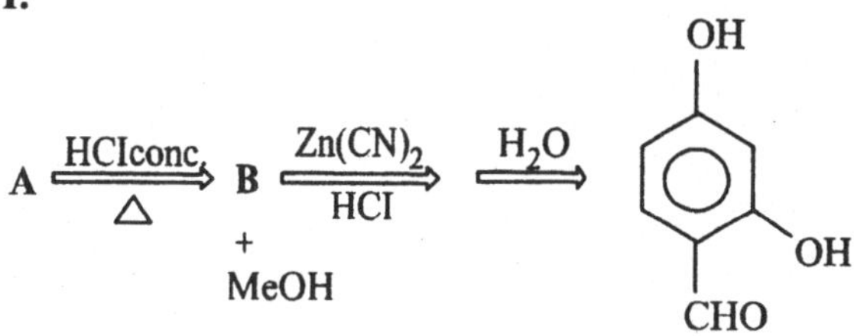

A ?

B ?

Answer

A ▶ 1:3 - O - Methyl benzene

B ▶ Resorcinol

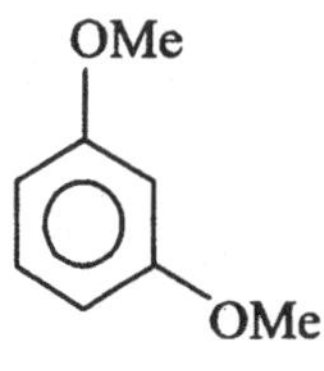

47

GATTERMANN-ALDEHYDE SYNTHESIS

Formylation of aromatic compound by HCN in the presence of anhy. $ZnCl_2$ is known as Gattermann-aldehyde synthesis.

This reaction is given by benzene, alkyl benzenes, phenols, phenolic ethers and some heterocyclic compounds. Nitrobenzene, amines and compounds containing deactivating groups do not give this reaction.

In this reaction, $ZnCl_2$ and HCN are conveniently generated from $Zn(CN)_2$ and HCl.

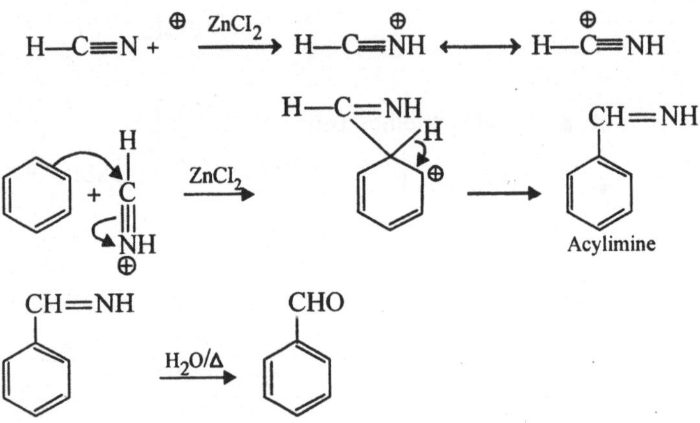

48

GATTERMANN -AMIDE SYNTHESIS

Benzene, alkyl benzene and anthracene gives this reaction in presence of anhy.

$AlCl_3/\Delta$. The reagent of the reaction is carbamoyl chloride and N, N-dimethyl carbamoyl chloride. Product of the reaction is amide.

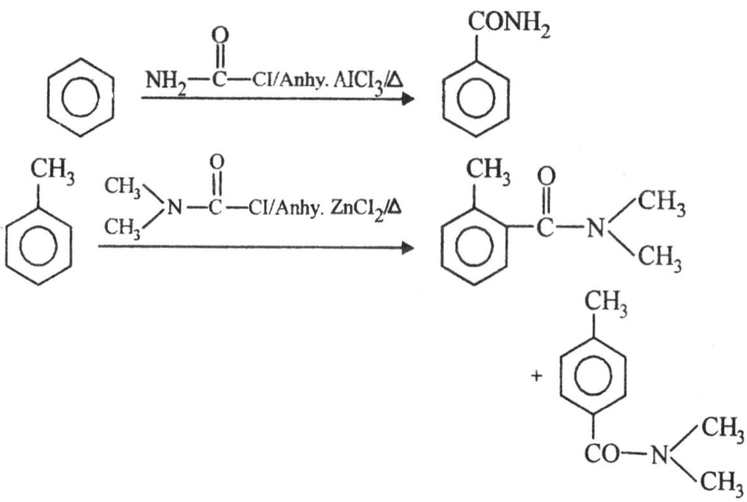

This reaction is an extension of Friedal-Craft acylation.

49

GATTERMANN -KOCH ALDEHYDE SYNTHESIS OR GATTERMAN-KOCH FORMYLATION

Formylation of benzene and alkyl benzene by carbon monoxide and hydrogen chloride in the presence of anhydrous $AlCl_3$ as catalyst and Cu_2Cl_2 as co-catalyst is known as Gattermann-Koch formylation.

The mixure of carbon monoxide and HCl is obtained by dropping chlorosulphuric acid on HCOOH.

$$HCOOH + HSO_3Cl \longrightarrow CO + HCl + H_2SO_4$$

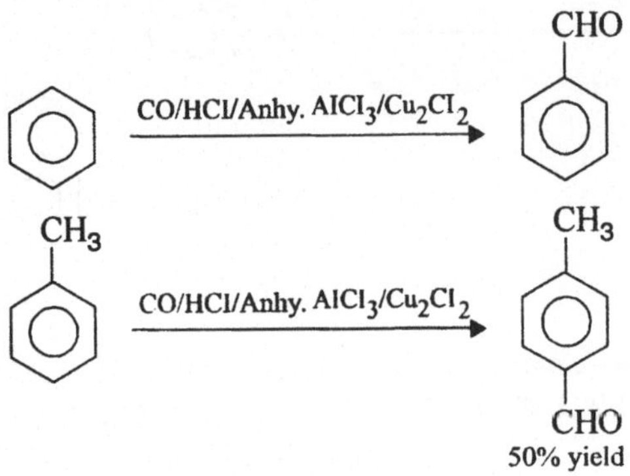

Alkyl benene gives only *p*-derivative.

MECHANISM:

$$HCl + \overset{\ominus}{C}\equiv\overset{\oplus}{O} + AlCl_3 \xrightleftharpoons{Cu_2Cl_2} H-\overset{\oplus}{C}\equiv O + \overset{\ominus}{AlCl_4}$$

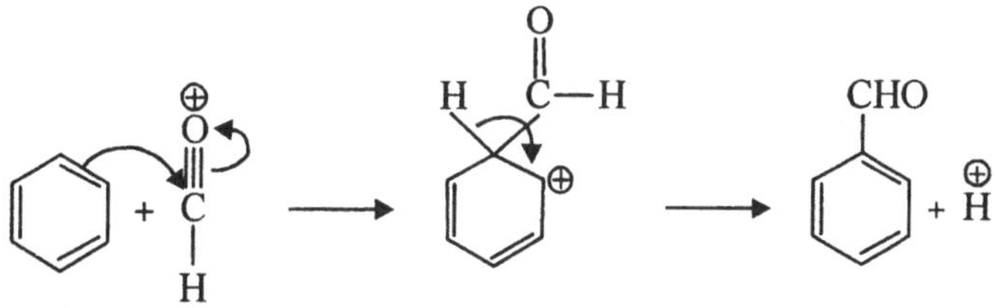

PROBLEMS

Problem

A compound **A** subjected to Gattermann-Koch reaction yielded **B** which on treatment with ethanolic solution of aqueous potassium cyanide produced **C.**

C upon mild oxidation by nitric acid produced **D.**

D condensed with ***ortho***-phenylenediamine to produce 2, 3-diphenylquinoxaline.

Identify **A, B, C & D**

Solution

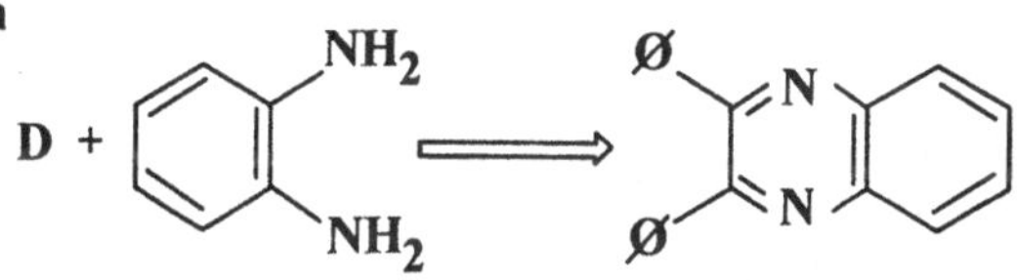

2,3- Diphenylquinoxaline

The only compound that forms 2,3-diphenylquinoxaline with ***ortho***-phenylenediamine is benzil.

Hence **D** is benzil.

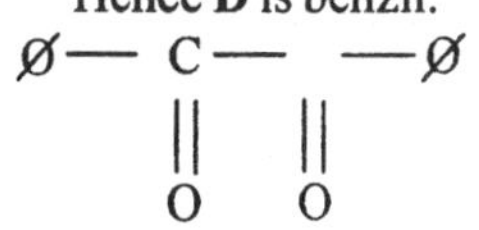

D is the mild oxidation product of **C.**

So **C** is benzoin.

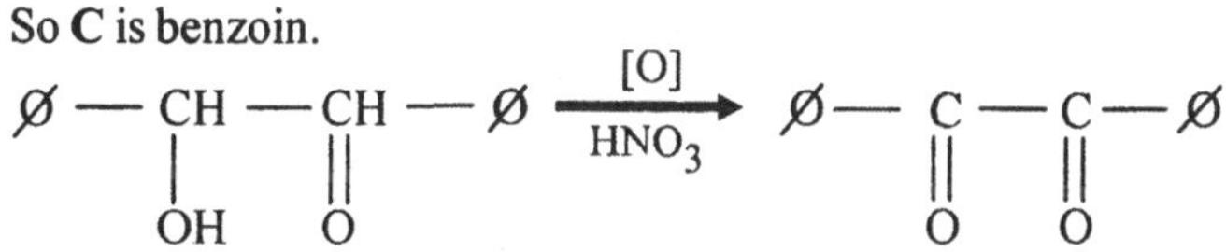

[C] [D]

C i.e benzoin is produced on treatment of **B** with ethanolic aq. KCN

Clearly **C** is the benzoin condensation product of benzaldehyde, since this reaction is typical of benzaldehyde. Therefore, **B** is benzaldehyde

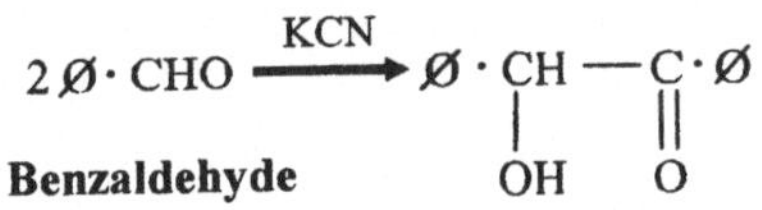

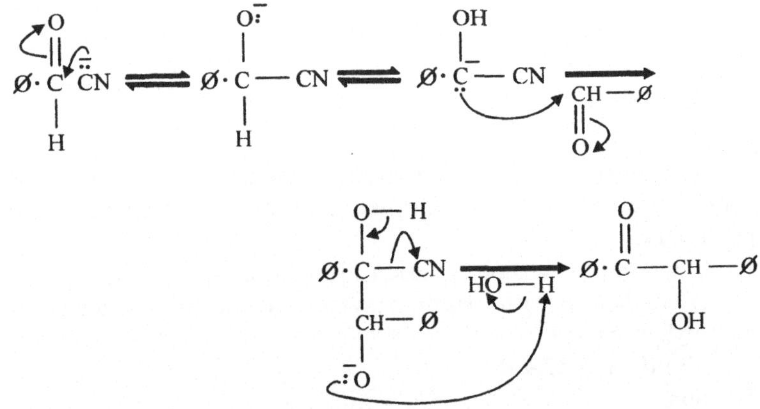

B i.e. benzaldehyde is the Gattermann-Koch reaction product of **A**. Hence, **A** is purely benzene

$$\text{Ø—H} + \text{CO} \xrightarrow[\text{CuCl / Et}_2\text{O}]{\text{AlCl}_3} \text{Ø—CHO}$$

Conclusion **A** Benzene **B** Benzaldehyde
C Benzoin **D** Benzil

Problem 2.

$$\underset{[A]}{C_7H_8} \xrightarrow[\text{HCl / CO/ 1 atm}]{\text{AlCl}_3\text{ /CuCl}} \underset{[B]}{C_8H_8O} \xrightarrow[\text{OH / }\Delta]{\text{CH}_3\text{CHO}} \underset{[C]}{C_{10}H_{10}O} \xrightarrow[\text{Pt}]{H_2} \underset{[D]}{C_{10}H_{14}O}$$

$$\text{D} \xrightarrow[\Delta]{-H_2O} \text{3 - (}p\text{-Methylphenyl) - propene}$$

Answer

A▶ Toluene

B▶ ***para-*** Tolualdehyde 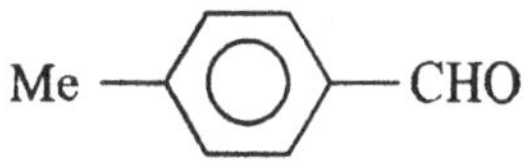

C▶ ***para-*** Methylcinnamaldehyde

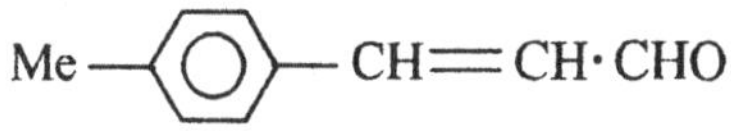

D▶ 3-[*para*-Methylphenyl]-propanol 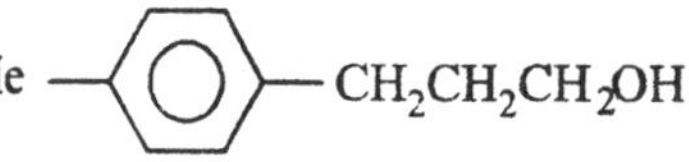

Problem -3

$$\underset{[A]}{C_7H_8} \xrightarrow[CO/HCl]{AlCl_3/CuCl} \underset{[B]}{C_8H_8O} \xrightarrow[CH_2O/EtOH]{NaOH} \underset{[C]}{C_8H_{10}O} \xrightarrow{PBr_3} \underset{[D]}{C_8H_9Br} \xrightarrow[\Delta]{Na/Et_2O}$$

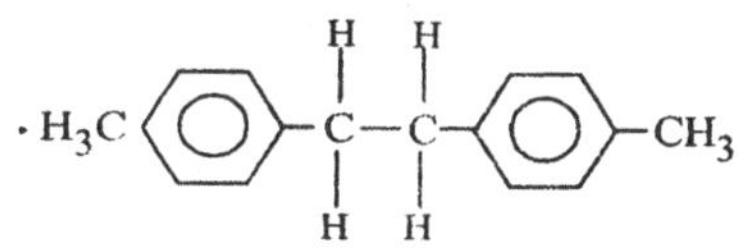

Identify **A, B, C, & D**

Answer

A▶ Toluene

B▶ ***para-*** Tolualdehyde

C▶ ***para***-Methylbenzyl alcohol

D▶ ***para***-Bromomethyl toluene 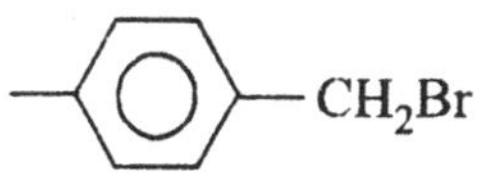

50

GABRIEL PHTHALIMIDE SYNTHESIS

A synthesis that gives primary amines without secondary and ter. amines isthe Gabriel phthalimide synthesis.

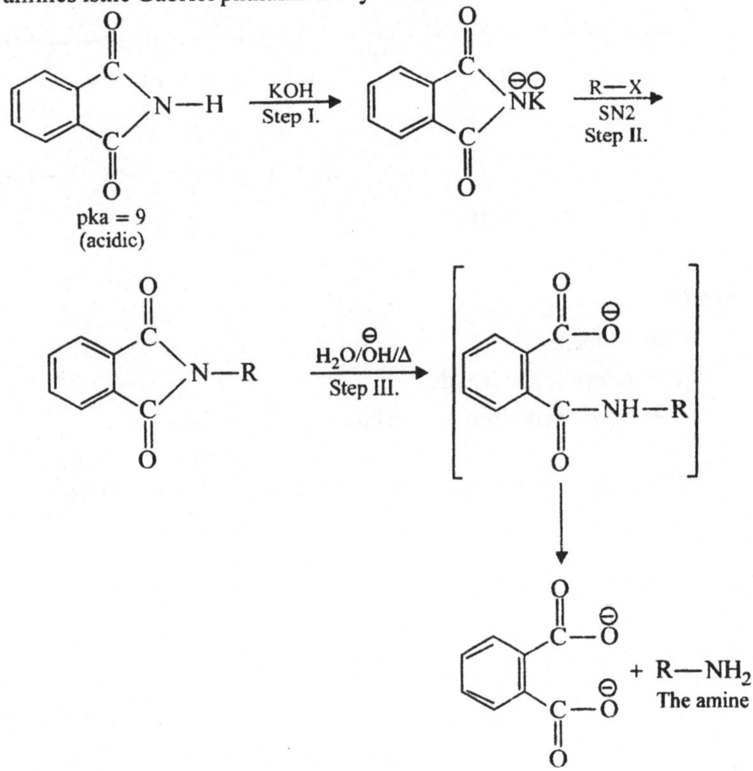

Hydrolysis of *N*-alkyl phthalimide is often difficult. It is often more convenient to treat *N*-alkyl phthalimide with hydrazine in refluxing ethanol to give *p*-amine and phthalazine-1, 4-dione.

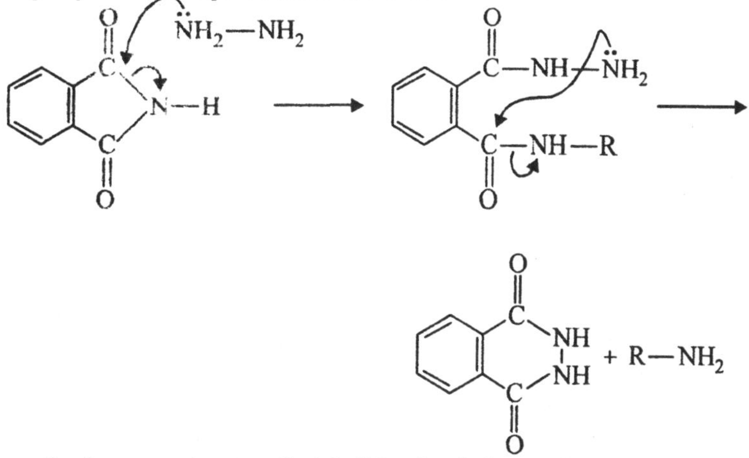

In the second step, alkyl halide shouls be methyl, primary and secondary alkyl halide because reaction is SN2. Ter alkyl halide leads almost exclusively to elimination due to strong base.

MECHANISM

Imides have acidic properties because the negative charge of the conjugate base is delocalized over both oxygen and the mitrogen. The pK_a of phthalimide is 8.3. In aqueous basic solution the compound is converted almost completely into the anion.

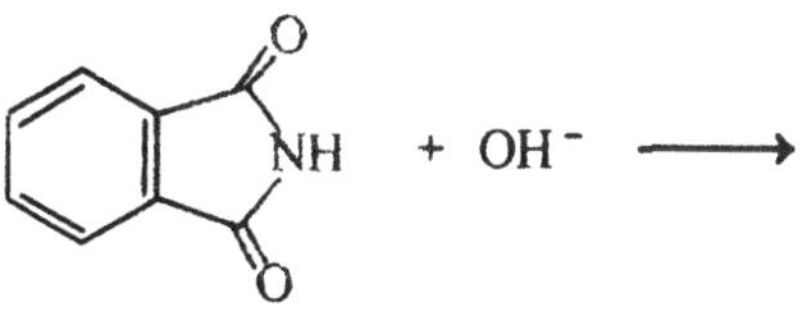

phthalimide

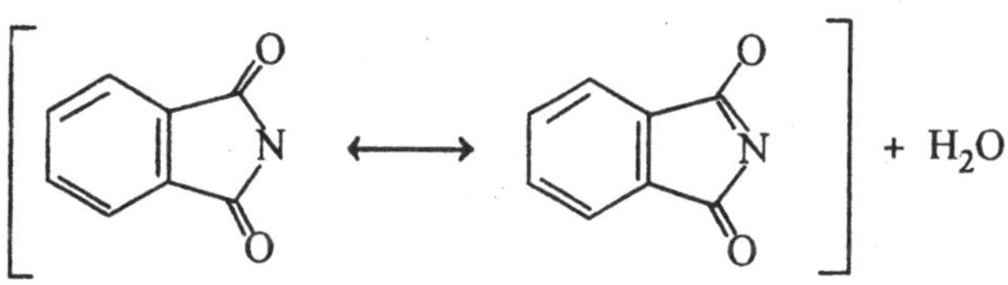

The phthalimide anion has nucleophilic properties and can enter into displacement reactions with alkyl halides. Reaction could in principle take place on either oxygen or nitrogen, but since nitrogen is more nuclephilic, it occurs mostly on mitrogen. Further alkylation cannot occur because there are no acidic protons. The product is an N-alkylphthalimide, and hydrolysis, and hydrolysis gives the amine and phthalic acid.

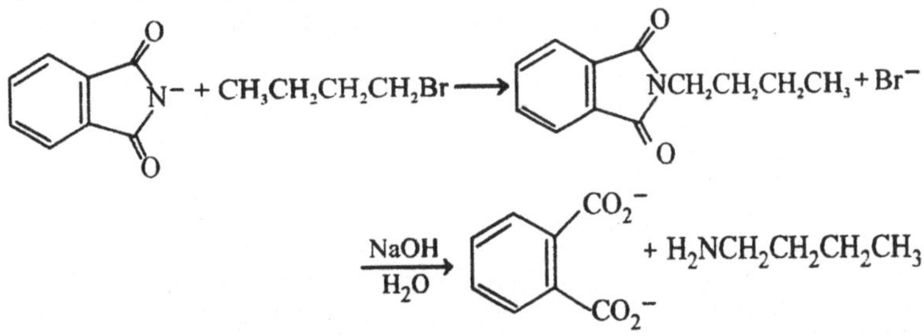

The best solvent for the alkylation appears to be dimethylformamide $HCON(CH_3)_2$.

The hydrolysis of N-alkyl phthalimide can be hydrolysed either with acid or alkyl. Usually a prolonged heating is required in both the mediums. The mechanism of hydrolysis is similar to ester hydrolysis.

(a) Base-catalysed hydrolysis :

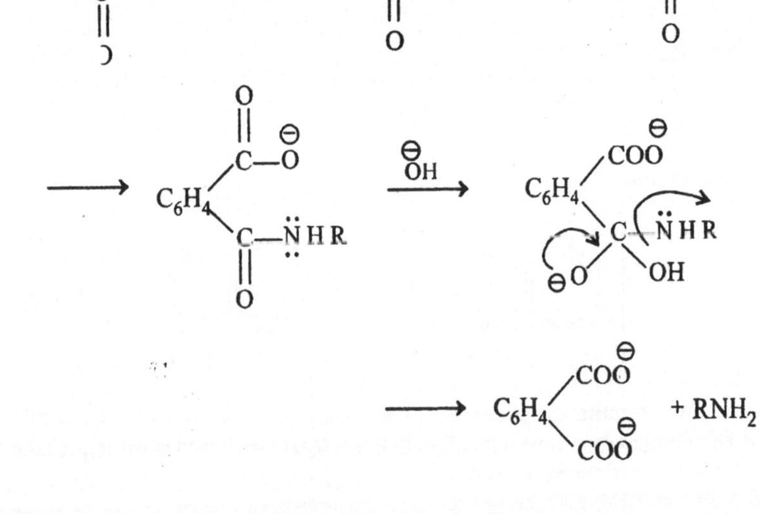

(*b*) *Acid-catalysed hydrolysis* :

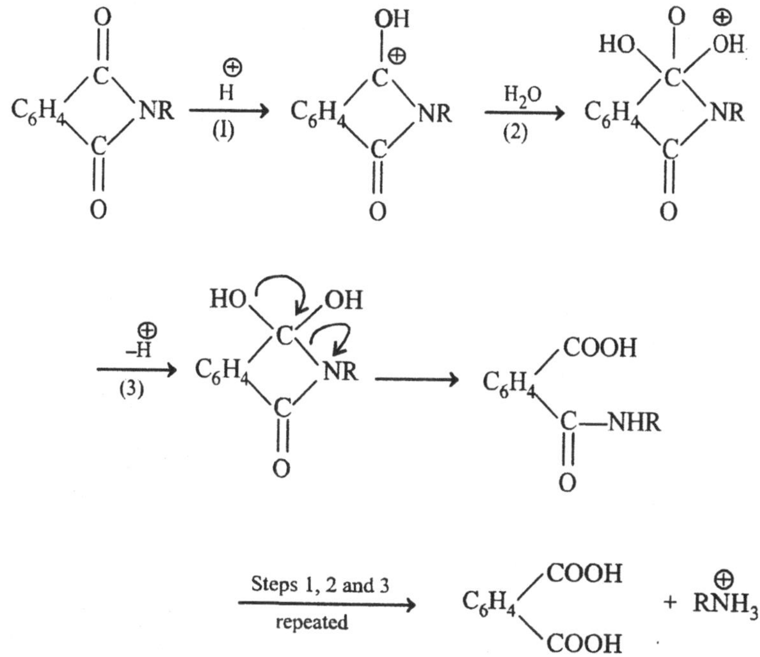

If substituted phthalimide is difficult to hydrolyse, hydrazine hydrate maybe used to get free amine and phthalohydrazide.

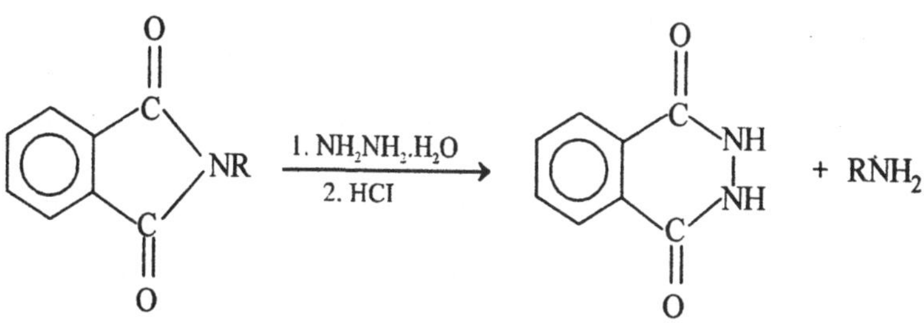

Phthalohydrazide

The reaction, called *hydrazinolysis,* takes place under milder conditions than alkaline hydrolysis. **Probable mechanism for hydrazinolysis** is as follows :

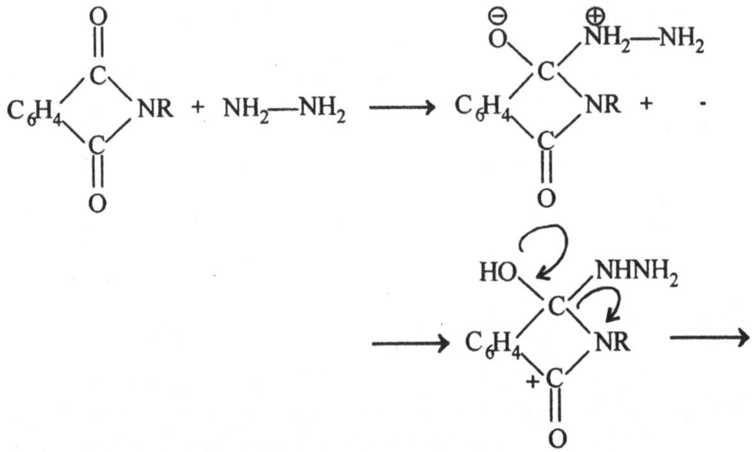

Dimethylformamide is a better solvent for Gabriel's synthesis.

APPLICATIONS

Gabriel synthesis is useful for the synthesis of complex compounds as phthalimide group is fairly unreactive and is readily separated at the end by hydrolysis. Hence,

(*a*) A dihalide is able to form monohthalamide which can undergo the usual reactions of alkyl halides. Finally the phthaloyl group can be removed to get the free amine.

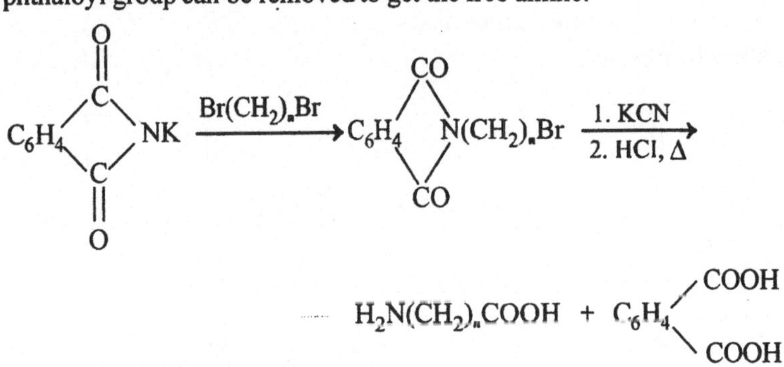

(*b*) Phthalimide derivative formed by reacting with halogeno esters can be converted to acid chlorides which find use in Friedel-Crafts reaction. The products on hydrolysis. yield amines.

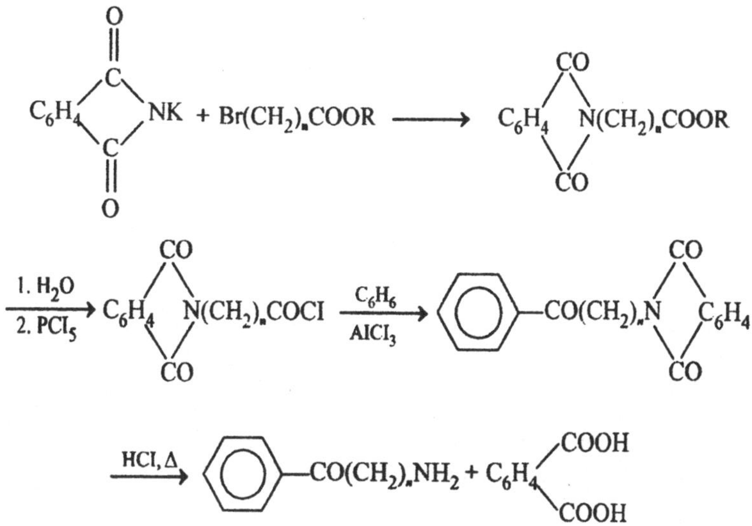

(*c*) Many amino acids can be obtained by using derivatives ofmalonic esters

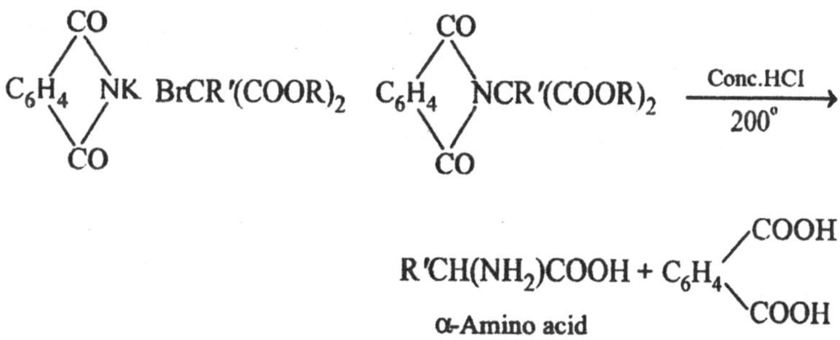

51

HALOFORM REACTION

From the standpoint of successful degradation and identification of methyl keto compounds, the Haloform Reaction typifies its fundamental importance in the domain of carbonyl compounds.

REACTION

The reaction is essentially the cleavage of ketones having the CH_3CO-group. The reaction is carried out by treating the methyl ketone dissolved in certain suitable solvent like dioxan with alkali and halogen taken directly or dissolved in a solvent (e.g. I_2 in KI solution).

$$CH_3 \cdot \underset{\underset{O}{\|}}{C} \cdot R + X_2 \xrightarrow{\bar{O}H} CHX_3 + R \cdot \underset{\underset{O}{\|}}{C} - \ddot{O}$$

MECHANISM

The reaction takes place in two distinctive stages :

STAGE - (I) ***Base Catalyzed Halogenation of the Methyl group attached to the Carbonyl group***

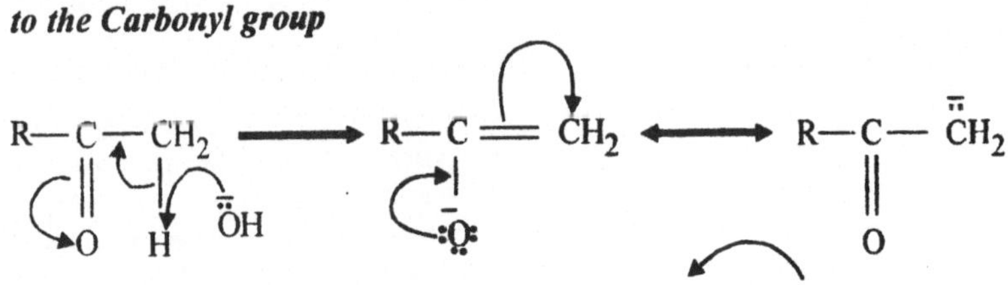

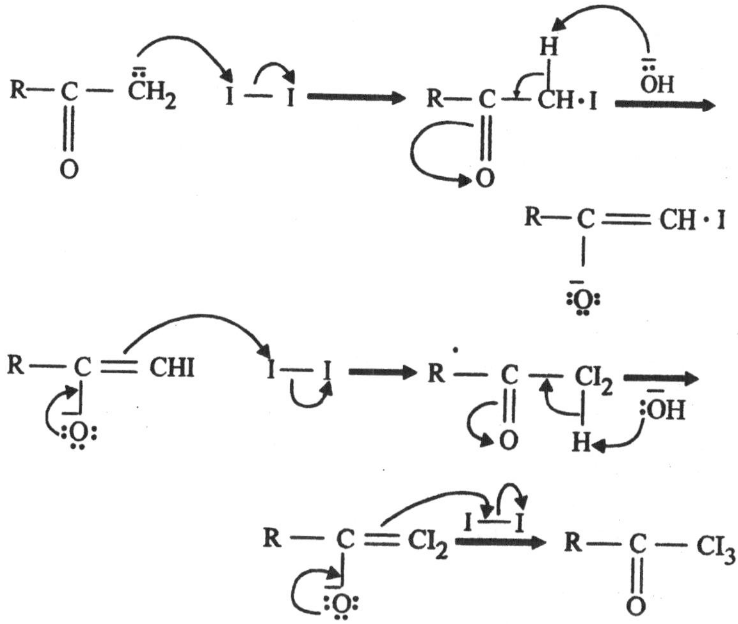

Stage - (II) ***Cleavage of Trihaloketone***

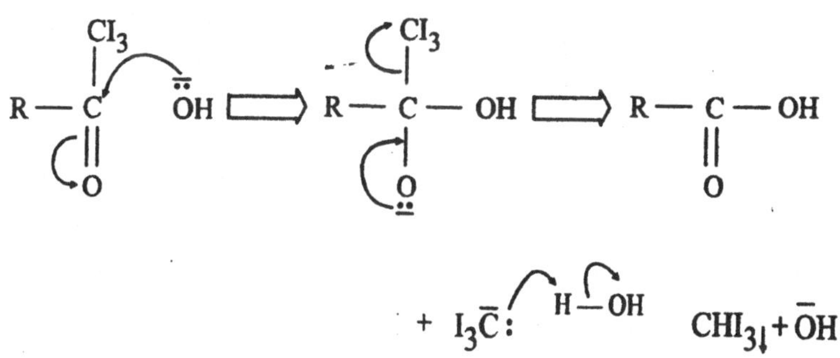

Note :

(I) The haloform reaction reaction is characteristic of methyl ketones and all other compounds containing CH_3CO·group.

(II) This reaction is also particular to secondary methyl carbinols Me·CHOH · R and ethanol for they are readily oxidized to methyl ketones and acetaldehyde respectively

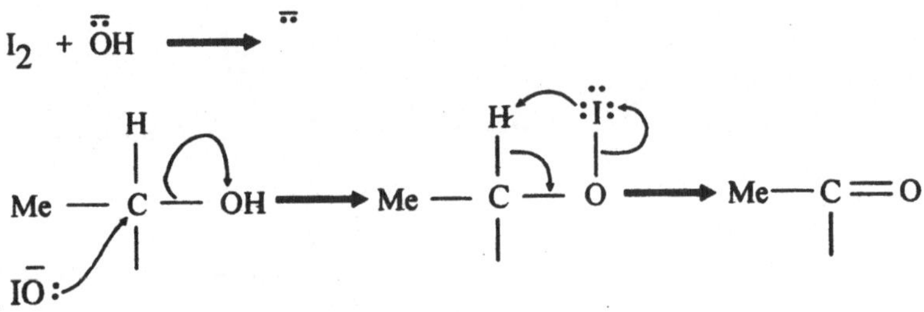

(III) The halogen may be Cl_2, Br_2 or I_2.
(IV) With iodine, sparingly soluble iodoform—a yellow solid—precipitates and the reaction is called IODOFORM

REACTION

With chlorine and bromine, the reaction results in chloroform (a volatile liquid) and bromoform (liquid, m.p. 149.5°C).

SIDE REACTION

α-Halogenation & cleavege :

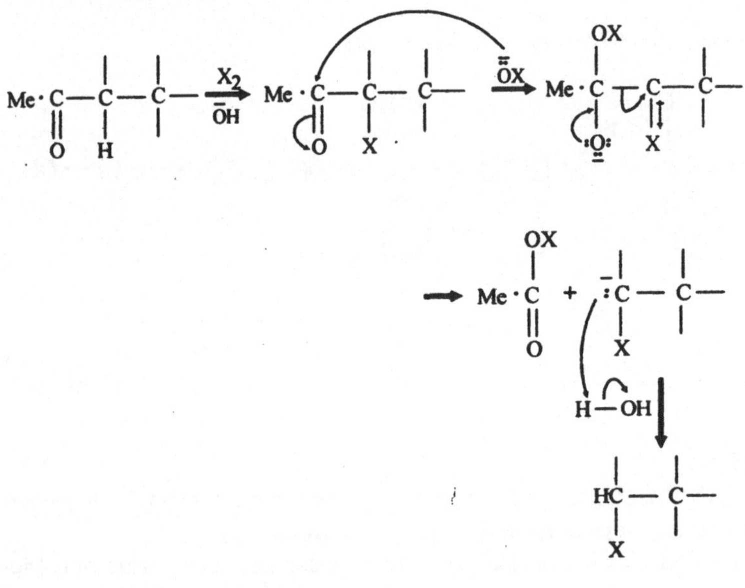

APPLICATION - Synthesis of Carboxylic acids from emthyl ketones

$$\text{Ø—Ø} \xrightarrow[AlCl_3\,/\,\Delta]{CH_3COCl} \text{Ø—Ø—}CO\cdot CH_3 \xrightarrow[\bar{O}H]{Cl_2} \text{Ø—Ø}\cdot CO_2H$$

Biphenyl

p- Biphenylcarboxylic acid

PROBLEMS

Problem 1.

All the three hydrogens are replaced by X_2/OH of carbonyl compound having a-CH_3. When the same compound is treated with X_2 in presence of acidic medium only one a-hydrogen is replaced. Why?

Solution : Let the medium be acidic

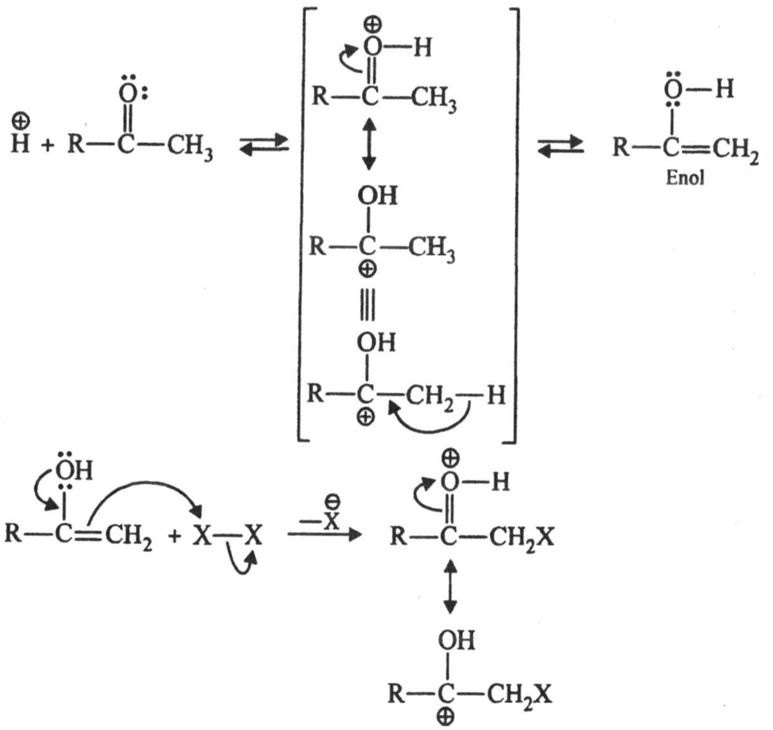

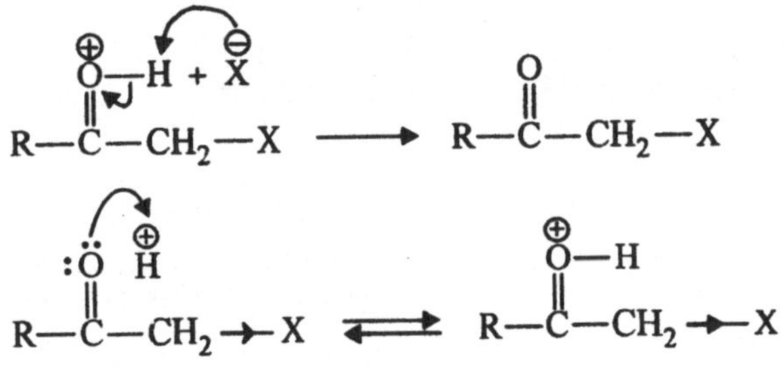

Presence of X (–I group) destabilises the protonation of carbonyl group and makes it difficutl to form and hence reaction does not take further.

Problem 2.

Explain with mechanism

$$\text{Me}\cdot\text{CO}\cdot\text{Et} \xrightarrow[\text{NaOH}]{\text{Cl}_2} \text{Et}-\text{Cl}$$

Solution

Look for the side reaction of the HALFORM reaction.

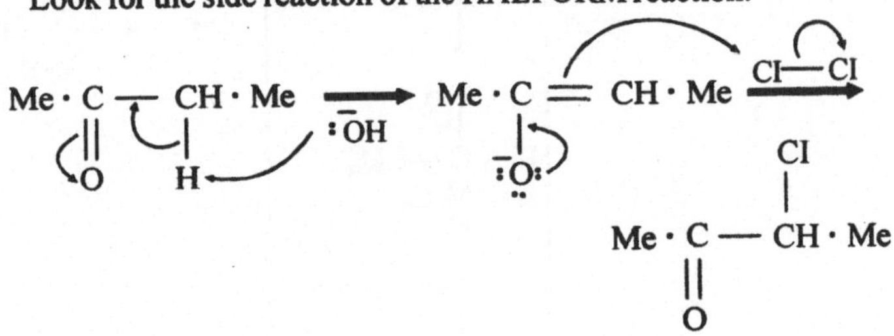

$$\text{Cl}_2 + \overset{-}{\text{O}}\text{H} \longrightarrow \overset{-}{\text{O}}\text{Cl} + \text{HCl}$$

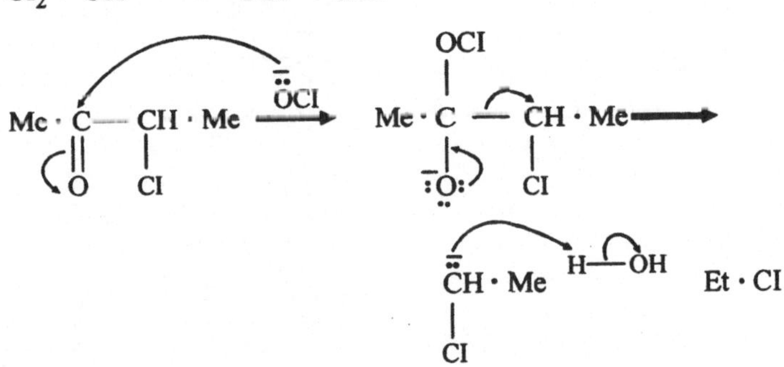

Problem 3.

$$\mathrm{A} \xrightarrow[\bar{O}H]{Br_2} \mathrm{B} \xrightarrow[\Delta]{} \textbf{Glutaric acid}$$

Identify **A & B**

Answer A ▶ Hexane-2,5-dione; **B ▶** lutaric acid

Problem 4.

Carry out following transformation Acetophenone ⟶ Diaryl

Solution

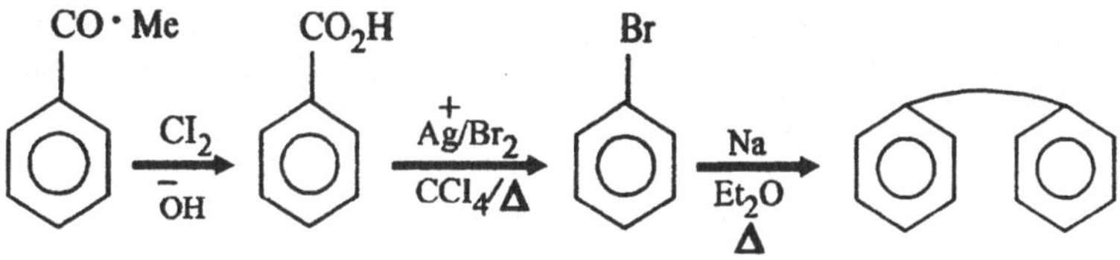

Problem 5.

$$\mathrm{A} \xrightarrow{HBr} \mathrm{B} \xrightarrow{KOH} \underset{\displaystyle OH}{\text{Ø·}\overset{}{CH}\cdot CH_3} \xrightarrow[{[O]}]{HNO_3}$$

$$\mathrm{C} \xrightarrow[\bar{O}H]{I_2} \underset{\displaystyle +\ CHI_3}{\mathrm{D}} \xrightarrow[CCl_4/\Delta]{\overset{+}{Ag}/Br_2} \mathrm{E} \xrightarrow[\Delta]{Na} \mathrm{F}$$

Answer

A ▶ Styrene Ø·CH═CH$_2$;

B ▶ 1 - Phenyl ethylbromide Ø·CH(Br)·CH$_3$

C ▶ Acetophenone Ø·CO·CH$_3$;

D ▶ Benzoic acid Ø·CO$_2$H

E ▶ Phenyl bromide Ø·Br

F ▶ Biphenyl Ø—Ø

52

HALLER-BAUER REACTION

The conversion of a phenacyl compound to a tri-substituted acetic acid is known as Haller-Bauer reaction. The intermediate of the reaction is the amide of tri-substituted acetic acid.

The reaction takes place as follows :

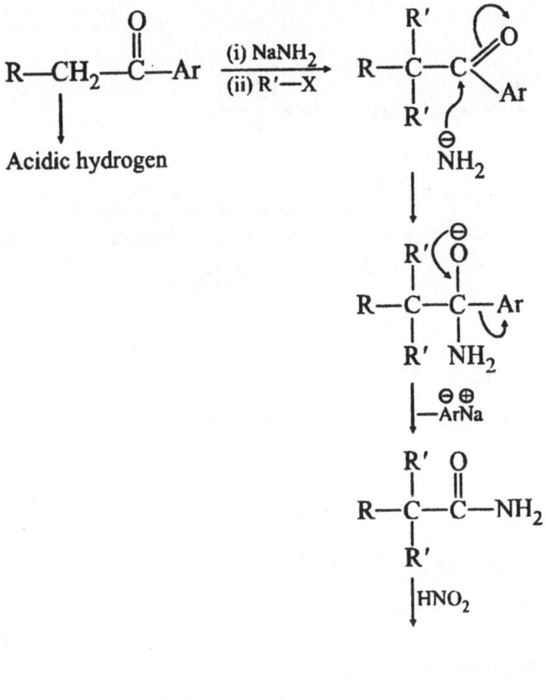

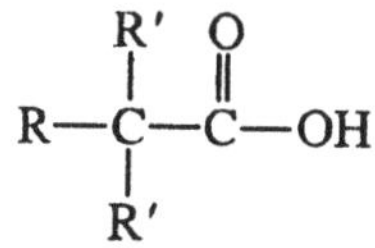

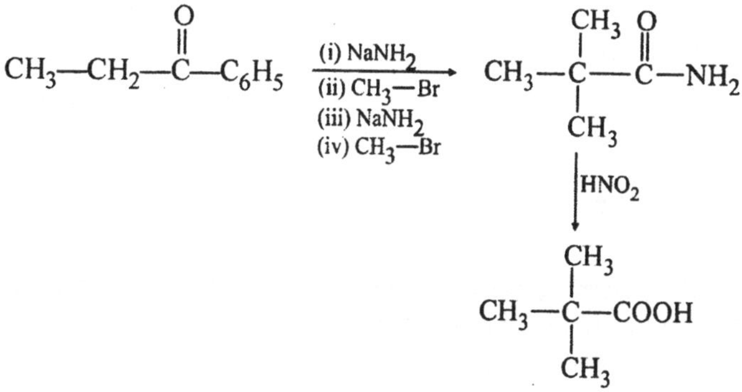

53

HYDROBORATION

Addition reaction of alkenes and alkynes with bronohydride is known as hydroboration. Hydroboration is carried out by diborane in THF. When diborane dissolves in THF, it dissociates into 2 molecules of BH_3.

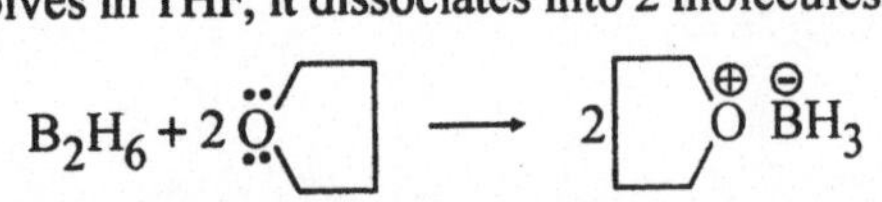

When alkene is treated with BH_3 : THF, one mole of boronhydride reacts with three moles of alkene to give trialkyl borane.

$$\overset{+\delta-\delta}{R—CH{=}CH_2 + BH_3} \longrightarrow R—CH_2—CH_2—BH_2 \xrightarrow{R—CH=CH_2}$$

$$\begin{array}{l} R—CH_2—CH_2—B—H \\ \qquad\qquad\qquad\quad | \\ \qquad\qquad\qquad CH_2—CH_2—R \end{array} \xrightarrow{R—CH=CH_2} \underset{\text{Trialkyl borane}}{(R—CH_2—CH_2)_3B}$$

In this addition reaction, boron atom of BH_3 gets attached to less hindred olifinic carbon. This reaction is regioselective.

MECHANISM

Alkene is nucleophile and BH_3 is electrophile, and hence the first step of the reaction is formation of π complex.

Step I. $R—CH{=}CH_2 + BH_3 \rightleftharpoons \underset{\pi\ \text{complex}}{\underset{BH_3}{R—CH \mp CH_2}}$

This π complex then gives cyclic four-centre transition state.

Step II.

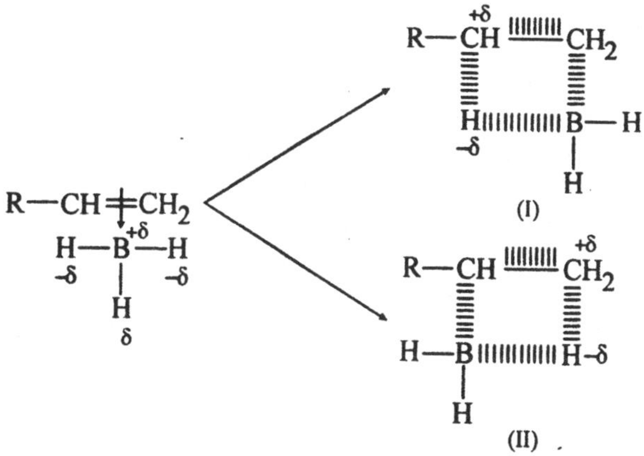

Cyclic four centre transition state has some carbocation character, hence T.S. (I) will be more stable than T.S. (II).

Formation of four center tranisition state suggests that addition of boron atom of BH_3 is on less hindred or less substituted olifinic carbon.

Step III. This four center transtion state then converts into addition product.

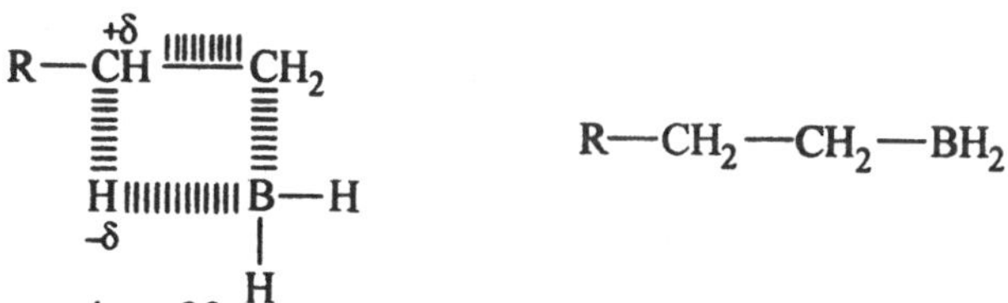

Formation of four centre TS also confirms that addition is syn.

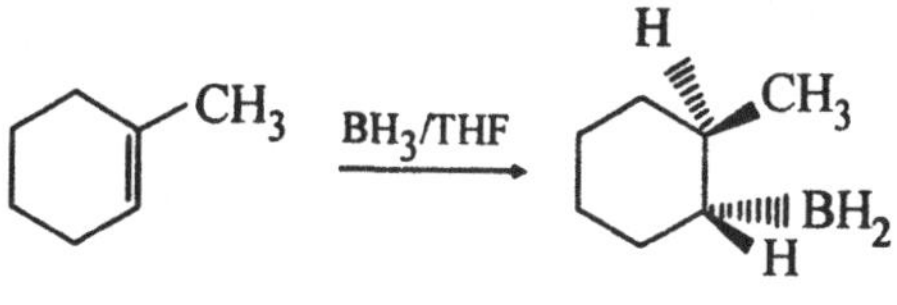

cis-addition

Trialkyl borane is a useful reagent for the preparation of the following compounds.

(A) Preparation of Alcohol : Trialkyl borane on oxidation with hydrogen peroxide in basic medium gives alcohol.

$$(R—CH_2—CH_2)_3B \xrightarrow{H_2O_2/\overset{\ominus}{O}H} 3R—CH_2—CH_2—OH$$

$$R—CH=CH_2 \xrightarrow[(ii)H_2O_2/OH^{\ominus}]{(i)BH_3/THF} R—CH_2—CH_2OH$$

Because hydroboration reactions are regioselective, the net result of hydroboration-oxidation is an apparent **anti-Markovnikov's addition of water.**

$$(CH_3)_2C=CH_2 \xrightarrow[(ii)H_2O_2/OH^{\ominus}]{(i)BH_3/THF} CH_3—\underset{\underset{CH_3}{|}}{CH}—CH_2OH$$

90% yield

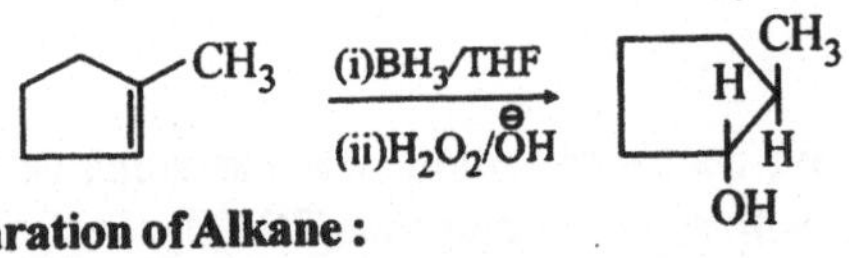

(B) Preparation of Alkane :

(1) Trialkyl borane reacts with organic acid, generally acetic acid, to give alkane corresponding to alkene.

$$(R—CH_2—CH_2)_3B \xrightarrow{CH_3COOH} 3R—CH_2—CH_3$$

$$CH_3—CH=CH_2 \xrightarrow[(ii)CH_3COOH]{(i)BH_3} CH_3—CH_2—CH_3$$

(2) Trialkyl borane undergoes coupling reaction to give higher alkane in presence of $AgNO_3/NaOH$.

$$(R—CH_2—CH_2)_3B \xrightarrow{AgNO_3/NaOH} R—CH_2—CH_2—CH_2—CH_2—R$$

$$\underset{\text{Propene}}{CH_3—CH=CH_2} \xrightarrow[(ii)AgNO_3/NaOH]{(i)BH_3/THF} n\text{-Hexane}$$

(C) Preparation of Amines :

$$(R—CH_2—CH_2)_3B \xrightarrow[(ii)\ HOH\ \Delta]{(i)NH_2Cl} R—CH_2—CH_2—NH_2$$

Addition reaction with alkyne : Alkyne also reacts with BH_3/THF to give addition product.

$$3R—C\equiv CH \xrightarrow{BH_3/THF} \underset{\text{Trivinyl borane}}{(R—CH=CH)_3B}$$

This trivinyl borane further reacts with BH_3 to give trialkyl borane.

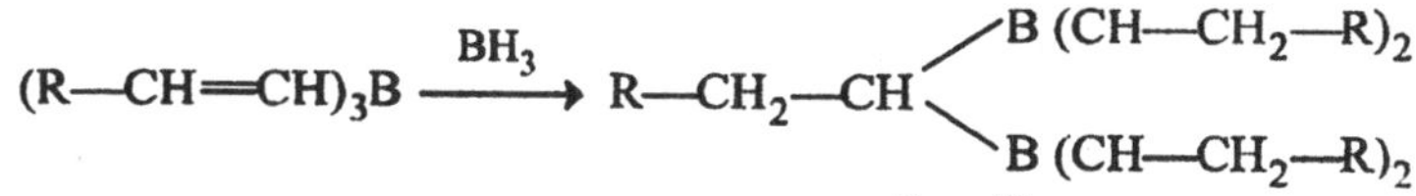

Gem diborane

A sterically hindred dialkylborane, disecisoamylborane,

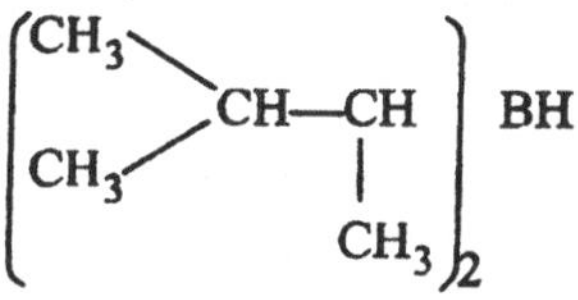

(abbreviated Sia_2BH) react with alkyne to give only vinyl drivative.

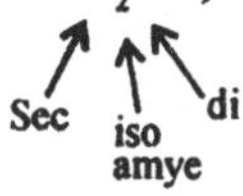

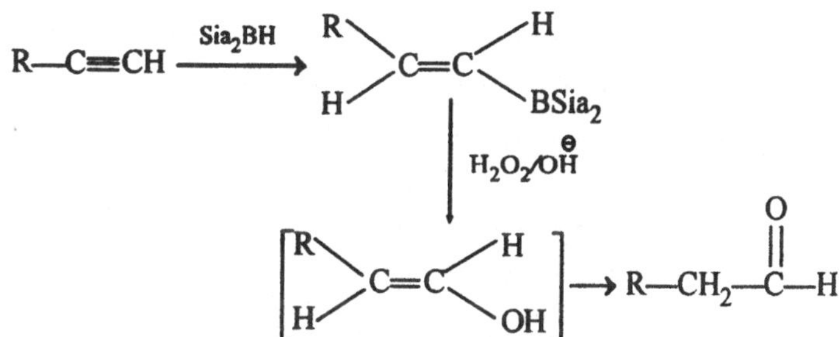

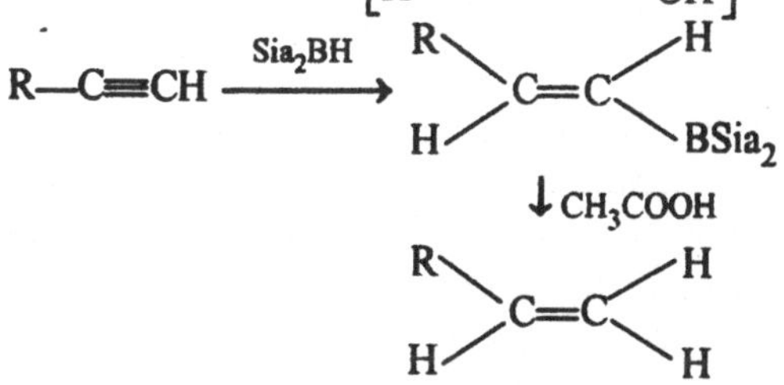

Alkene

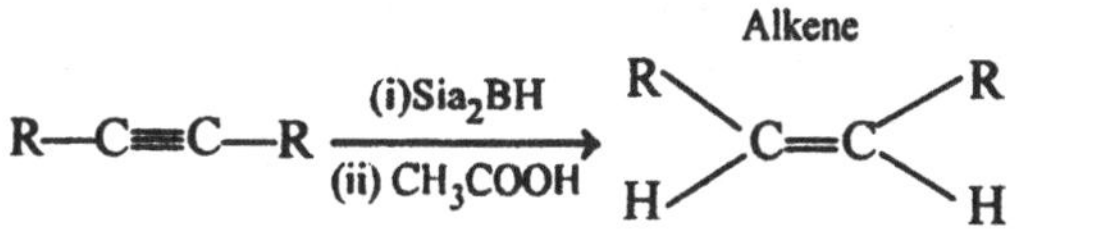

cis-alkene

54

HARRIES OZONOLYSIS REACTION

Ozonolysis reaction is given by alkenes and alkynes.

This reaction is a two step reaction.

First step is 1, 3-dipolar addition of ozone across C, C π bond to give adduct known as molozonide. This molozonide undergoes rearrangement to give ozonide.

Second step is hydrolysis of ozonide, either in the presence of reducing agents (Zn or CH_3—S—CH_3) or in the absence of reducing agents.

OZONOLYSIS IS OF TWO TYPES :

(1) **Reductive ozonolysis** : Addtion of ozone followed by hydrolysis in the presence of reducing agent is known as reductive ozonolysis. This ozonolysis is mainly carried out in the case of alkenes. Reductive ozonolysis of alkenes is simply known as ozonolysis.

(2) **Oxidative ozonolysis** : Additon of ozone followed by hydrolysis in absence of reducing agent is known as oxidative ozonolysis. This ozonolysis is mainly carried out in case of alkynes and ozonolysis is simply known as ozonolysis of alkyne.

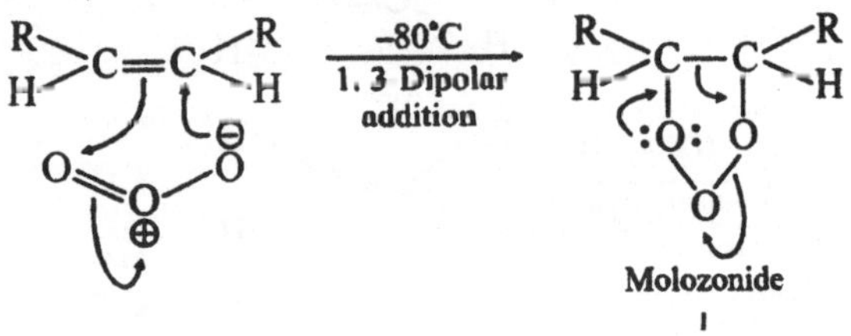

I

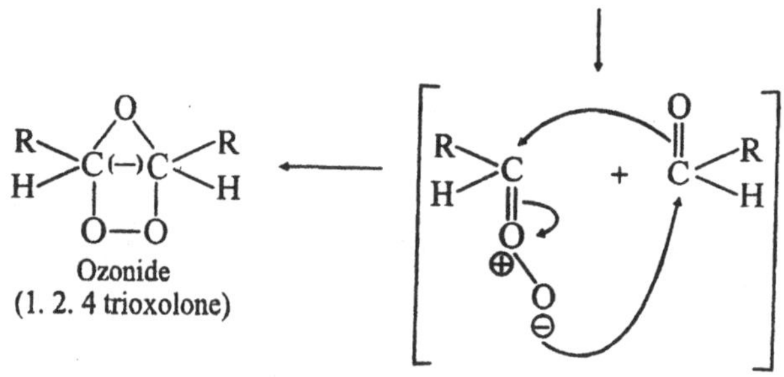

Ozonolysis of Alkene :

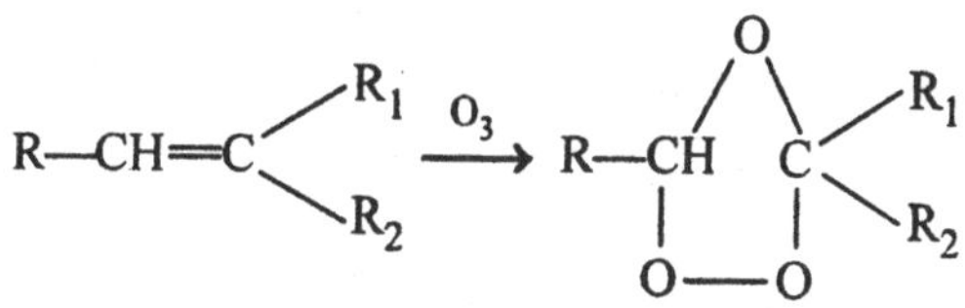

Reductive work up :

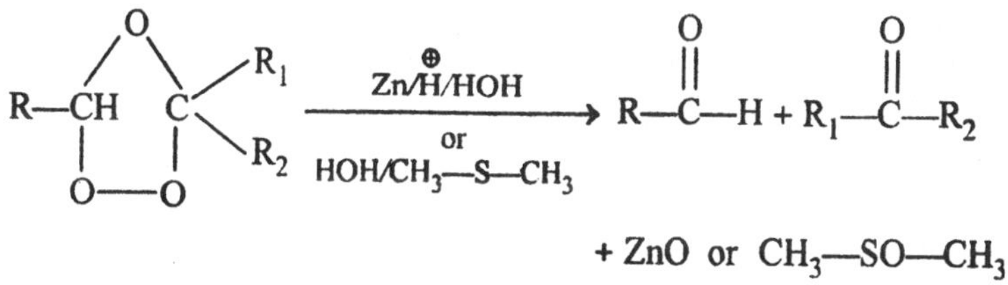

In reductive ozonolysis, breaking of C=C bond takes place and monosubstituted (or unsubstituted) olifinic carbon yields aldehyde and disubstituted olifinic carbon ketone.

Oxidative work up :

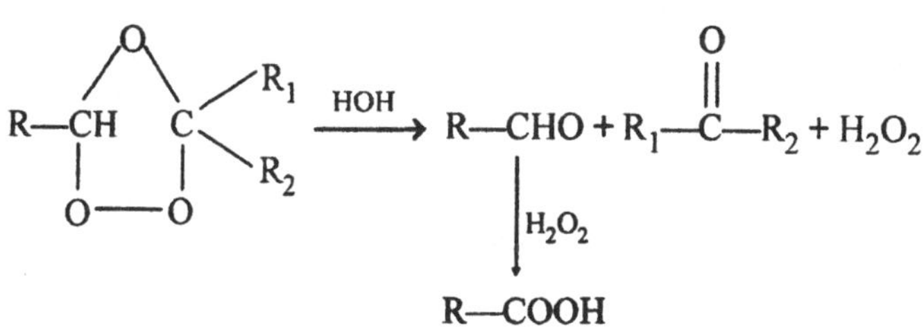

In oxidative ozonolysis, breaking of C=C double bond takes place and monosubstituted olifinic carbon yields carboxylic acid and disubstituted olifinic carbon yields a ketone.

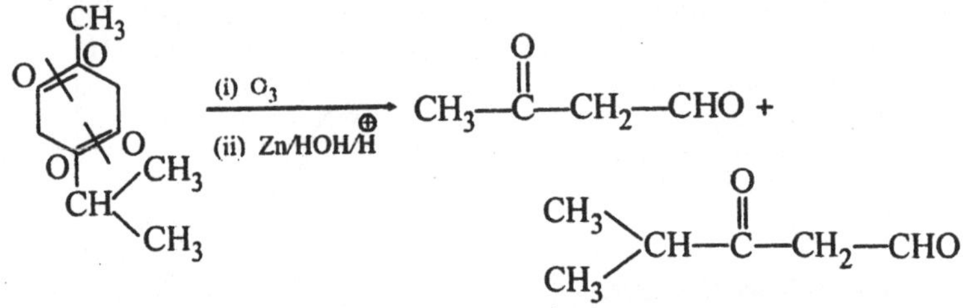

Ozonolysis of Alkyne :

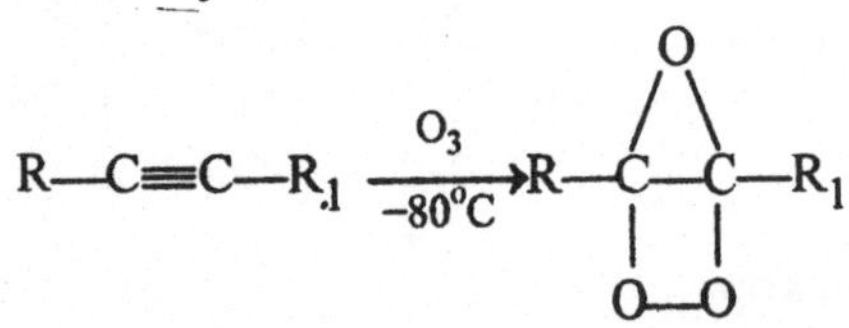

Ozonide

Reductive work up :

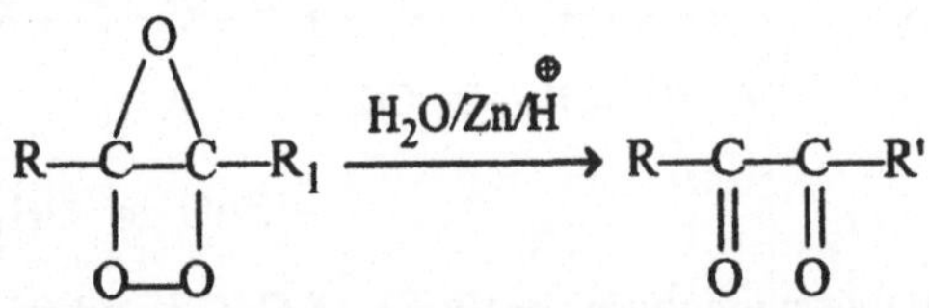

In this case triple bond converts into single bond and both the acetylenic carbons convert into carbonyl group and product is vic dicarbonyl compound.

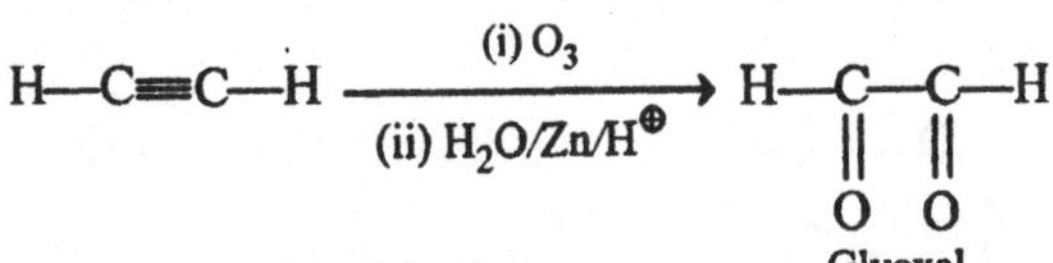

Glyoxal

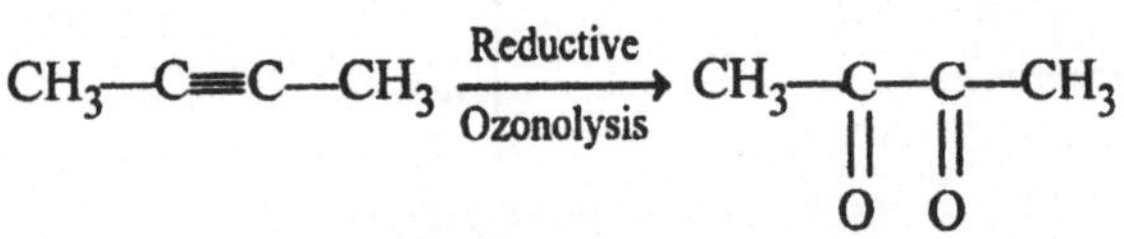

Dimethyl glyoxal

Oxidative ozonolysis or ozonolysis of alkyne

$$R{-}C{\equiv}C{-}R_1 \xrightarrow[\substack{\text{(ii) } H_2O/H^{\oplus} \\ \text{or} \\ CH_3COOH}]{\text{(i) } O_3} R{-}COOH + R_1{-}COOH$$

In this case, breaking of C, C triple bond takes place and acetylenic carbons convert into carboxylic group.

$$H{-}C{\equiv}C{-}H \xrightarrow[\text{(ii) AcOH}]{\text{(i) } O_3} HCOOH + HCOOH$$

$$CH_3{-}C{\equiv}C{-}H \xrightarrow[\text{(ii) AcOH}]{\text{(i) } O_3} CH_3{-}COOH + HCOOH$$

55

HELL-VOLHARD-ZELINSKY REACTION

Carboxylic acids having at least one α-hydrogen react with X_2 (Cl_2 or Br_2) in presence of phosphorus (or phosphorus halide) to give α-halo acids. This reaction is known as Hell-Volhard Zelinsky reaction (HVZ reaction)

$$R{-}CH_2{-}COOH \xrightarrow[\text{(ii) HOH}]{\text{(i) } Br_2/P} R{-}\underset{\alpha\text{-Halo acid}}{\overset{\overset{Br}{|}}{CH}{-}COOH}$$

If more than one molar equivalent of X_2 is used in the reaction, the products obtained are α, α-dihalo acids or α, α, α-trihalo acids.

MECHANISM:

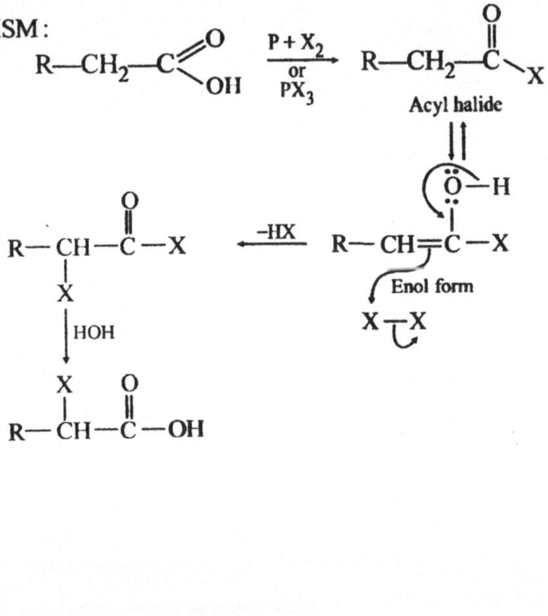

The HVZ reaction is limited to the formation of α-chloro and α-bromo acids and is, in truth, sometimes awkward to carry out. The reagents X_2 and phosphorous are noxious and the reaction time is often long and the reaction conditions are harsh.

Methods have evolved to extend and replace the classic HVZ process. In this method, acid chlorides are treated with the appropriate *N*-halosuccinamide and a trace of HX to produce α-chloro and α-bromo acyl chlorides.

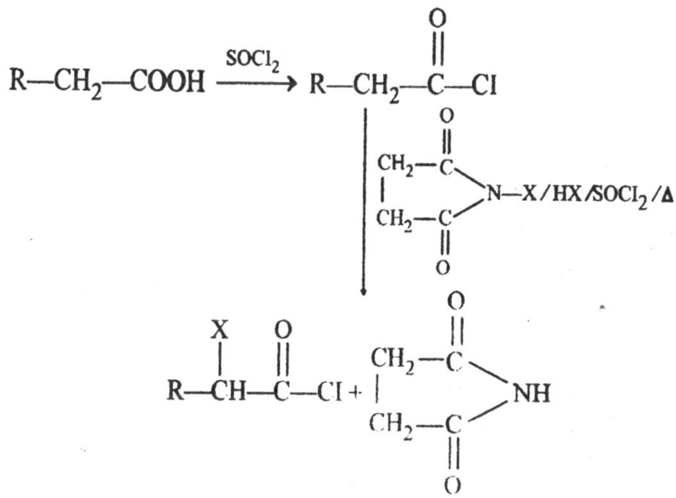

α-iodo acyl chlorides can be obtained by using molecular Iodine.

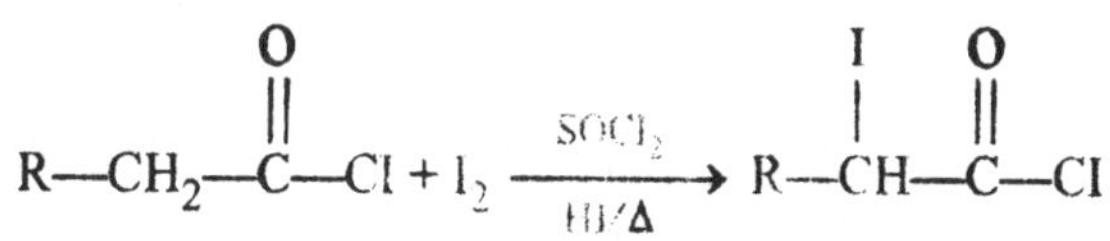

α-halo acyl chloride on hydrolysis gives α-halo acid.

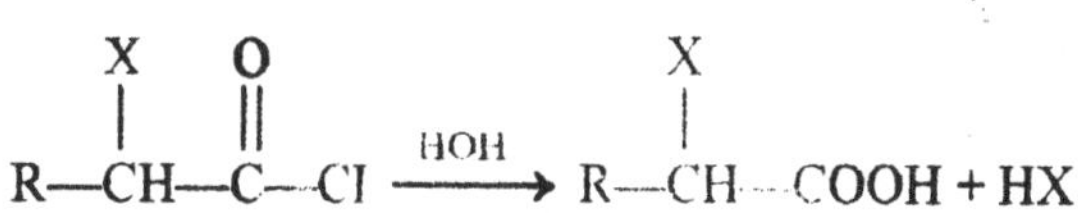

This modification is knwon as **Harpp modification** of HVZ reaction.

α-halo acids are important synthetic intermediates because they are capable of reacting with different nucleophiles.

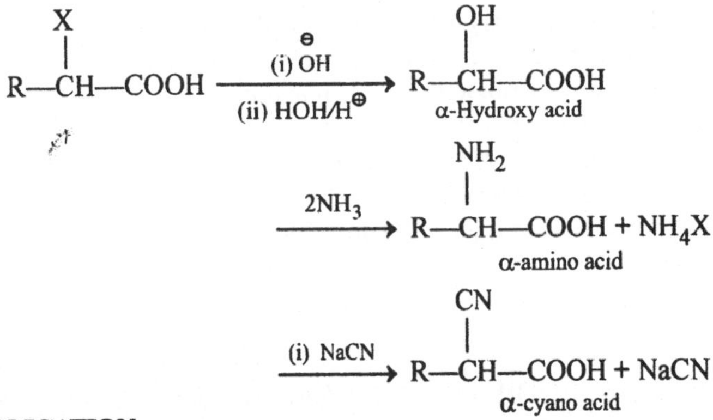

APPLICATION

The reaction can be employed for structure determination and in the synthesis of α-amino acids. The reaction (HVZ) can be used to detect α-hydrogen in acids because the reaction takes place exclusively with acids having α-hydrogen.

The reaction can be used to prepare α-monobromo derivatives of dicarboxylic acids from their half esters.

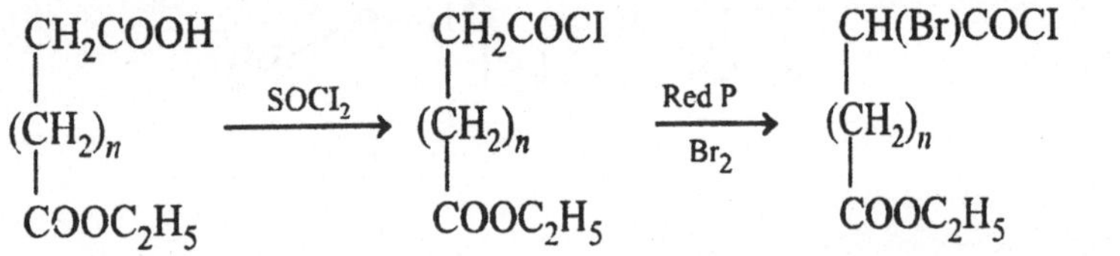

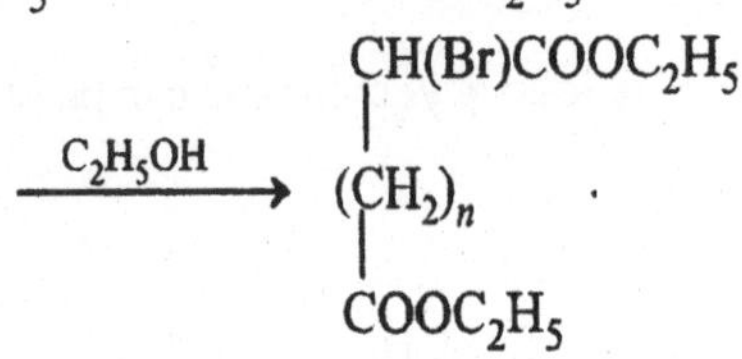

The yield is usually very high.

The α-halogen can be readily replaced by nucleophilic reagents to yield various substituted products

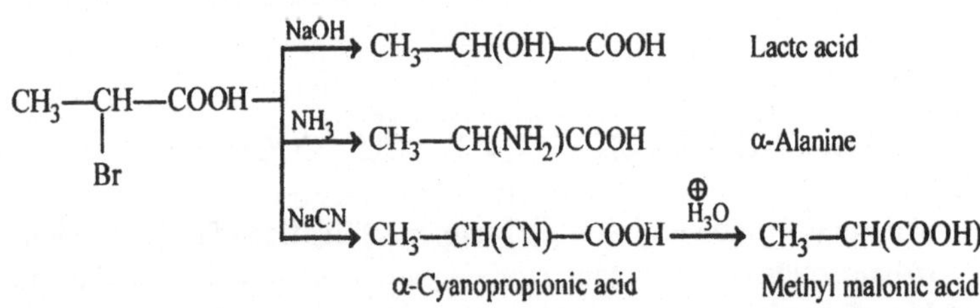

56

HYDROCARBOXYLATION REACTION (Koch-Haaf Reaction)

Alkene gives carboxylic acid when treated with carbonmonoxide at high temperature (300°atm) in presence of conc. H_2SO_4 followed by reaction with water. This reaction is called hydrocarboxylation reaction.

Reaction occurs at room temperature under atmospheric pressure if HCOOH is taken in place of CO and H_2O.

$$CH_3—CH{=}CH_2 + CO + H_2O \xrightarrow[300^{\circ}C/500\ atm.]{\overset{\oplus}{H}} CH_3—\underset{\underset{CH_3}{|}}{CH}—COOH$$

$$CH_3—CH{=}CH + \underset{\underset{CO + H_2O}{\downarrow}}{HCOOH} \xrightarrow{\overset{\oplus}{H}} CH_3—\underset{CH_3}{CH}—COOH$$

MECHANISM :

$$R—CH{=}CH_2 + \overset{\oplus}{H} \longrightarrow R—\overset{\oplus}{C}H—CH_3 \longrightarrow$$

$$\overset{-\delta}{C}{\equiv}\overset{+\delta}{O}$$

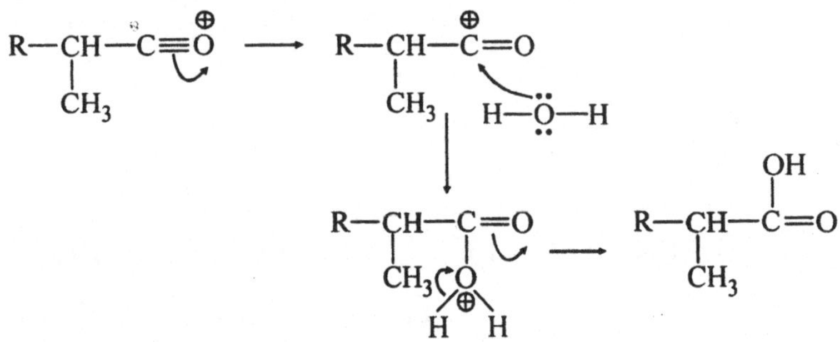

In this reaction R—OH, R—S—H and amines (*P* or Sec) can be used in palce of water.

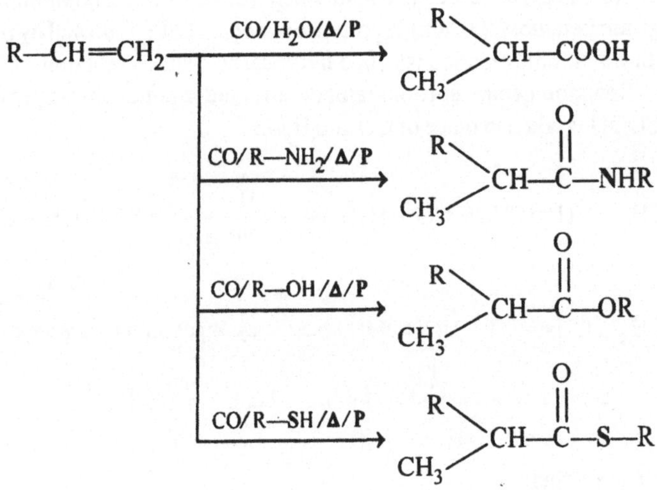

(P = Pressure)

57

HOUBEN-HOESCH REACTION

This reaction was dicovered independently by Hoesch (1927) and Houben (1926). It is a reaction in which acylation of polyhydroxy phenols (especially metasubstituted), phenolic ethers and some heterocylic compounds such as pyrrole is done with nitrile and hydrogen chlōride in the presence of a Lewis acid (*e.g.* zinc chloride, ferric chloride, dry hydrogen chloride, in dry ether). In reality, Houben-Hoesch reaction is simply the Friedel-Crafts acylation with nitriles.

The reaction is carried out by passing dry HCl gas through an equimolecular mixture of nitrile and phenol in dry ether contraining $ZnCl_2$. The ketimine chloride is formed which on hydrolysis yields the aromatic hydroxy ketone. For example, resorcinol yields 2, 4-dihydroxyacetophenone.

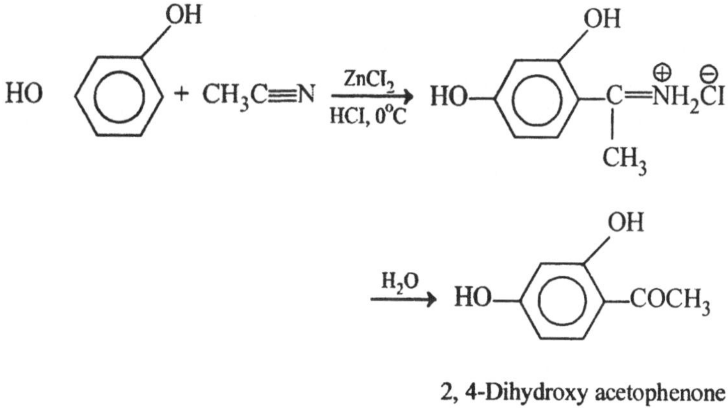

2, 4-Dihydroxy acetophenone

Another example is that phloroglucinol condenses with methyl cyanide to form phloroacetophenone (m.pt. 492 K).

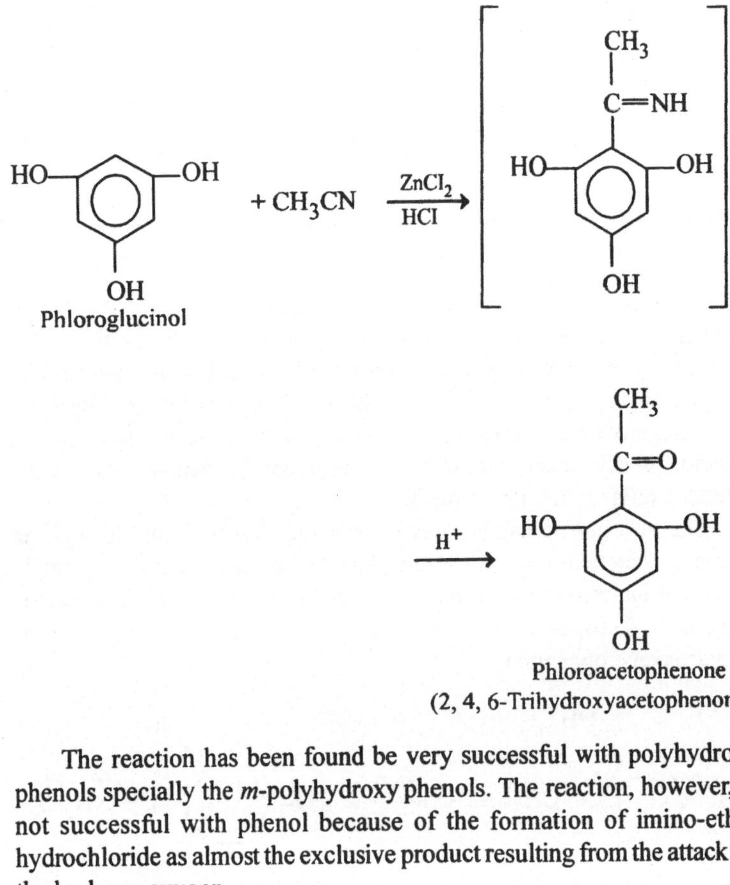

Phloroglucinol

Phloroacetophenone (2, 4, 6-Trihydroxyacetophenone)

The reaction has been found be very successful with polyhydroxy phenols specially the *m*-polyhydroxy phenols. The reaction, however, is not successful with phenol because of the formation of imino-ether hydrochloride as almost the exclusive product resulting from the attack on the hydroxy oxygen.

$$C_6H_5OH + RCN \xrightarrow[HCl]{ZnCl_2} C_6H_5O{-}\underset{}{\overset{R}{\overset{|}{C}}}{=}\overset{\oplus}{N}H_2\overset{\ominus}{Cl}$$

Hydrolysis yields phenyl ester.

In addition to a variety of aliphatic nitriles (*e.g.*, acetonitrile, mono- and trichloro-aceto nitriles), aromatic nitriles have also been found to give good yield. Ether, chloroform,ethyl acetate, etc., may be used as solvent.

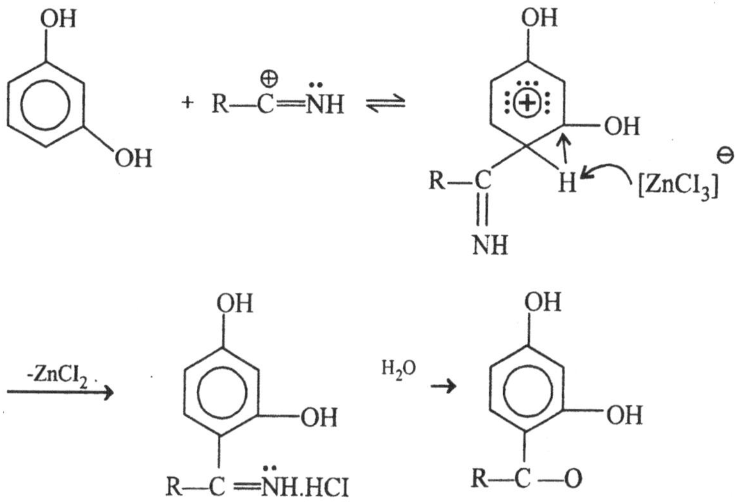

(II)

If hydrogen cyanide is used in place of nitrile, aromatic aldehyde may be obtained. This modification is known as Gattermann synthesis of aldehydes. By this method aldehydes of various aromatic compounds could be prepared.

APPLICATION

The reaction is generally used for the synthesis of many polyhydroxy acetophenones and polyhydroxy benzophenones. This reaction is also used for the synthesis of many natural products of medicinal value.

1. **Synthesis of w-substituted acetophenones :** With benzyl cyanide, w-phenylacetophenones are formed

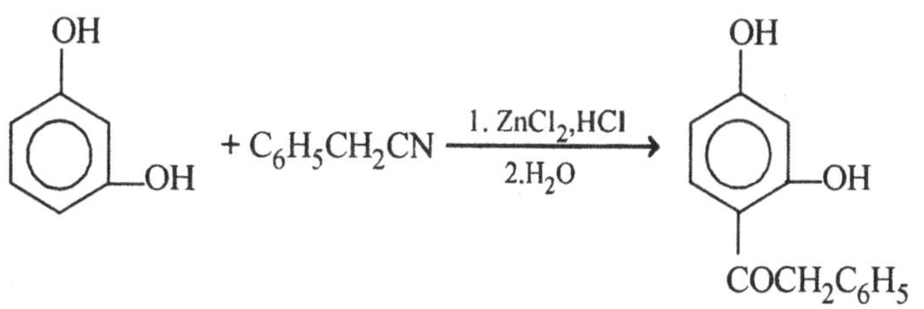

2, 4-Dihydroxy-ω-phenylacetophenone

Similarly, with chloromethyl cyanide ($ClCH_2CN$), 2, 4-dihydroxy-ω-chloroacetophenone is formed

2. **Synthesis of heterocyclic ketones :** A few pyrrol ketones are also prepared.

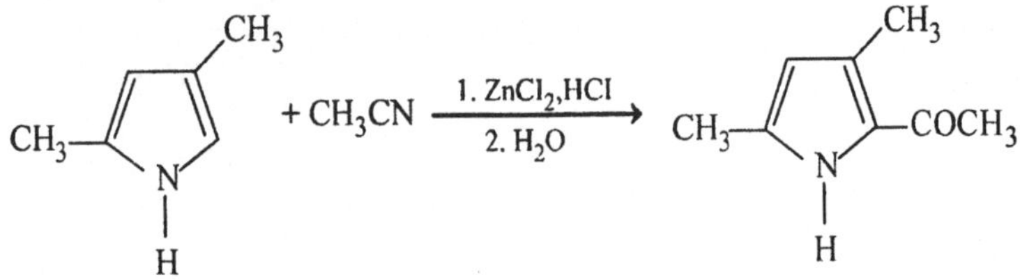

2, 4-Dimethyl-5 acetyl pyrrole

3. **Synthesis of coumarins :** Oxocoumarin is formed from malononitrile and resorcinol.

58

HOFMANN REARRANGEMENT

Apart from its outstanding importance in the synthesis of aliphatic (primary) and aromatic amines, **Hofman's Rearrangement** finds its extensive application in the synthesis of carbamates and ureas.

REACTION

The reaction takes place between an acid and hypohalite ion, predominantly hypobromite ion generated by bromine in presence of alkali

$$Br_2 \xrightarrow{NaOH} \overset{-}{\ddot{O}}Br$$

$$R \cdot CO \cdot NH_2 \xrightarrow{\overset{-}{\ddot{O}}Br} \underset{\text{alkyl isocyanate}}{RNCO} \xrightarrow{H_2O} R \cdot NH_2$$

The alkyl isocyanate produced due to rearrangement of acid amide on treatment with hypobromite ion is the main product which hydrolyzes under the reaction condition to produce amine.

IMPROVED TECHNIQUE

The improved version of this method being : Addition of bromine rapidly to a methanolic solution of the amide containing sodium methoxide

$$R \cdot CO \cdot NH_2 + 2Me\overset{-}{\ddot{O}}\overset{+}{Na} + Br_2 \longrightarrow \underset{[A]}{R \cdot NH \cdot CO_2Me} + 2NaBi + MeOH$$

and hydrolyzing the product [A] i.e. N-alkyl-urethan in presence of alkali

$$R \cdot NH \cdot CO_2Me \ + \ NaOH \longrightarrow RNH_2 \ + \ Na_2CO_3 \ + \ MeOH$$

MECHANISM

The mechanism of HOFMANN REARRANGEMENT follows via formation of an electron-deficient-nitrogen [**nitrene,** —N] which rearranges to alkyl isocyanate

STEP - [I] ***N-Halogenation***

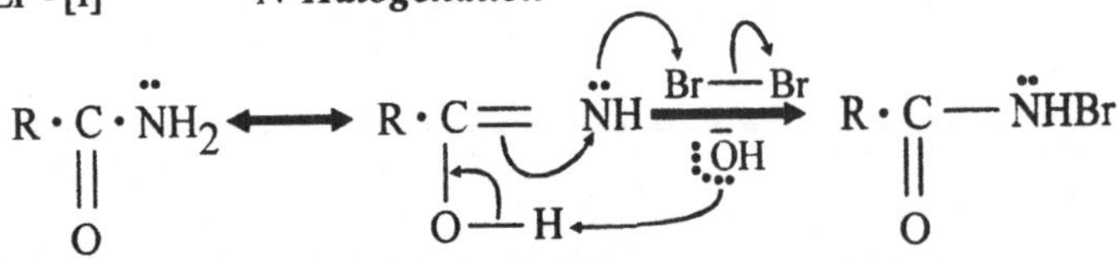

STEP - [II] ***Generation of Nitrene***

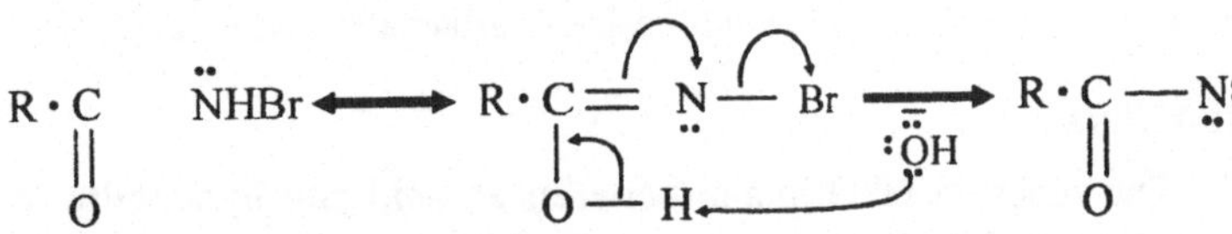

STEP - [III] ***Rearrangement***

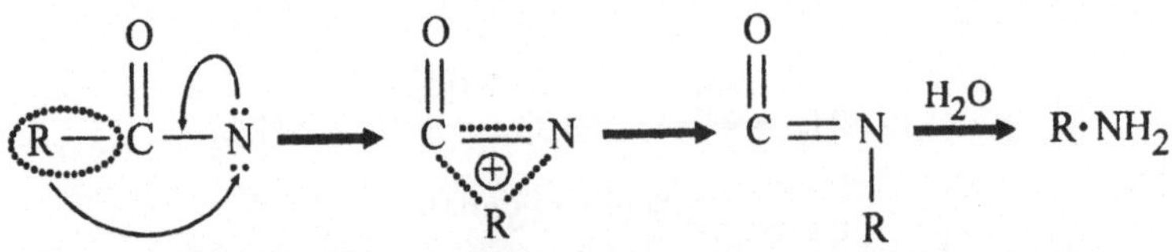

[B]

SUPPORT OF THE MECHANISM

The foregoing mechanism has been supported by a great deal of experimental works :

(1) The first evidence comes from the isolation of the following intermediates :

N-Bromamidues	R·CO·NHBr
Salts of Bromamides	$[R \cdot CO \cdot NBr]^- Na^+$
Alkyl isocyl	R·N=C=O

(2) The formation of bridge intermediate **B** has been supported by the works of Hauser *et. al.* who showed that an electron releasing group attached to the ***para***-position of phenyl nucleus, if it is a migrating one, accelerates the reaction rate while the presence of an electron-withdrawing

group in the same position retards the reaction rate. Had the migrating phenyl group be free, in time, this rate difference would unlikely to result. (3) The last but not the least evidence comes from the works of Wallis *et. al.* who showed that optically active **R**-group retained its configuration even after migration—this means that the R-group is linked to the carbon atom of the carbonyl group and the nitrogen atom through the same point of contact and it is never free.

This conclusively proves that **B** is a bridged intermediate which is also true if the migrating group a phenyl one.

Note :

(I) Although alkyl isocyanate is the main product, efforts to isolate it has been met with failure since it rapidly hydrolyzes to amine under the reaction condition.

(II) The R group may be alkyl or aryl.

(III) The amide containing less than seven carbon atoms yields amine [70 - 80%] as shown above while with more than seven carbon atoms, alkyl cyanide becomes the main product

$$R \cdot CH_2 \cdot CONH_2 \xrightarrow[KOH]{Br_2} RCN$$

(IV) This reaction can be directed to synthesize ureas and carbamates also [See below].

(V) Since amides can be resulted easily from carboxylic acids, this rearrangement offers a convenient method to convert carboxylic acids into ***primary*** amines.

With the amides containing carbon atoms upto seven, the yield of ***primary*** amine ranges from 70 - 90 percent.

SIDE PRODUCTS

(1) Formation of acylurea

$$R \cdot CO \cdot NH_2 + R \cdot NCO \xrightarrow[NaOH]{Br_2} R \cdot CO \cdot NH \cdot CO \cdot NH \cdot R$$

(2) Formation of cyanides

$$R \cdot CH_2NH_2 \xrightarrow{\bar{O}Br} R \cdot CN$$

That the rearrangement is intramolecular has been supported by the following evidence :

1. When T.J. Prosser and E.L. Ellel carried out the Hofmann reaction with a mixture of *m*-deteriobenzamide and benzamide -^{15}N, they found only *m*-deuterioaniline and aniline-^{13}N are formed. This shows that the reaction is

intramolecular. Had the reaction been intermlecular then there would have been crossproduct such as aniline and *m*-deutereo aniline-^{15}N.

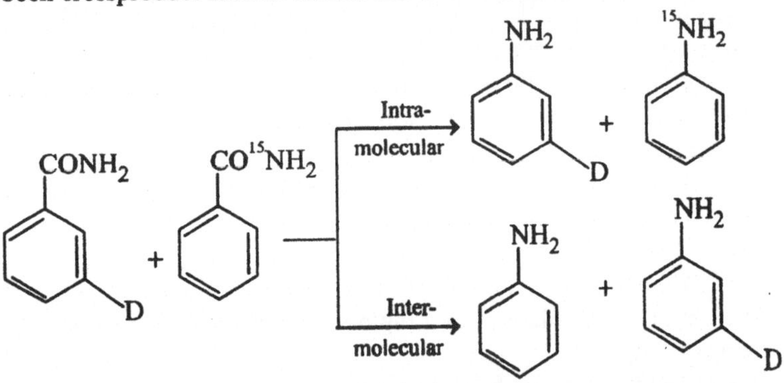

Cross-product not formed

2. When optically active α-phenyl propionamide undergoes Hofmann rearrangement, then α-phenylethylamine having the same configuration is formed thereby revealing that the migrating group is never free and the reaction is an intramolecular rearrangement.

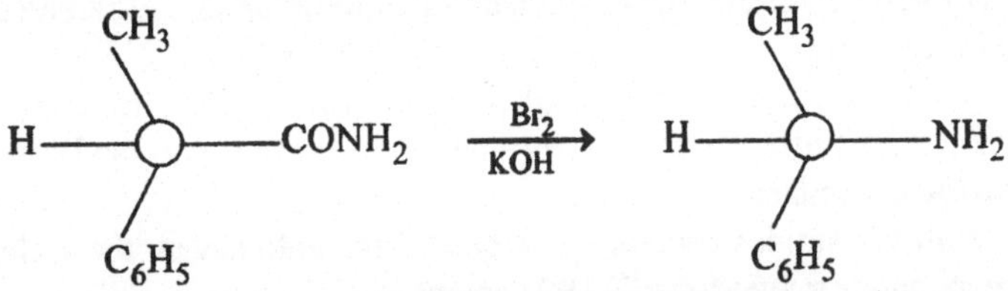

Rate of Hofmann Reaction : In Hofmann's rearrangment N-bromamide rearranges to isocyanate in one step with a loss of bromide ion. This is a slow step and proceeds through the easier S_N2 like path.

Further, in arylamides, *i.e.*, when the migrating group is aryl, then the rate of Hofmann reaction gets increased by the presence of electron releasing substituents in the aromatic ring. For example, in the case of substituted benzamide as :

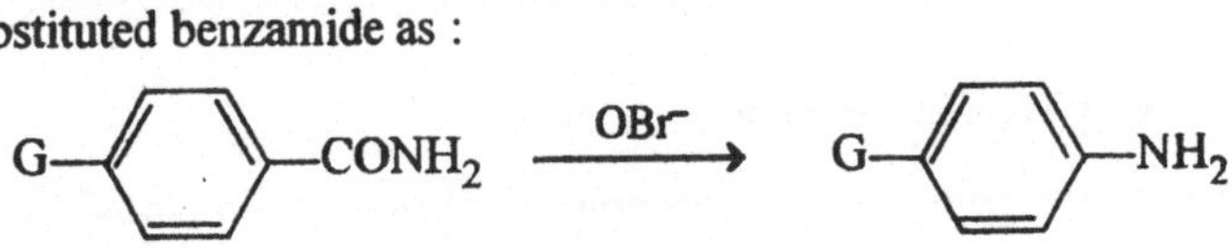

the reactivity of G is of the following order :

$$—OCH_3 > —CH_3 > —H > —Cl > —NO_2$$

APPLICATION S OF HOFMANN REACTION

(*a*) *Primary aliphatic and aromatic amines* : The Hofmann reaction is useful for conversion of carboxylic acids and their derivatives into primaryamines having one carbon atom less than the starting material. The amines so obtained are free from secondary and tertiary amines.

$$\underset{\text{Isobutyric acid}}{(CH_3)_2CHCOOH} \xrightarrow[\text{(ii) } NH_3]{\text{(i) } PCl_5} (CH_3)_2CHCONH_2 \xrightarrow[KOH]{Br_2} \underset{\text{Isopropylamine}}{(CH_3)_2CHNH_2}$$

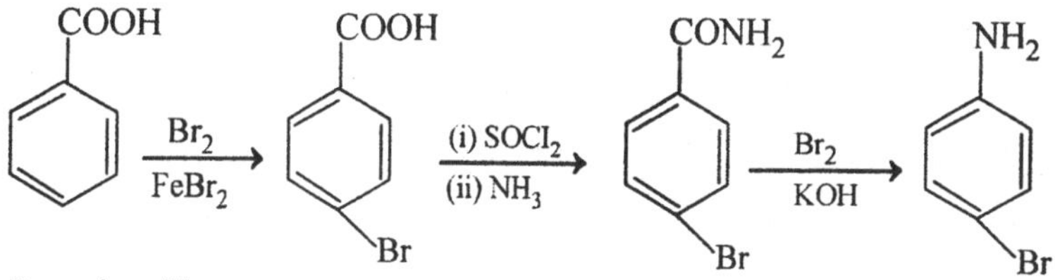

Benzoic acid *m*-Bromoaniline

The following primary amines are perpared from the reaction in the following way :

(*i*) *Preparation of methylamine, aniline, benzylamine, etc.*

$$CH_3CONH_2 \xrightarrow{Br_2/KOH} CH_3NH_2$$
$$C_6H_5CONH_2 \longrightarrow C_6H_5NH_2$$
$$C_6H_5CH_2CONH_2 \longrightarrow C_6H_5CH_2NH_2$$
$$\underset{\text{Urea}}{H_2N.CO.NH_2} \longrightarrow \underset{\text{Hydrzine}}{H_2N.NH_2}$$

(*ii*) *Preparation of* β*-aminopyridine* : β-Aminopyridine is produced from the nicotianmide (available from natural sources), because it can not be produced in good yield *viz* the nitration of pyridine.

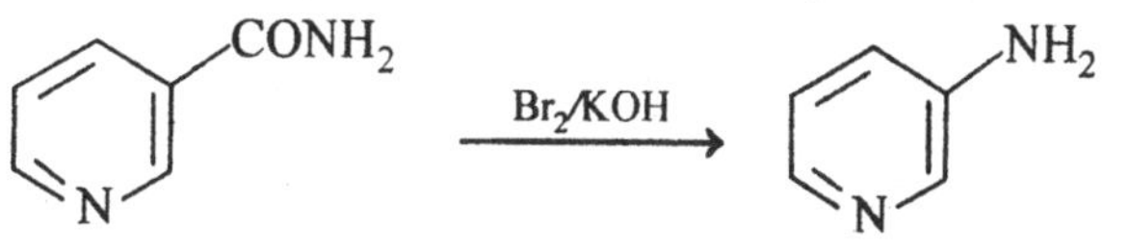

(65-70%)

(*iii*) *Preparation of amino acids* : β-Alanine is prepared (about 45% yield) by treating succinimide with bromine and aqueous caustic potash; reaction takes place through the haloamide of succinic acid.

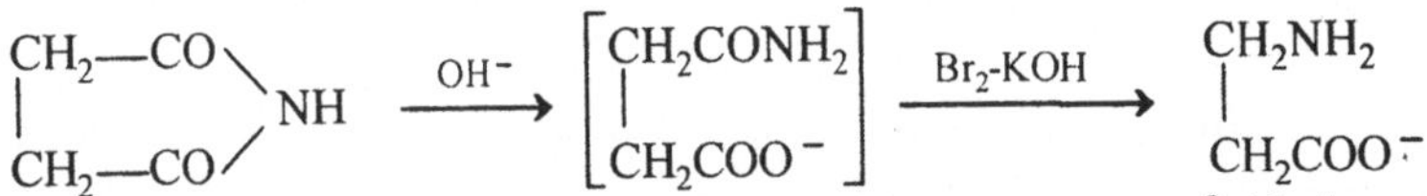

β-Alanine

One of the most important industrial application of this reaction is the synthesis of anthranilic acid which is basic starting material for the preparation of *ortho*-disubstituted benzene derivatives.

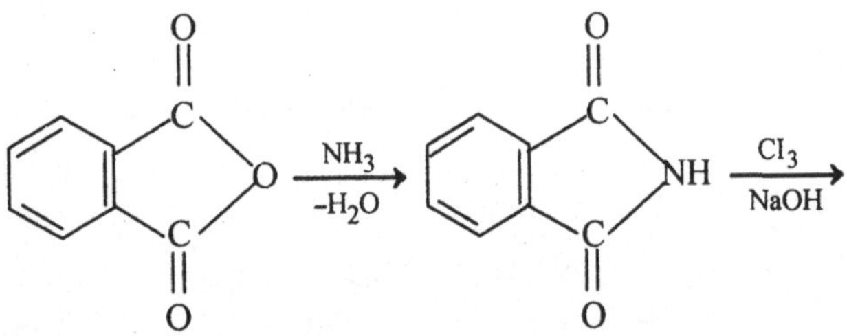

Phthalic anhydride Phthalimide

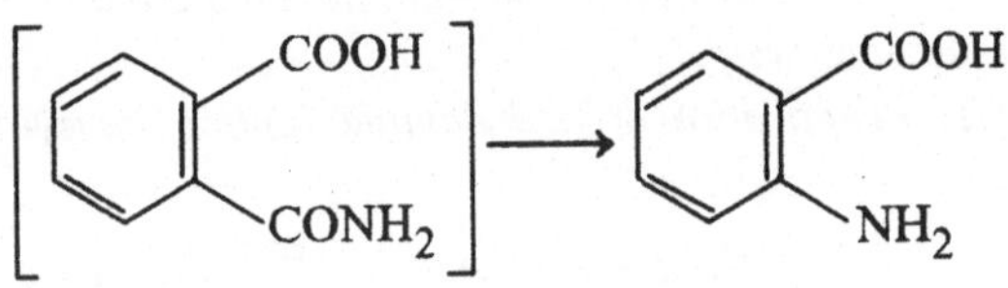

Anthranilic acid

(b) *Preparation of aldehydes* : Hofmann reaction is used to convert α, β-unsaturated acids and α-hydroxy acid amides into aldehydes according to the following steps.

$$\text{R CH=CH.CONH}_2 \xrightarrow{\text{Cl}_2/\text{NaOH}} \text{RCH=CHNHCOCH}_2 \xrightarrow{\text{HCl}} \text{RCHO}$$

$$\underset{\text{Mandelamide}}{\text{C}_6\text{H}_5\text{—}\overset{\text{OH}}{\overset{|}{\text{CH}}}\text{—CONH}_2} \xrightarrow[\text{NaOH}]{\text{Cl}_2} \left[\text{C}_6\text{H}_5\text{—}\overset{\text{OH}}{\overset{|}{\text{CH}}}\text{—NH}_2\right] \longrightarrow \text{C}_6\text{H}_5\text{HO} + \text{NH}_3$$

(c) Synthesis of ***meta*- bromoaniline**

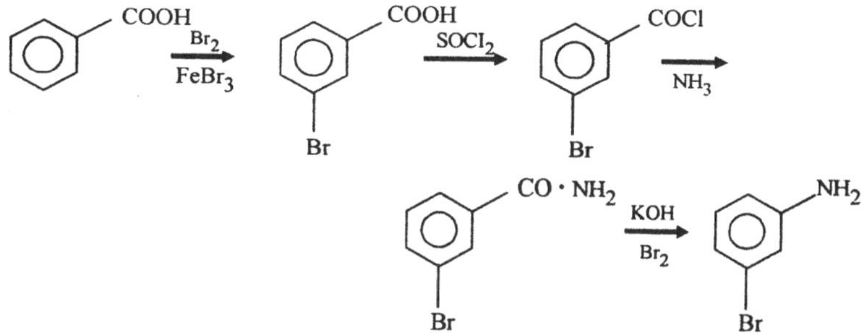

(d) Synthesis of **ureas and carbamates**

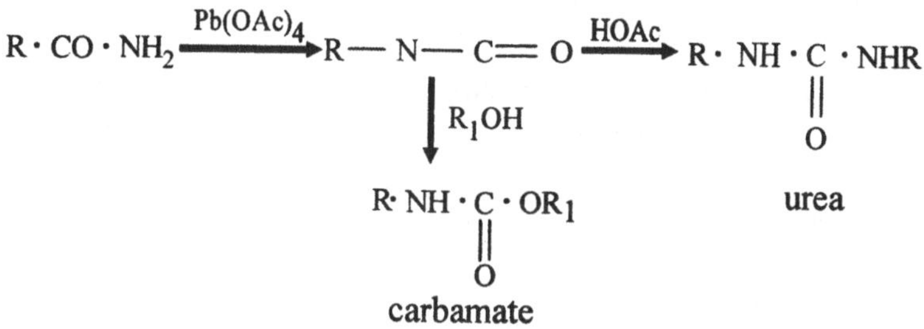

Problem-1

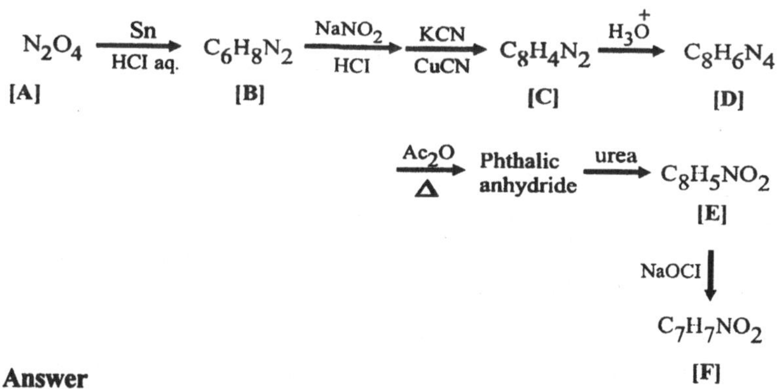

Answer

A▶ ***ortho*-**Dinitrobenzene; **B▶** ***ortho*-**Phenylene diamine
C▶ ***ortho*-**Phenylene dicyanide; **D▶** Phthalic anhydride
E▶ Phthalamide; **F▶** Anthranilic acid

59

HOFMANN MARTIUS REARRANGEMENT

When hydrochlorides or hydrobromides of the N-alkyl substituted anilines or quaternary ammonium halides are heated strongly, (200–300°C), they undergo rearrangement in which the alkyl group migrates preferentially from nitrogen atom to para carbon atom or if this is occupied, then to the ortho carbon atom. This reaction is known as the Hofmann Martius rearrangement (1871).

This rearrangement is best illustrated by taking the following example of phenyltrimethyl ammonium chloride (I).

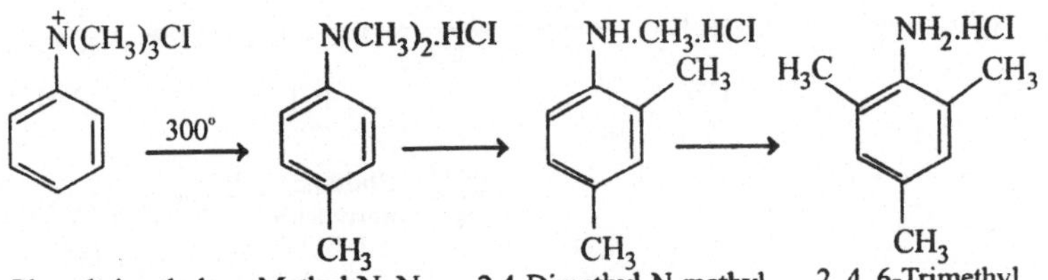

Phenyltrimethyl ammonium chloride; *p*-Methyl-N, N-dimethyl aniline hydrochloride; 2,4-Dimethyl-N-methyl aniline hydrochloride; 2, 4, 6-Trimethyl aniline hydrochloride

MECHANISM

The mechanism of this rearrangement is found to be similar to that of Orton rearrangement. However, Hughes and Ingold in 1956 postulated the following mechanism :

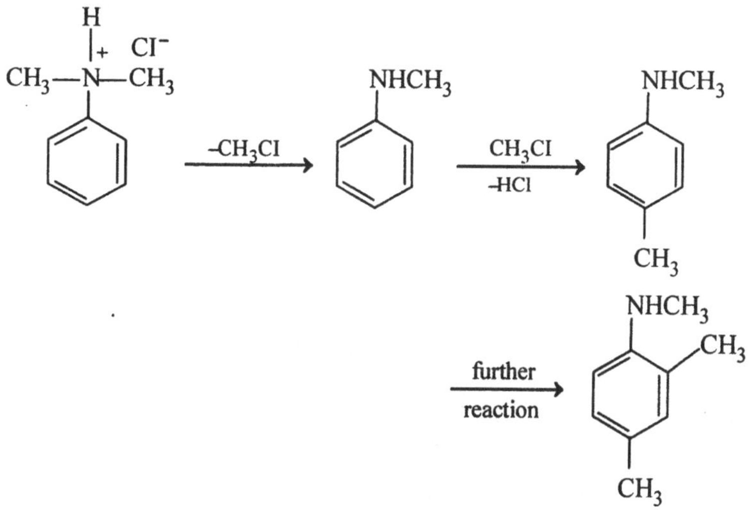

The above mechanism was based on the suggestion given by Hicknbottom (1934) that the rearrangement occurs *via* the formation of an alkyl carbonium ion. This is evidenced by the following facts :

During the rearrangement, there also occurs the formation of alkyl halide, olefin and alkylated aniline with rearranged alkyl group. The formation of these products could only be explained on the basis of formation of alkyl halide as an intermediate which undergoes elimination and rearrangement reaction *via* the carbonium ion. For example, isobutyl anilinium bromide gives rise the formation of isobutyl bromide (IV) as an intermediate which in turn explains the formation of isobutene (III) and *p*-*t*-butyl aniline (II) in addition to normal product, the *p*-isobutylaniline (I) as follows :

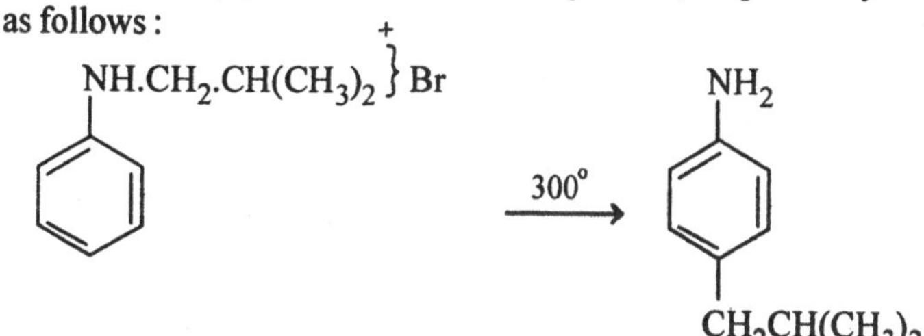

Isobutyl anilinium bromide

p-Isobutyl aniline, (I)

+

$(CH_3)_2C{=}CH_2$ +

Isobutene

(III)

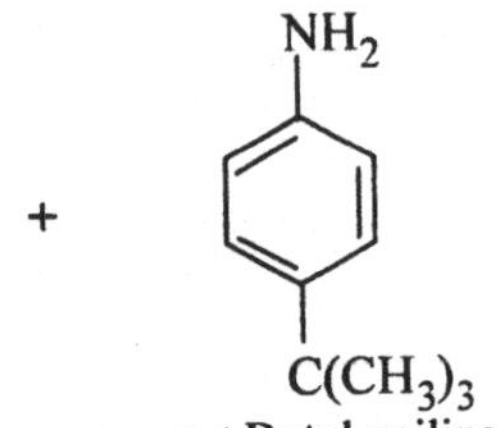

p-t-Butyl aniline, (II)
(having rearranged alkyl group)

$(CH_3)_2CH.CH_2Br$
Isobutyl bromide
(IV)

Formation of (II) and (III) products may be explained as follows :

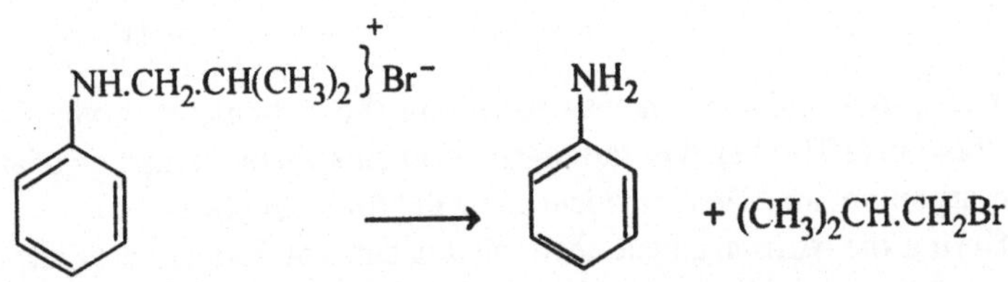

$$\underset{\text{Isobutyl bromide}}{(CH_3)_2CH.CH_2Br} \rightleftharpoons \underset{\text{Isobutyl carbonium ion}}{(CH_3)_2CH.\overset{+}{C}H_2 + Br^-}$$

$$\underset{\text{Isobutene, (III)}}{(CH_3)_2C{=}CH_2} \xleftarrow{-\overset{+}{H}} \underset{\text{Isobutyl caropnium ion}}{(CH_3)_2CH.\overset{+}{C}H_2} \xrightarrow[\text{to}]{\text{rearranges}} \underset{t\text{-Butyl carbonium ion}}{(CH_3)_3\overset{+}{C}}$$

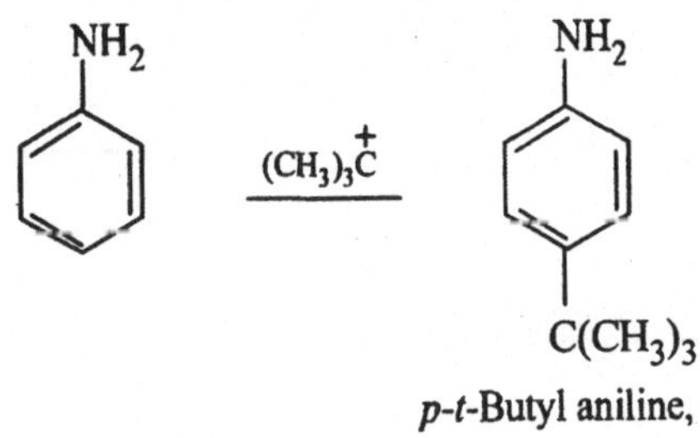

p-t-Butyl aniline, (II)

60

HUNSDIECKER REACTION

Treatment of silver salt of acid with bromine in presence of carbon tetra chloride to give an alkyl or aryl bromide is known as Hunsdiecker reaction.

$$RCOONa \xrightarrow{AgNO_3} RCO\overset{\ominus}{O}Ag^{\oplus} \xrightarrow{Br_2/CCl_4} R\text{—}Br + CO_2 + AgBr$$

MECHANISM

The mechanism of Hunsdicker reaction is uncertain and one of the favoured theories being the free-radical mechanism :

Step - [I] ***Formation of Hypohalite***

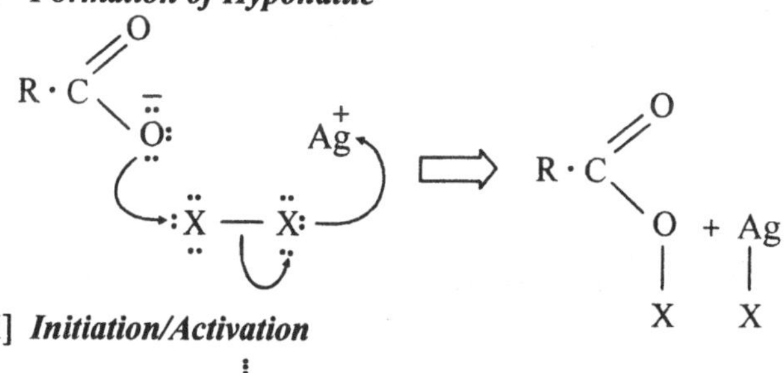

Step - [II] ***Initiation/Activation***

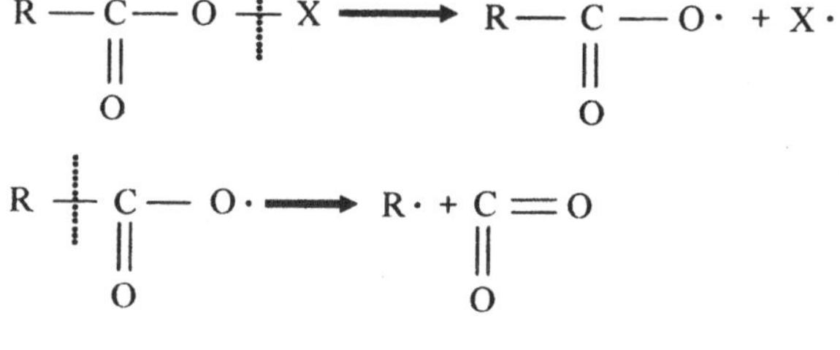

Step - [III] ***Growth or Propagation***

$$R\cdot + R\cdot\underset{\underset{O}{\|}}{C}-OX \longrightarrow R-X + R\cdot\underset{\underset{O}{\|}}{C}-O\cdot$$

The first step is, of course, not a free-radical one but its actual mechanism being unknown, it may possibly the one mentioned here.

The hypohalite is the intermediate though it has not been isolated. the hypohalite decomposes into free-radicals to facilitate propagation—the details of which are yet to be investigated.

SUPPORT OF THE ABOVE MECHANISM

(1) Optical activity of R is loss if R asymmetric

(2) No rearrangement takes place if R is neopentyl. This rearrangement would invariably take place if the reaction proceeds via generation of R^+, instead of R.

(3) The side products R—R has been isolated. No two R^+ would form this coupling, but two R· free-radicals can. The foregoing evidences outweigh the ionic mechanism in favour of free-radical mechanism.

SIDE REACTIONS

Two main side reactions reduce the final yield :

[I] Generation fo Alkane

$$R\cdot + \cdot R \longrightarrow R-R$$

[II] Generation of Ester

$$R\cdot + R\cdot\underset{\underset{O}{\|}}{C}-O\cdot \longrightarrow R\cdot\underset{\underset{O}{\|}}{C}-OR$$

With 2:1 ratio of acid salt to iodine, the ester is the main product.

Note -1

(I) The Hunsdicker Reaction endows a convenient method to reducte the carbon number by one unit step.

(II) Although bromine is most frequently used, chlorine and iodine have also been employed.

(III) When iodine is taken in the same ratio as the salt, the alkyl halide is the main product. If the concentration is twice that of iodine, the ester is the main product. If the concentrations ratio of silver salt of the acid to iodine is 3:2, the ester and the alkyl halide is obtained almost in equal amount.

(IV) The R group may be aryl or alkyl—cyclic or straight chained.
(V) R may be branched. In these cases secondary or tertiary alkyl halides are obtained.
(VI) Good results have been found with R groups having carbon number ranging from 2 to 18.

Note-2
(I) The yield of alkyl halide is in the order

***primary- > secondary- > tertiary* - R**

(II) Chlorine gives poorer yield than bromine
(III) The yield is also affected by solvent used. Trichloroethane is better solvent than CCl_4 or CS_2.

ALTERNATIVE ROUTES

The Hunsdicker Reaction can be carried out more conveniently by adding bromine to refluxing carbon tetrachloride solution of the acid in presence of mercuric oxide taken in excess. This eliminates much trouble in the preparation of pure & dry silver salts.

Another method is to treat the acid with lead tetracetate in presence of halide ions.

APPLICATIONS

(*i*) One useful modification of Hunsdiecker reaction is the use of HgO and Pb(IV) salt in place of silver salt.

(*ii*) In presence of Pb(IV) salt, $Pb(OAc)_4$ Iodide derivative can also be prepared.

$$RCOOH \xrightarrow{Pb(OAc)_4/I_2/hv} R—I$$

Silver salt of acid (excess) is treated with I_2 leading to the formation of ester. This reaction is known as **Simonini reaction.**

$$2RCOOAg \xrightarrow{I_2/\Delta} RCOOR + CO_2 + 2AgI$$

$$CH_3COOH \xrightarrow{AgNO_3} CH_3CO\overset{\ominus}{O}Ag^{\oplus} \xrightarrow{I_2/\Delta} CH_3—\overset{\overset{\displaystyle O}{\|}}{C}—OCH_3$$

(iii) A free radical is generated at bridgehead position *via* the Hunsdiecker reaction to show that a free radical need not be planar.

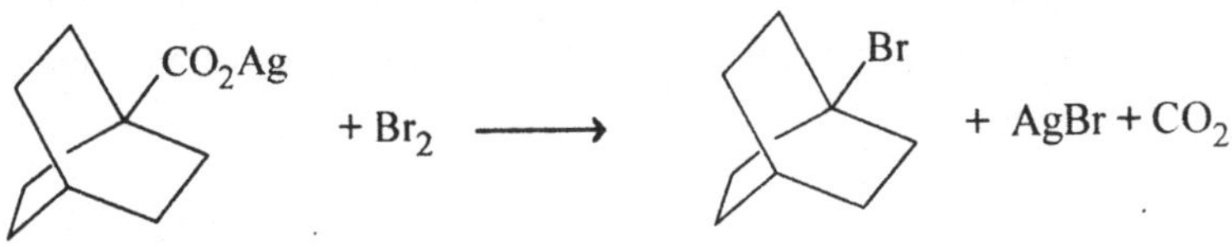

87%

Problem -1

$$\underset{[A]}{C_8H_{10}} \xrightarrow{\overset{+}{H} - KMnO_4} \underset{[B]}{C_8H_6O_4}$$

$$\underset{[B]}{C_8H_6O_4} \xrightarrow{AgNO_3} C_8H_6O_4Ag_2 \xrightarrow[CCl_4/\Delta]{Br_2} \underset{[C]}{C_6H_4Br_2}$$

(I) **A** does not declorize bromine

(II) **B** is soluble in alkali but insoluble in acids. It does not react with phenyl hydrazine

(III) **C** forms only one mononitroderivative.

Identify **A, B, & C**

Answer

A ▶ *para*-Cresol **B** ▶ Terephthalic acid

C ▶ *para*-Dibromobenzene

Problem -2

▶ $$\underset{[A]}{C_7H_5N} \xrightarrow{\overset{+}{H_3O}} \underset{[B]}{C_7H_6O_2} \xrightarrow[CCl_4/\Delta]{\overset{+}{Ag}/Br_2} [C] + CO_2$$

$$[C] \xrightarrow[\Delta]{Cu}$$

Hints

C upon heating with Cu yields Ar—Ar.

So **C** is probably and aryl halide

B sports two **O** atoms.

B forms a silver salt which when heated with Br_2 in CCl_4 produces **C**.

So **B** is a carboxylic acid that undergoes Hunsdicker reaction giving rise to **C** which is therefore ArBr.

Answer	**A**	Phenyl cyanide ArCN
	B	Benzoic acid $ArCO_2H$
	C	Phenyl bromide Ar Br

Problem -3

$$\underset{[A]}{C_8H_9NO} \xrightarrow{P_2O_5} \underset{[B]}{C_8H_7N} \xrightarrow{H_3\overset{+}{O}} \underset{[C]}{C_8H_8O_2} \xrightarrow[CS_2/\Delta]{\overset{+}{Ag}/Br_2} \underset{[D]}{C_7H_7Br}$$

$$\underset{[D]}{C_7H_7Br} \xrightarrow[\Delta]{Cu} \underset{[E]}{C_{14}H_{14}} \xrightarrow[\overset{+}{H}]{KMnO_4} \text{Diphenic acid}$$

Identify **A, B, C, D & E.**

Answer

A▶ *ortho*-Methyl benzamide 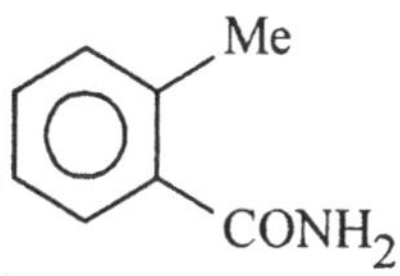

B▶ *ortho*-Methyl phenyl cyanide 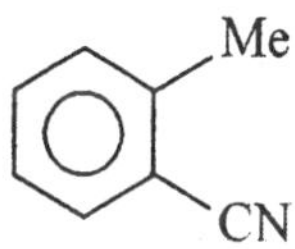

C▶ *ortho*-Methyl benzoic acid 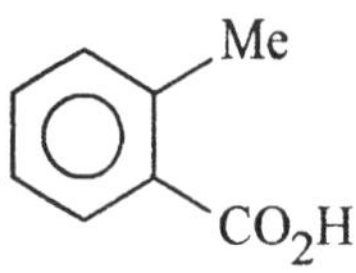

D▶ *ortho*-Bromotoluene 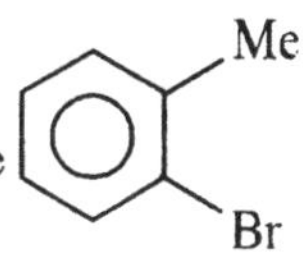

E▶ 2,2'-Dimethylbiaryl 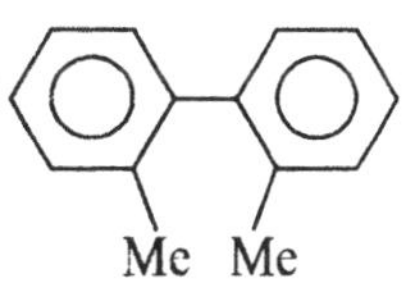

61

KNOEVENAGEL REACTION

Condensation reaction of compound having active hydrogen with carbonyl compounds in presence of a weak base is known as Knoevenagel reaction.

In this reaction, base as a catalyst is amine, mainly tertiary and sec. amine but best result is obtained with sec amine.

Carbonyl compounds are usually those which have no α-hydrogen.

Compounds having active hydrogen are mainly those compounds which have active methylene group. However, terminal alkynes, haloform, cyclopentadiene and nitro alkanes also undergo this reaction.

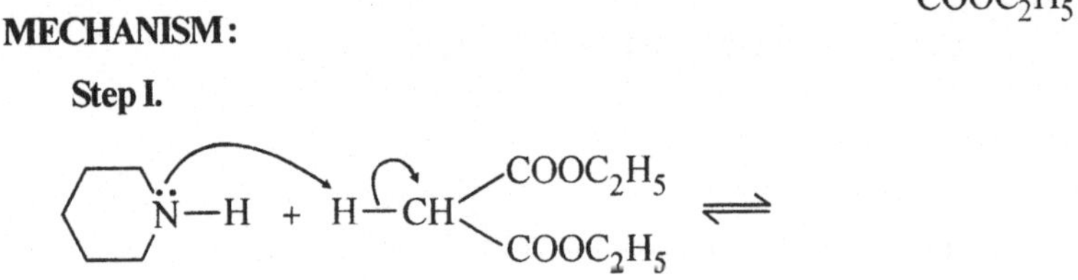

MECHANISM:

Step I.

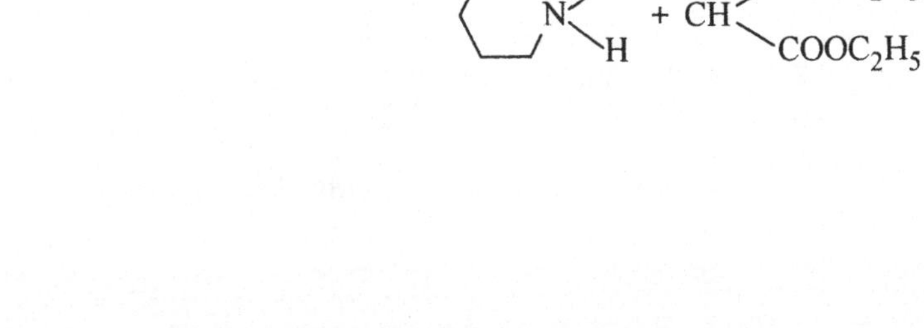

Step II.

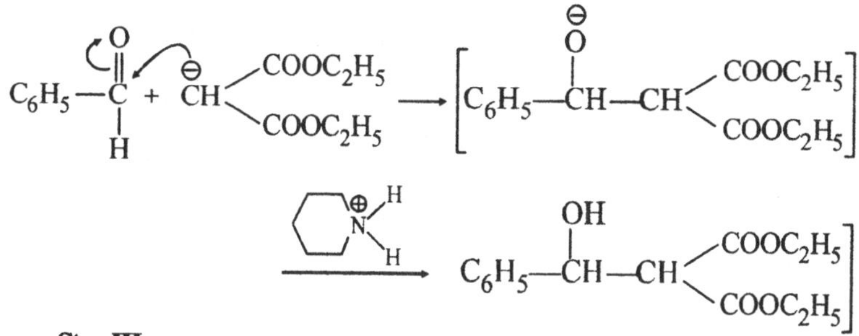

Step III.

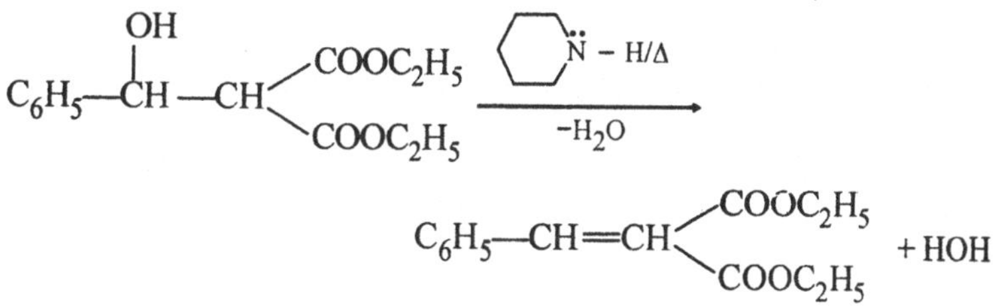

89% yield

Ketones undergo this reaction only with those compounds which contain highly activated α-hydrogen like those compounds which have active methylene group.

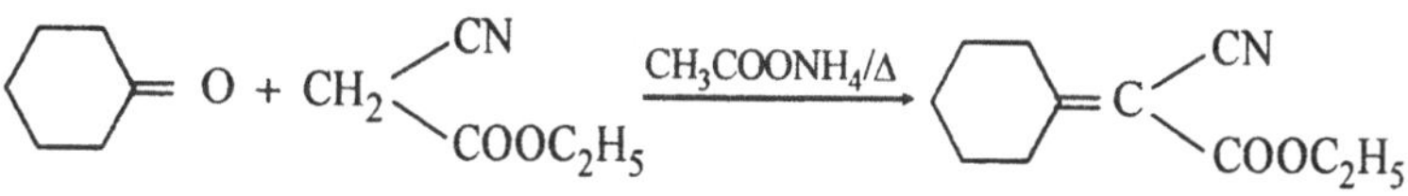

80% yield

Cyclopentadiene (pKa = 16) is enough acidic to give an anion when treated with base and cyclopentadienide ion can add to carbonyl compound.

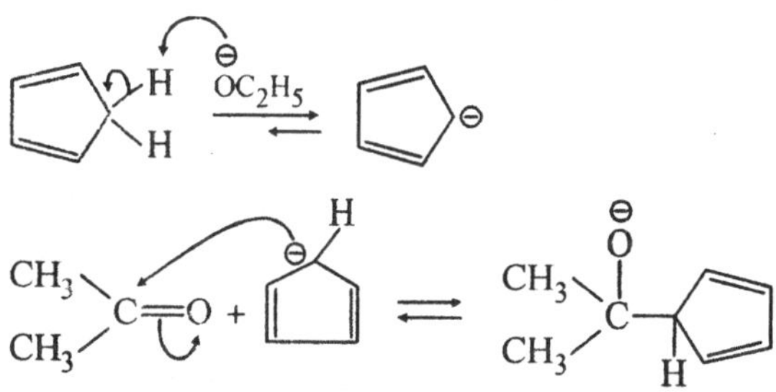

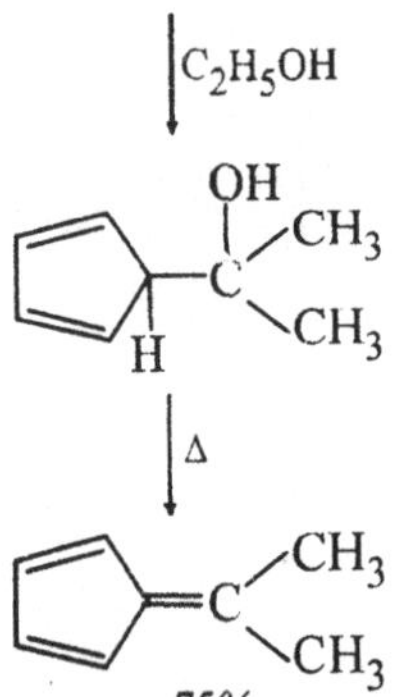

75%
6,6-Dimethylfulvene

Both aromatic and aliphatic aldehydes undegoes this reaction. For steric and electronic reasons ketones have been less reactive than aldehydes. Hence,ketones react with compounds containing powerful active methylene group only *e, g.,* ethyl acetoacetate, cyanoacetic acid and its ester but not with malonic

$$(CH_3)_2C{=}O + H_2C\begin{matrix} CN \\ COOR \end{matrix} \xrightarrow[CH_3COOH]{CH_3CONH_2} (CH_3)_2C \quad C\begin{matrix} CN \\ COOR \end{matrix}$$

$$\xrightarrow[2.\ \Delta,\ -CO_2]{1.\ H_2O} (CH_3)_2C{=}CHCOOH$$

3-Methylcrotonic acid

The intermediate unsaturated product formed during the course of reaction with aldehydes tends to undergo Michael reaction specially in the presence of excess of the active methylene compound. The Michael condensation product may get hydrolysed and decarboxylated to yield dibasic acid.

$$CH_3CHO + H_2C(COOR)_2 \xrightarrow{\text{Piperidine}} CH_3{-}CH{=}C(COOR)_2$$

$$\xrightarrow[\text{Michael reaction}]{CH_2(COOR)_2} CH_3{-}CH\begin{matrix} CH(COOR)_2 \\ CH(COOR)_2 \end{matrix} \xrightarrow[2.\ \Delta\ -CO_2]{1.\ H_2O} CH_3{-}CH\begin{matrix} CH_2COOH \\ CH_2COOH \end{matrix}$$

3-Methylglutaric acid

Hence, Knoevenagel reaction best proceeds with aromatic aldehydes

SCOPE OF THE REACTION

Many active methylene compound, like diethyl malonate, ethyl acetoacetate, ethyl cyanoacetate, as well as phenylacetonitrile, benzyl ketones and aliphatic nitro-compounds are used in the reaction. Though pyridine is a satisfactory catalyst, other catalysts like ammonia, piperidine, aminoacids, ammonium acetate and acetamide are also very effective.

Diethyl malonate, readily condenses with aldehydes and reactive ketones like acetone and cyclohexanone, but less reactive ketones require the use of cyanoacetic esters for obtaining good yield of the product. Further, aliphatic aldehydes react with excess of the active methylene compound to undergo Michael reaction and thus Knoevenagel reaction is more useful for the condensation of aromatic rather than aliphatic aldehydes.

APPLICATIONS

The Knoevenagel reactions has more synthetic value aromatic aldehydes than aliphatic aldehydes. The addition to the C=C of the product from an aromatic aldehyde should be less likely due to loss of conjugation to the aromatic system. Knoevenagel reaction has been generally used for the synthesis of α, β-unsaturated compounds :

1. Different α, β-unsaturated acids like crotonic, cinnamic, fumaric and β-piperonly acrylic acids can be prepared.

(*a*) $$CH_3CHO + H_2C(COOR)_2 \xrightarrow{\text{Pip}} CH_3{-}CH{=}C(COOR)_2$$

$$\xrightarrow[\text{2. }\Delta,\ -CO_2]{\text{1. }H_2O} \underset{\text{Crotonic acid}}{CH_3{-}CH{=}CH{-}COOH}$$

(*b*) $$C_6H_5CHO + H_2C(COOR)_2 \xrightarrow{\text{Pip}} C_6H_5{-}CH{=}C(COOR)_2$$

$$\xrightarrow[\text{2. }\Delta,\ -CO_2]{\text{1. }H_2O} \underset{\text{Cinnamic acid}}{C_6H_5{-}CH{=}CH{-}COOH}$$

(*c*) $$\underset{\text{Glyoxalic acid}}{HOOC{-}CHO} + H_2C(COOR)_2 \xrightarrow{\text{Pip}}$$

$$HOOC{-}CH{=}C(COOR)_2 \xrightarrow[\text{2. }\Delta,\ -CO_2]{\text{1. }H_2O}$$

$$\underset{\text{Maleic acid}}{HOOC{-}CH{=}CH{-}COOH}$$

(d) 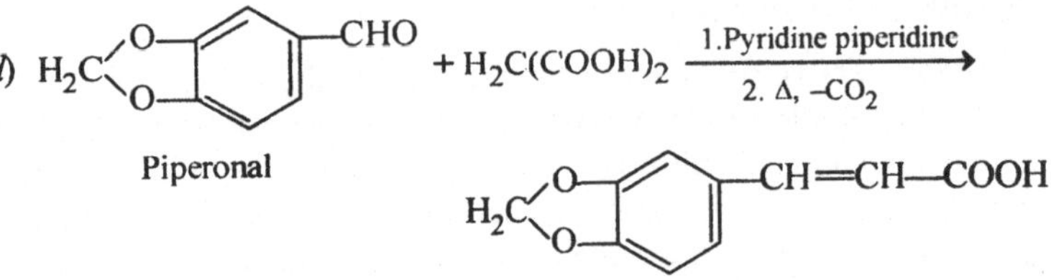

Piperonal

β—Piperonyl acrylic acid

2. By reacting aldehyde with active methylene compound in 1:2 molar proportions, different dibasic acids and diketones can be synthesised

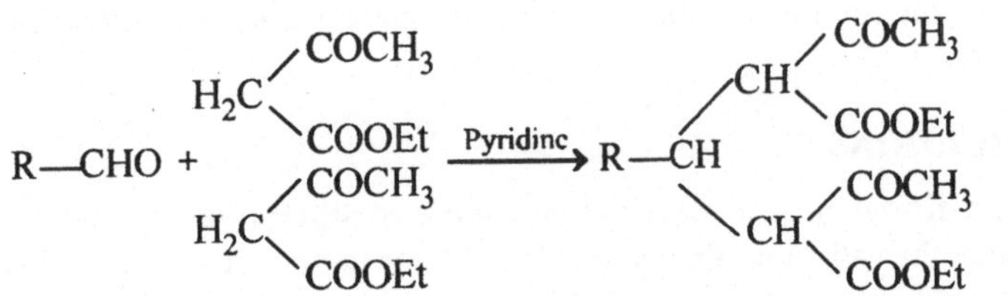

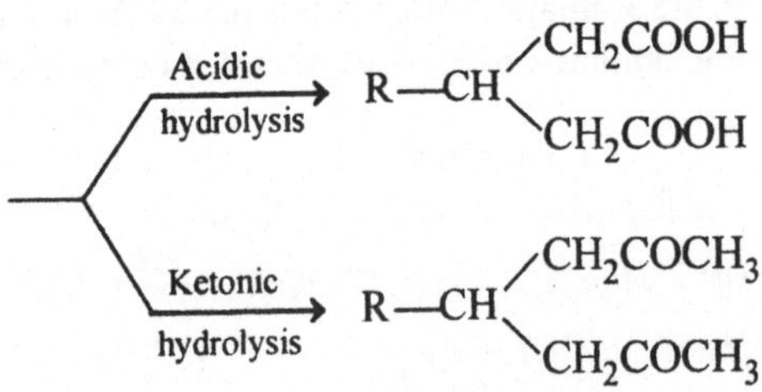

4-Alkyl-heptan-2, 6 dione
(1, 5-Diketone

The 1 : 5-diketones may be cyclised by treating with alcoholic alkali to obtain cyclohexene derivatives.

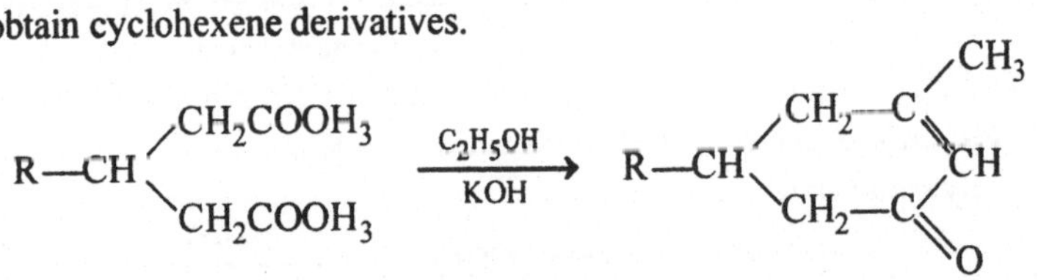

3. Cyclic compounds can also be prepared from suitably selected compounds possessing one active hydrogen. The reaction proceeds upto aldol stage and then the aldol gets cyclised to paraconic acid.

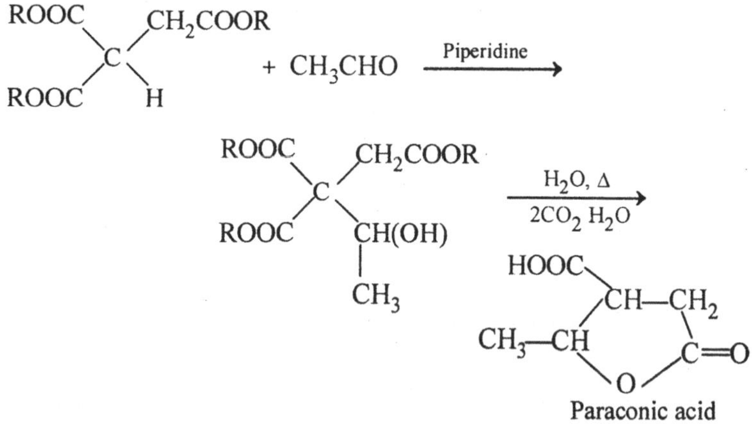

If paraconic acid is heated to 150°C, it gives β, γ-unsaturated acid

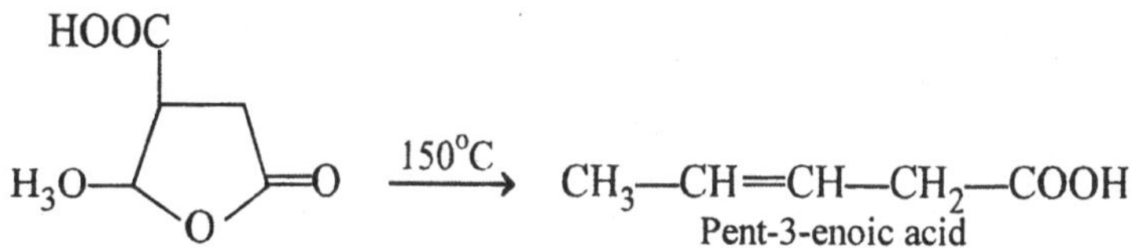

62

KOLBE REACTION

Phenol, in presence of base NaOH, reacts with CO_2 at high temperature and pressure to yield sodium salicylate which on acidification yields salicylic acid. This reaction is electrophilic aromatic substitution reaction in which CO_2 behaves as electrophile. The reaction is known as Kolbe reaction.

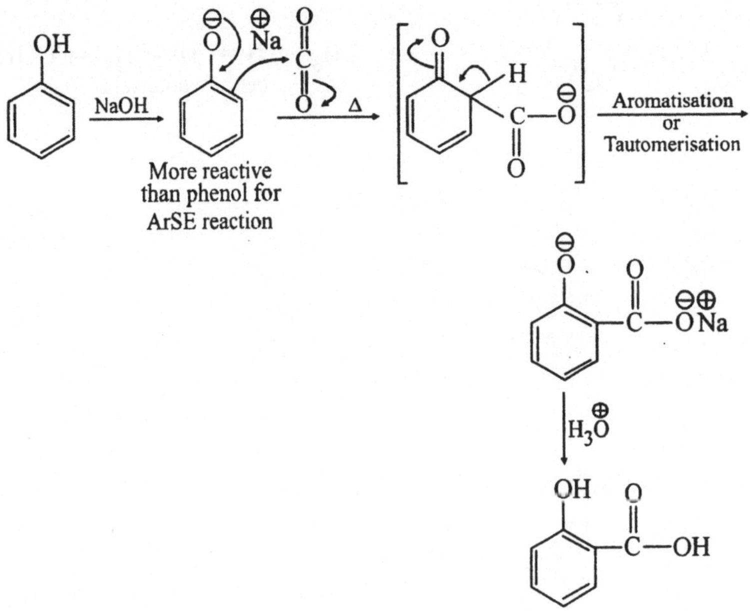

63

KOLBE HYDROCARBON SYNTHESIS

Electrolysis of sodium salt of carboxylic acid to give coupled product, alkane, is known as Kolbe hydrocarbon synthesis.

$$RCO\overset{\ominus\oplus}{ONa} + RCO\overset{\ominus\oplus}{ONa} \xrightarrow{\text{Electrolysis}} R\text{—}R + 2CO_2 + 2H_2 + 2NaOH$$

MECHANISM:

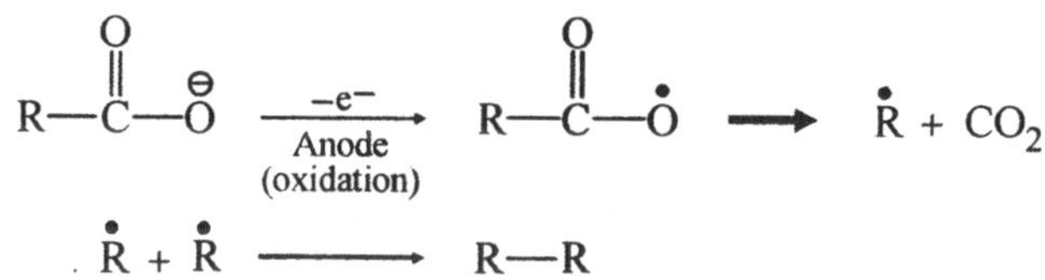

1, 4-saturated dibasic acid gives alkene.

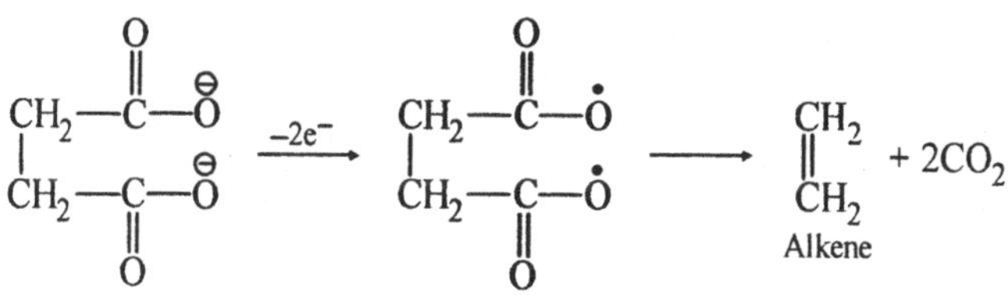

1, 4-unsaturated dibasic acid gives alkyne.

$$\begin{array}{l} CH{-}CO\overset{\ominus}{O} \\ \| \\ CH{-}CO\overset{\ominus}{O} \end{array} \xrightarrow{\text{Electrolysis}} CH{\equiv}CH + 2CO_2$$

$$C_6H_5COONa \xrightarrow{\text{Electrolysis}} C_6H_5{-}C_6H_5$$

$$C_6H_5CH_2COONa \xrightarrow{\text{Electrolysis}} C_6H_5{-}CH_2{-}CH_2{-}C_6H_5$$

$$\begin{array}{l} CH_3{-}CH{-}COONa \\ \quad\quad | \\ \quad\ CH_2{-}COONa \end{array} \xrightarrow{\text{Electrolysis}} \begin{array}{l} CH_3{-}CH \\ \quad\quad\ \| \\ \quad\quad CH_2 \end{array}$$

$$\begin{array}{l} CH_3{-}C{-}COONa \\ \quad\quad \| \\ CH_3{-}C{-}COONa \end{array} \xrightarrow{\text{Electrolysis}} CH_3{-}C{\equiv}C{-}CH_3$$

64

KOLBE REACTION (Electrolytic Method)

In the synthesis of symmetrical alkanes, the importnace of Kolbe Reaction flows parallelly with that of Wurtz Reaction.

RECATION

It is carried out by electrolyzing concentrated aqueous solution of sodium or potassium salt of a fatty acid or a mixture of fatty acids whereupon it brings about the union of two alkyl groups.

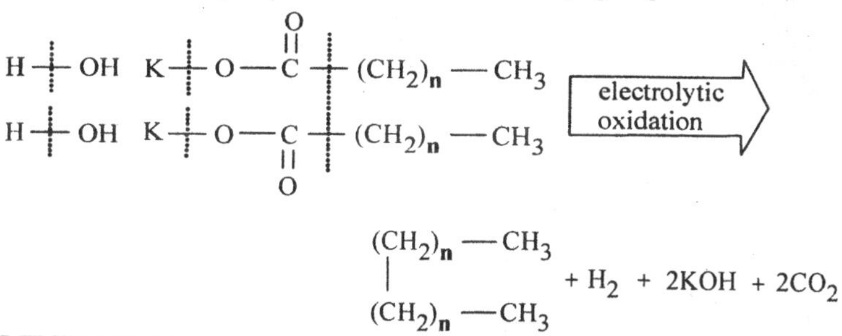

MECHANISM

Several mechanisms have been proposed for Kolbe Reaction. The ***free-radical*** mechanism is the one now favoured.

STEP - [I] ***Formation of a Carboxylate Free-radical at the anode***

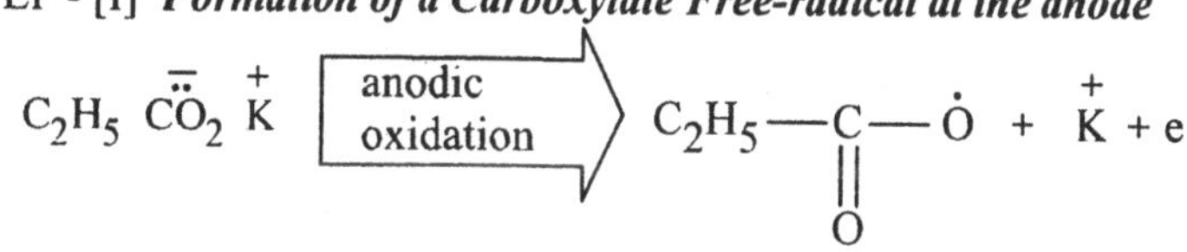

STEP - [II] ***Breaking up of the Carboxylate Free-radical to yield the alkyl free-radical &carbon dioxide***

$$C_2H_5 - \overset{\overset{O}{\|}}{C} - O\cdot \longrightarrow \dot{C}_2H_5 + CO_2$$

Step - [III] ***Combination of the two alkyl Free-radicals***

$$C_2H_5\cdot \; + \; \cdot C_2H_5 \longrightarrow C_2H_5 - C_2H_5$$

SIDE REACTIONS

Several side reactions reduce the yield of the desired product. These reactions produce alkenes, alcohols and esters as by-products.

(1) Disproportion of the two alkyl free-radicals to generate equal amount of an olefin and a paraffin

$$\underset{H_2C\,-\,H}{\overset{H_2C\cdot}{|}} \;\cdot CH_2 - CH_3 \Longrightarrow \underset{H_2C}{\overset{H_2C}{\|}} + CH_3 - CH_3$$

(2) Combination of the alkyl free-radical with a hydroxyl free-radical group ×OH to yield an alcohol

$$H - OH \longrightarrow H\cdot + \cdot OH$$

$$\dot{C}_2H_5 + \dot{O}H \longrightarrow C_2H_5 - OH$$

(3) Formation of a lighter parafin

$$\dot{C}_2H_5 + \dot{H} \longrightarrow C_2H_6$$

(4) Formation of an ester

$$\dot{C}_2H_5 + C_2H_5 - \underset{\underset{O}{\|}}{C} - O\cdot \longrightarrow C_2H_5 - \underset{\underset{O}{\|}}{C} - O - C_2H_5$$

LIMITATIONS

Kolbe Reaction is particularly useful in the synthesis of Symmetrical alkanes viz. ethane [CH_3—CH_3]; *n*-butane [C_2H_5—C_2H_5].

Applied to the mixture of two sodioderivatives of two different fatty acids, KOLBE REACTION, though will generate unsymmetrical alkanes, yet the yield of unsymmetrical alkenes is poor. Nevertheless, the mixtures R—R′, R—R, & R′—R′ can often be separated readily.

$$\begin{array}{l} R\cdot C\bar{O}_2\ \overset{+}{N}a \\ R'\cdot C\bar{O}_2\ \overset{+}{N}a \end{array} + 2H_2O \xrightarrow{\text{electrolysis}} \begin{array}{c} R\cdot \\ | \\ R'\cdot \end{array} + \begin{array}{c} R—R \\ \\ R'—R' \end{array} + 2CO_2 + 2H_2 + 2NaOH$$

Finally to say, Kolbe Reaction fails to induce coupling between two aromatic groups such as Ar—Ar *i.e.* sodio or potasio derivative of aromatic carboxylic acid in which the $-CO_2H$ group is directly attached to the phenyl nucleus will not give rise to this reaction.

Using different straight-chain fatty acids containing 2—18 carbon atoms, the yields of alkanes have been found to range within 50—90 percent. Alkyl groups in the ***alfa***-position decrease the yield usually below 10 percent. This is ascribed to steric hindrance exerted bythe alkyl group in the ***alfa***-position.

APPLICATIONS

(I) The KOLBE electrolytic method is now being employed in the synthesis of natural products, particularly, lipids.

(II) For the preparation of ethylene

$$\begin{array}{l} CH_2.C\bar{O}_2\ \overset{+}{Na} \\ | \\ CH_2.C\bar{O}_2\ \overset{+}{Na} \end{array} + 2H_2O \Longrightarrow \begin{array}{c} CH_2 \\ \| \\ CH_2 \end{array} + 2NaOH + 2CO_2 + 2H_2$$

Problem

$$\underset{[A]}{C_6H_{10}O_2} \xrightarrow{Br_2} C_6H_{10}O_2Br_2$$

$$[A] \xrightarrow[CaO/\Delta]{NaOH} C_5H_{10} \xrightarrow{Br_2} C_6H_{10}Br_2$$

$$[A] \xrightarrow[KOH]{} \underset{[B]}{C_6H_9O_2K} \xrightarrow{\text{electrolysis}} \underset{[C]}{C_{10}H_{12}}$$

$$[C] \xrightarrow{\text{Ozonolysis}} \underset{(2\ \text{molecules})}{\text{Acetone}} + \underset{[D]}{C_4H_6O_4}$$

$$[D] \xrightarrow[\Delta]{} \text{succinic anhydride}$$

Identify **A, B, C, & D.**

Solution

$$\underset{[A]}{C_6H_{10}O_2} \xrightarrow[\text{1mole}]{Br_2} C_6H_{10}O_2Br_2$$

Hence **A** contains one C=C

$$\underset{[A]}{C_6H_{10}O_2} \xrightarrow[\Delta]{NaOH / CaO} C_5H_{10}$$

Since C_5H_{10} is less then $C_6H_{10}O_2$ by one CO_2, **A** is a carboxylic acid that undergoes decarboxylation in this process. Therefore, **A** can be written as C_5H_9COOH.

A on treatment with KOH produced its salts $C_5H_9CO_2K$, **B** which on electrolysis produces C i.e. $C_{10}H_{18}$.

B undergoes electrolytic oxidation as

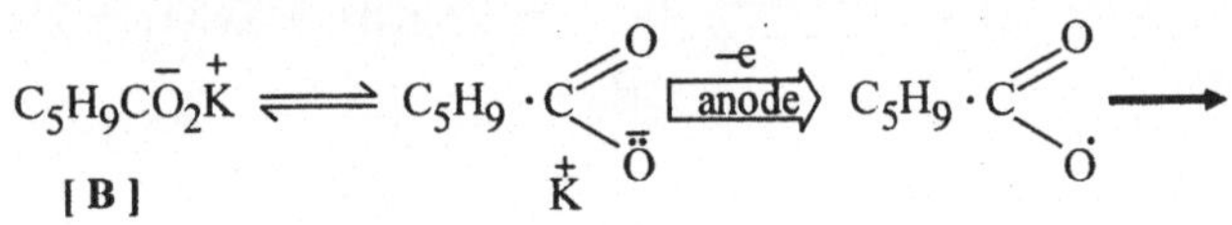

$$\mathbf{D} \xrightarrow[\Delta]{} \text{Succinic anhydride}$$

$$\dot{C}_5H_9 + CO_2 \xrightarrow{\cdot C_5H_9} \underset{[C]}{H_9C_5-C_5H_9}$$

Hence **D** is succinic acid

C upon ozonolysis produces 2 molecules of acetone and one molecule of succinic acid which is **D.**

So **C** is 2,7-dimethyloct-2,6-diene

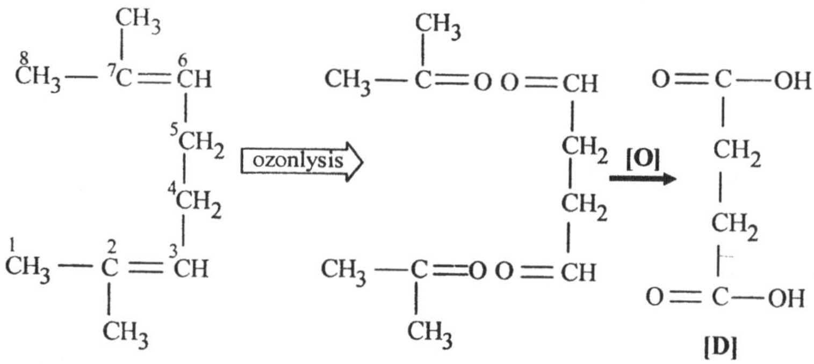

[C]

Now that **S** is a symmetrical compound having a plane of symmetry,

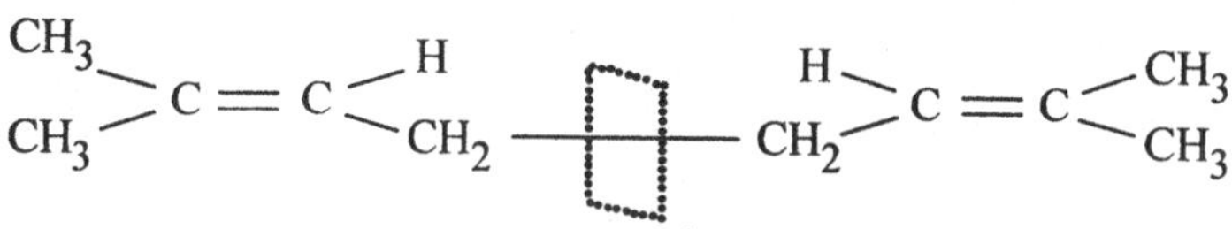

Plane of symmetry

so its each C_5H_9 unit comes from **B** as 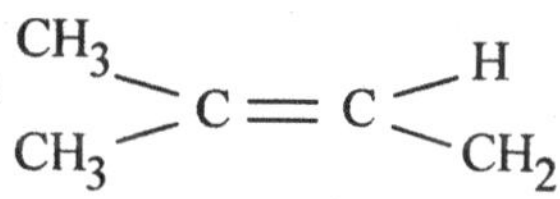

Therefore A is 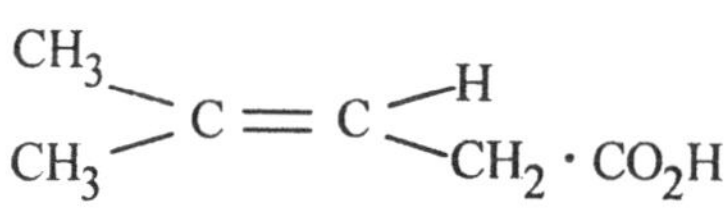

Conclusion

A▶ 4-Methylpent-3-enoic acid

B▶ 4-Methylpent-3-ene-1-potassium carboxylate

C▶ 2,7-Dimethyloct-2, 6-diene

D▶ Succinic acid

65

KNORR-PAAL SYNTHESIS

Conversion of 1,4-diketones into five membered heterocyclic compounds in presence of suitable inorganic dehydrating agent is known as Knorr-Paal synthesis.

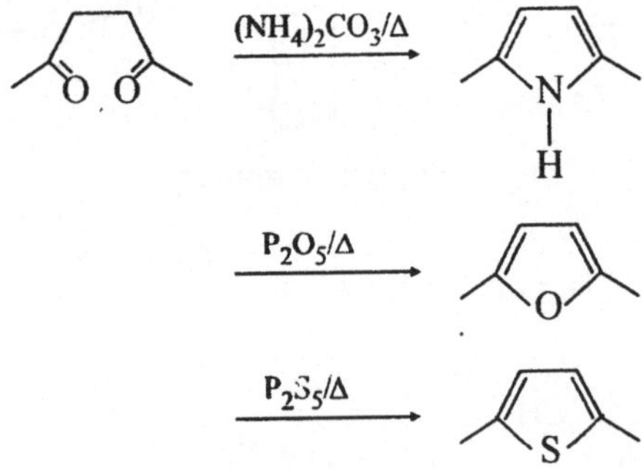

66

KOLBE-SCHMIDT REACTION

One of the technically successful methods in carboxylating phenolic rings is the Kolbe-Schmidt Reaction.

REACTION

The reaction is carried out by heating the phenolate salt with carbon dioxide under pressure

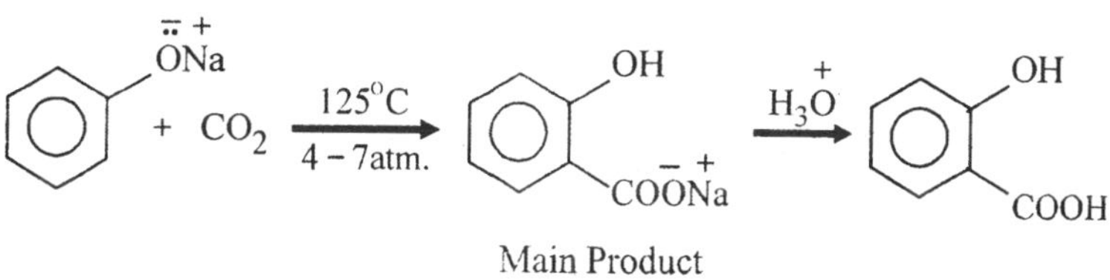

Main Product

MECHANISM

The mechanism of this reaction is uncertain. One possibility being :

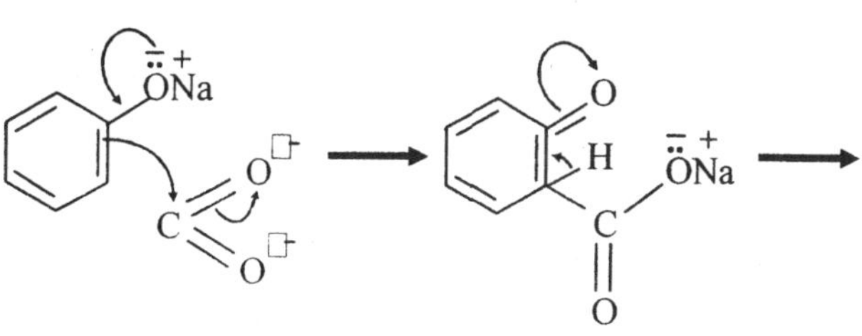

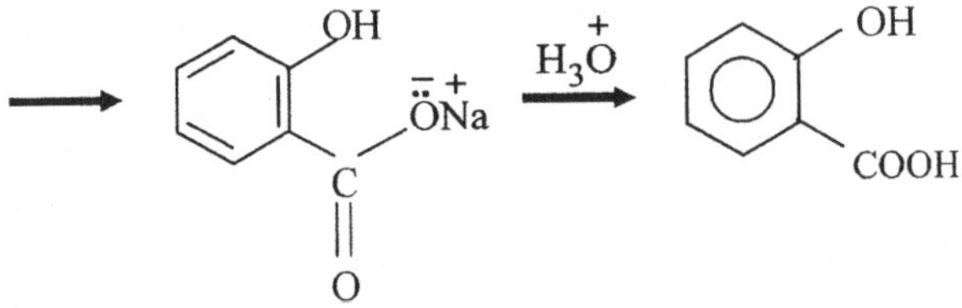

Note :

(I) The main product is the ***ortho***-hydroxybenzoic acid if the temperature is not raised above 140 °C. Above this temperature, ***para***-derivative is the main product.

(II) For polyhydric phenols, good yields are resulted in relatively mild conditions.

(III) A possible explanation why the ***ortho***-carboxylation occurs in preference to ***para***-isomer at low temperature is that CO_2 complextates with the phenoxide

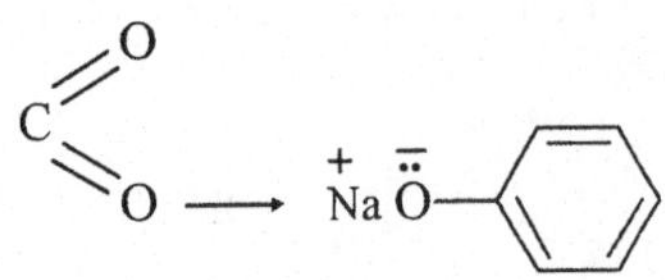

APPLICATIONS

1. 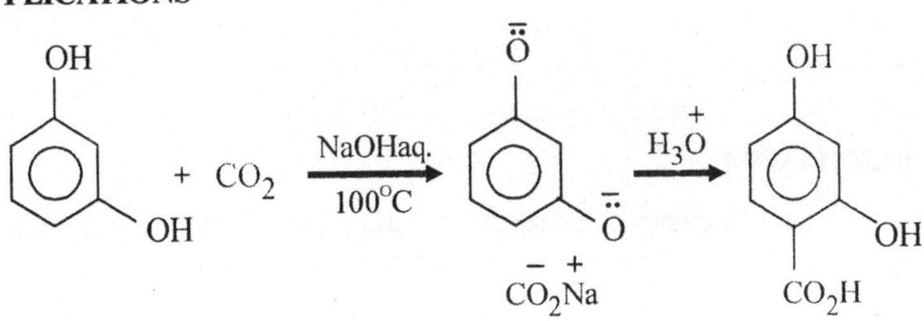

2. 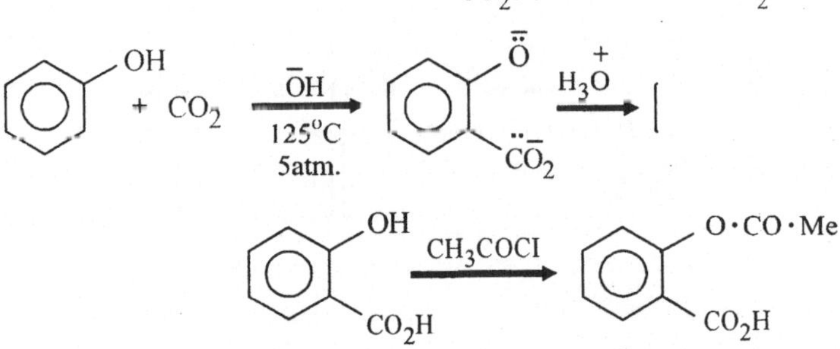

aspirin

Problem-1

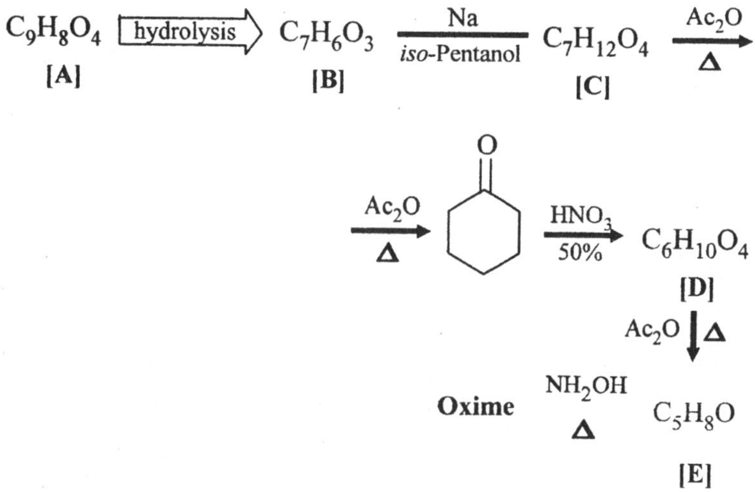

A, B, C, D are all precipitated from solution be mineral acid. Identify **A, B, C, D & E.**

Answer **A▶** ***ortho***-Acetoxybenzoic acid

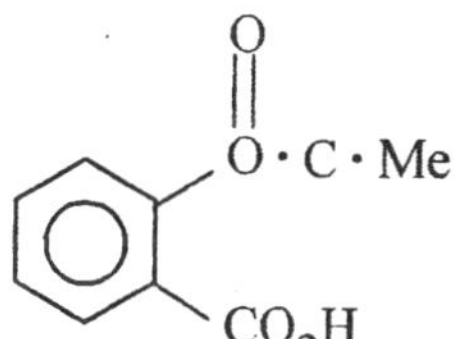

B▶ ***ortho***-Hydroxybenzoic acid
C▶ Pamitic acid
D▶ Adipic acid
E▶ Cyclopentanone

Problem-2

A, B, & C are isomeric—each having the formula $C_6H_6O_2$.

$$\mathbf{A} \xrightarrow[CO_2/100^\circ C]{\text{NaOH aq.}} \xrightarrow{H_3\overset{+}{O}} C_7H_6O_4 \qquad \text{[Three isomers]}$$

$$\mathbf{B} \xrightarrow{\text{NaOH aq.}} \xrightarrow{H_3\overset{+}{O}} C_7H_6O_4 \qquad \text{[Two isomers]}$$

$$\mathbf{C} \xrightarrow[CO_2/100^{\circ}C]{\text{NaOH aq.}} \xrightarrow{H_3\overset{+}{O}} C_7H_6O_4 \qquad \text{[One isomers]}$$

Identify **A, B, C.**

Hint

A, B, & C are since they undergo Kolbe-Schmidt reaction. If one oxygen comes from phenolic OH group, the other oxygen must be of phenolic OH, otherwise it would not fit the molecular formula.

Answer

A ▶ ***meta***-Dihydroxyphenol
B ▶ ***ortho***-Dihydroxyphenol
C ▶ ***para***-Dihydroxyphenol

67

LEDERER-MANASSE REACTION

Since its very incipiation, the Lederer-Manasse Reaction has gained much industrial importance.

REACTION

The reaction takes place at room temperature in the presence of dilute acid or alkali

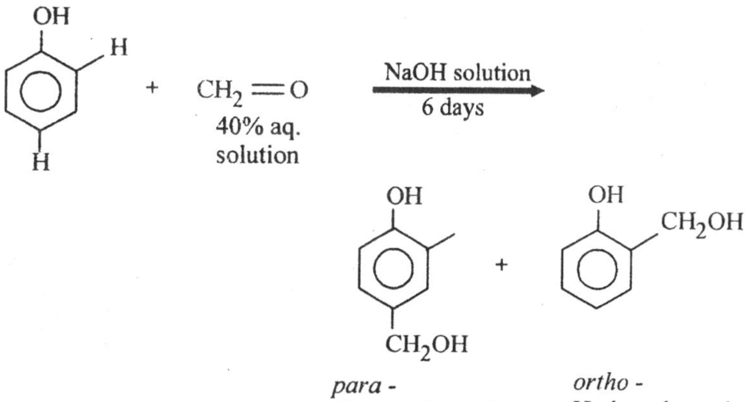

para - Hydroxybenzyl alcohol

ortho - Hydroxybenzyl alcohol

MECHANISM

[I]*Acid* ***Catalysis***

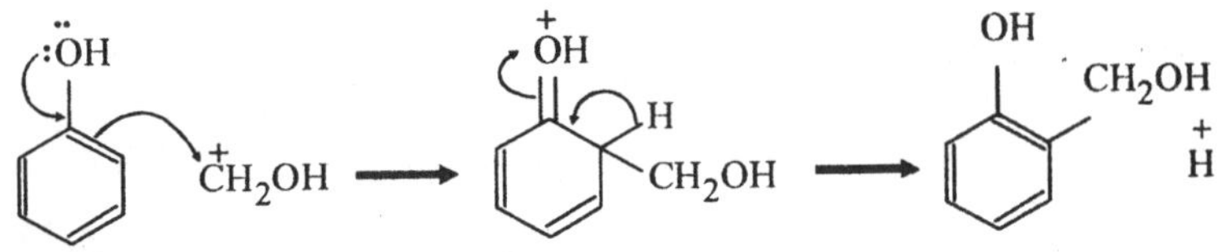

[II] ***Base Catalysis***

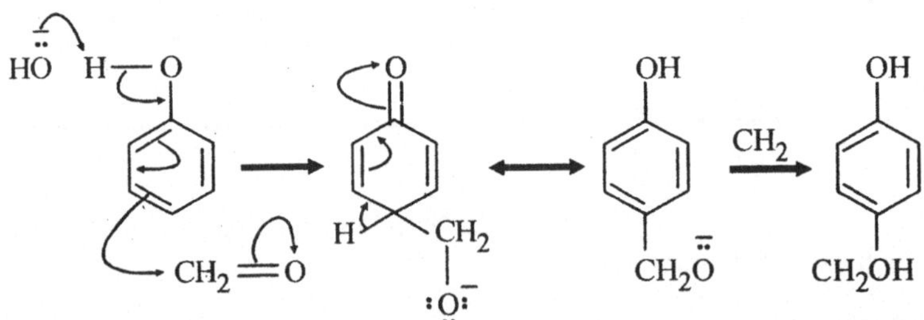

Note :

(I) The condensation takes place predominantlyin the para-position and if it is occupied, the condensation is entirely followed at the ortho-position. In usual case, the main product is the para-isomer with small of ortho-isomer.

(II) The reaction is slow and takes place at low temperature.

(III) Both dilute acid and alkali can give rise to this reaction. They act as catalyst.

(IV) The nature of the product depends upon the quantity of reacting meterials involved:

[A] With large amount of formaldehyde, the products are 2, 4-dilhydroxymethylphenol and 2,6-dihydroxomethylphenol.

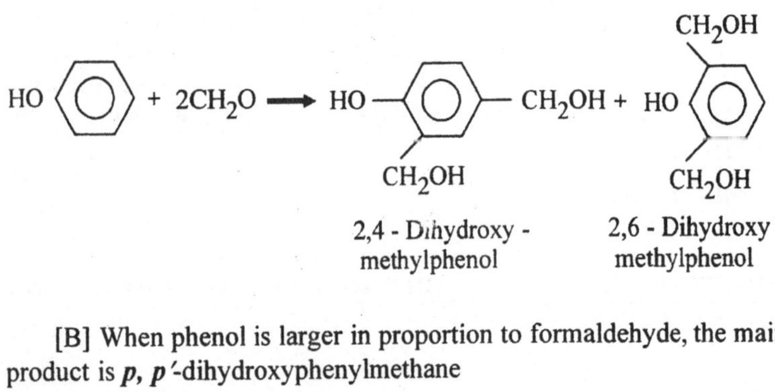

2,4 - Dihydroxy - methylphenol   2,6 - Dihydroxy - methylphenol

[B] When phenol is larger in proportion to formaldehyde, the main product is ***p, p***′-dihydroxyphenylmethane

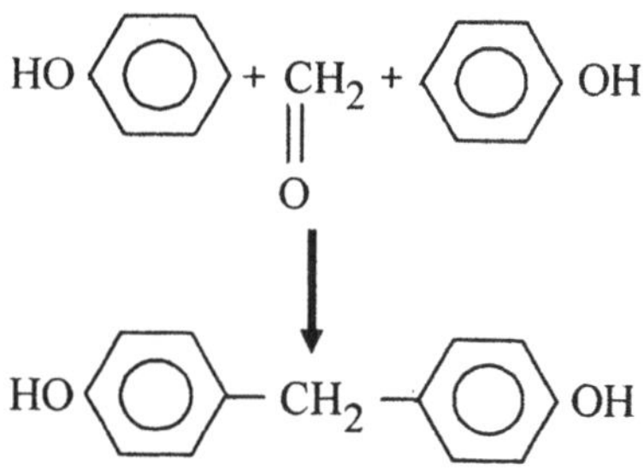

APPLICATIONS

Synthesis of phenol-formaldehyde plastic

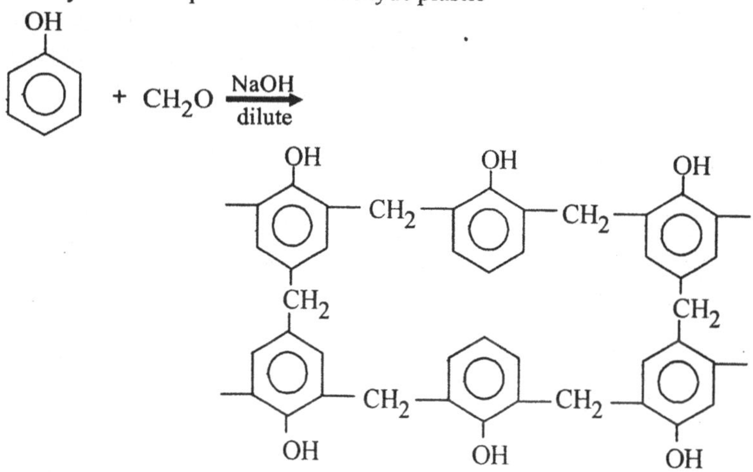

68

LEUCKART REACTION

This reaction is a modification of Eschweiler-Clark reaction. In this reaction, reductive amination of carbonyl compounds is carried out with ammonium formate or formamide at 180-200°C.

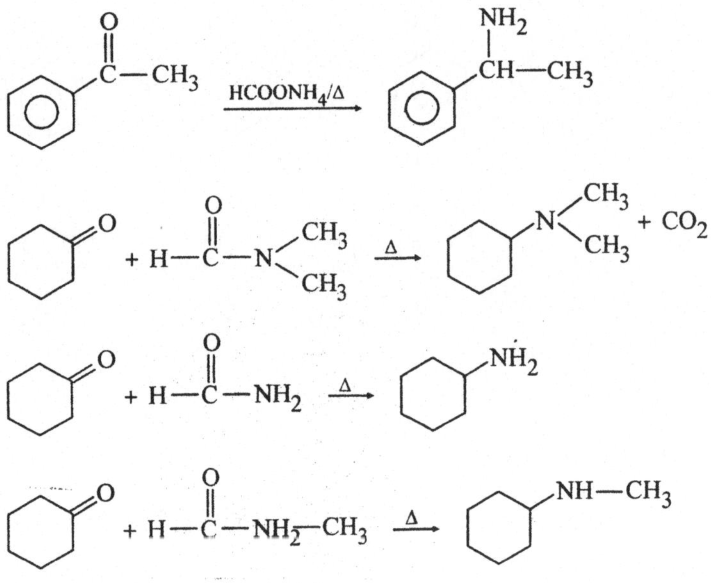

MECHANISM

The mechanism of this reaction is not known with certainty. However, it appears that an amine is first formed and this is then reduced to an amine by hydride ion transfer from formic acid.

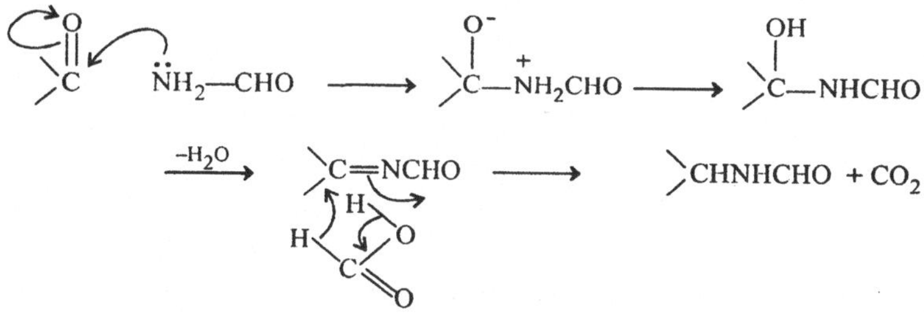

APPLICATIONS

This reaction may be used for the preparation of primary amines. It has been applied to synthesis riboflavin.

69

LOSSEN REARRANGEMENT

Though not of synthetic importance,
Lossen Rearrangement is one of academic importance.

REACTION

It involves treatment of hydroxamic acid* with a base or strong inorganic acid whereupon hydroxamic acid rearranges to amine.

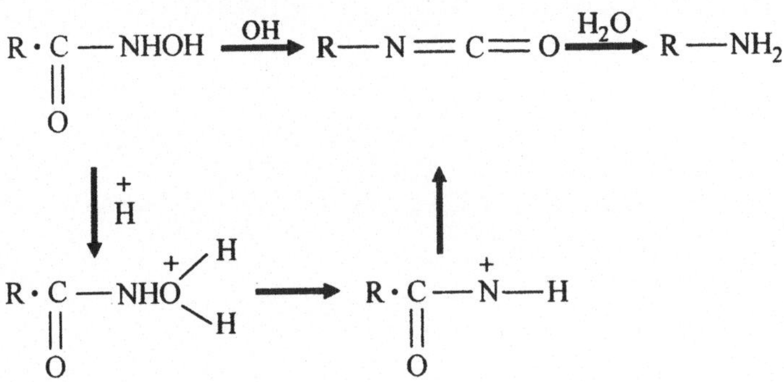

MECHANISM

It runs very close that of Hofmann Rearrangement.

[I] ***Acid Catalyzed***

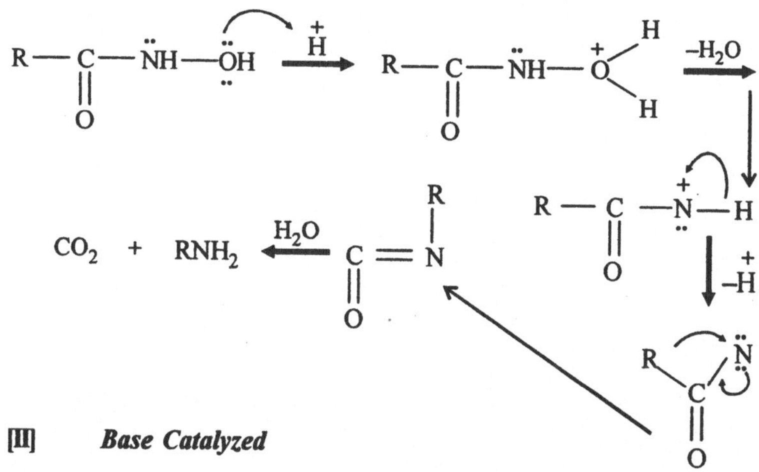

[II] ***Base Catalyzed***

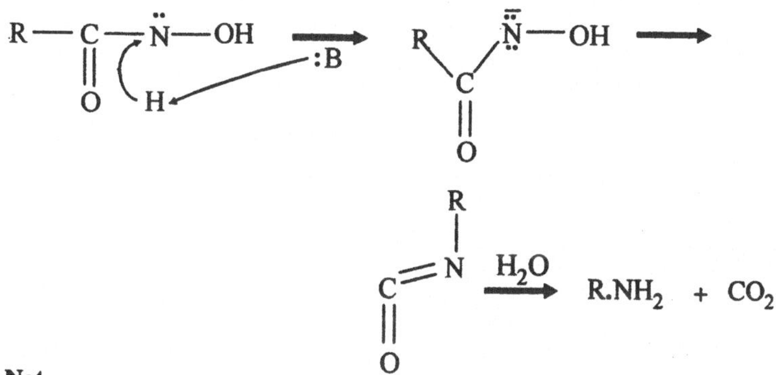

Note:

1. The rearrangement can occur in the presence of bases or strong inorganic acids; sometimes just heating completes the reaction.
2. Not only the hydroxamic acids but also their derivatives undergo this rearrangement

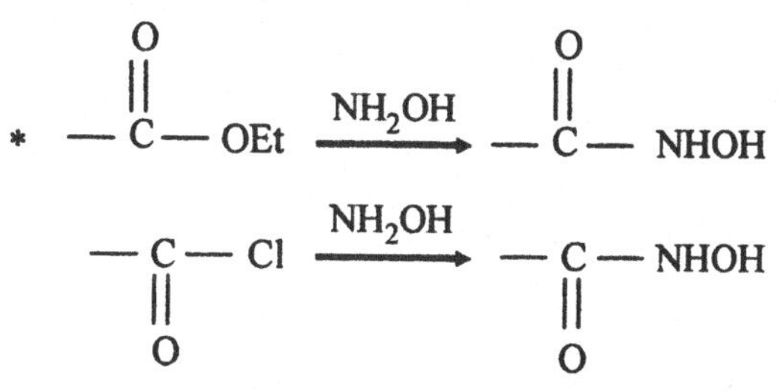

Hydroxainic acid

70

MICHAEL REACTION

This is the addition reaction between an α,β-unsaturated keto compound (acceptor) and a compound with an active methylene group (addendum) in the presence of a base.

$$\underset{\text{Acceptor}}{C_6H_5.\,CH=CH.\,COOC_2H_5} + \underset{\text{Addendum}}{CH_2(COOC_2H_5)_2} \xrightarrow{C_2H_5ONa} C_6H_5\,\underset{\substack{|\\ CH(COOC_2H_5)_2}}{CH}-CH_2.\,COOC_2H_5$$

EXAMPLES:

(1) $$CH_3-NO_2 + CH_2=CH-\overset{\overset{\displaystyle O}{\|}}{C}-CH_3 \xrightarrow{iPr_2NH/CHCl_3} NO_2-CH_2-CH_2-CH_2-\overset{\overset{\displaystyle O}{\|}}{C}-CH_3$$

(2) $$I-CH_2-CH_2-CH_2-CH_2-CH=CH\ \ COOC_2H_5 + CH_3-NO_2$$

$$\downarrow K_2CO_3/DMF$$

$$I-CH_2-CH_2-CH_2-CH_2-\underset{\substack{|\\ CH_2\\ |\\ NO_2}}{CH}-CH_2-COOC_2H_5$$

(3) 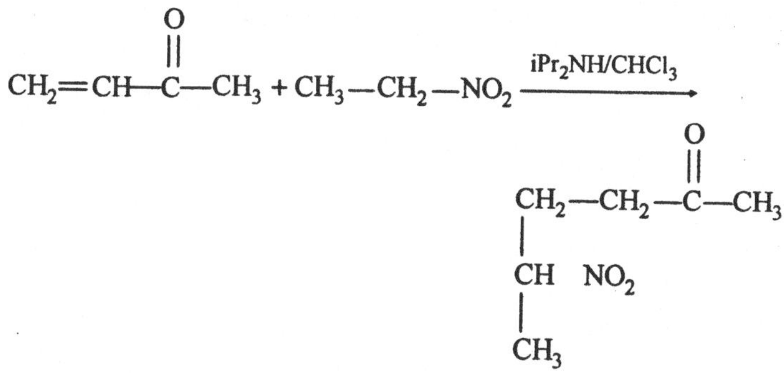

(4) 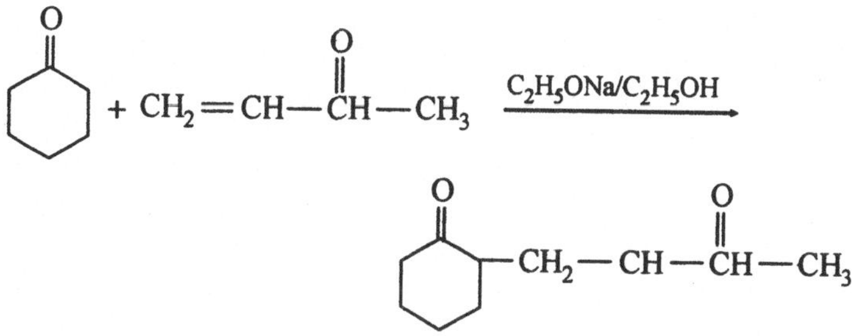

(5) 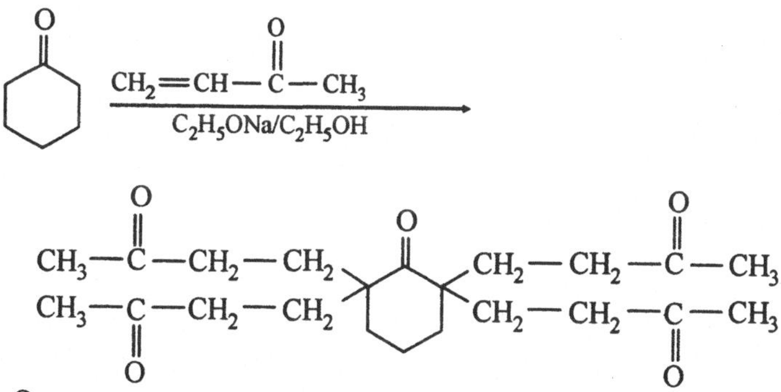

(6) 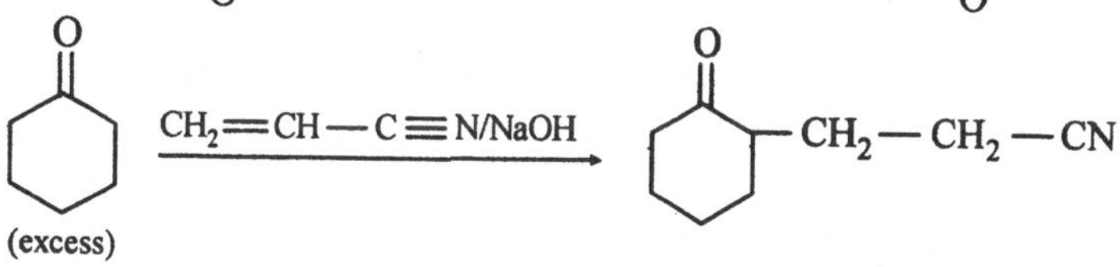

(7) 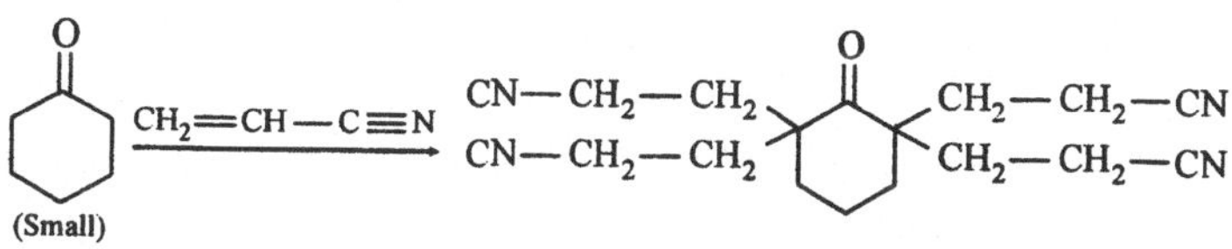

COMPONENTS OF THE MICHAEL REACTION

1. **Acceptors :** All the α, β-unsaturated compounds possessing a carbonyl or nitrile group in conjugation with carbon-carbon double bond are referred to as acceptors of Michael condensation. Some examples of acceptors are as follows :

(*i*) **Ketones :** Examples of these are chcone ($C_6H_5CH=CH\,COC_6H_5$), ethyl cinnamate ($C_6H_5CH = CH.COOC_2H_5$), benzylidene acetone ($C_6H_5CH = CH.CO.CH_3$), mesityl oxide [$(CH_3)_2C=CH.COCH_3$), etc.
(*ii*) **Esters :** Examples of these are acrylic ester ($CH_2 = CH.COOCH_3$), fumaric esters ($C_2H_5OOC.CH=CHCOOC_2H_5$), etc
(*iii*) **Nitriles :** An example is acrylonitrile, ($CH_2=CH.CN$).
(*iv*) **Aldehydes :** An example is acraldehyde, ($CH_2=CH.CHO$).

2. **Addenda :** There are the compounds having an active methylene grup *i.e.* these can ionise to a proton and a carbanion. Examples of addenda are malonic, acetoacetic, cyanoacetic and phenylacetic ester, aliphatic nitro compounds, benzylcyanide, cyanacetamide, cyclopentadiene, indene, fluorene, sulphones, etc.

Cyclopentadience

Indene

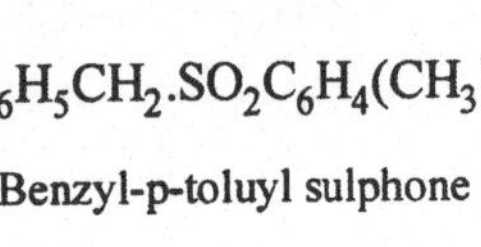
Fluorene

$C_6H_5CH_2.SO_2C_6H_4(CH_3)$-*p*

Benzyl-p-toluyl sulphone

3. **Catalyst :** Michael reaction is generally carried out in the presence of basic catalyst. When catalysts like aqueous and alcoholic caustic alkali and alkoxides (like sodium ethoxide or potassium *t*-butoxide, etc.) are used the addition product formed very often undergoes cyclisation. This can be avoided by using mild catalysts, such as, secondary amines like dialkylamines, piperidine, or tertiaryamines like trialkylamines, pyridine, etc.

Acid catalysts like $AlCl_3$, $ZnCl_2$, BF_3, H_2SO_4, etc., have also been used to bring about Michael reaction.

MECHANISM

It involves nucleophilic addition of enolate anions (carbanion) to the carbon-carbon double bond of the Michael acceptor. The reaction has been initiated by the base (catalyst) which withdraws a proton from the active methylene group, thereby forming the corresponding carbanion (or enolate anion).

$$C_2H_5ONa \rightleftharpoons C_2H_5\bar{O} + Na$$

$$CH_2(COOC_2H_5)_2 + C_2H_5\bar{O} \rightleftharpoons :\bar{C}H(COOC_2H_5)_2 + C_2H_5OH$$

Carbanion

Then, the carbanion adds to the β-carbon of the α, β-unsaturated ester acceptor, ethyl cinnamate to yield the anion (I) which accepts a proton from alcohol to produce an enol. The enol then automerises to yield the more stable product , ketone.

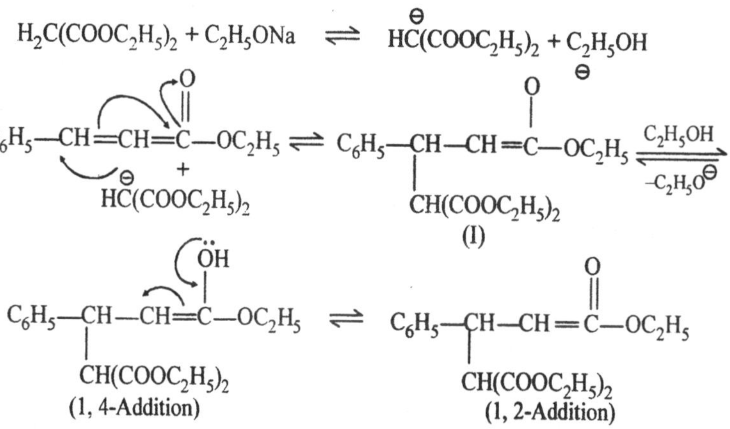

The electron-attracting group —$COOC_2H_5$ (Z) tends to facilitate the attack by stabilizing the intermediate anion (I) by dispersal of the charge. It can be seen that although 1, 4-addition takes place intially, the final result is 1,2-addition to the α, β-unsaturated carbons. The reason is that the enol reverts to the more stable ketone (recall that vinyl alcohol is unknown). In the presence of strong base, the product may undergo cyclisation. No cyclisation takes place with mild bases such as 2° or 3° amines and piperidine.

As the ethoxide ion is regenerated, only a catalytic amount of the base catalyst is required. Also, the reactions are reversible because the product is a weaker acid than the reactant because C—H bond is converted into C—C bond. therefore, a smaller concentration base will favour the product.

In order to minimise the side reactions and to get a better yield, the following conditions are to be kept in mind :

(*i*) Use of excess of active methylene compound,

(*ii*) Use of weaker basic catalyst (like piperidine, pyridine, etc.).

(*iii*) Carrying out the reaction at low temperature (25°C or below),

(*iv*) Shorter reaction time, and

(*v*) Carrying out the reaction in the presence of a protic solvent (*e.g.*, water or alcohol).

If a compound is having two double bonds in conjugation with electron-with drawing group it may undergo nucleophilic attack at the β-carbon or δ-carbon to give three products. Hence,

(a) *On attack at β-carbon* :

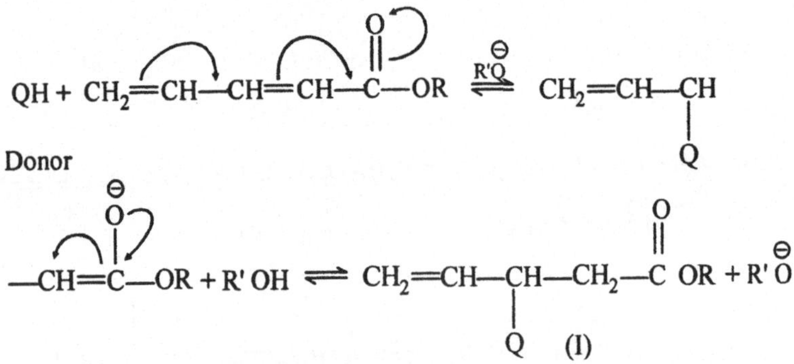

(b) *On attack at δ-carbon* :

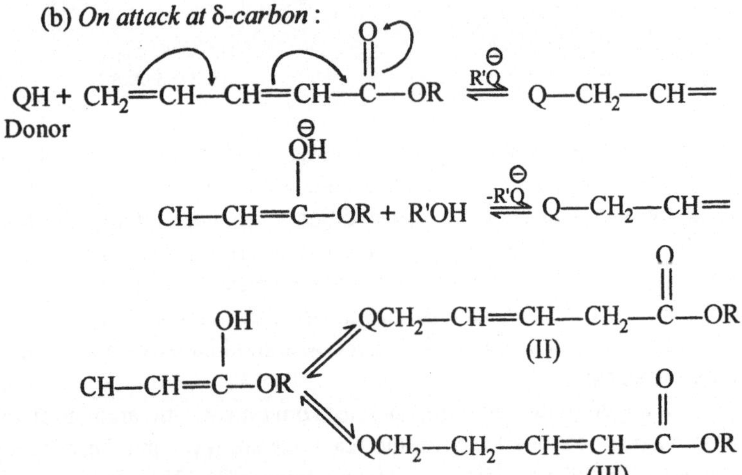

As the double bond is conjugated in (III), its the most stable of the three products. Hence it is the predominant product.

APPLICATIONS

This reaction is of great synthetic importance because a large number of organic compounds can be synthesised by using this reaction.

1. Synthesis of tricarboxylic acids

(*a*) Propane—1, 2, 3-tricarboxylic acid (*Tricarballylic acid*)

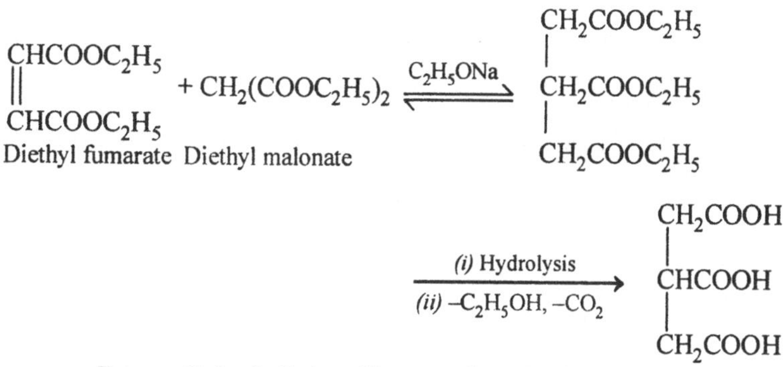

Esters of tricarballyic acid are used as plasticiser.

(*b*) Propene-1, 2, 3-tricarboxylic acid (*Aconitic acid*)

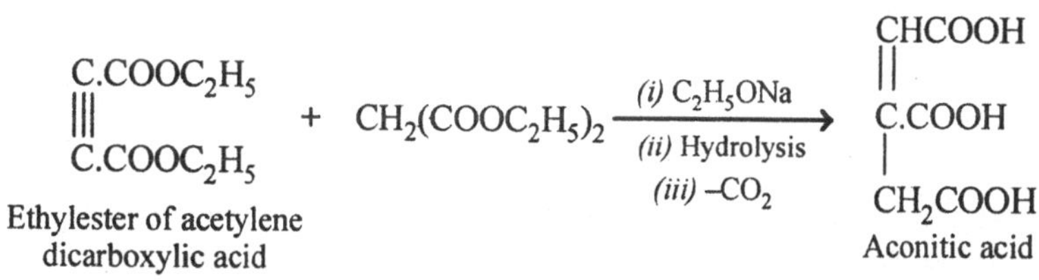

Aconitic acid finds use as a pain reliever and a fever reducer medicine. Its esters are used as plasticisers and in the manufacture of wetting reagents.

(*c*) Synthesis of β,β-Dimethylglutaric acid :

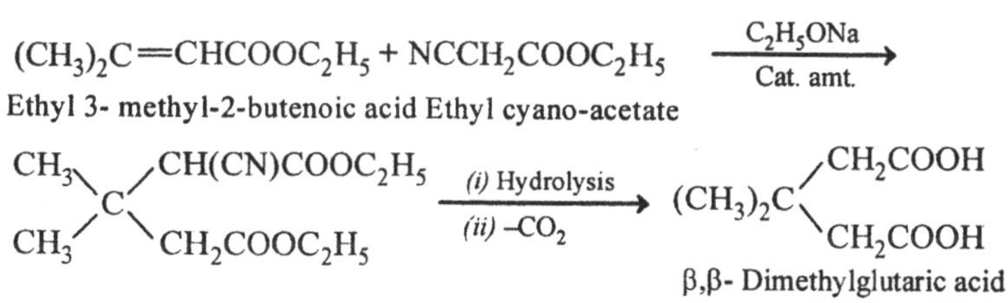

2. Synthesis of dicarboxylic acids

$$\begin{array}{l} COOH \\ | \\ CHO \end{array} + CH_2(COOC_2H_5)_2 \xrightarrow[(ii)\ H_2O]{(i)\ \text{Pyridine}} \begin{array}{l} CH.COOH \\ \| \\ CH.COOH \end{array}$$

Fumaric acid

3. **Preparation of ketocarboxylic acids :** From ketones and compounds containing reactive methylene group, keto-acids can by synthesised.

(*a*)

$$\underset{\text{Methyl vinyl ketone}}{CH_2COCH{=}CH_2} + CH_2(COOC_2H_5)_2 \xrightarrow[C_2H_5OH\ -\ 10/25°C]{C_2H_5ONa\ (\text{cat. amt.})}$$

$$CH_3COCH_2CH_2CH(COOC_2H_5)_2 \xrightarrow[(ii)\ -CO_2]{(i)\ \text{Hydrolysis}} \underset{\text{5-Oxohexanoic acid}}{CH_3COCH_2CH_2COOH}$$

(*b*)

$$\begin{array}{c} CH_2COCH_2 \\ | \\ CH_3 \end{array} + 2CH_2{=}CH{-}CN \xrightarrow[(CH_3)_3COH]{KOH}$$

Butanone

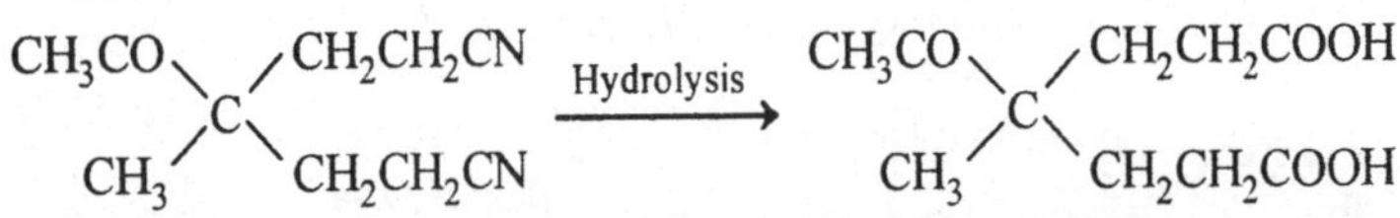

4-Acetyl-4-methylhe ptanedioic acid

(*c*) Enamines, i.e. β-unsaturated amines, such as *N*-cyclohexenopyrrolidine can undergo Michael condensation thereby forming cyclohexanone drivative.

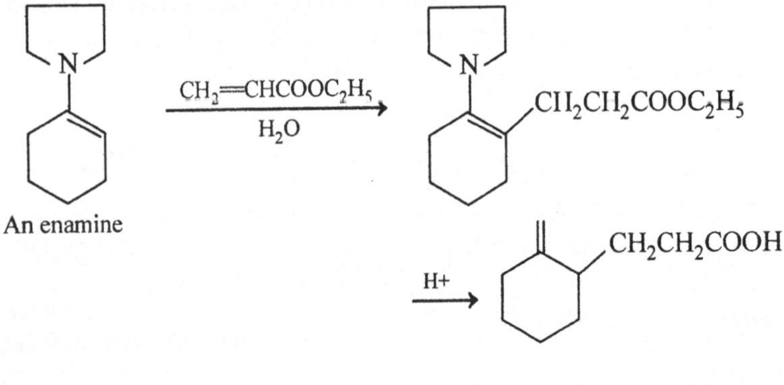

4. **Preparation of nitro-derivatives**

(*a*) $(CH_3)_2CH-NO_2 + CH_2{=}CHCOOCH_3 \xrightarrow[\text{H}_2\text{O, Dioxane, 70-100°C}]{C_6H_5CH_2\overset{+}{N}(CH_3)_3\overset{-}{O}H}$

2-Nitropropane — Methyl propeonate (Methyl acrylate)

$$O_2N-\underset{CH_3}{\overset{CH_3}{\underset{|}{\overset{|}{C}}}}-CH_2CH_2COOCH_3 \xrightarrow[(ii)\ HCl]{(i)\ KOH} O_2N\underset{CH_3}{\overset{CH_3}{\underset{|}{\overset{|}{C}}}}CH_2CH_2CH_2COOH$$

4-Methyl-4-nitropentanoic acid

(*b*) Nitromethane undergoes Michael reaction with mesityl oxide, in the presence of alcoholic potassium hydroxide, to form 4, 4-dimethyl-5-nitropentane-2-one.

$$\underset{\text{Nitromethane}}{CH_3NO_2} + KOH \rightleftharpoons CH_2{=}N\overset{-}{O_2}\overset{+}{Na}$$

$$\underset{\text{Mesityl oxide}}{(CH_3)_2C{=}CHCOCH_3} + \overset{-}{C}H_2{=}NO_2\overset{+}{Na} \longrightarrow$$

$$(CH_3)_2\underset{CH=N\overset{-}{O_2}\overset{+}{Na}}{\underset{|}{C}}CH_2COCH_3 \xrightarrow{HCl} (CH_3)_2\underset{CH_2NO_2}{\underset{|}{C}}CH_2COCH_3$$

4, 4-Dimethyl-5-nitrobutan-2-one

5. **Preparation of Cyano derivatives,** *e.g.,*

$$HCN + \underset{\text{2-Methyl-1-nitropropene}}{(CH_3)_2C{=}CH.NO_2} \xrightarrow[(CH_2OCH_3)_2]{KOH} (CH_3)_2\underset{CN}{\underset{|}{C}}.CH_2NO_2$$

2, 2-Dimethyl-3-nitropropanonitrile

6. **Preparation of cyclic compounds :** In certain the addition products, formed during Michael reaction, undergo further intramolecular condensation thereby forming cyclic products.

Synthesis of Dimedone

(*a*) $\underset{\text{Mesityl oxide}}{(CH_3)_2C{=}CH-\overset{\overset{O}{||}}{C}-CH_3} + H_2C(COOC_2H_5)_2 \xrightarrow{C_2H_5ONa}$

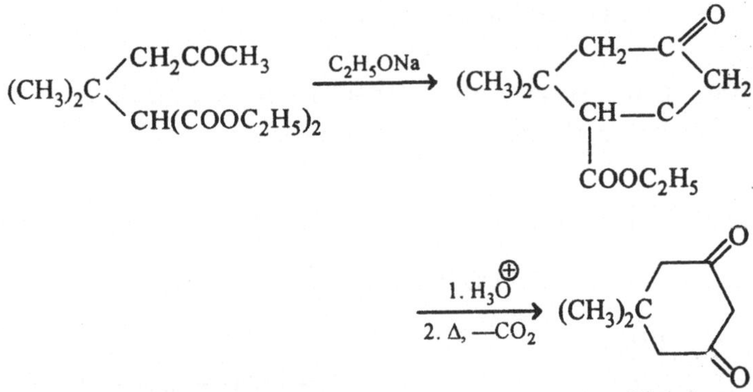

Dimedone

Dimedone is also employed for the separation of aldehydes from ketone because it reacts only with aldehydes.

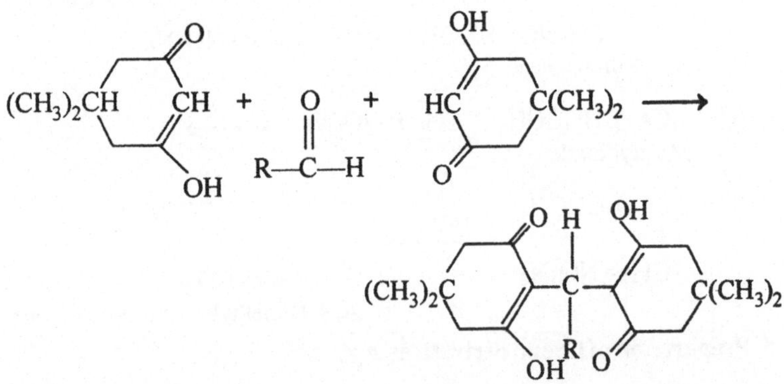

This reagent has been found to be very useful for detecting the presence of an aldehyde group, even in the presence of a ketogroup, because the reagent attacks only the aldehyde group in the following manner :

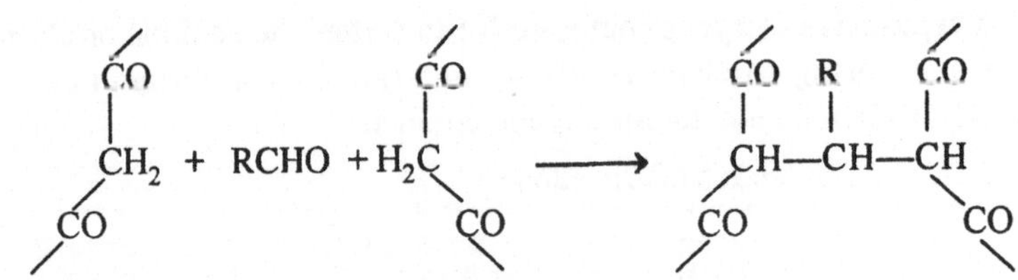

Dimedone   Dimedone

(*b*) *Synthesis of an octalone derivative*

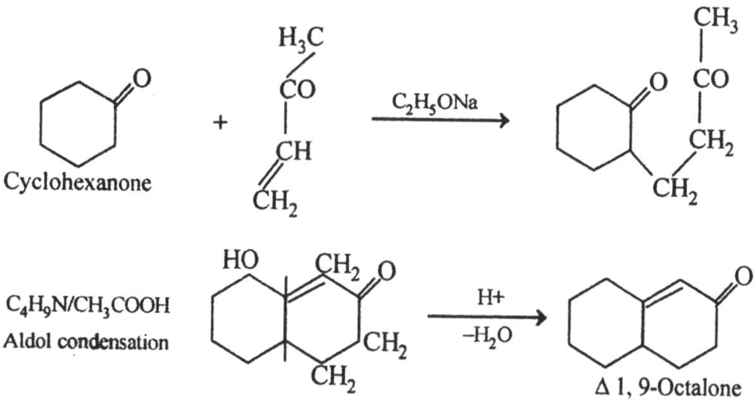

(*c*) *Double Michael addition results in ring formation*

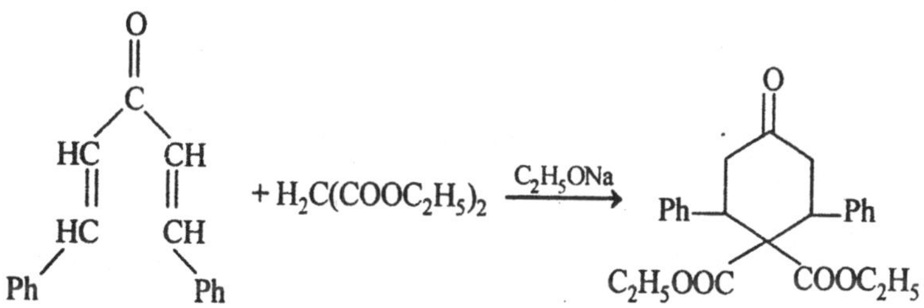

7. **Synthesis of cyclopropane derivatives,** *e.g.*, **carnoic acid.**

$$(CH_3)_2C{=}CH{-}COOC_2H_5 + CH_2(CN){-}COOC_2H_5 \xrightarrow{C_2H_5ONa}$$

Ethyl-3-methylcrotonate　　Ethyl cynoacetate

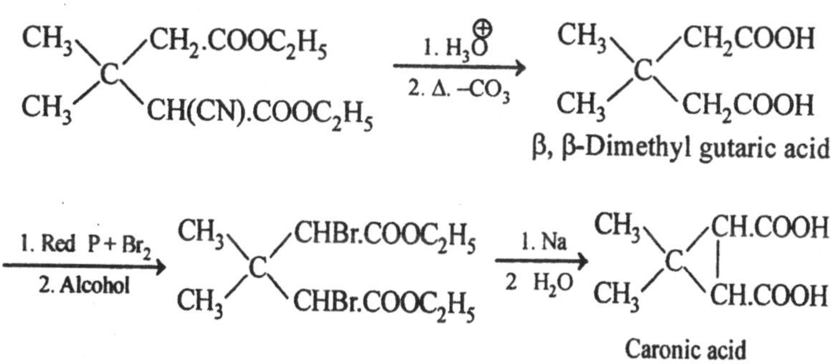

β,β-Dimethylglutaric acid can also be synthesised from dimedone, However, is one of the starting materials camphoric acid synthesis

8. Alicyclic System of Steroids : Michael addition is frequently used in building up the alicyclic system of steroids, *e.g.*, in the synthesis of cholesterol.

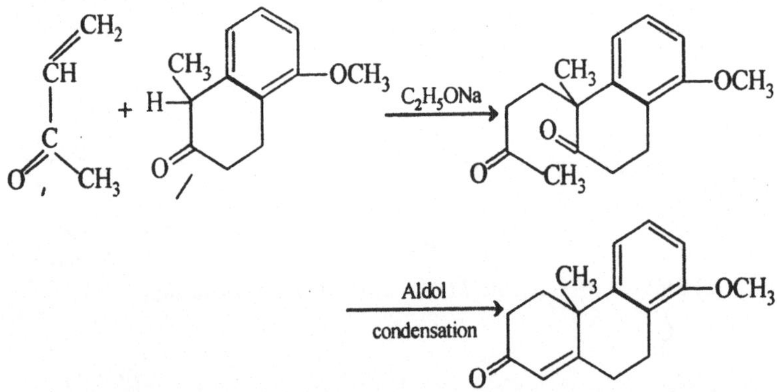

This method has been employed during the synthesis of chloesterol to build the ring system.

9. Synthesis of cincholoiponic acid :

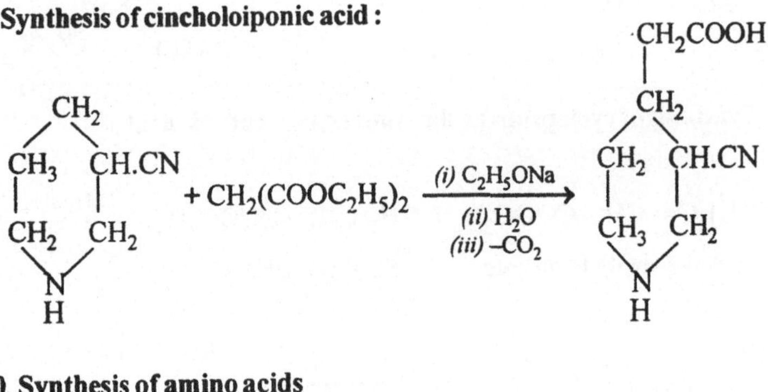

10. Synthesis of amino acids

$$CH_2{=}CH{-}COOCH_3 + CH_3CONHCH(COOC_2H_5) \xrightarrow{C_2H_5ONa}$$

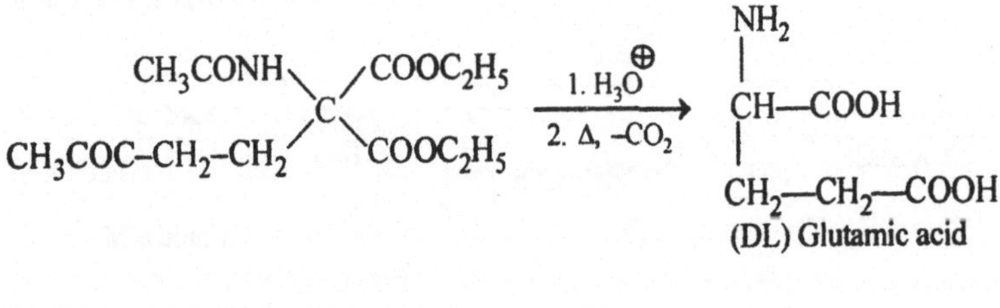

(DL) Glutamic acid

71

MANNICH REACTION

This is the condensation between a compound containing at least one active hydrogen atom, formaldehyde and ammonia, a primary or a secondary amine (preferably as the hydrochloride).

$$C_6H_5.COOH_3 + CH_2O + R_2NH.HCl \rightarrow C_6H_6.CO.CH_2CH_2NR_2.HCl + H_2O$$

So the *net change, during this reaction, is the replacement of the active hydrogen atom by a aminomethyl group or substituted aminomethyl group.* In practice the base, commonly known as "**Mannich base**" is usually isolated as the hydrochloride.

MECHANISM

[A] *Base Catalyzed Reaction*

Much controversy brewed up over the years regarding the base catalyzed, especially as to whether the aldehyde is first attacked by the compound having active hydrogen or by the ammonia/amine. The evidences available so far go in favour of the latter

STEP - [I]

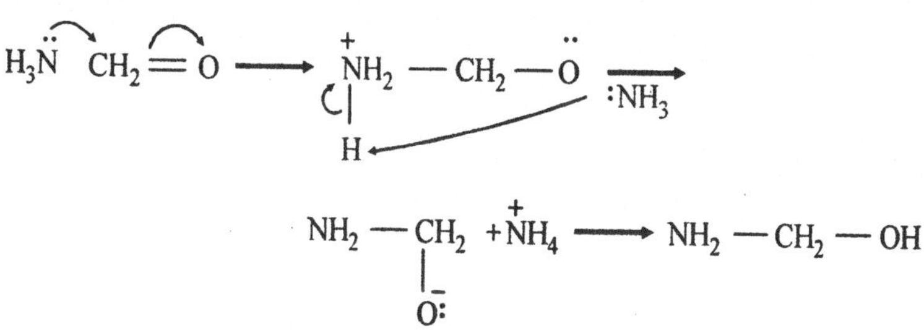

STEP - [II]

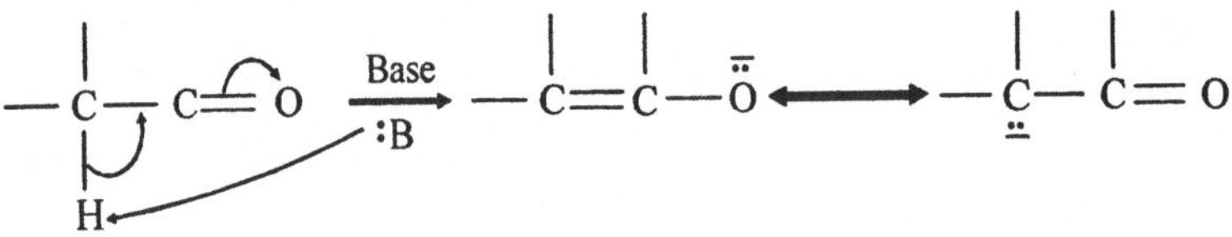

STEP - [III] Nucleophilic Attack

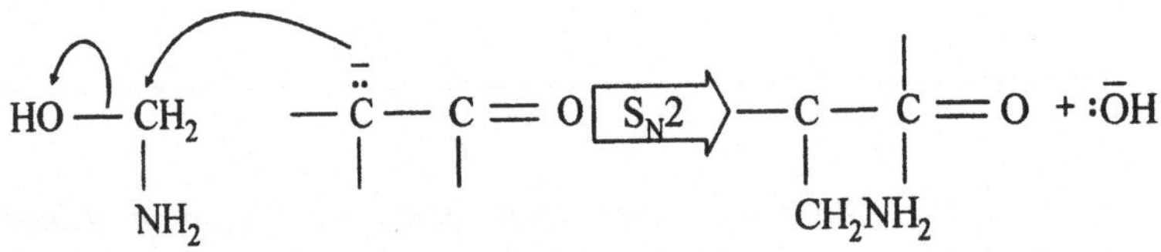

[B] ***Acid Catalyzed Reaction***

STEP - [I]

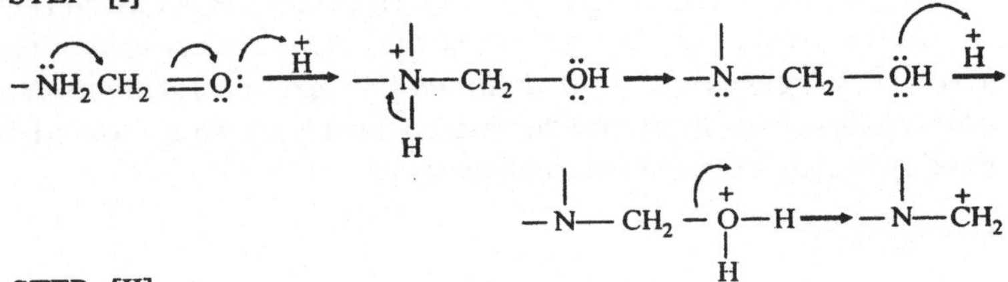

STEP - [II]

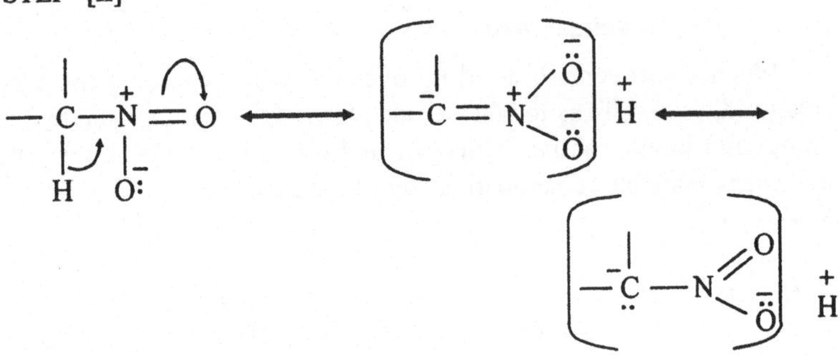

STEP - [III] Nucleophilic Attack

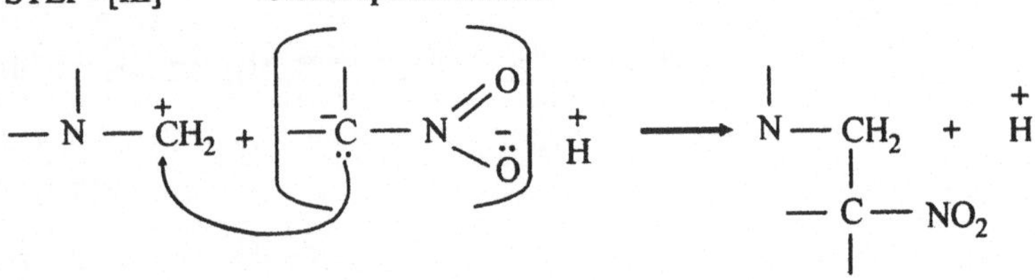

Note :

Ammonia is furnished in the form of its salt which formally forms the MANNICH base as NH_2CH_3OH

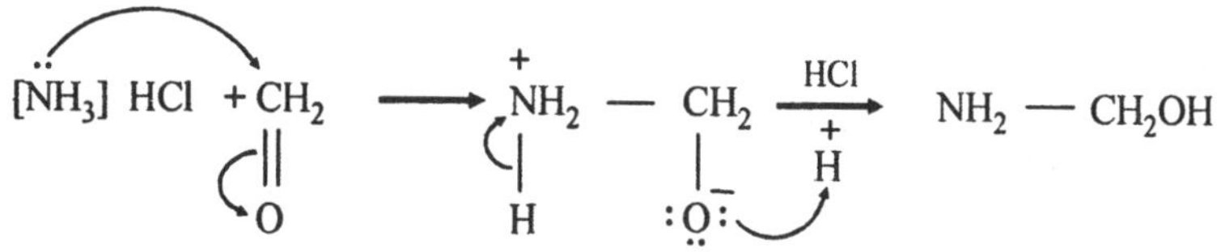

(I) ***Primary*** as well as ***secondary*** amines can replace amonia

(II) Among those compounds which contain active hydrogen, mention may be made of the followings

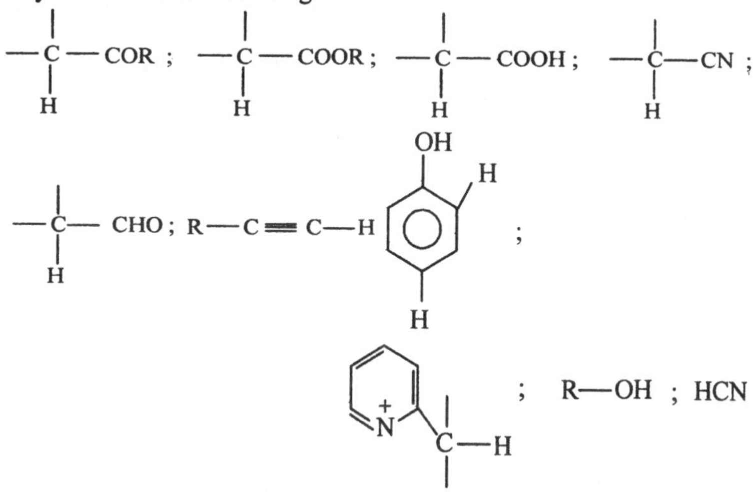

The negative inductive groups like CN, CHO, COR, COOR, NO_2 etc. attached to the carbon atom having at least one hydrogen atom render the latter labile as H^+

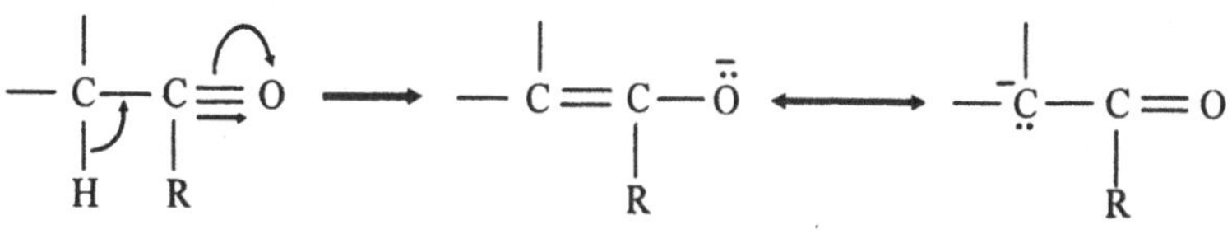

resonance stabilizes the anionic structure

Resonance stablization of the resulting structures renders proton mobilization easy.

COMPONENTS OF THE MANNICH'S REACTION

In this reaction there are three important reactants.

1. **Compounds having reactive hydrogen atom :** the various examples of such compounds are as follows :

(*i*) *Ketone* : All the ketones having the grouping >COOH< undergo Mannich reaction. Examples of such ketones are acetophenone, acetone, etc.

(*ii*) *Acid and their esters* : Acids and their esters especially acetoacetic, malonic, β-cyano and β-ketoacids undergo Mannich reaction. For example, ethylmalonic acid undergoes Mannich reaction.

$$C_2H_5.CH.(COOH)_2 + CH_2O + (CH_3)_2NH \longrightarrow C_2H_5C(COOH)_2.CH_2.N(CH_3)_2$$

(*iii*) *Phenols* : Phenols undergo Mannich reaction. In phenols, the basic group enters *o*-and/or *p*-to the phenolic group, *e.g.*,

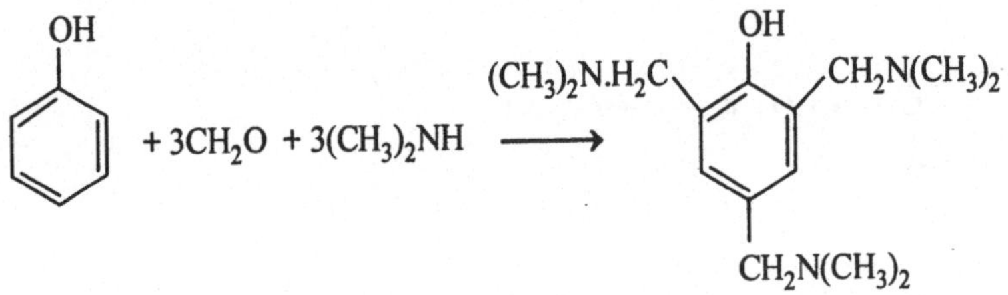

(*iv*) *Furan, pyrrole, indole and their derivatives* : These compounds undergo Mannich reaction. However, these compounds are sensitive to acids. Therefore, Mannich reaction with such compounds must be carried out in the presence of a weak acid, *e.g.*, CH_3COOH_2.

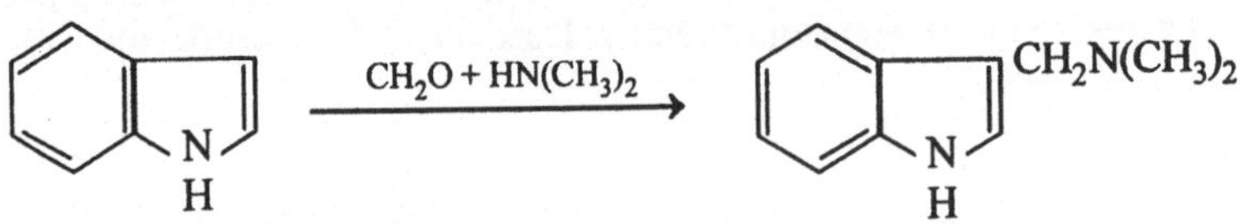

(*v*) *Methylated pyridines and quinolines* : These compounds undergo Mannich reaction. For example, α-picoline.

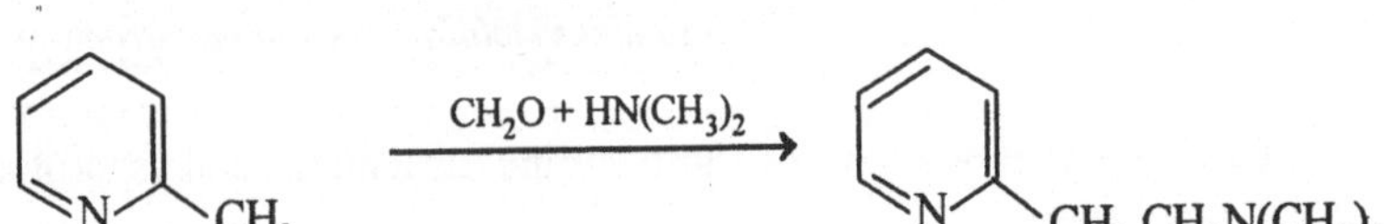

(*vi*) *Nitro compounds* : Primary and secondary nitro compounds also undergo the Mannich reaction.

$$R_2CH.NO_2 + CH_2O + NH_4Cl \longrightarrow R_2\text{—}\overset{\displaystyle NO_2}{\overset{|}{C}}\text{—}CH_2.NH_2.HCl + H_2O$$

(*vii*) *Acetylenes* : Phenylacetylenes and some of its unclear substituted derivates may undergo Mannich reaction.

$$C_6H_5C{\equiv}CH + CH_2O + (CH_2)_2NH \xrightarrow[H_2O]{H+} C_6H_5C{\equiv}C.CH_2N(CH_3)_2$$

It is to be noted that if the compound has a second reactive hydrogen atom, further reaction may take place with the introduction of another—$CH_2.NR_2$ group *e.g.*, with acetophenone.

$$C_6H_5.CO.CH_3 \xrightarrow{CH_2O + HN(CH_3)_2} C_6H_5.COCH_2.CH_2N(CH_3)_2$$

$$\xrightarrow{CH_2O + HN(CH_3)_2} C_6H_5.COCH.[CH_2N(CH_3)_2]_2$$

2. **Amines :** The amine is usually added as its hydrochloride and several drops of hydrochloric acid are often added to ensure that the reaction mixture is not basic. The amine may be ammonia, primary or a secondary amine but usually a secondary amine, such as, dimethylamine, diethylamine, piperidine, pyrrolidine, etc., is employed to avoid side reactions. The secondary amine contains one replaceable hydrogen and the Mannich base formed is a tertiary base. When a primary amine is used, the product is a secondary base and can further react to give teritary amine.

$$CH_3.CO.CH_3 + CH_3O + H_2NCH_3 \xrightarrow[-H_3O]{H+} \underset{\text{Secondry amine}}{CH_3.CO.CH_2CH_2NH.CH_3}$$

$$\xrightarrow{CH_2O + CH_3.COCH_3} \underset{\text{Tertiary amine}}{CH_3.CO.CH_2.CH_2\overset{\displaystyle CH_3}{\overset{|}{N}}.CH_2.CH_2.COCH_3}$$

Similarly, with ammonia three successive reactions take place in the following manner :

$$3CH_3.CO.CH_3 + 3CH_2O + NH_3 \longrightarrow \underset{\text{Tertiary amine}}{(CH_3.CO.CH_2CH_2)_3N}$$

3. **Aldehyde :** Formaldehyde which is the most common aldehyde used is usually introduced as an aqueous solution (as formalin), as solid trimer trioxymethylene, $(CH_2O)_3$, or as the solid ploymer (polyoxymethylene), the latter two reagents are converted into formaldehyde by acid added to reaction mi xture. Higher aldehydes have also been used.

EXAMPLES:

(*i*)

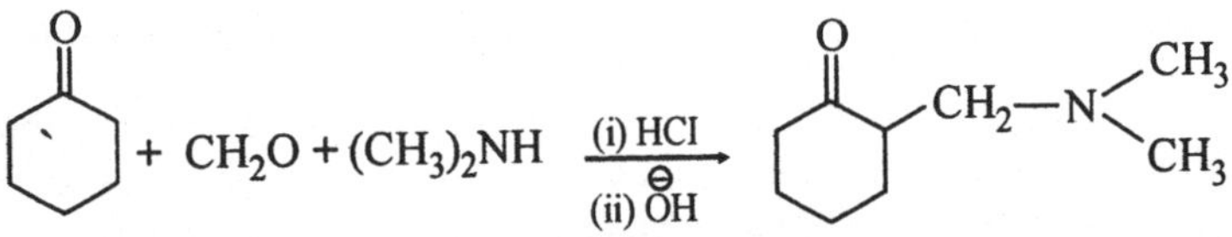

(*ii*)

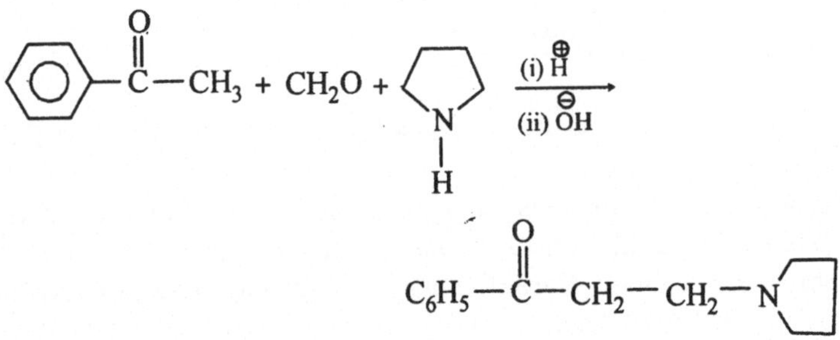

(*iii*)

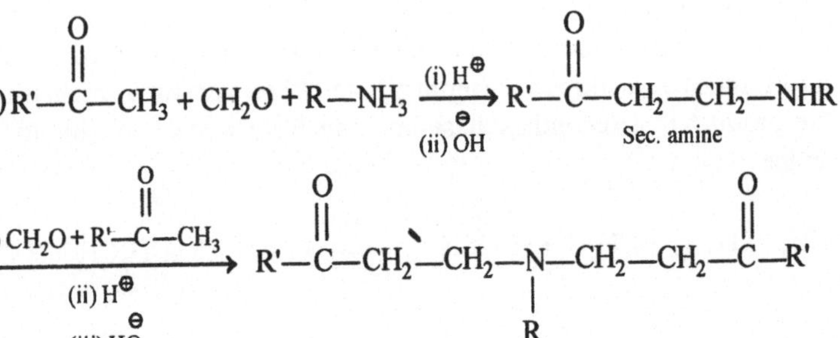

Mannich reaction is given by those carbonyl compounds which undergo enolisation. Apart from these compounds, 2-methylpyridine and related compounds also give this reaction because in the presence of acid these compounds undergo nitrogen-analogue of an enol.

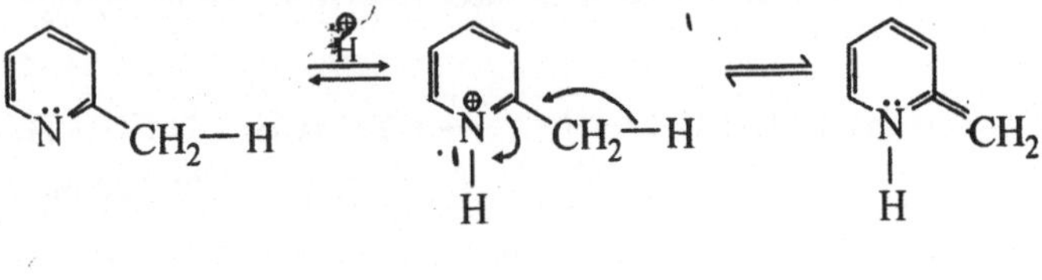

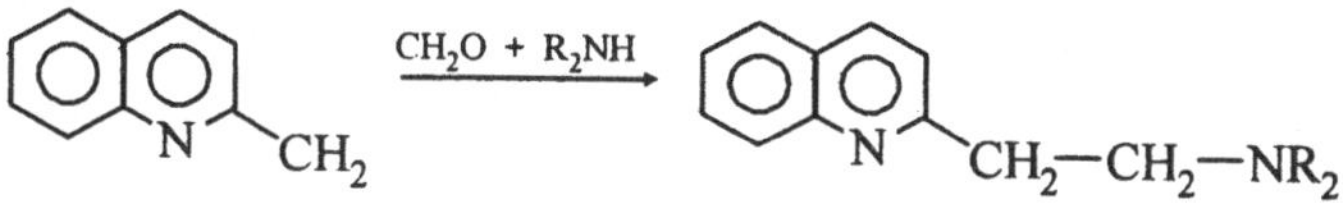

Phenols, Furan, Pyrrole, Indole also gives this reaction because intermediate of the reaction is Iminium cation which is an electrophile and these compounds give ArSE reaction.

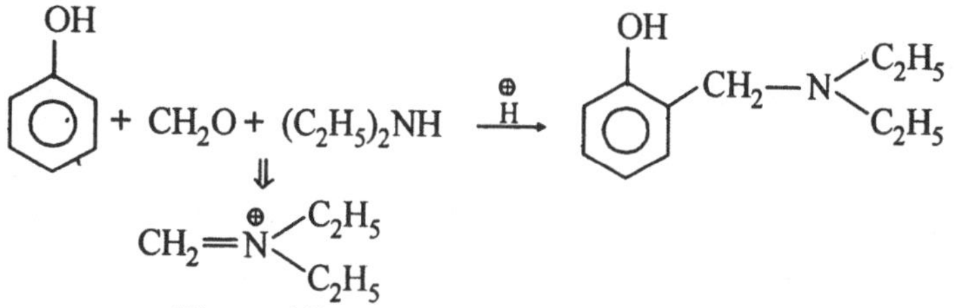

Electrophile

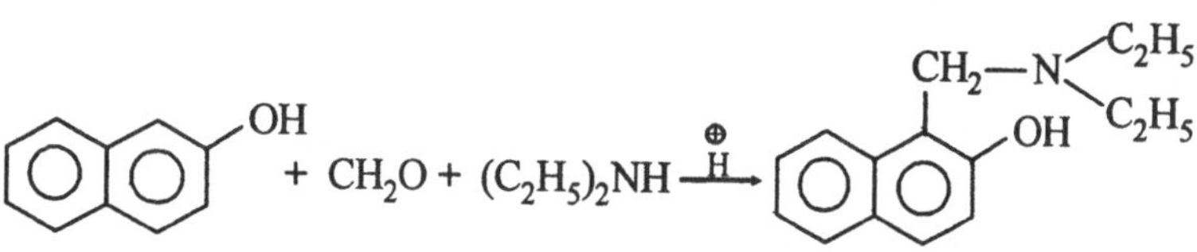

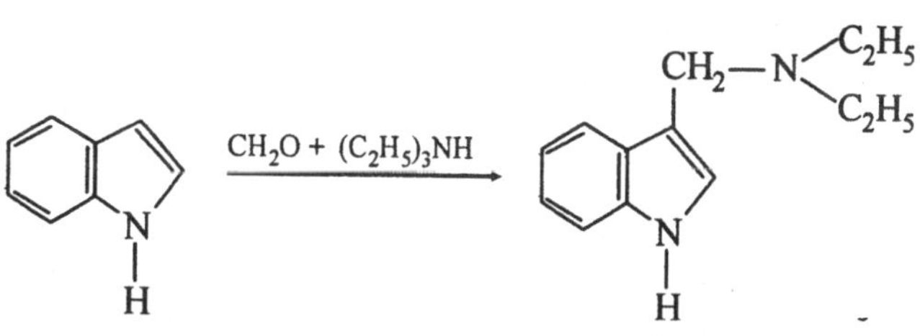

Gramine

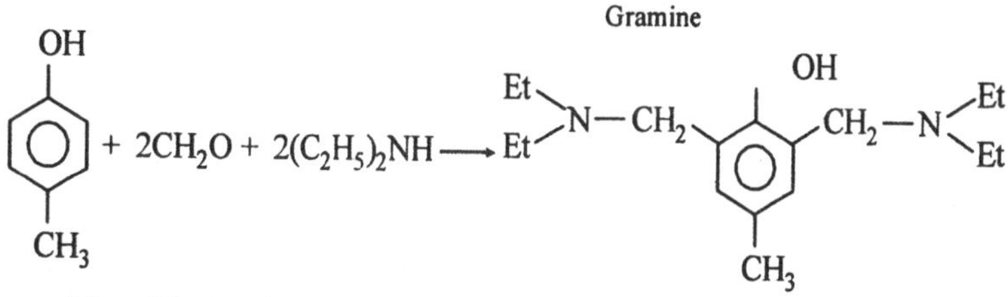

Mannich reaction is carried out in the presence of acid and the product of the reaction before basidification is Mannich salt. *i.e.* hydrochloride of Mannich base. This salt is stable at room temperature but undergoes β-elimination on heating.

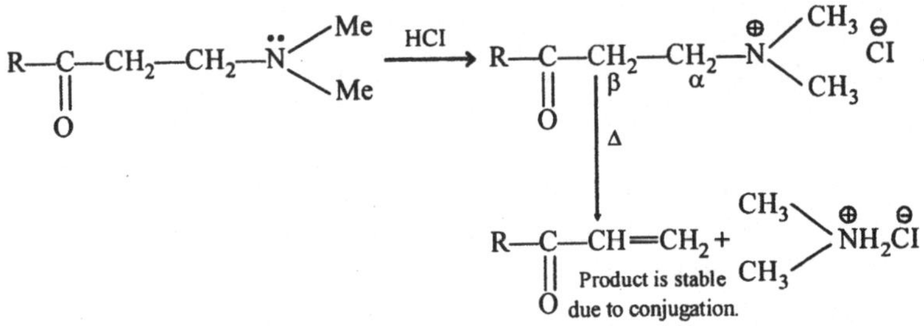

Mannich base also undergoes elimination reaction is presence of strong base,

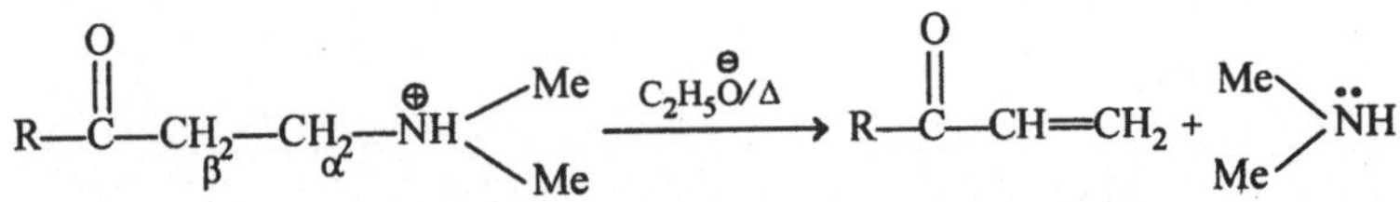

SYNTHETIC APPLICATIONS

From the above considerations, it follows that the reaction is widely used to solve synthetic problems, *viz.*

1. *Synthesis of alkaloids* : Many alkaloids are synthesised by Mannich reaction; the most important of which is Robinson's synthesis of tropinone which is used for the synthesis of atropine.

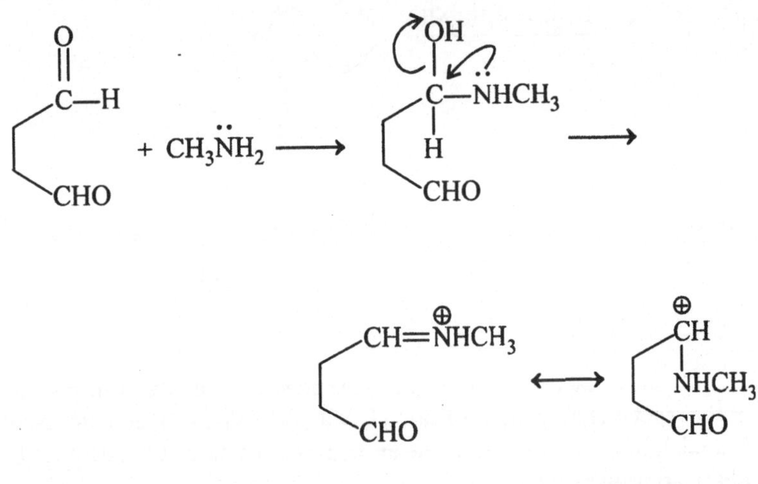

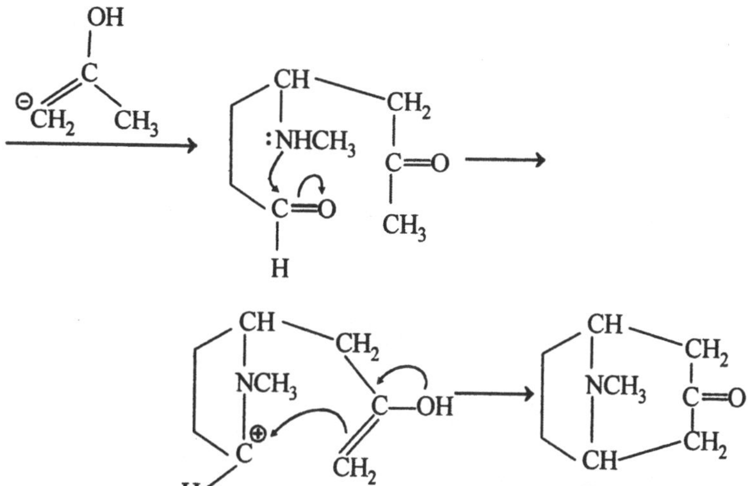

Tropinone

Better yield of tropinone is obtained by the use of calcium salt of acetone dicarboxylate in place of acetone.

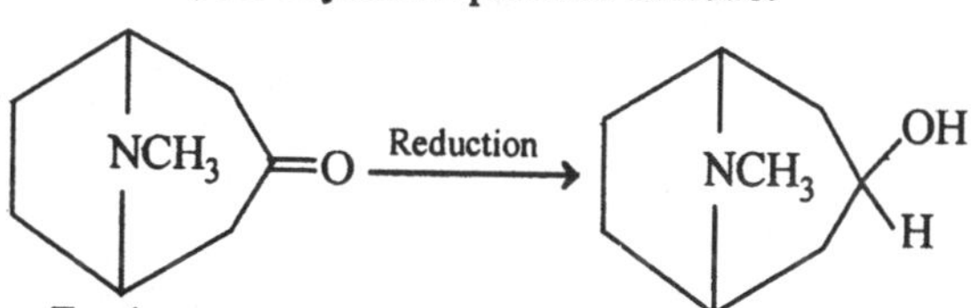

Tropinone

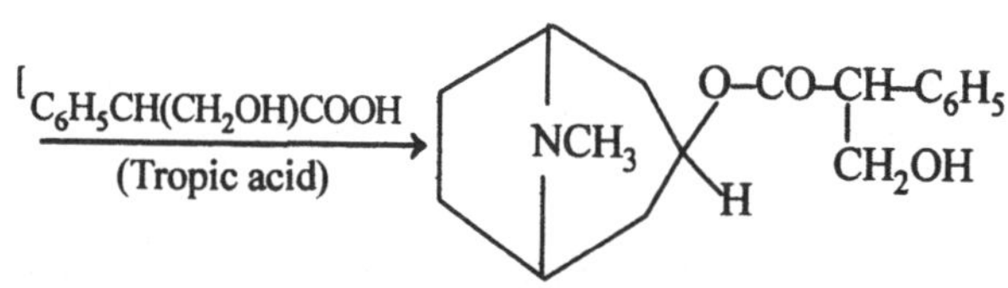

Atropine

Other alkaloids synthesised by Mannich reaction are cocaine, pseudopelletierme, hygrine, etc. Their syntheses are as follows :

Sythesis of cocaine :

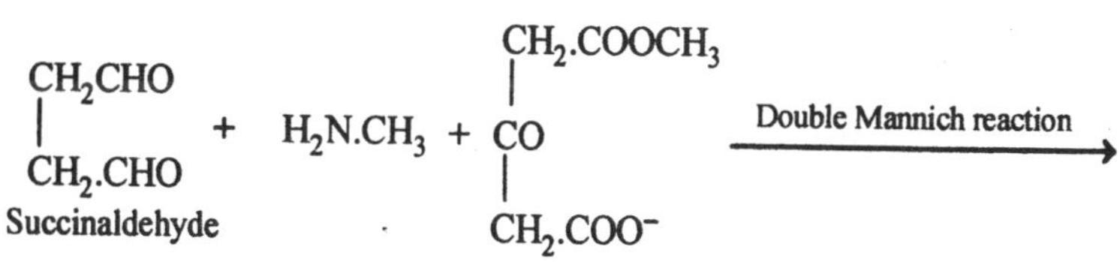

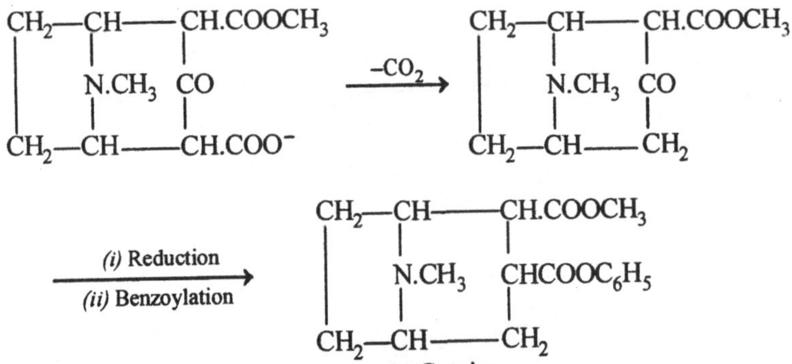

Cocaine

Synthesis of pseudopelletierine.

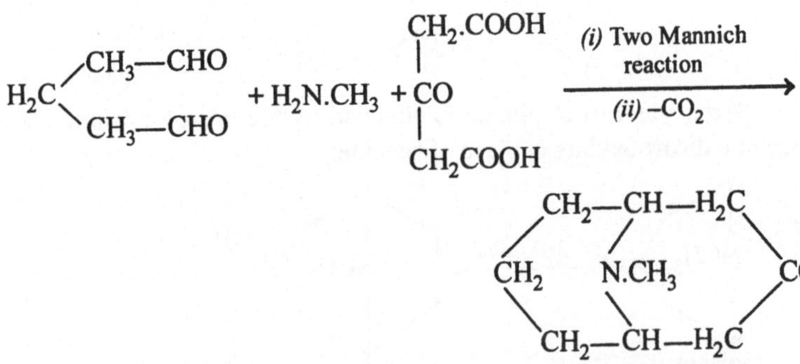

Pseudopelletierine

2. *Synthesis of local anaesthetics* : Important local anaesthetics may be synthesised by using Mannich reaction. It can be exemplified by the synthesis of tutocaine.

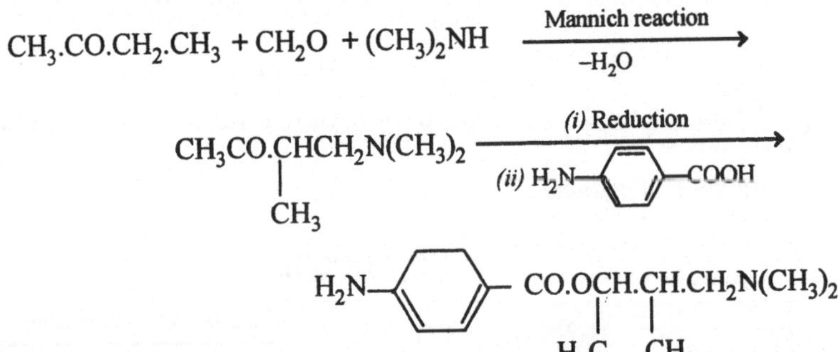

Tutocaine

3. *Other applications* : As the Mannich base formed can rapidly eliminate nitrogen as R_2NH in the following manner,

$$C_6H_5.CO.CH_2.CH_3 \xrightarrow[\text{Ni/H}_2]{\text{Reduction}} \underset{\text{Mannich base}}{C_6H_5.COCH_2.CH_2NR_2}$$

$$\downarrow \text{Heat}$$

$$R_2NH + C_6H_5.CO.CH{=}CH_2$$

the Mannich reaction is of synthetic importance. Some important uses involving such a reaction are as follows :

(*a*) *Preparation of an* α, β*-unsaturated carbonyl compounds* : Mannich base hydrochloride which is especially derived from aliphatic compounds eliminates an amine hydrochloride forming unsaturated product, *e.g.*

$$[C_6H_5.CO.CH_2CH_2\overset{+}{N}H(CH_3)_2]\overline{C}l \xrightarrow{\text{Heat}}$$

$$C_6H_5.CO.CH{=}CH_2 + (CH_3)_2\overset{+}{N}H\overline{Cl}$$

This elimination reaction finds important uses :

(*i*) Catalytic reduction of the unsaturated product yields the next higher homologue of the ketone from which the Mannich base was derived, *e.g.*

$$\underset{\text{Phenyl methyl ketone}}{C_6H_5.CO.CH_3} \xrightarrow[\text{(Mannich reaction)}]{CH_2O + HN(CH_3)_2} C_6H_5.CO.CH_2.CH_2N(CH_3)_2$$

$$\xrightarrow{\text{Heat}} C_6H_5.CO.CH{=}CH_2 \xrightarrow{\text{Ni/H}_2} \underset{\text{Phenylethyl ketone}}{C_6H_5.CO.CH_2.CH_3}$$

(*ii*) α, β-Unsaturated carbonyl compounds produced in *situ* are used for the condensation reaction (Michael reaction) as it is better to obtain the unsaturated carbonyl compound from the quaternized Mannich base than to use the free ketone, because the latter readily gets polymerized. For example, for condensation reaction using methyl vinyl ketone it is convenient ot generate the ketone in *situ* from quaternized base than to take it as a starting material. The best example of this is Robinson's ring extension which finds use in building up the ring system of steroids.

(I) $CH_3COCH_3 + HCHO + (CH_3)_2NH \longrightarrow CH_3COCH_2CH_2N(CH_3)_2I$

$$\xrightarrow[\Delta]{CH_3I} \underset{\text{Quanteinary salt}}{CH_3COCH_2CH_2\overset{\oplus}{N}(CH_3)_2\overset{\ominus}{I}} \xrightarrow{\Delta} CH_3COCH{=}CH_2 + (CH_3)_3NHI$$

(II) 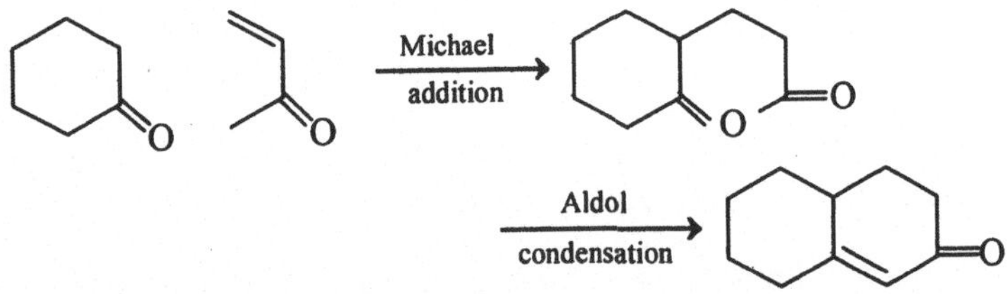

The reaction finds uses in Woodward's synthesis of chlolesterol and Jonson's synthesis of epiandrosterone.

(*b*) *Replacemant of amino Group* : It is possible to replace the amino group of the Mannich base by other function for which the Mannich base is first quaternised. Some examples are as follows :

(*i*) *Synthesis of heteroauxin (indole-β-acetic acid)*

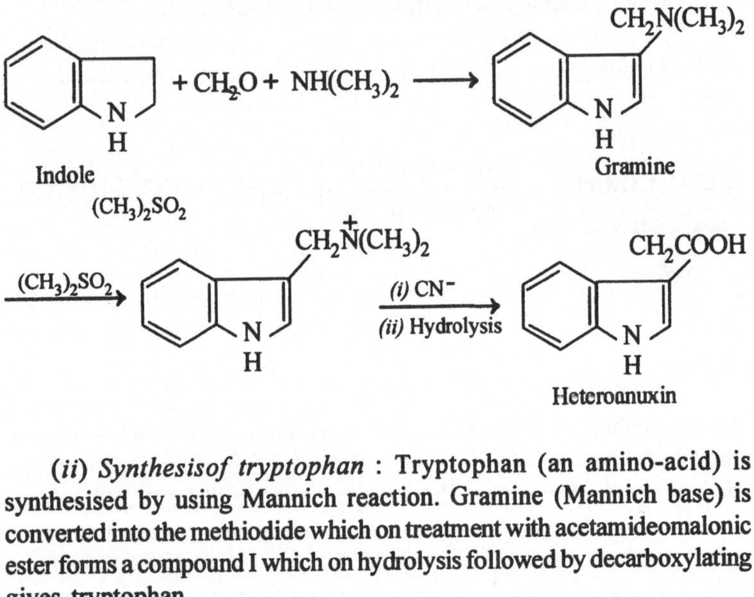

(*ii*) *Synthesisof tryptophan* : Tryptophan (an amino-acid) is synthesised by using Mannich reaction. Gramine (Mannich base) is converted into the methiodide which on treatment with acetamideomalonic ester forms a compound I which on hydrolysis followed by decarboxylating gives tryptophan.

$$CH_3CONH{-}CH(COOC_2H_5)_2 \xrightarrow{C_2H_5ONa} CH_3CONH{-}\bar{C}(COOC_2H_5)_2$$

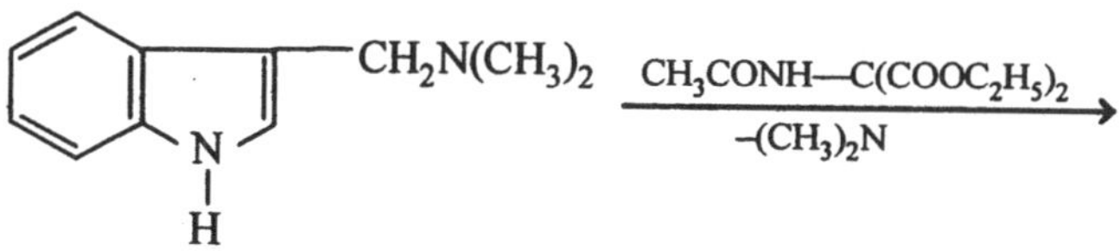

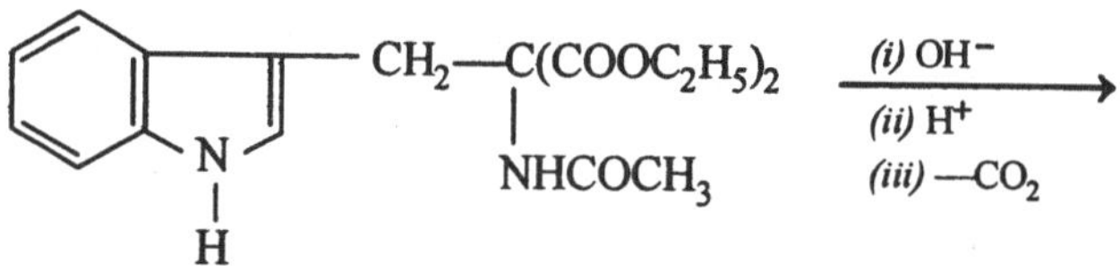

(I)

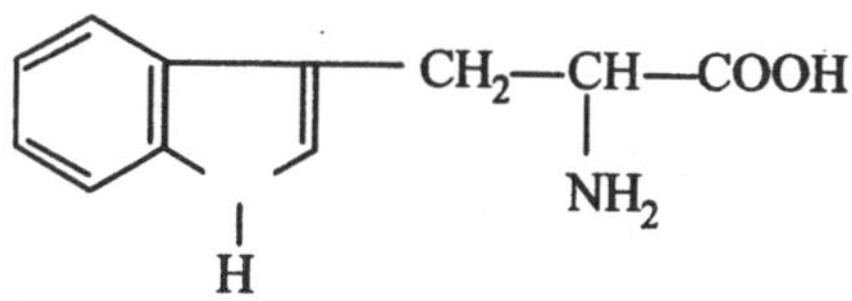

Tryptophan

(*iii*) *Synthesis of quinine* : Mannich reaction is used in the synthesis of quinine to introuduce a methyl group into the 8-position of 7-hydroxysioquinoline. Mannich base is obtained from 7-hydroxyisoquinoline, formaldehyde, and piperidine. Now the amino group of the base has been replaced by hydrogen from methoside ion (hydride ion transfer).

Problem-1

$$\underset{[A]}{C_7H_7O_2N} + CH_2O + NH.Me_2 \longrightarrow \underset{[B]}{C_{10}H_{14}O_2N_2} \xrightarrow{\Delta} \text{1-Nitrostyrene}$$

$$\xrightarrow{Zn/\overset{+}{H}} \xrightarrow{HNO_2} \underset{[C]}{C_8H_{10}O} \xrightarrow[\Delta]{} \underset{[D]}{C_8H_8}$$

$$[D] \xrightarrow{Br_2 / H_2O} \underset{[E]}{C_8H_8Br_2}$$

Answer

A ► $\varnothing.CH_2.NO_2$; B ► $\varnothing.CH_2.NO_2$ with $CH_2{-}NMe_2$ on the CH_2 ; C ► $\varnothing.CH.OH$ with CH_3 on the CH

$$\text{B: } \varnothing.\underset{\displaystyle |\atop \displaystyle CH_2{-}NMe_2}{CH}.NO_2 \qquad \text{C: } \varnothing.\underset{\displaystyle |\atop \displaystyle CH_3}{CH}.OH$$

D ▶ Ø.CH$=$CH$_2$; **E** ▶ Ø.CH(Br)—CH$_2$(Br)

Problem-2

$$C_2H_4 + HBr \longrightarrow [A] \xrightarrow[\text{EtOH}]{\text{2KCN}} [B] \xrightarrow{\text{LiAlH}_4} \underset{[C]}{C_4H_6O_2}$$

$$[C] + HN.Me_2 + C_2H_2O_2$$

Identify **A, B, C, & D.**

Answer

A ▶ CH$_2$Br — CH$_2$Br ; **B** ▶ CH$_2$(CN)—CH$_2$(CN)

C ▶ CH$_2$(CHO)—CH$_2$(CHO) ; **D** ▶ CHO—CHO

72

MEERWEIN-PONNDORF-VERLEY REDUCTION (MPV REDUCTION)

Meerwein-Ponndorf-Verley reduction (MPV Reduction)

Conversion of carbonyl compound into alcohol by the action of Aluminium isopropoxide in isopropyl alcohol is called MPV reduction. This reduction takes place under very mild condition. The reaction is highly specific for carbonyl compounds; other groups which are susceptible to reduction, such as C = C, C ≡ C, —NO_2, COOR do not get reduced by this reagent.

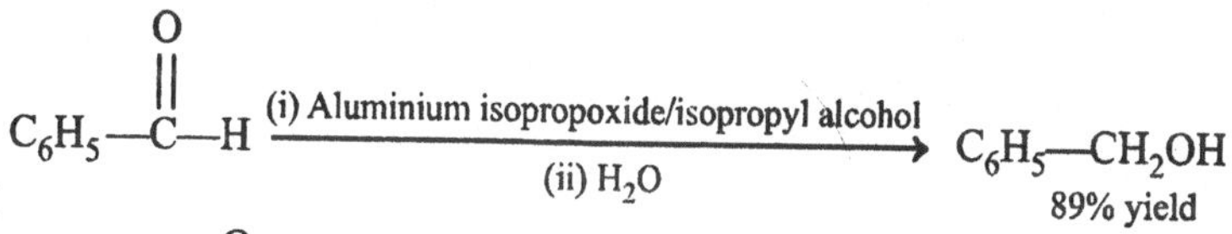

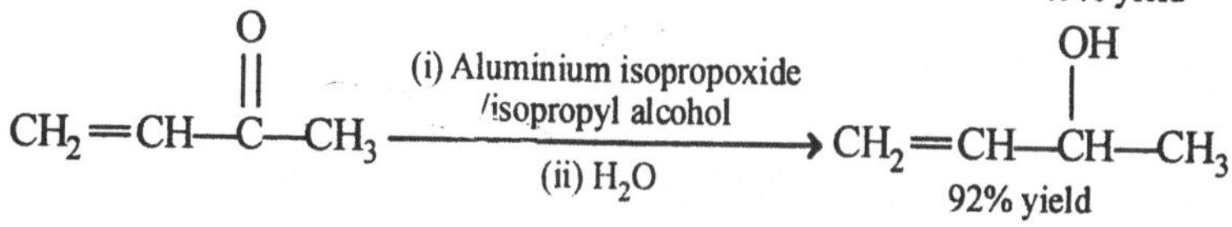

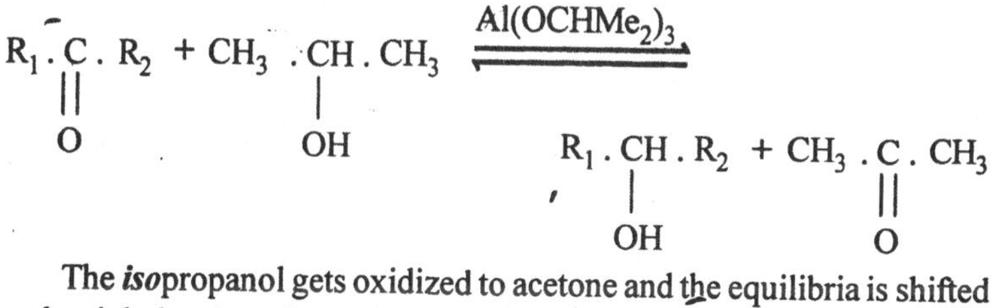

The *iso*propanol gets oxidized to acetone and the equilibria is shifted to the right by slow distillation.

MECHANISM

The MEERWEIN-PONNDORF-VERLEY REDUCTION is believed to proceed via a cyclic transition state and can be visualized as follows :

[I] ***Co-ordination of the ketone with the Alkoxide***

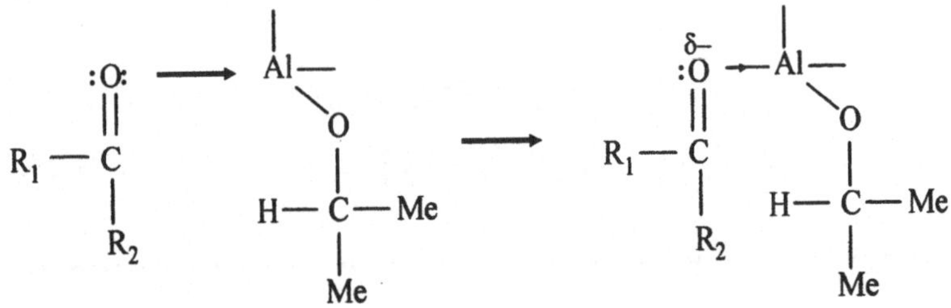

[II] ***Hydride Transfer***

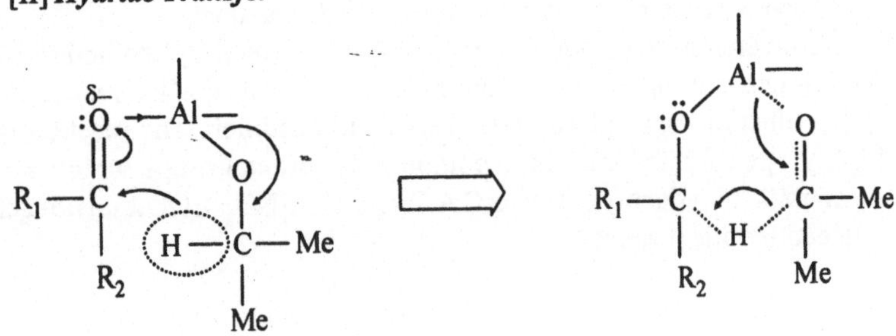

[III] ***Fission of cyclic complex***

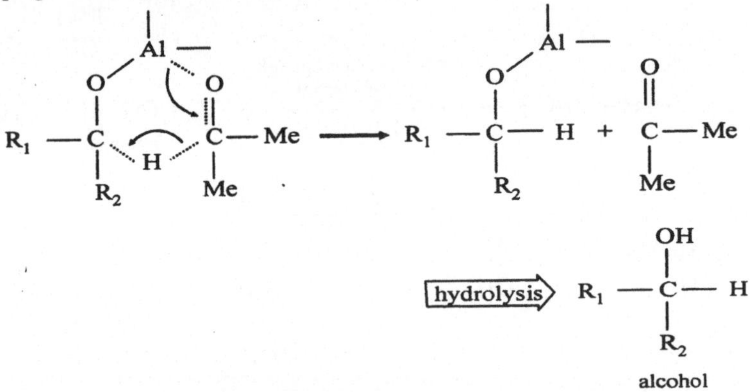

SUPPORT OF THE AVBOVE MECHANISM

Employing the tagging technique using deuterium as the labelled atom, Williams, Kreiger & Day have shown [**J. Am. Chm. Soc.** vol. 75/P-

2404/1953] that it is the ***alfa***-hydrogen of the alkoxide that attaches itself, during the reaction course, with the carbonyl carbon of the starting aldehyde or ketone. Basing on this observation, they have concluded that the reaction proceeds via formation of cyclic complex.

Note :

(I) The reaction takes place under very mild conditions.

(II) This reaction is particularly specific for carbonyl compounds and as such other functional groups may be present with out being attacked.

LIMITATIONS

Beta-keto ester, ***beta***-diketones and other ketones & aldehydes with high enolic percentage do not give this reaction

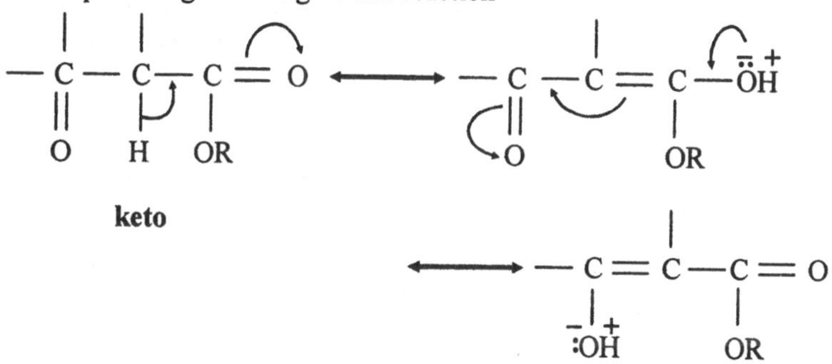

keto

enol

APPLICATIONS

This method of reduction is specific for carbonyl group and therefore it can be used for reducing aldehydes and ketones containing some other reducible group, such as, a double bond, a nitro or an ester group, which are not reduced under these conditions.

1. The reaction has been employed to *reduce* α, β-unsaturated aldehydes to α, β-unsaturated alcohols.

$$\underset{\text{Acraldehyde}}{CH_2{=}CH{-}CHO} \xrightarrow[(CH_3)_2CHOH]{Ai(OCHMe_2)_3} \underset{\text{Allyl alcohol}}{CH_2{=}CH{-}CH_2OH}$$

$$\underset{\text{Crotonaldehyde}}{CH_3{-}CH{=}CH{-}CHO} \xrightarrow[(CH_3)_2CHOH]{Ai(OCHMe_2)_3} \underset{\text{Crotyl alcohol}}{CH_3{-}CH{=}CH{-}CH_2OH}$$

2. *Reduction of aromatic ketones, e.g.,*

$$\underset{\text{Acetophenone}}{C_6H_5COCH_3} \xrightarrow[\text{Me}_2\text{CHOH}]{\text{Ai(OCHMe}_2)_3} \underset{\alpha\text{-Methyl benzylalcohol}}{C_6H_5CHOHCH_3}$$

3. *Reduction of alicyclic ketones, e.g.,*

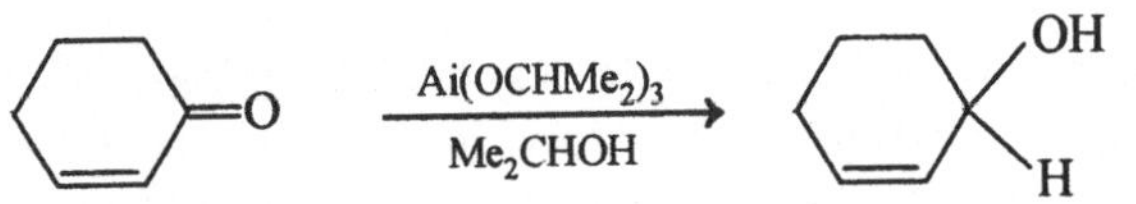

2-Cyclohexene-1-one 2-Cyclohexene-1-ol

4. *Reduction of α-halogenated ketone, e.g.,*

$$\underset{\text{Trichloroacetone}}{CH_3COCCl_3} \xrightarrow[\text{Me}_2\text{CHOH}]{\text{Ai(OCHMe}_2)_3} \underset{\text{Trichloropropanol-2}}{CH_3CHOHCCl_3}$$

5. *Reduction of keto esters, e.g.,*

$$\underset{\text{Ethyl trimethylpyruvate}}{Me_3C.CO.COOC_2H_5} \xrightarrow[\text{Me}_2\text{CHOH}]{\text{Ai(OCHMe}_2)_3} \underset{\text{Ethyl teimethyllactate}}{Me_3C.CH(OH).COOC_2H_5}$$

6. As the reduction requires mild conditions, it is accepted widely as synthetic procedre. However, it is used in the synthesis of chloromycetin and oestradiol.

(*a*) Synthesis of chloromycetin is as follows :

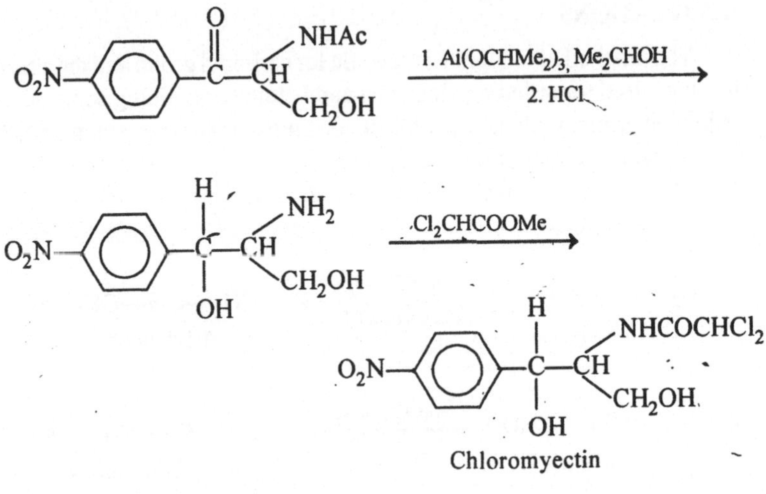

Chloromyectin

(*b*) Synthesis of oestradiol is as follows :

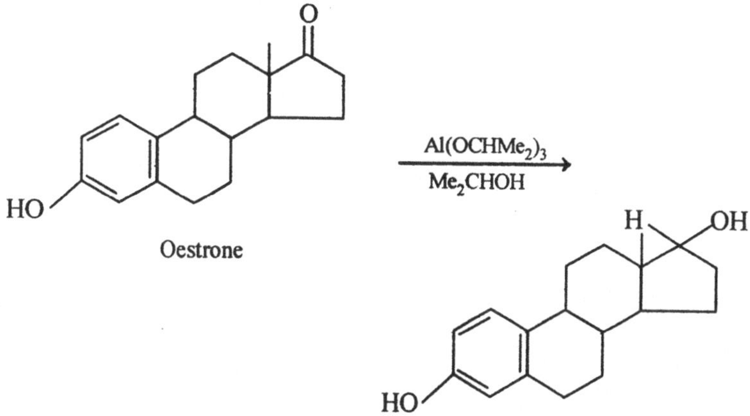

Oestrone

Oestradiol

(7) $CH_3{-}C{\equiv}C{-}\overset{O}{\overset{\|}{C}}{-}CH_3 \xrightarrow[\text{isopropyl alcohol}]{\text{Aluminium isopropoxide}} CH_3{-}C{\equiv}C{-}\overset{OH}{\overset{\|}{C}H}{-}CH_3$

(8) $C_6H_5{-}\overset{NO_2}{\overset{|}{C}H}{-}CH{=}CH{-}\overset{O}{\overset{\|}{C}}{-}CH_3 \xrightarrow[\text{isopropyl alcohol}]{\text{Aluminium isopropoxide}}$

$C_6H_5{-}\overset{NO_2}{\overset{|}{C}H}{-}CH{=}CH{-}\overset{OH}{\overset{|}{C}H}{-}CH_3$

(9) 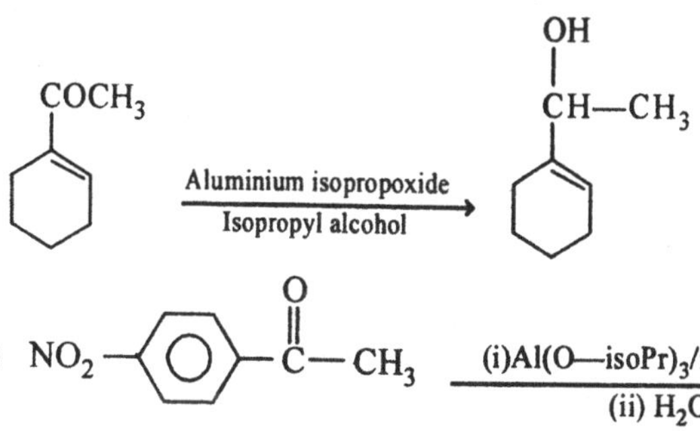

(10) 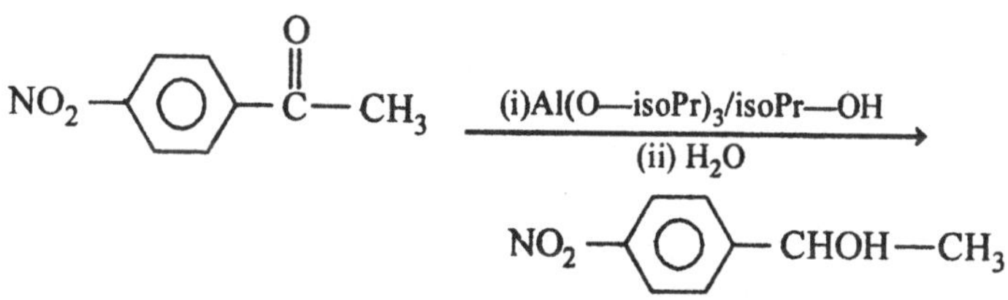

74% yield

(11) 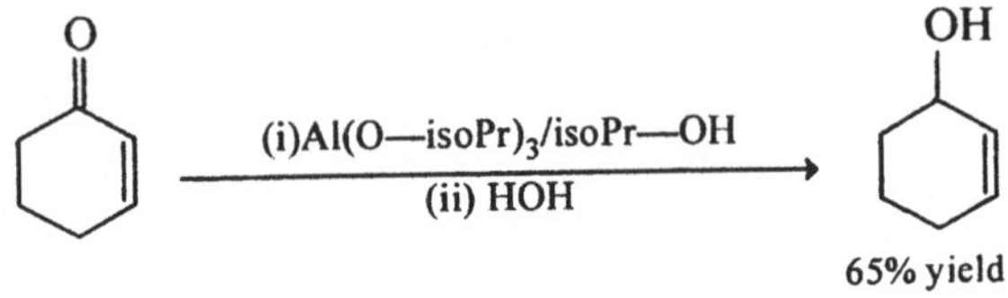

(12) 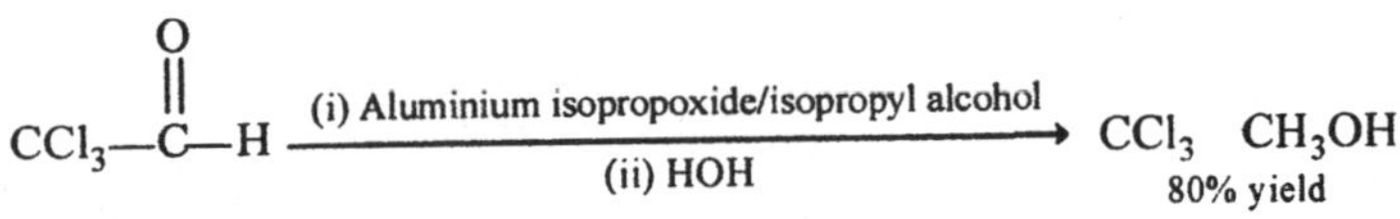

(13) 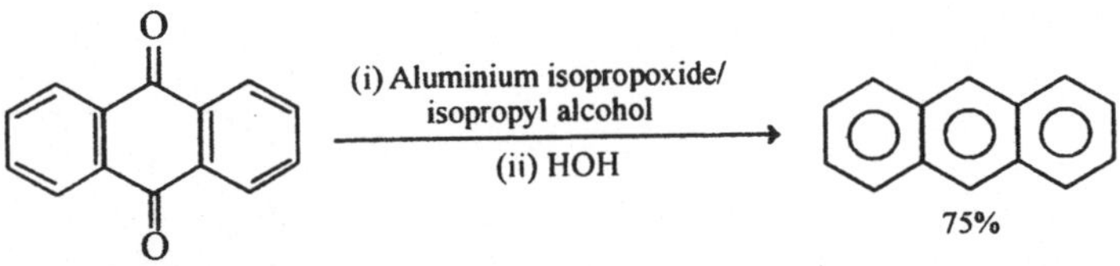

MODIFICATIONS

Among the various by-products obtained during the M.P.V. reduction, hydrocarbons are formed. For example, Hoffsommer *et al.* reduced aromatic ketones at 250°C with an excess of aluminium isopropoxide forming hydrocarbons wiht about 95% yield.

$$\underset{\text{Benzophenone}}{C_6H_5.CO.C_6H_5} \longrightarrow \underset{\text{Diphenylmethane}}{C_6H_5.CH_2.C_6H_5}$$

73

MEERWEIN ACYLAION REACTION

If vinyl carbon has —I group (—NO_2, —CN, CHO, C_6N_5 etc), then hydrogen present on this vinylic carbon is known as activated vinylic hydrogen.

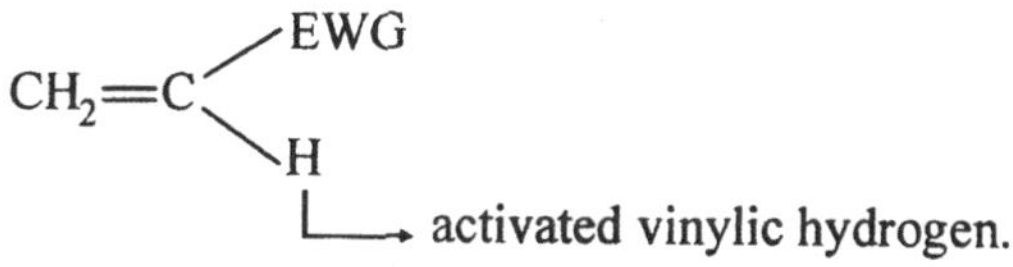

The replacement of an activated hydrogen by an aryl group achieved by the reaction of a diazonium salt in the presence of Cu^{++} salt is known as Meerwein-Acylation reaction.

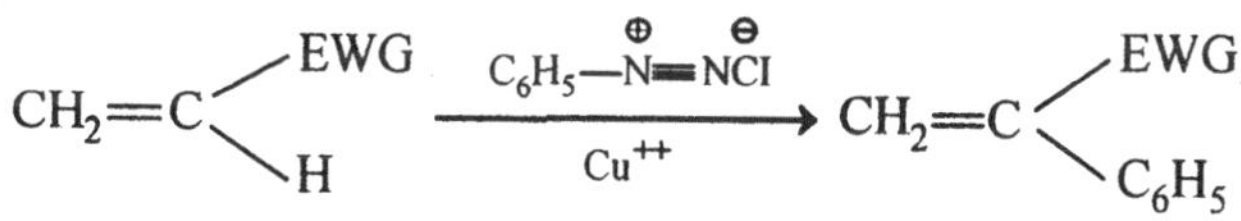

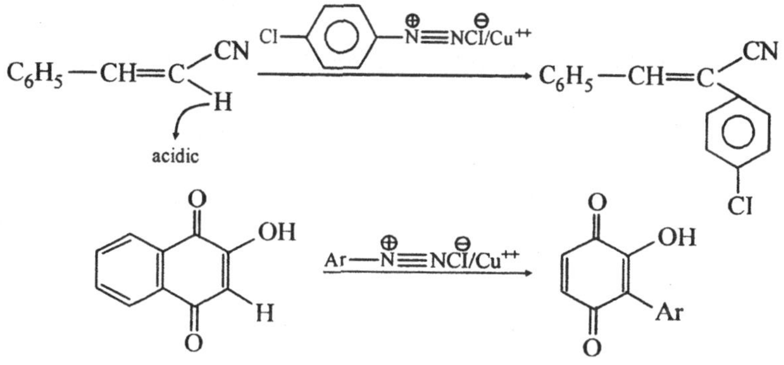

74

OPPENAUER OXIDATION

Oppenauer Oxidation is of paramount importance in the synthesis of ketones from ***secondary*** alcohols.

REACTION

The ***secondary*** alcohol is refluxed with a base viz. aluminum *iso*propoxide, potassium-***tert***-butoxide—used as an oxidizing agent —in the presence of a ketone —used as a refluxing agent which gets reduced.

$$R-\underset{\underset{OH}{|}}{CH}-R' + CH_3\cdot\underset{\underset{O}{\|}}{C}\cdot CH_3$$

$$R-\underset{\underset{O}{\|}}{C}-R' + CH_3-\underset{\underset{OH}{|}}{CH}-CH_3$$

MECHANISM

STEP - [I] ***Formation of Alkoxide of the secondary alcohol***

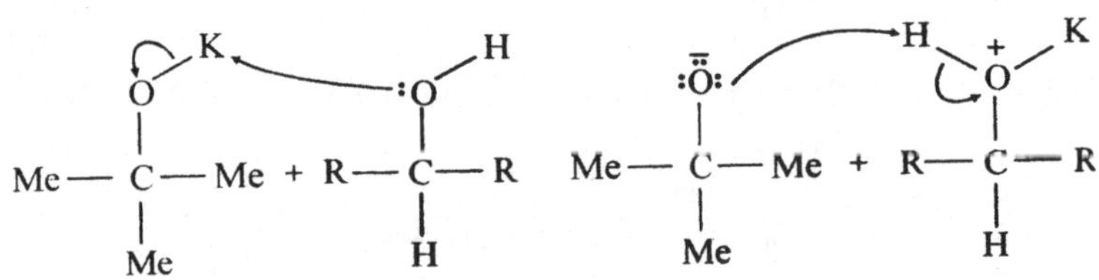

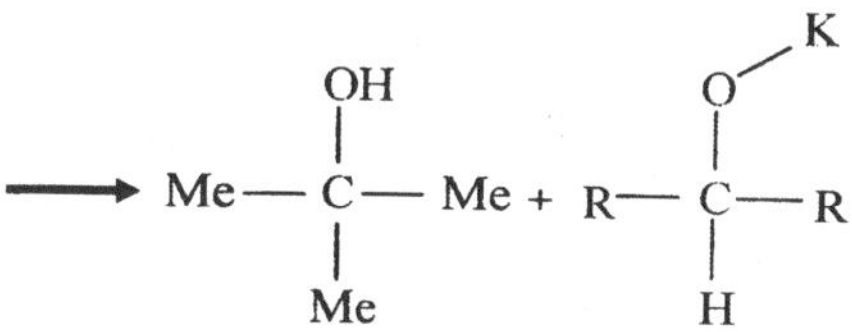

STEP - [II] ***Formation of a Cyclic Intermediate & Its Decomposition***

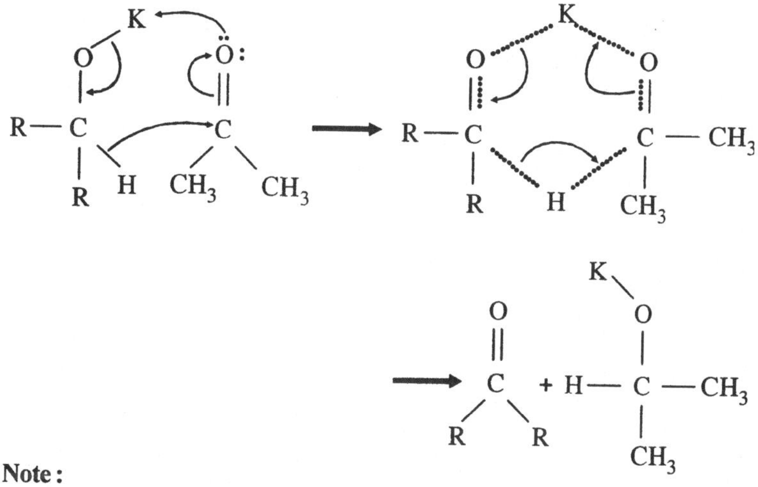

Note :

(I) Among the ketones most widely used are acetone, methylethyl ketone, cyclohexanone etc. as refluxing agent.

(II) The common base being potassium-***tert***-butoxide and alumium ***iso***propoxide

The reaction is highly selective. Usually ketones are prepared through this method but it may be useful for the preparation of aldehyde also. When aldehyde is prepared from primary alcohol, *p*-benzoquinone is used as oxidant, *i.e.* hydrogen acceptor, instead of aliphatic ketones. In this reaction no other oxidising group undergoes oxidation.

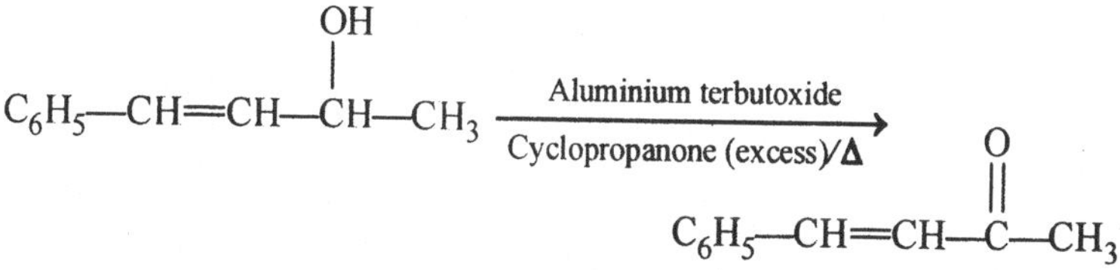

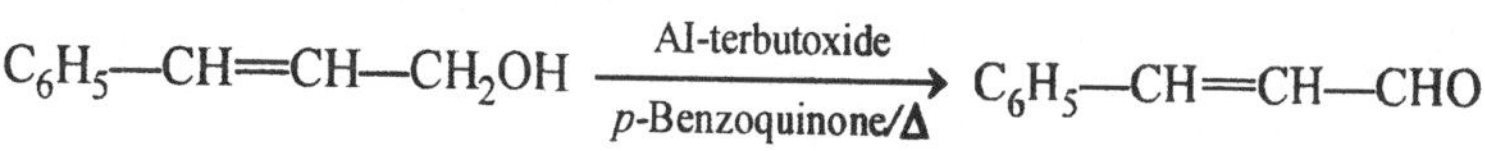

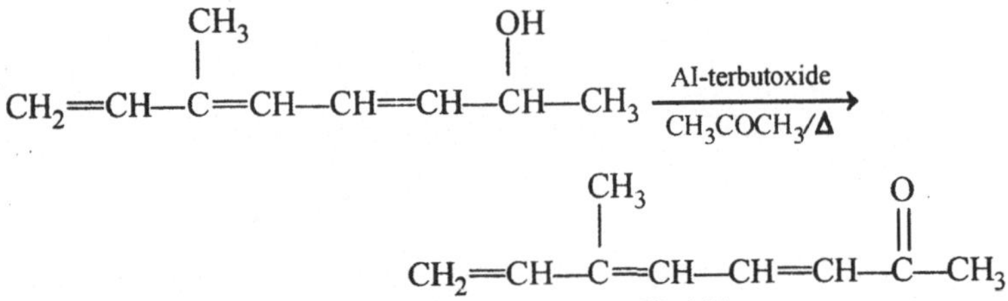

80 yield

Note : Aldehydes are not satisfactorily prepared by this reaction in the presence of aliphatic ketone because the basic nature of reagent induces reaction between aldehyde and ketone. In presence of *p*-benzoquinone.

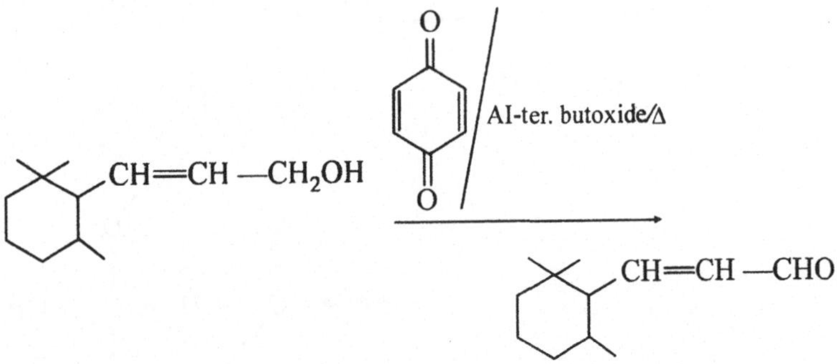

APPLICATIONS

1. Oppenauer oxidation is specific for alcohols. Therefore, the reaction is specially used to prepare a ketone from a secondary alcohol having other oxidizable functions (*e.g.* olefinic double bond, phenolic group, etc.).

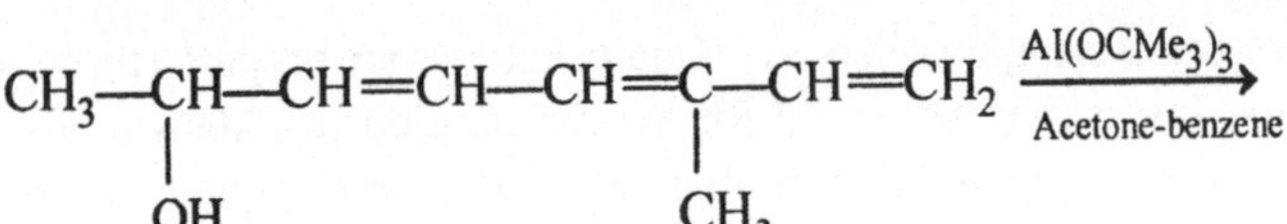

6-Methyl- Δ 3, 5, 7-octatriene-2-ol.

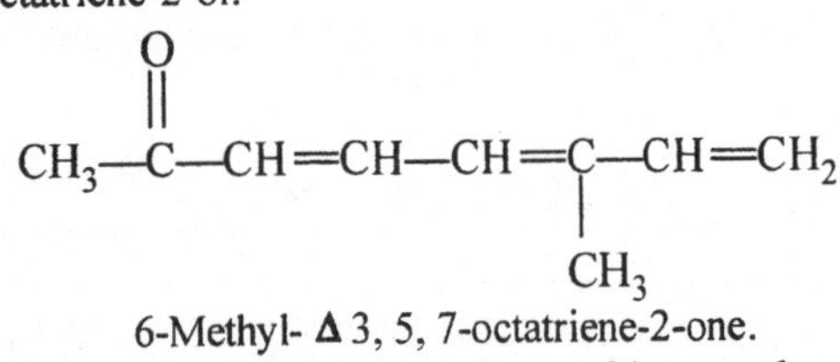

6-Methyl- Δ 3, 5, 7-octatriene-2-one.

When Ruzicka *et al.* were elucidating the structure of ionone they employed aluminium isopropoxide in acetone and benzene for converting α-and β-ionols to the corresponding ionones.

Organic Reactions : Mechanism with Problems

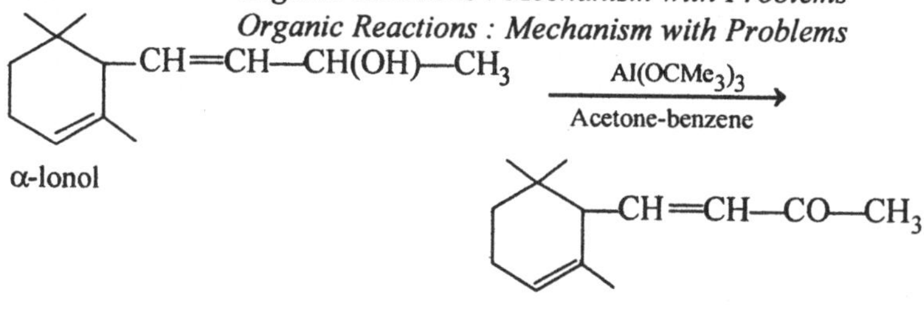

α-lonol

α-lonone

It is not possible to use this reaction for the synthesis of all aldehydes as most of the aldehydes undergo condensation reaction with ketones (acetone). However, if acetone has been replaced by *p*-benzoquinone, aldehydes may be prepared from primary alcohols by employing this reaction. Thus, quinones and aromatic ketones are better hydrogen acceptors than acetone.

2. This method was originally introduced by Oppenauer for preparing steroid ketones and since then it is still used in the steroid chemistry, mainlyin the synthesis. It finds use in the synthesis of cholesterol, testosterone, cholanic acid, progesterone, some of the corticoids, etc. The fundamental changes taking place in the Oppenaurer oxidation may be illustrated by the conversion of cholesterol into cholestenone.

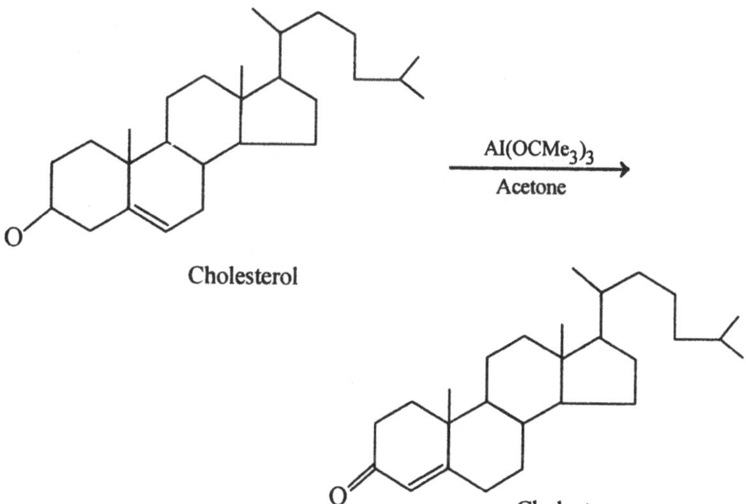

Cholesterol

Cholestenone

3. Another important application of Oppenauer oxidation involves oxidation of formates to carbonyl group [Ringold, 1956].

$$R_2CHO.\overset{\overset{O}{\|}}{C}\text{—}H \xrightarrow{(O)} R_2CHO.\overset{\overset{O}{\|}}{C}\text{—}O\text{—}H \xrightarrow{-CO_2} R_2CHOH$$

$$\xrightarrow[(CH_3)_2C{=}O]{Al(OCMe_3)_3} R_2C{=}O$$

4. Primary unsaturated alcohols are oxidized to aldehydes in the presence of gold hydrogen acceptors, *e.g.*, *p*-benzoquinone. In some cases acetaldehyde or cinnamaldehyde are used.

(*a*) 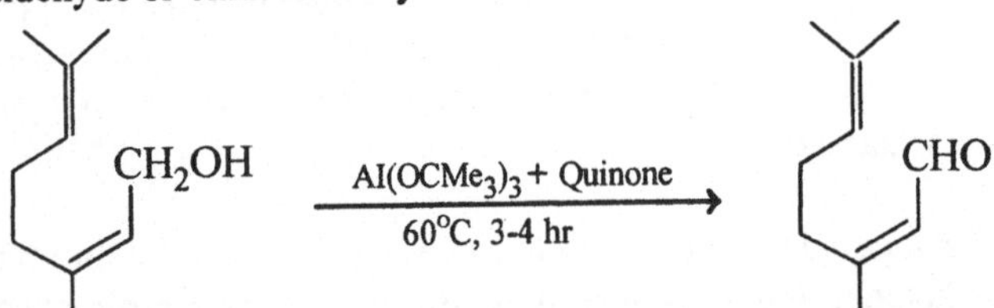

Geraniol — Citral

(*b*) $(CH_3)_2C{=}CH\text{—}CH_2\text{—}CH_2\text{—}CH_2OH + C_6H_5\text{—}CH{=}CH\text{—}CHO$
5-Methyl-4-hexene-1-ol — Cinnamaldehyde

$$\xrightarrow{Al(OCMe_3)_3}$$

$(CH_3)_2C{=}CH\text{—}CH_2\text{—}CH_2\text{—}CHO + C_6H_5\text{—}CH{=}CH\text{—}CH_2OH$
5-Methyl-4-hexenal — Cinnamyl alcohol

The reaction has been successful with low-boiling aldehydes which can distilled off as they get formed.

(*c*) The terminal CH_2OH group of the unsaturated side chain in vitamin A_1 is oxidized to aldehyde group by using aluminium isopropoxide in the presence of acetaldehyde and benzene and prolonged heating at 70°C.

5. The β, γ-unsaturated alcohols undergo oxidation with migration of double band to yield α, β-unsaturated ketones.

$$R\text{—}CH{=}CH\text{—}CH_2\text{—}CH(OH)\text{—}R \xrightarrow[p\text{-Benzoquinone}]{Al(OCMe_3)_3}$$

$$R\text{—}CH_2\text{—}CH{=}CH\text{—}CO\text{—}R$$

6. Alicyclic alcohols are oxidized to alicyclic ketones.

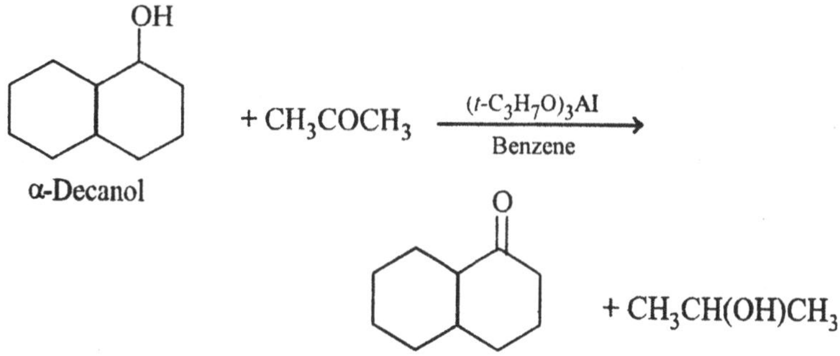

α-Decanol

α-Decanol

MODIFICATION AND LIMITATIONS

This reaction is mainly applicable to secondary alcohols. However, some primary alcoholic groups are also oxidised, giving a low yield (50-90%) of aldehydes.

The Oppenauer oxidation does not work with molecules having basic groups which are in close proximity to hydroxyl group. However, Woodward was successful in oxidising aromatic alcohols having basic groups in the benzene solution with benzophenone as the hydrogen accptor.

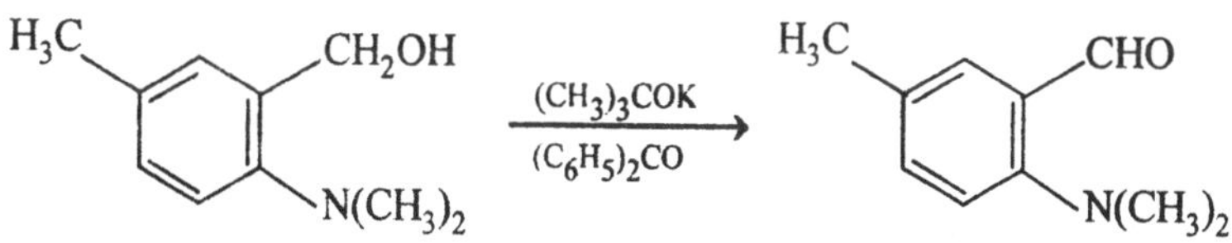

2-Dimethylamino-5-methyl benzyl alcohol

2-Dimethylamino-5-methyl benzaldehyde (40%)

Fluorenone is a rapid hydrogen acceptor and is use for carrying oxidations of sensitives alcohols under mild conditions.

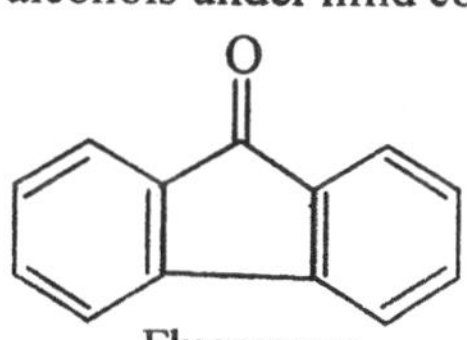

Fluorenone

Problem - 1

$$\underset{[A]}{C_4H_8O} \xrightarrow[p\text{- Benzoquinone}\ \Delta]{\text{Al-}isopropoxide} \underset{[B]}{C_4H_6O}$$

$$\underset{[B]}{C_4H_6O} \xrightarrow{O_3} C_4H_6O_4 \xrightarrow{Na/NH_3(liq)} CH_2O + \textit{alfa}\text{ - Ketopropionic acid}$$

[I]A absorbs one molar equivalent of bromine & forms sodioderivative
[II] **B** reduce ammoniacal silver nitrate solution.
Identify **A** and **B.**

Solution

A absorbs one molar equivalent of bromine.

Hence it contains one olefinic double bond.

A forms sodioderivative & it contains one oxygen atom.

So **A** is most probably an alcohol

A on refluxing with ***p***-benzoquinone in presence aluminum-***iso***propoxide produces **B.**

Clearly A undergoes Oppenaur oxidation to produce B. So A is an alcohol & B is a Carbonyl compound B, just proved to be a carbonyl compound, reduces ammoniacal silver nitrate.

Hence B is an aldehyde

B reacts with ozone to form an ozonide which undergoing decomposition by the reducing agent [Na/NH_3(liq)] produces one molecule of formaldehyde and one molecule of ***alfa***-ketoproponic acid.

Clearly the alfa-ketogroup of this acid has been resulted from the ozonolytic fission of B whose aldehyde group has been simultaneously oxidized to COOH group of the ***alfa***-ketopropionic acid.

Hence the structures of A & B can be derived as follows:

Going backwards

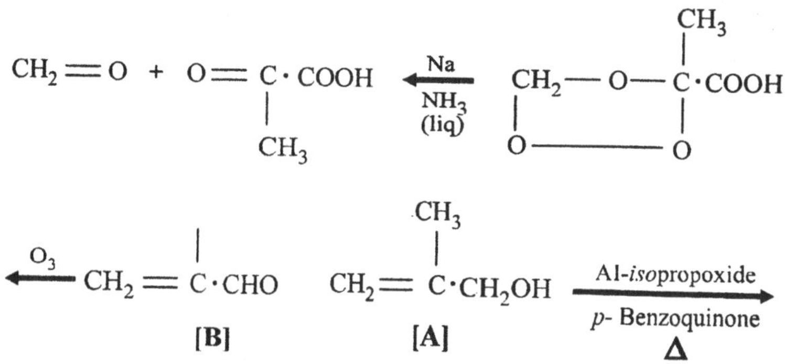

2-Methyl acraldehyde 2-Methyl allyl alcohol

Conclusion **A**▶ 2-Methyl allyl alcohol **B**▶ 2-Methyl acraldehyde

Problem - 2

$$\underset{[A]}{C_4H_8O} \xrightarrow[p\text{- Benzoquinone}\ \Delta]{\text{Al-}isopropoxide} \underset{[B]}{C_4H_6O}$$

$$\underset{[B]}{C_4H_6O} \xrightarrow[H_2O]{Na_2O_2} \xrightarrow{HIO_4} \underset{[C]}{C_2H_4O} + \text{Oxalic acid}$$

$$\underset{[C]}{C_2H_4O} \xrightarrow[\Delta]{NaOH} \text{Aldol}$$

Identify **A, B & C.**

Answer **A** But-2-ene-1-ol **B** But-2-ene-1-al **C** Acetaldehyde

Problem - 3

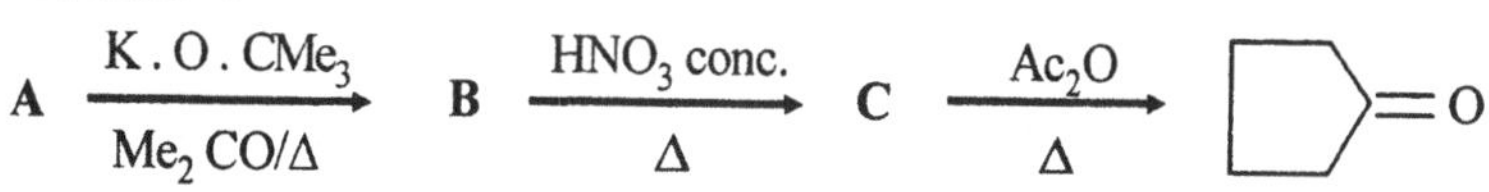

m.p. 24°C

Solution

Working from the end product

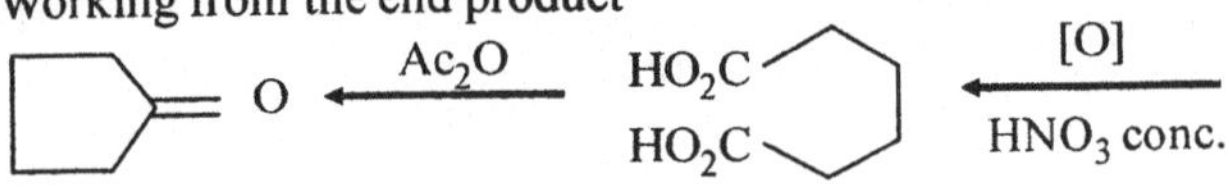

cyclopentanone adipic acid

[C]

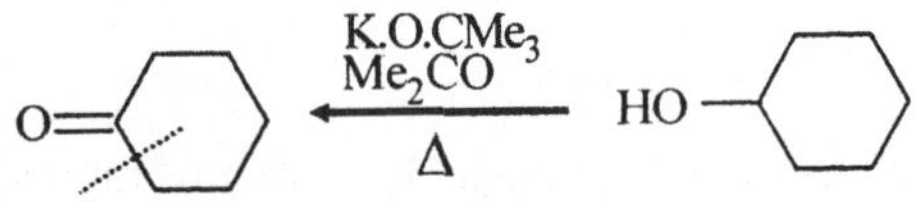

cyclohexanone cyclohexanol

[B]

Conclusion **A** Cyclohexanol; **B** Cyclohexanone
C Adipic acid; **D** Cyclopentanone

75

PERKIN REACTION

Perking reaction consists in the condensation of an aromatic aldehyde and aliphatic acid anhydride containing at least two α hydrogen atoms, viz $(RCH_2CO)_2O$, in the presence of the sodium or potassium salt of the corresponding acid to form α, β-unsaturated acid.

$$C_6H_5CHO + (CH_3CO)_2O \xrightarrow{CH_3COONa}$$

$$\underset{\text{Cinnamic acid}}{C_6H_5CH = CHCOOH} + CH_3COOH$$

It must be noted that during the condensation only α-hydrogen atoms of the anhydride are involved, *e.g.*

$$C_6H_5CHO + (\overset{\alpha}{CH_3}\overset{\beta}{CH_2}.CO)_2O \xrightarrow{CH_3CH_2COONa}$$

$$\underset{\alpha\text{-Methylcinnamic acid}}{C_6H_5CH = \overset{\overset{\displaystyle CH_3}{|}}{C}.COOH}$$

Since a weak base is used with a weak acidic reagent high temperature is required during the Perkin reaction. The catalyst (sodium or potassium salt of the acid) may be replaced by other bases, such as sodium carbonate, quinoline, pyridine and triethylamine.

MECHANISM

The base (sodium acetate or the other base) abstracts the α-hydrogen atom of the anhydride to form a carbanion which then attacks the carbonyl carbon of the aldehyde. Now the protonation of the carbanion followed by the removal of water yields an unsaturated anhydride which gets hydrolysed forming the α, β-unsaturated acid.

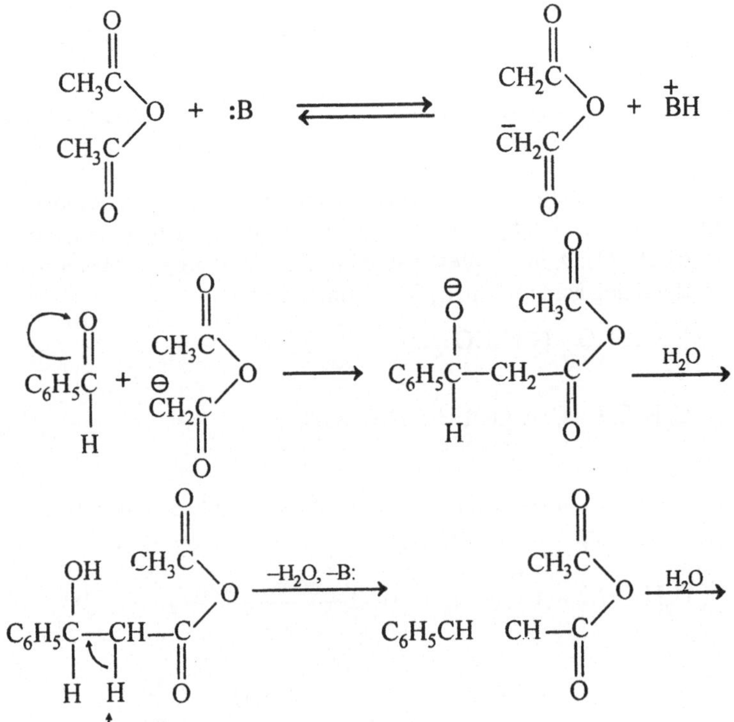

$$C_6H_5CH{=}CHCOOH + CH_3COOH$$

When sodium acetate is used as a base, the mechanism of the Perkin reaction may be put as follows:

$$CH_3COONa \rightleftharpoons CH_3CO\overset{\ominus}{O} + \overset{\oplus}{Na}$$

$$CH_3CO\overset{\ominus}{O} + \underset{\text{Acetic anhydride}}{H—CH_2—CO—O—CO—CH_3} \rightleftharpoons$$

$$CH_3COOH + \overset{\ominus}{C}H_2—CO—O—CO—CH_3$$
(I)

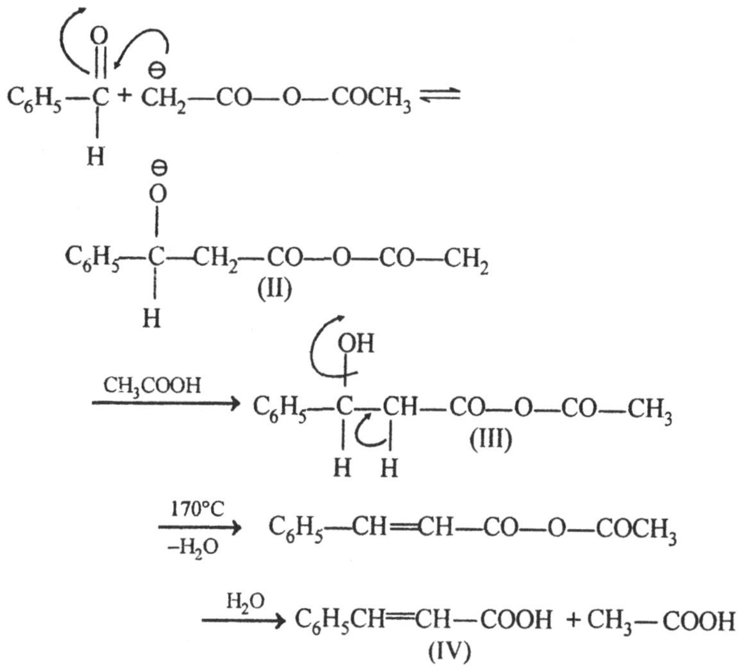

Prolonged heating (about 5h) and high temperature are required because a weak base (acetate ion) has to react with a weak acid (anhydride).

That the anhydride adds to the aldehyde has been proved by the fact that when benzaldehyde, acetic anhydride and trimethyl amine or pyridine are heated, cinnamic acid is obtained, but when benzaldehyde, sodium acetate and pyridine are heated, it is not obtained. Hence, the alkali salt acts only as the catalyst.

Meta-directing groups in the atomic ring promotes the reaction. When anhydride with one α-H atom is used, no dehydration would be possible and an aldol-type product is isolated.

The reaction has been found to be specific for aromatic aldehydes which cannot undergo self-condensation in the presence of the basic catalyst.

In some cases, when triethylamine is used as a base, in place of sodium salt of the acid, it gives better yield. Further, aliphatic aldehydes donot usually undergo Perkin reaction.

Aromatic aldehydes having an electron attracting group (*i.e.*, having —I effect) undergo Perkin reaction more readily, *e.g.*,

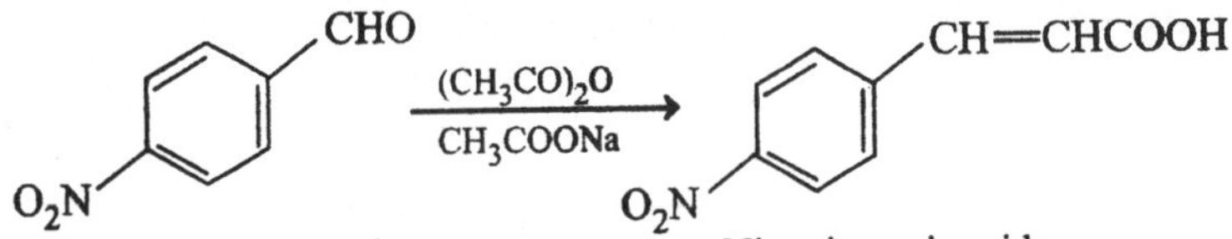

p-Nitrobenzaldehyde *p*-Nitrocinnamic acid (85% yield)

o-Hydroxybenzaldehyde has been found to undergo Perkin reaction readily and forms *o*-hydroxycinnamic acid, which spontaneously undergoes dehydration to form coumarin.

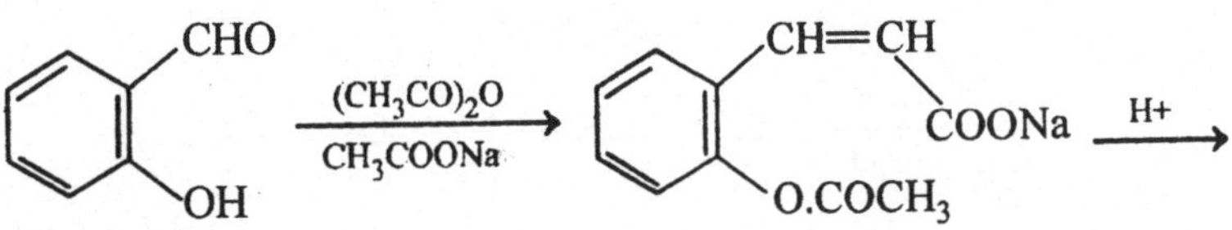

o-Hydroxybenzaldehyde

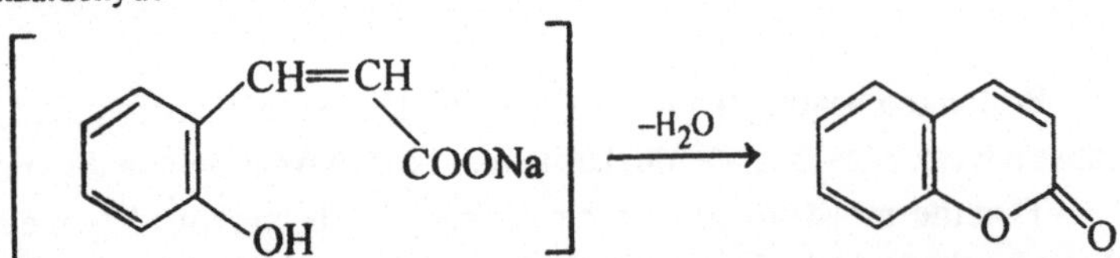

unstable Coumarin

On the other hand, the presence of an electron releasing group has been found to reduce the yield of the unsaturated acid, *e.g.*, *o*-methylbenzaldehyde yields *o*-methylcinnamic acid only in 15% yield white *p*-dimethylaminobenzaldehyde does not undergo Perkin reaction at all.

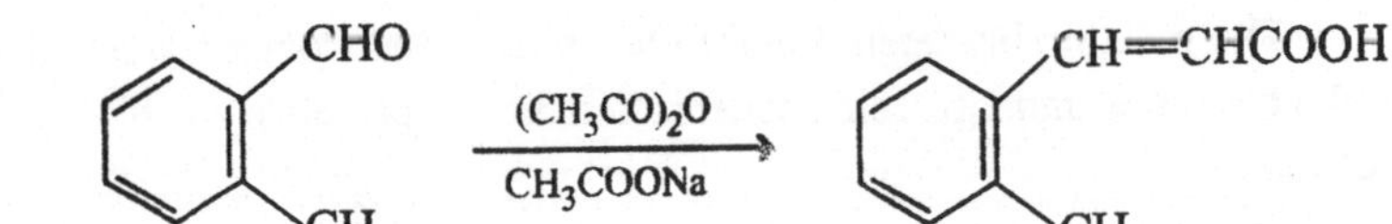

o-Methylbenzaldehyde *o*-Methylcinnamic acid (15% yield)

EXTENSION AND APPLICATIONS

1. Furfural may undergo Perkin reaction forming furylacrylic acid in 65-70% yield.

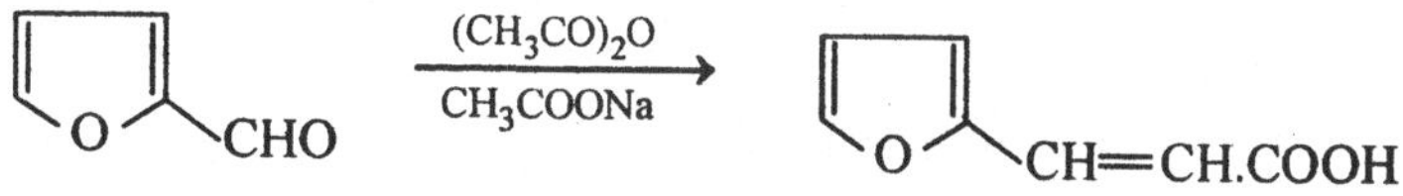

Furfural — Furylacrylic acid

2. Pschorr synthesis of phenanthrene is based upon Perkin's reaction.

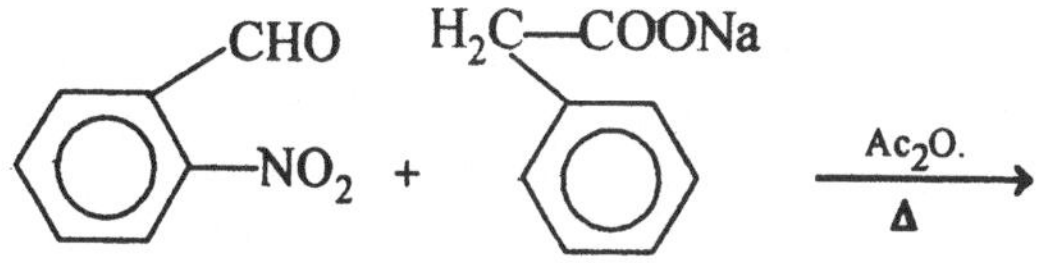

o-Nitrobenzaldehyde — Sodium phenyl acetate

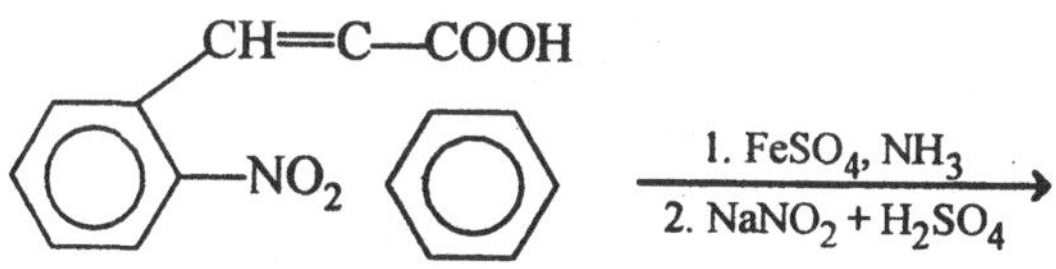

α-Phenyl-o-nitro cinnamic acid

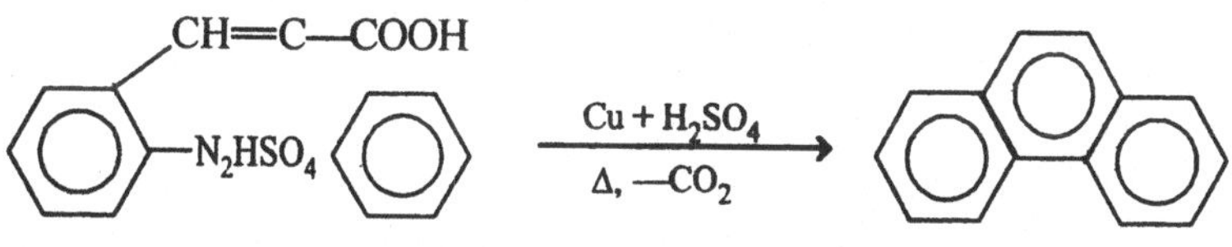

Phenanthrene

3. When salicylaldehyde undergoes Perkin's reaction, if forms *trans-o*-hydroxycinnamic acid or coumaric acid which on dehydration gives coumarin.

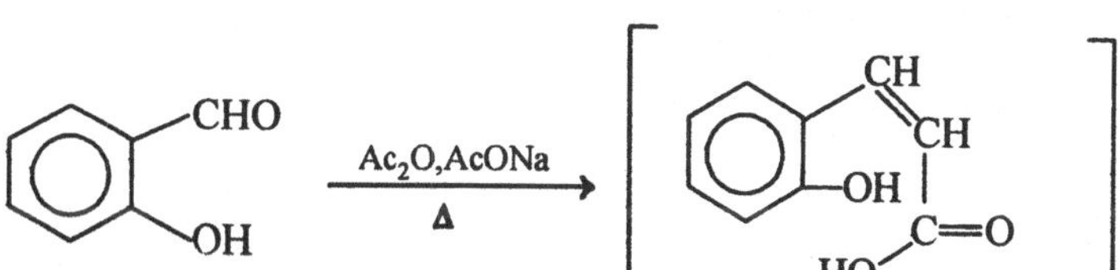

Salicylaldehyde

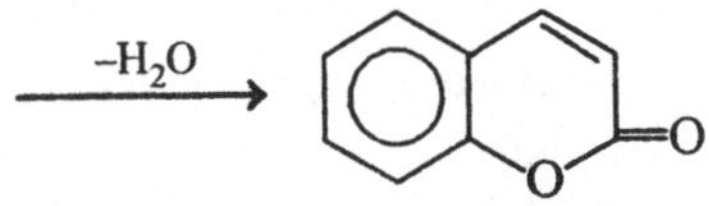

Coumarin

Synthesis of carbostyril (nitrogen analogue of coumarin may be put as follows :

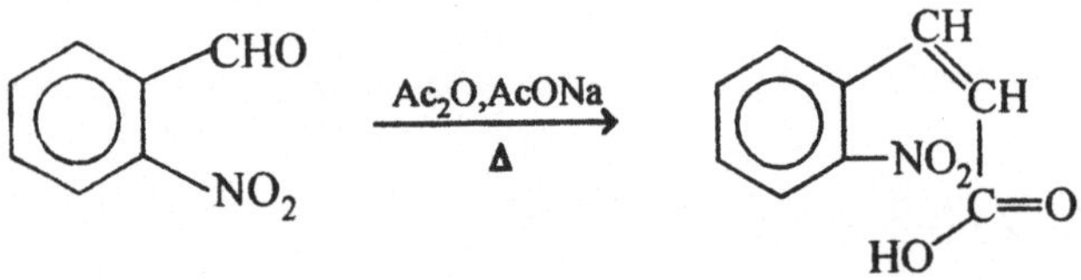

o-Nitrobenzaldehyde *o*-Nitro cinnamic acid

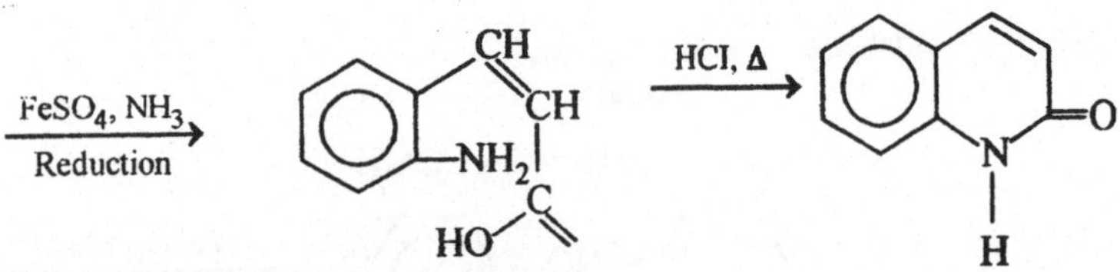

o-Amino cinnamic acid

Carbostyril

4. When cinnamaldehyde undergoes Perkin's condensation, if forms doubly unsaturated acid, *e.g.*,

$$\underset{\text{Cinnamaldehyde}}{C_6H_5.CH = CH.CHO} + (CH_3CO)_2O \xrightarrow{CH_3COONa}$$

$$\underset{\text{Cinnamylacrylic acid}}{C_6H_5.CH = CH.CHCOOH}$$

This reaction finds application in the synthesis of piperic acid and hence piperine (an alkaloid).

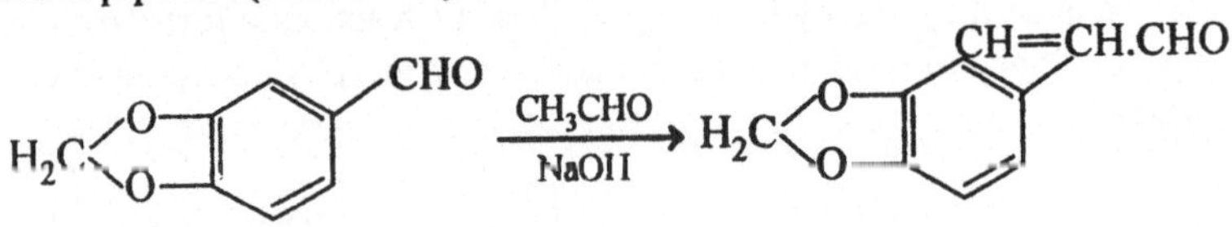

Piperonal

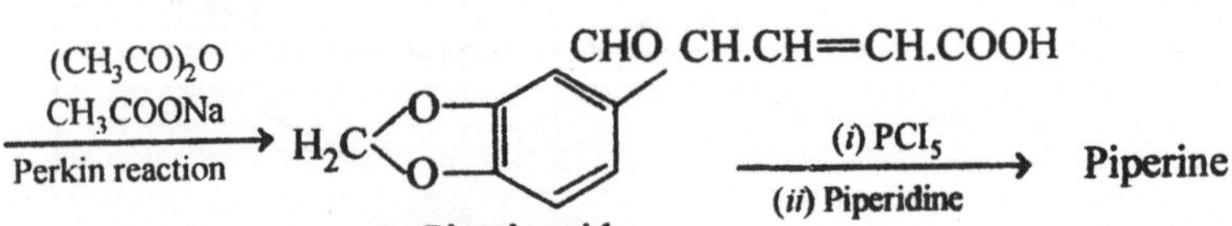

Piperic acid

5. In place of aldehydes, it is also possible to use anhydrides in Perkin's reaction. For example, phthalic anhydride undergoes Perkin's reaction with acetic anhydride and sodium acetate at 150-160°C to form phthalyl acetic acid.

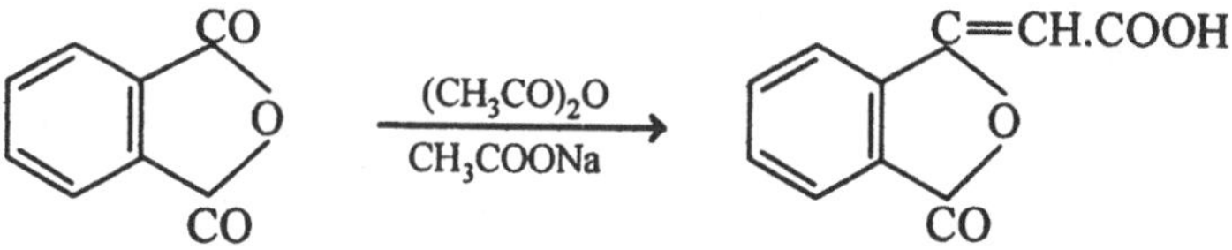

Phthalic anhydride Phthalylacetic acid

Phthalylacetic acid is further converted into many useful compounds, *e.g.*,

(*a*) When treated with CH_3ONa followed by heating with hydrochloric acid, phthalylacetic acid forms 1, 3-indandione.

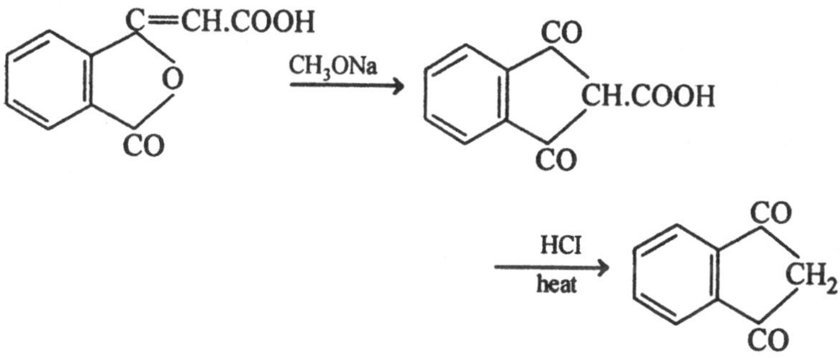

1, 3-Indandione

(*b*) When treated with cold aqueous alkali and HCl, it yield 2 carboxybenzoylacetic acid which on losing CO_2 forms *o*-acetylbenzoic acid.

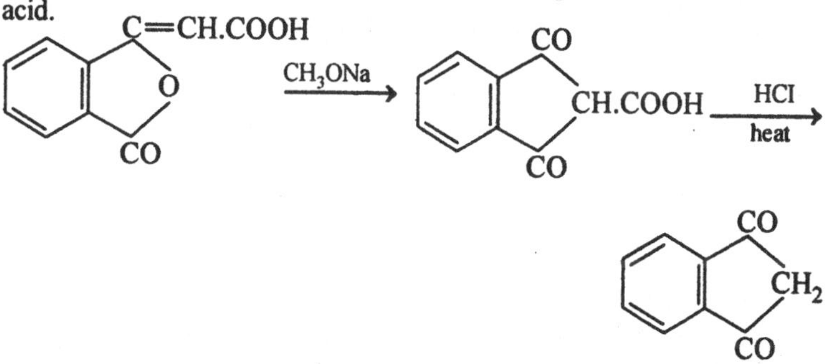

1, 3-Indandione

6. Kuhn employed Perkin's reaction to synthesis diphenylpolyenes,*e.g.*,

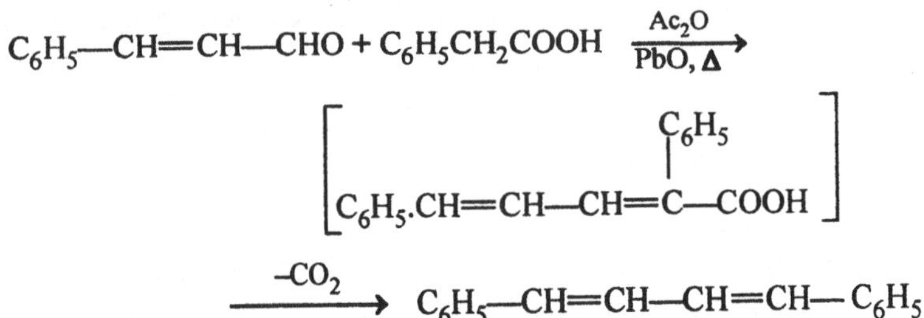

1, 4-Dipenyl 1, 3-butadiene

7. An important modification of Perkin's reaction involves the condensation of benzaldehyde, succinnic anhydride and sodium acetate.

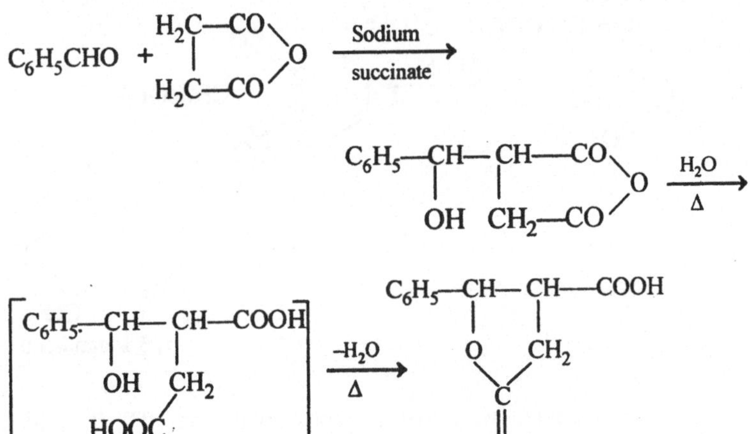

Phenylparaconic acid

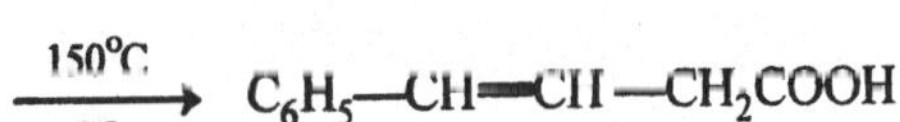

4-Phenyl-3-butenoic acid
(β, γ-unsaturated acid)

8. An important application of Perkin's reaction is Erlenmeyer's azlactone synthesis in which an acylglycine
gets condensed with aromatic aldehyde in the presence of acetic anhydride and sodium acetate.

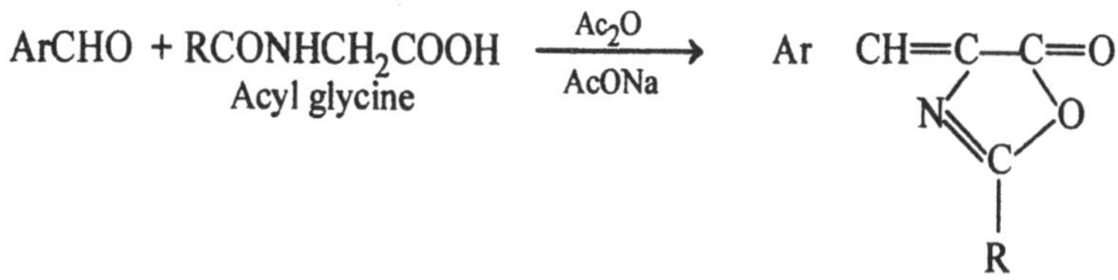

Azlactone

In this process the first step involves the conversion of acylglycine (*e.g.* benzoylglycine) into azlactone by the dehydration of its enolic tautomer with acetic anhydride. Now the azlactone which is having active methylene group undergoes Perkin reaction with the aldehyde, *e.g.*, benzaldehyde.

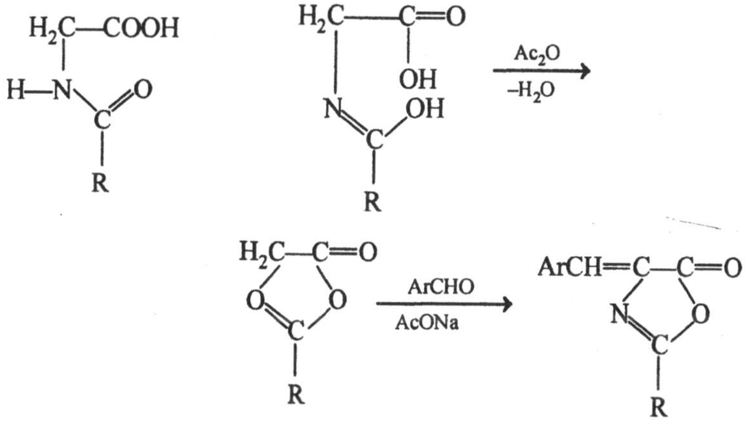

Substituted azlactone

The resulting substituted azlactone is further used in the following two ways.

(*i*) *Synthesis of α-amino-acids :* The above formed azlactone on reduction followed by hydrolysis yields α-amino-acids, *e.g.*, *dl*-phenylalanine may be formed in the following way :

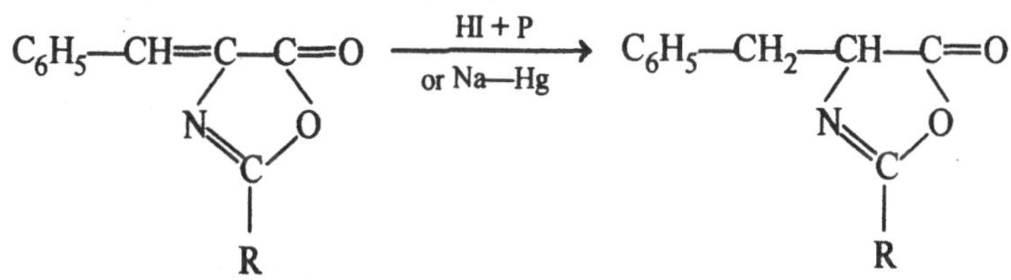

Azlactone

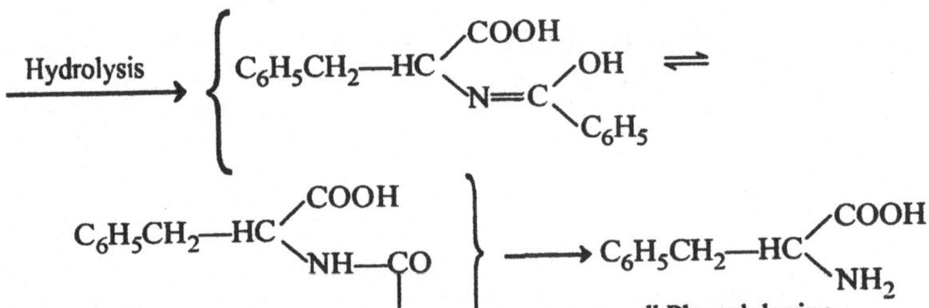

dl-Phenylalanine

(*ii*) *Synthesis of α-keto-acids* : By avoiding the reduction step, α-keto-acids may be prepared from azlactones in the following way :

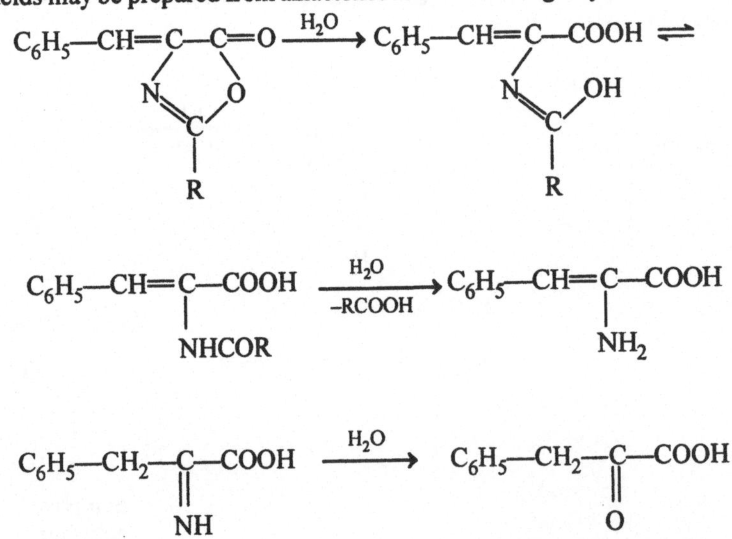

Phenylpyruvic acid

9. Aromatic anhydride in place of aldehydes is also used.

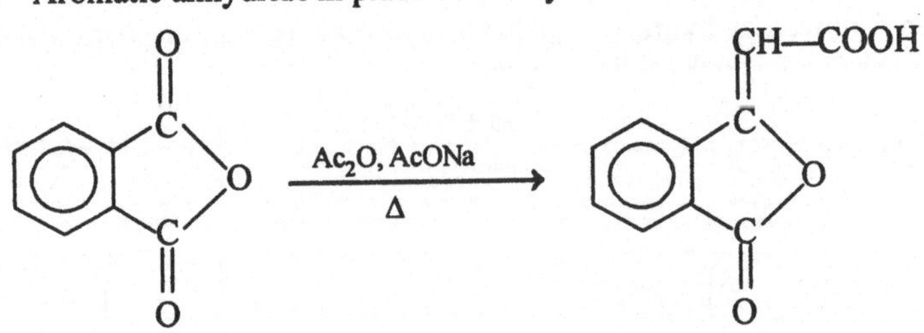

Phthalylacetic acid

Phthalylacetic acid can converted to different products, thus,

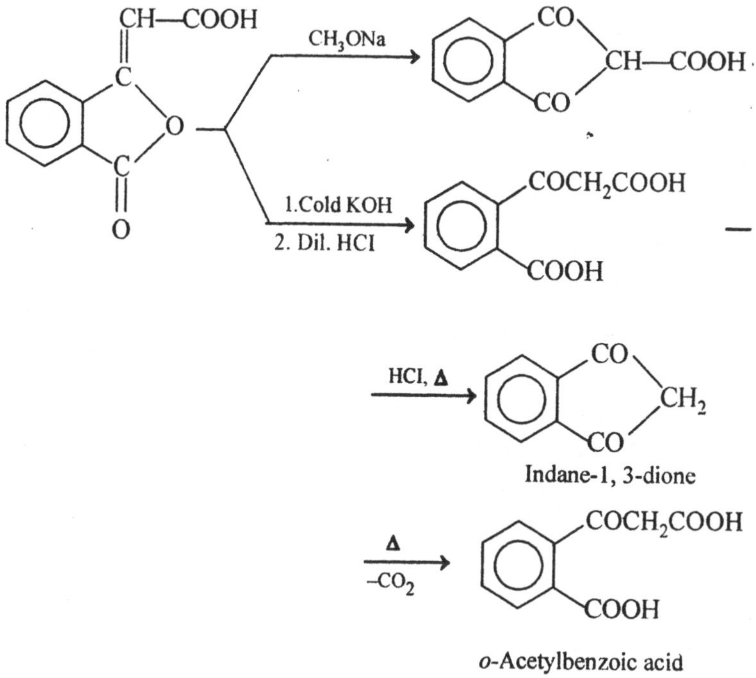

76

PERKIN MARKONIKOV KRESTOVNIKOV-FREUND SYNTHESIS

Condensation of α-w-dihalo alkanes with compound containing an active methylene group to produce alicyclic compounds in presence of sodiumethoxide and ethyl alcohol is known as PMKF reaction.

(i)

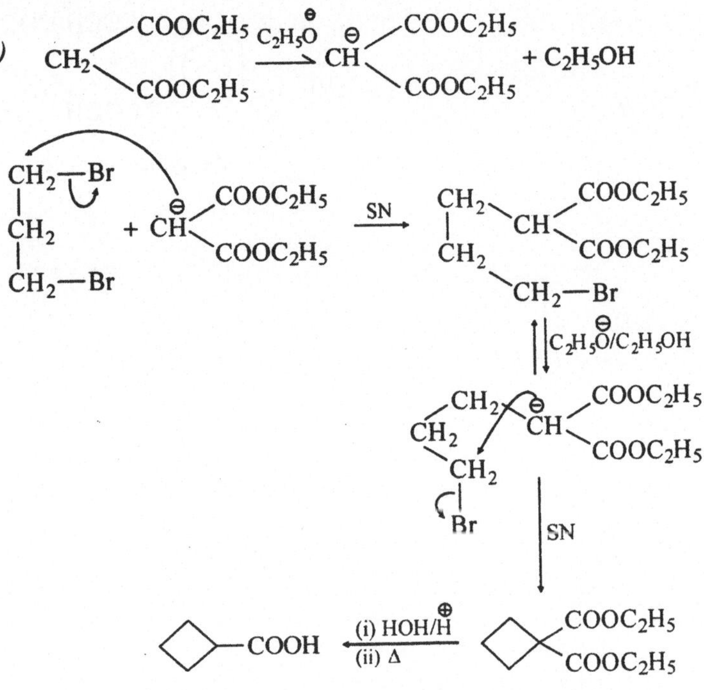

77

PECHMANN REACTION

Pechmann reaction is the condensation of phenols with malic acid or β-keto esters in the presence of conc.. H_2SO_4 or other condensing agent (discussed later on) to form coumarins, e.g.

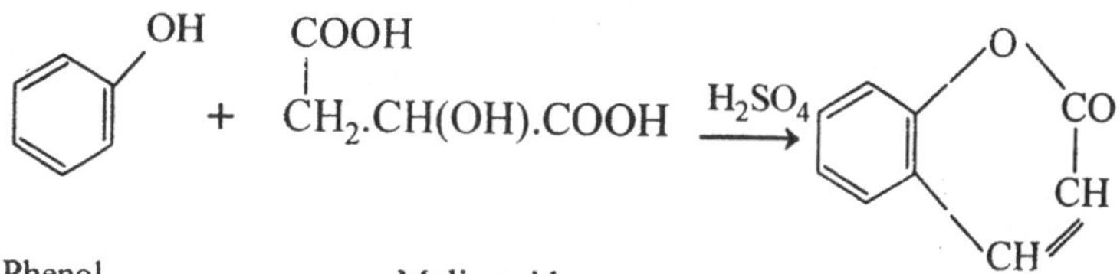

Phenol Malic acid Coumarin

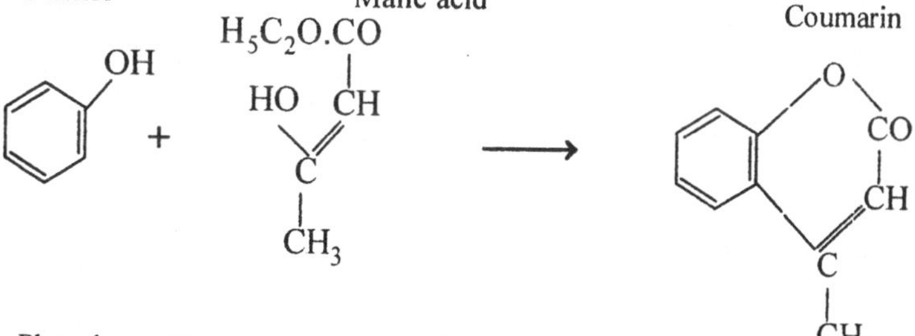

Phenol Ethylacetoacetate (enol form) 4-Methyl coumarin

By taking the appropriately substituted or unsubstituted phenols and β-ketone esters, coumarins having substituents either in the benzene ring and/or in the heterocyclic ring can be synthesised.

Simonis *et al.* (1913-1914) used phosphorus pentoxide as the condensing agent (in place of H_2SO_4) in the Pechmann reaction and obtained generally, but not always, chromones rather than coumarins as the final products.

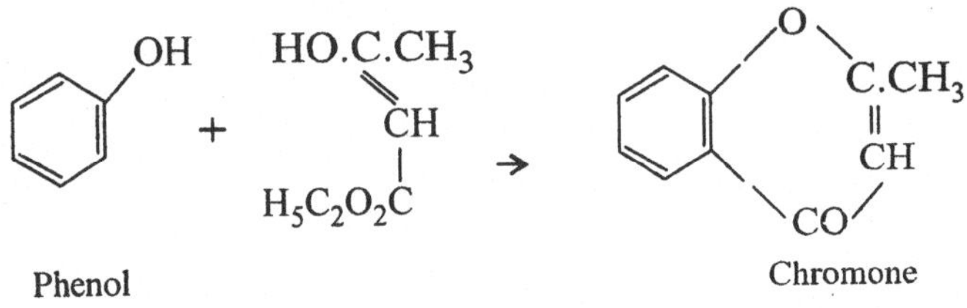

Hence the condensation of phenol and β-ketonic ester in the presence of P_2O_5 is sometimes called Simonis reaction, but actually it is merely a variation of the Pechmann reaction and hence will be discussed here.

Components of the Pechmann reaction : The course of this reaction has been found to depend on the nature of the phenol, the nature of the β-ketonic ester and the condensing agent.

1. Nature of the phenols. Among the simple mono-di- and trihydric phenols, resorcinol is the most reactive, followed by pholoroglucinol, α-naphthol and pyrogallol. However, the least reactive phenols are phenol, quinol and β-naphthol. For instance, when phenol is condensed with ethyl acetoacetate in the presence of H_2SO_4, 3% yield of methylcoumarin is obtained. Also, catechol does not condense even with the acetoacetic ester.

Furthermore, the reactivity of the substituted phenol is determined by the nature and the position of the substituent. It is found that the tendency of condensation of a particular phenol gets increased by the presence of electron-donating group such as CH_3, OH, OCH_3, NH_2, $NHCH_3$, $N(CH_3)_7$ and halogens in the *m*-position to the hydroxyl group. On the other hand, the tendency of condensation of a particular group gets decreased or becomes almost nil by the presence of electron-donating groups such as NO_2, SO_3H, COOH, $COOCH_3$, $COCH_3$, CN and CHO in the *m*-position to the hydroxyl group. The following is the order of the effect of the position of a particular group in monohydric phenol :

$$o > p > m$$

By the order it means that the substituent in the ortho position has the maximum inhibition followed the para and meta positions.

2. Reactivity of malic, maleic and fumaric acids and β-ketoesters : Malic esters generally do not condense with many of the substituted phenols. However, whenever malic esters condense, the yield of the product is very poor. Furthermore, malic acid will only condense in the presence of H_2SO_4 but not others.

It was found by Ponndorff and Thompson the *p*-cresol on condensation with fumaric or maleic acid in the presence of H_2SO_4 forms 6-methylcoumarin in quantitative yield.

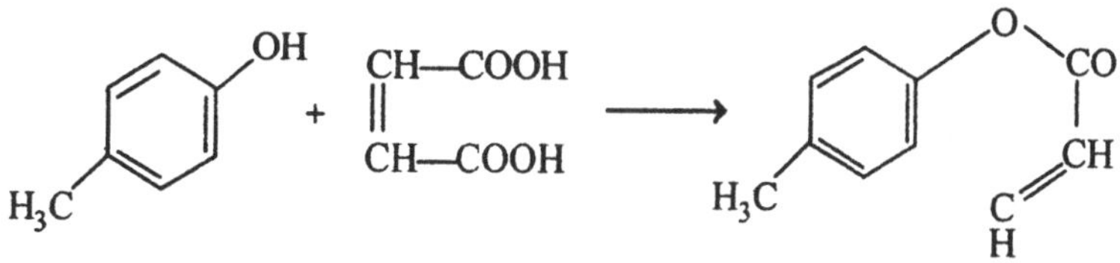

p-Cresol Maleic acid 6-Methyl coumarin

As β-keto ester has been found to react with phenols in its enolic form, it means that any substituent increasing the enolization or stabilising the enolic form should increase the reactivity than the β-keto ester itself (*e.g.* ethylacetoacetate) and *vice versa*. Furthermore, as β-keto ester may be substituted either at the α- or γ-carbon atom, it provides a means of getting coumarins with different substituents in the heterocyclic ring.

3. Nature of condensing agents : The role of the condensing agent in the Pechmann reaction is very important. The condensation between a phenol and β-keto ester is sometimes not effected by one reagent but may be effected by another reagent. The nature of the condensing reagent also controls the yield of the product. However, although rare, the two different condensing agents may sometimes lead to the formation of two different products. For example, resacetophenone and acetoacetic ester yield different products with $POCl_3$ and $AlCl_3$ as condensing agents.

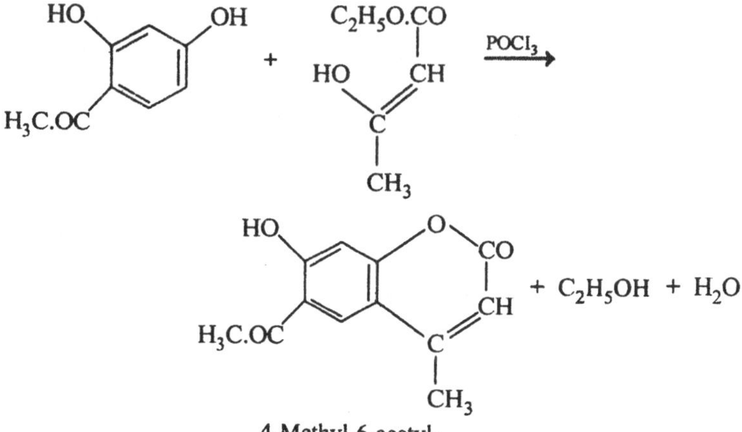

4-Methyl-6-acetyl-
7-hydroxy coumarin

The condensing agents used in Pechmann reaction are H_2SO_4, P_2O_5, $POCl_3$, $AlCl_3$, $ZnCl_2$, etc. The use of H_2SO_4 as the condensing agent favours the formation of mainly coumarins (except in the case of β-naphthol which yields a mixture of coumarin and chromone). However, P_2O_5 as the condensing agent favours the formation of chromone.

MECHANISM

The mechanism of this reaction can be understood by considering the following categories.

1. **Condensation of phenols with malic acid :** Pechmann postulated that malic acid is first of all converted into malonaldehydic acid and formic acid (which gets decomposed into H_2O and CO). Then, malonaldehydic acid reacts with the phenol, thereby forming an addition products which on eliminating two molecules of water forms the coumarin derivative.

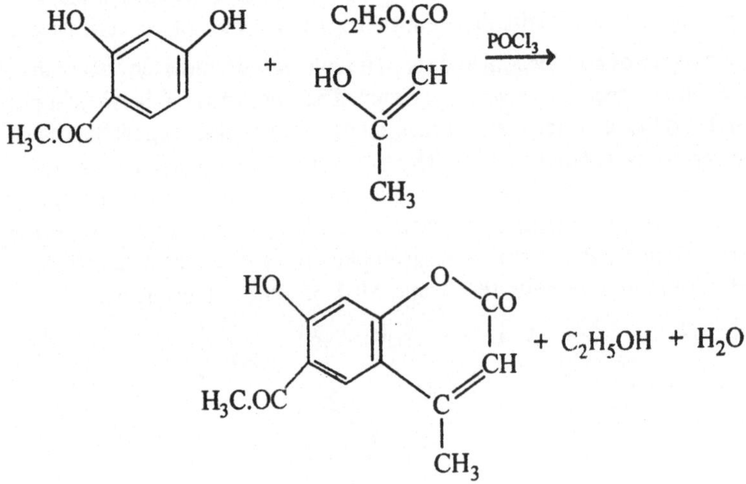

4-Methyl-6-acetyl-
7-hydroxy coumarin

2. **Condensation of Phenols with β-keto esters :** β-Keto esters been found to condense with phenols in either of the following ways :

(*a*) Pechmann and Duisbery (1883) postulated that the β-keto esters add on the hydrogen atom *ortho* to the hydroxyl group, thereby forming an intermediate hydroxy ester which on losing a molecule of water an ethyl alcohol cyclises.

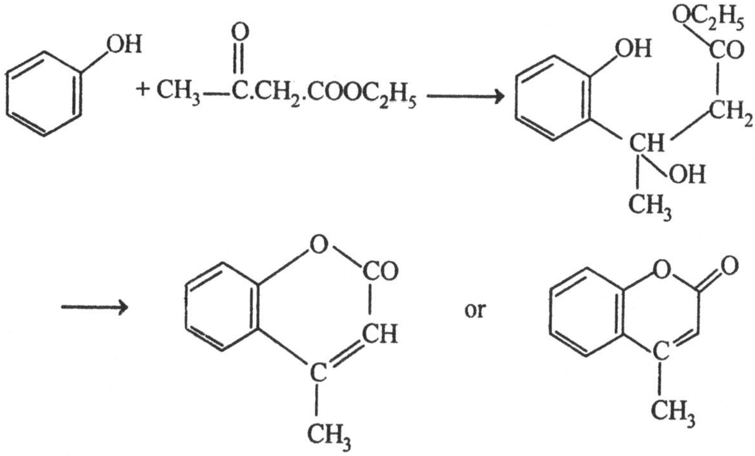

4-Methylcoumarin

As no intermediate could be isolated, this course of reaction is thus speculative.

(*b*) Robertson *et al* : (1932) postulated that the reaction proceeds through the following steps :

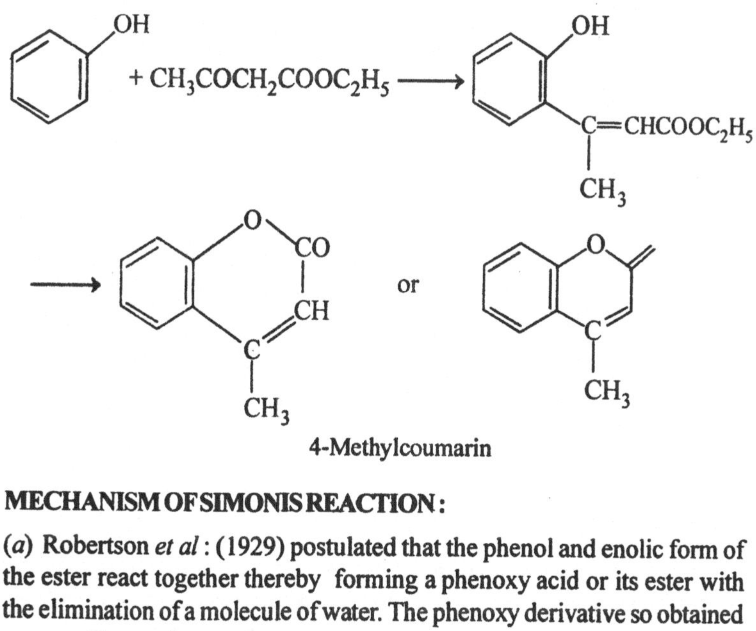

4-Methylcoumarin

MECHANISM OF SIMONIS REACTION :

(*a*) Robertson *et al* : (1929) postulated that the phenol and enolic form of the ester react together thereby forming a phenoxy acid or its ester with the elimination of a molecule of water. The phenoxy derivative so obtained now cyclises to form a chromone.

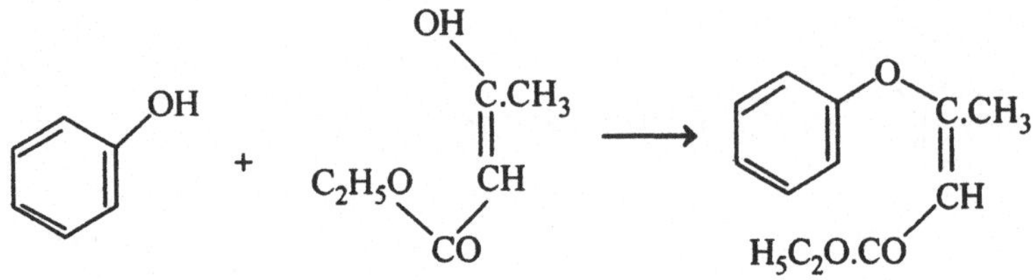

Phenol Acetoacetic ester

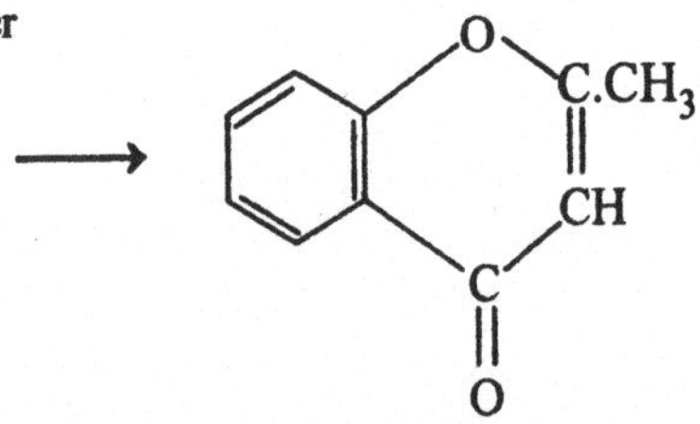

2-Methylchromone

(*b*) Ahmad and Desai (1937) postulated that the reactive hydrogen or the phenolic hydroxylic group reacts with the ethoxyl group of the β-keto ester thereby forming an aryl ester of the acid, (I). The aryl ester, (I) then gets isomerises to an *o*-hydroxybenzoylacetone, (II) that finally gets dehydrated to the chromone, (III).

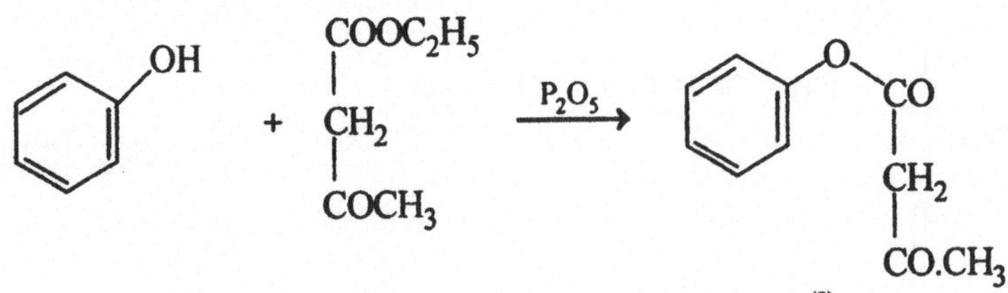

(I)

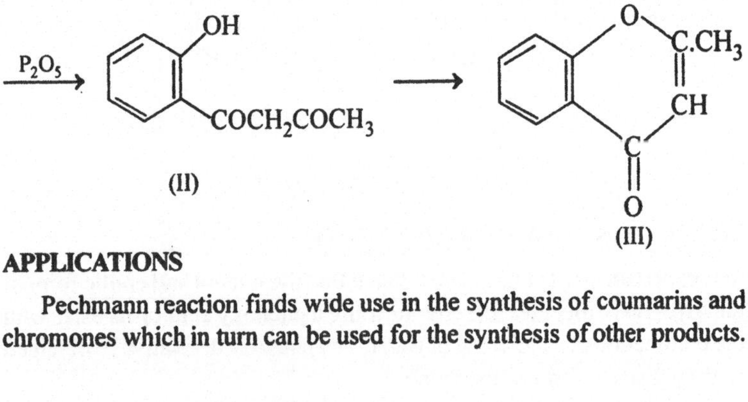

(II) (III)

APPLICATIONS

Pechmann reaction finds wide use in the synthesis of coumarins and chromones which in turn can be used for the synthesis of other products.

78

PINACOL-PINALCOLONE REARRANGEMENT (Pinacol Rearrangement)

The conversion of pinacols (1, 2-glycols) to ketones or aldehydes by means of mineral acids, acid chloride, zinc chloride or other electrophilic reagents is known as pinacol-pinacolone rearrangement.

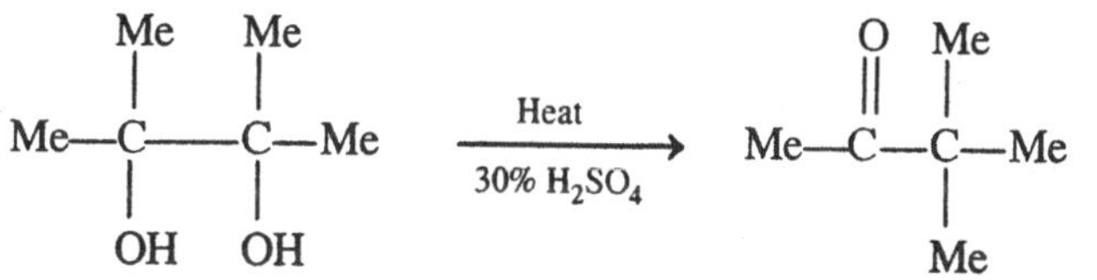

2, 3-Dimethyl-2, 3-butanediol (pinacol) → Methyl *t*-butyl ketone (pinacolone)

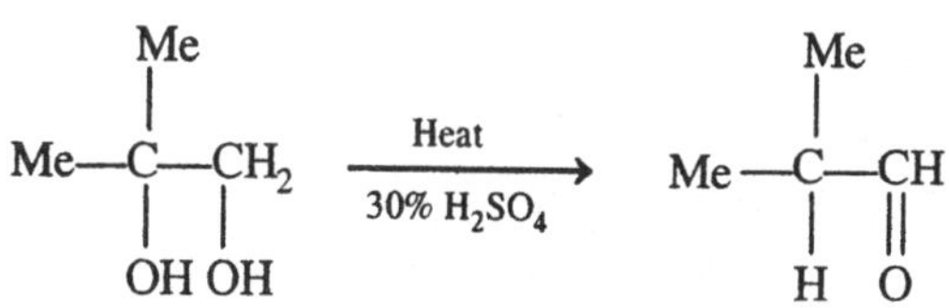

2-Methyl propane 2, 2-diol → Isobutyraldehyde

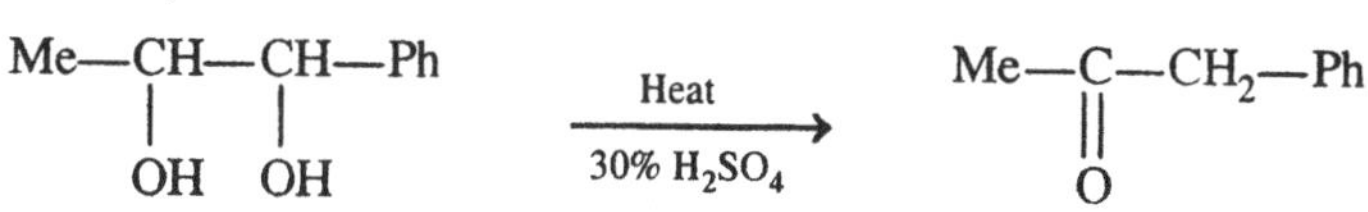

The pinacol-pinacolone rearrangement is general for 1, 2-glycols and the migrating groups may be alkyl or aryl.

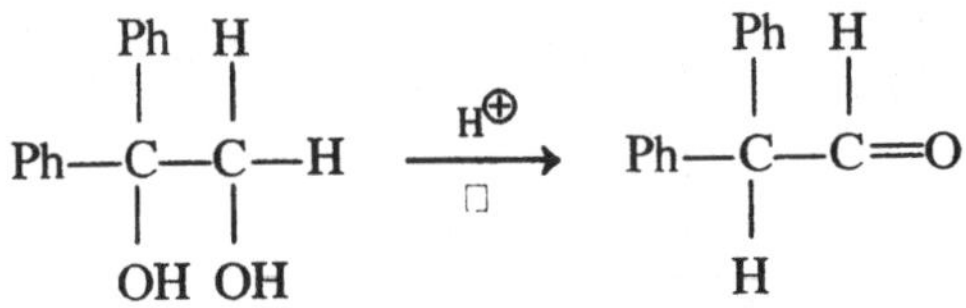

2, 2-Diphenyl-1, 2-glycol　　　　2, 2-Dipnenyl acetaldehyde

Elimination of water without rearrangement-the normal reaction of alcohols-may be carried out only under drastic conditions (Al_2O_3, 450°C).

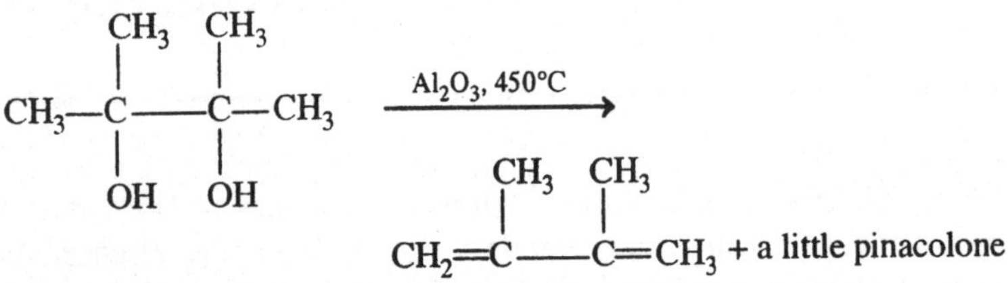

MECHANISM

The reaction mechanism is as follows :

[I]***Generation of Carbonium Ion***

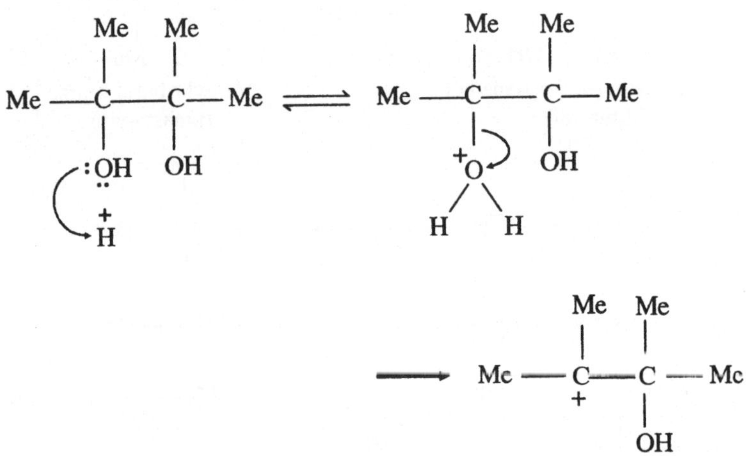

carbocation
[carbonium ion]

[II] ***1 : 2-shift***

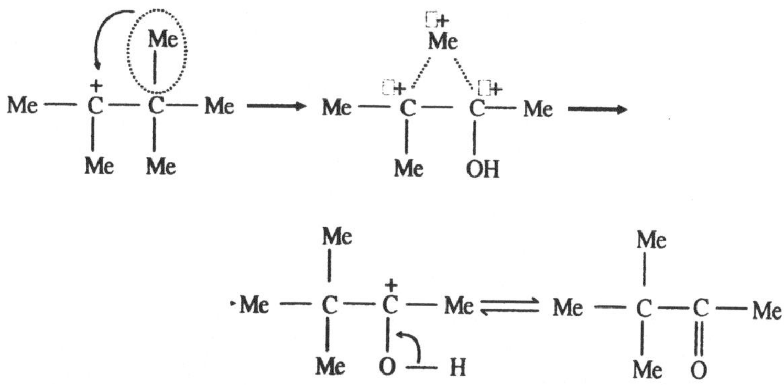

Support for the Mechanism of Pinacol Rearrangement

(*i*) The mechanism of pinacol rearrangement given above has been supposed by the Kinetic studies of the rearrangement which reveals that the rate determining step is the conversion of oxonium ion to carbonium ion, *i.e.*, the formation of a carbonium ion.

(*ii*) When the pinacol is subjected to rearrangement in $H_2{}^{18}O$ and the reaction is stopped before it gets completed, the recovered pinacol contains ^{18}O. This reveals the reversible formation of a carbonium ion.

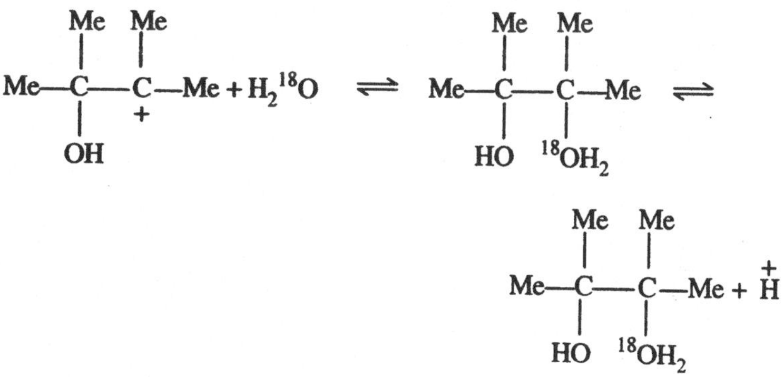

(*iii*) If the mechanism given above is correct, it means that wherever such a carbonium ion is involved, it should rearrange to form a pinacalone. This has been confirmed by the various experiments. For example, The reaction of the pinacol bromohydrin with silver oxide results in the formation of pinacolone. Also the deamination of 2, 3-dimethyl, 3-aminobutan 2-al with HNO2 results in the formation of pinacalone.

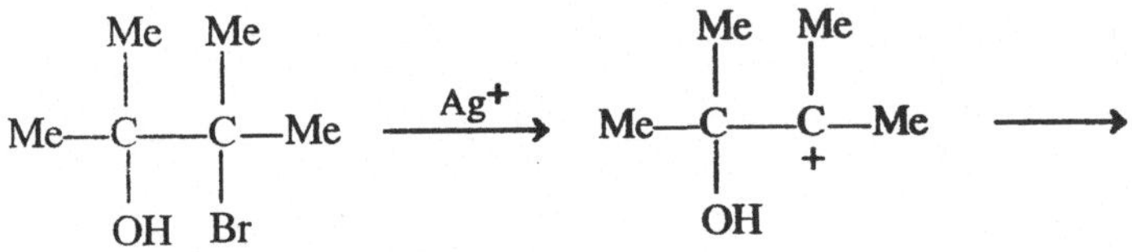

Pinacol bromohydrin

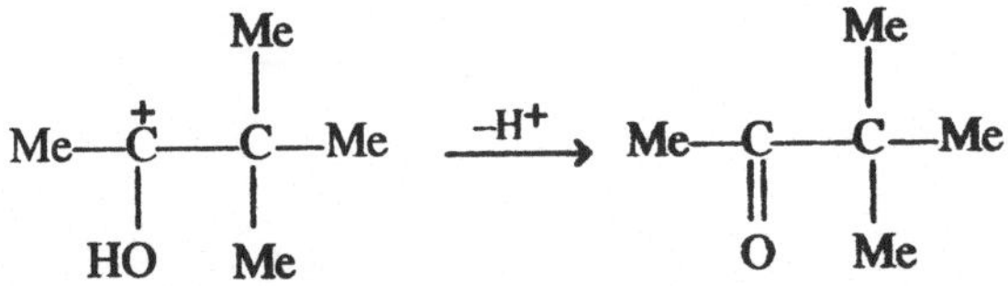

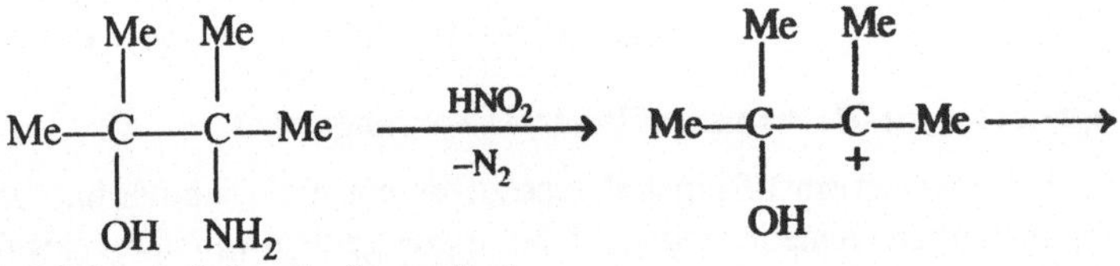

2, 3-Dimethyl-3-aminobutan-2-ol

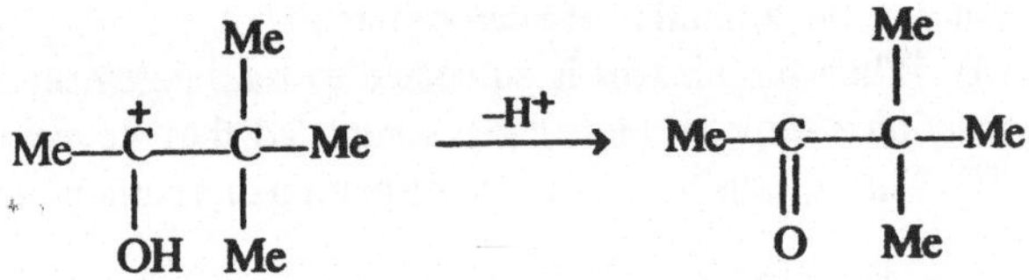

(*iv*) The mechanism is also supported by the fact that any carbocation having the positive charge on the carbon adjacent to the one having the hydroxyl group $\left(-\underset{\oplus}{\overset{|}{C}}-\underset{|}{\overset{|}{C}}-OH \right)$ also undergoes similar rearrangement.

For example,

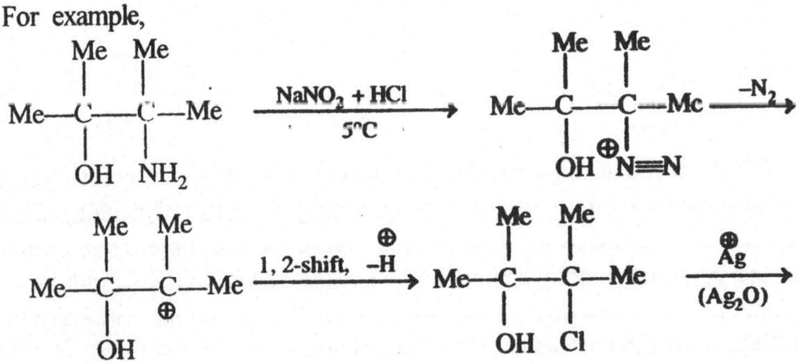

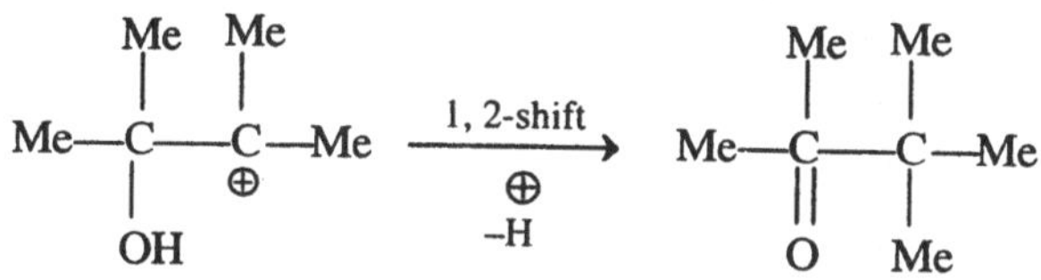

FEATURES OF PINACOL REARRANGEMENT

(*i*) **Stability of the carbonium ion :** If there is choice in removing hydroxyl group preferentially, *i.e.*, if two OH groups are different as in unsymmetrically substituted glycols, only that OH will be removed which produces the more stable carbonium ion (Difference from Wagner-Meerwein rearrangement). In the following example, it can be easily known that carbonium ion gets generated on the phenylated carbon atom rather than on the methylated carbon atom due to resonance stabilisation.

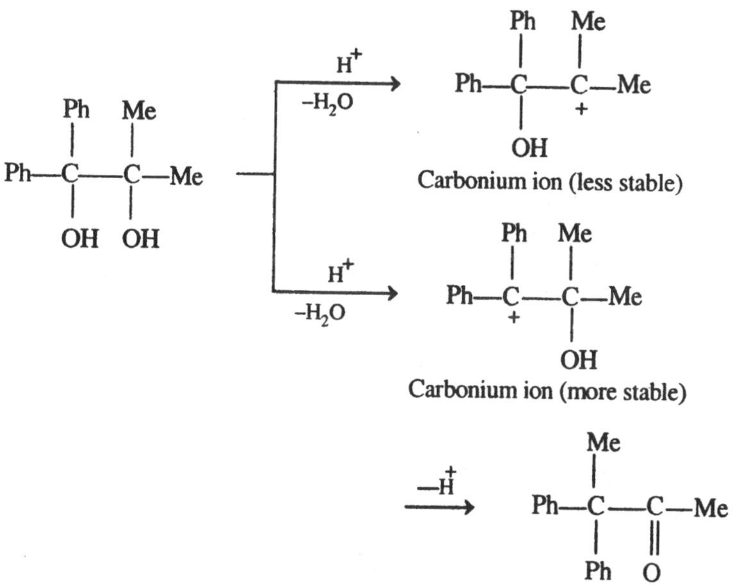

Carbonium ion (less stable)

Carbonium ion (more stable)

3, 3-Diphenyl-2-butanone

From the rearrangements of α, β-diol esters, it is possible to know the relative stabilities of carbonium ions. The partial positive charge on the carbonyl carbon of the ester group does not permit the formation of a carbonium ion at the α-carbon due to mutual repulsion between two similar charges. However, it is possible to generate the positive charge at the α-carbon by stabilizing it over a phenyl substituent (II) together with increasing the energy of the alternate β-carbonium ion by making it a primary one (I).

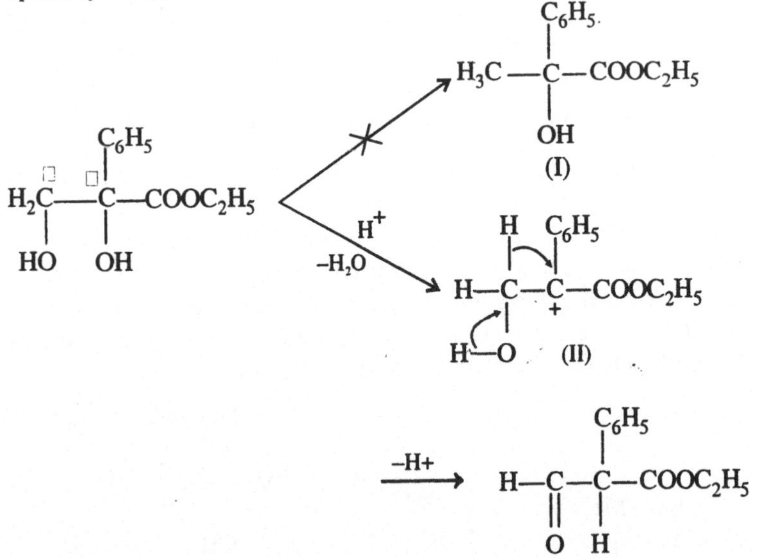

(*ii*) When both the hydrogen atoms are substituted by methyl groups, carbonium ion gets generated on the β-carbon. This shows that a tertiary carbonium ion (III) has somewhat more stability than a benzylic carbonium ions adjacent to a catbethoxy group.

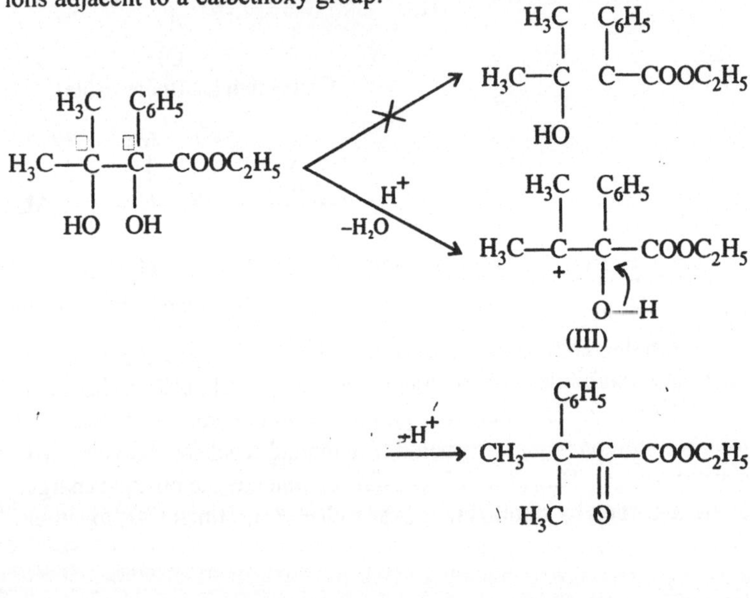

(*iii*) If the two groups present on the non-ionic carbon are different, then the better nucleophile (*i.e.* more powerful electron donor towards carbon) of the two gets migrated preferentially. In the following example, *p*-$MeOC_6H_4$ is a better nucleophile than phenyl group due to electron donating effects of its $—OCH_3$ group and it, therefore, migrates in preference to phenyl.

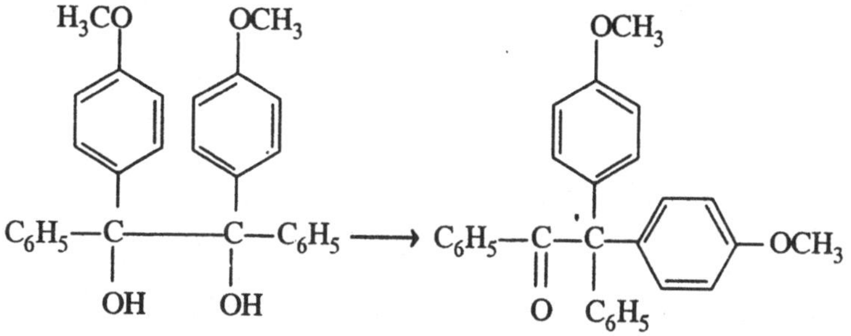

Generally, the phenyl group gets migrated in preference to alkyl groups.

(*iv*) **Migratory aptitude of the groups :** If the migratory aptitude of phenyl group is assigned the value 1, then the values for the migratory aptitudes of substituted phenyl groups with reference to this group are as follows :

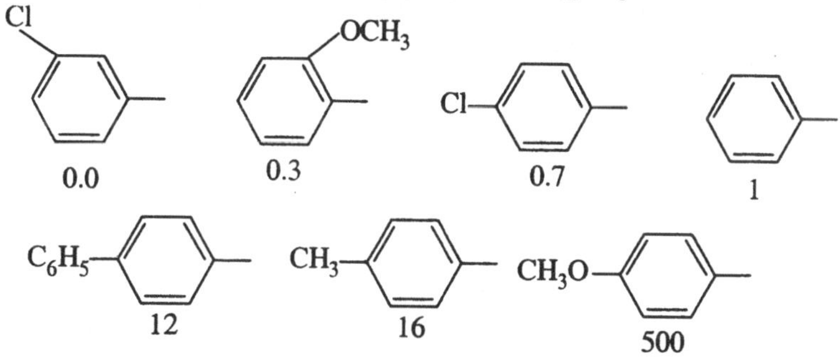

This is illustrated by the following example :

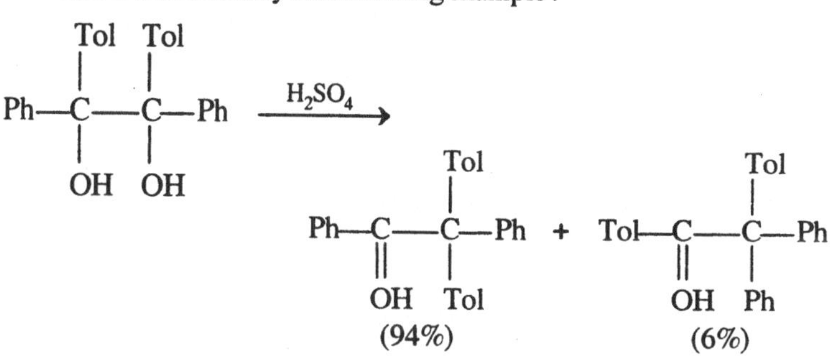

From the above data, it follows that the migratory aptitude of a group depends upon its ability to release electrons and this is also indicative of the fact that there occurs the formation of a bond of migrating group with the adjacent electron deficient centre even though the departing group (H_2O) has not completely broken from the molecule.

(*v*) **Steric effect :** The migratory aptitude of the group is also influenced by steric factors. For example, the *o*-$CH_3OC_6H_4$– group is migrated thousand times less than the *p*-$CH_3OC_6H_4$– group due to steric interference.

The migrating group attacks from the trans position (back side) to the leaving group. This exerts important effect in cyclic systems. Hence, the two isomers of 1, 2-dimethyl-cyclohexane-1, 2-diol yield

different products because of different orientations of the methyl and hydroxyl groups. The one (VI) in which the Me and OH groups are trans to each other yields 2, 2-dimethylcyclohexanone by methyl shift.

The other (VII) in which the Me and OH groups are *cis* to each other undergoes ring methylene group shift instead of Me-shift with consequent ring contraction to yield 1-acetyl-1-methylycyclopentane(VIII)

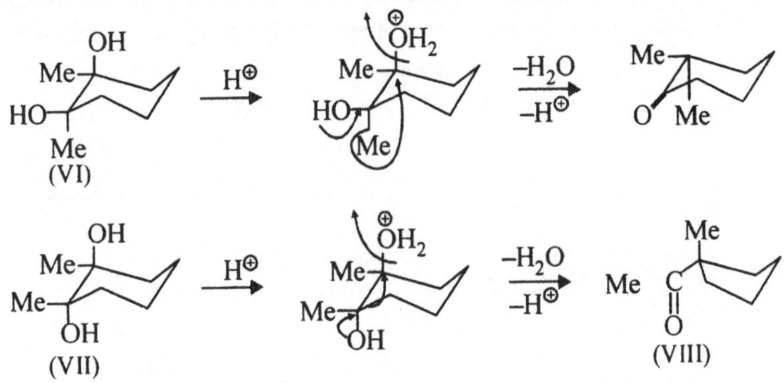

The rearrangement finds interesting applications in synthesis.

(*vi*) **Transmigration.** The pinacol rearrangement occurs in such a manner that the migrating group is always trans to the leaving hydroxyl group. It has important consequences.

APPLICATIONS OF PINACOL REARRANGEMENT

The pinacol rearrangement has important applications. It is used to prepare the pinacolones which are highly branched ketones. These cannot by easily prepared by other methods. Further, the cyclic ketones can be converted into pinacols which on rearrangement give spiro-ketones.

1. Synthesis of carbonyl compounds from alkenes : Isobutyraldehyde may be obtained on a large scale from isobutylene.

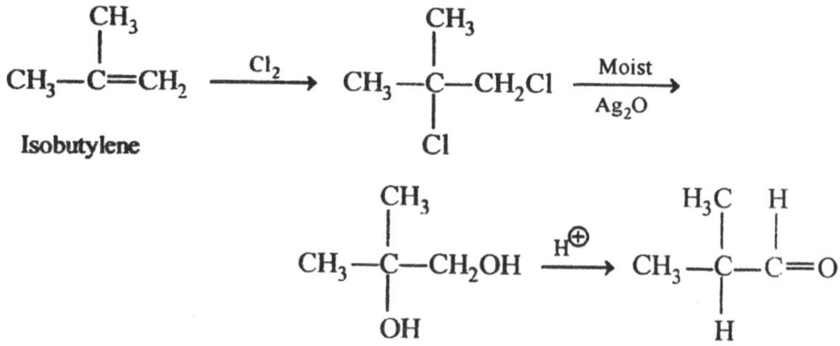

Isobutylene

Dimethyl acetaldehyde
(Isobutyraldehyde)

2. Ring expansion of cyclic ketones : Cyclohexanone can be converted to cycloheptanone. The yield is quite good.

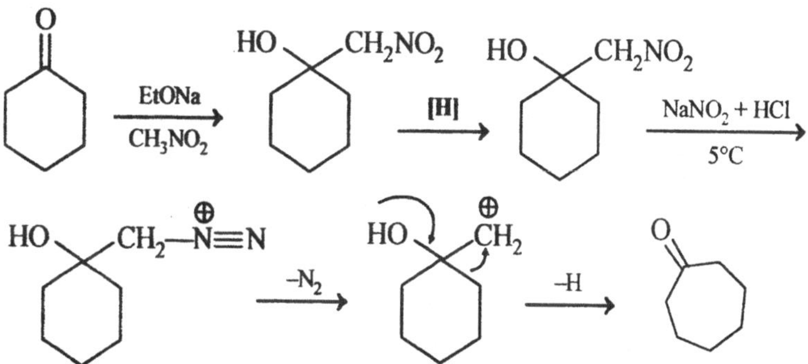

Cycloheptanone

3. Ketones from cyclic diols : Pinacol rearrangement is used to prepare ketones which are otherwise inaccessible or are very different to synthesize.

(*i*)

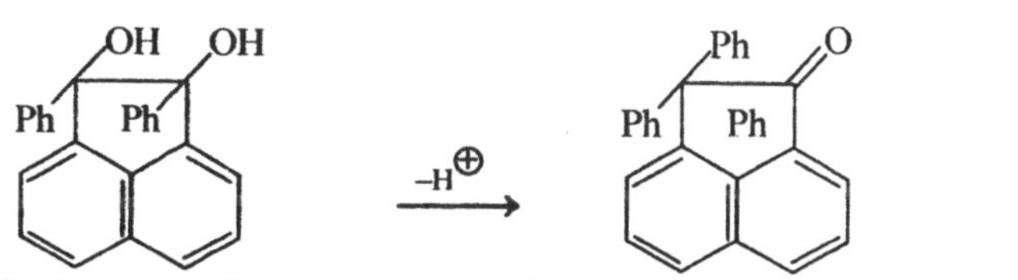

7, 8-Diphenylacenaphthene -7, 8-diol

7-Oxo-8-, 8-diphenylacenaphthene

(*ii*) 2 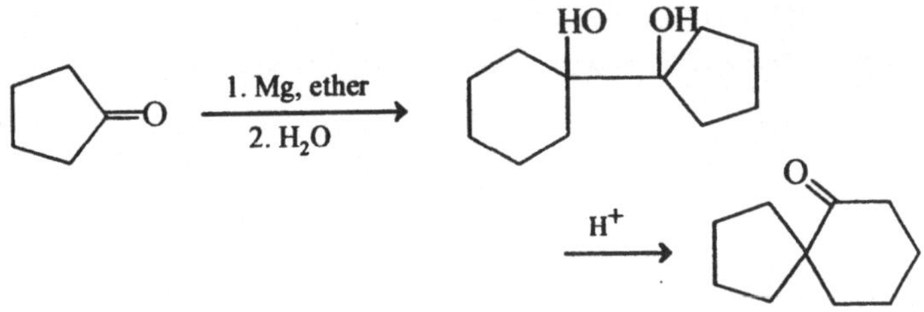

Cyclopentanone pinacol

4. 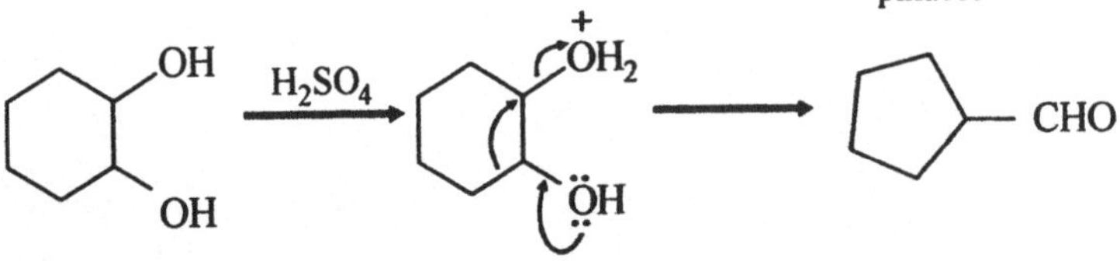

5. 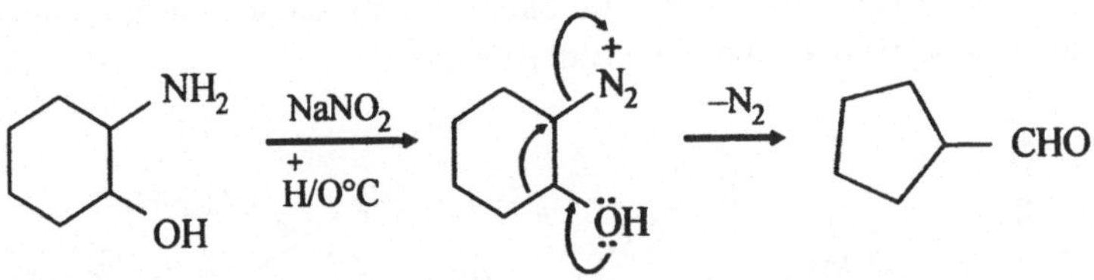

6. 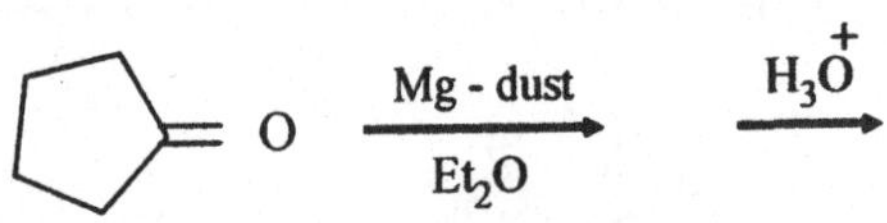

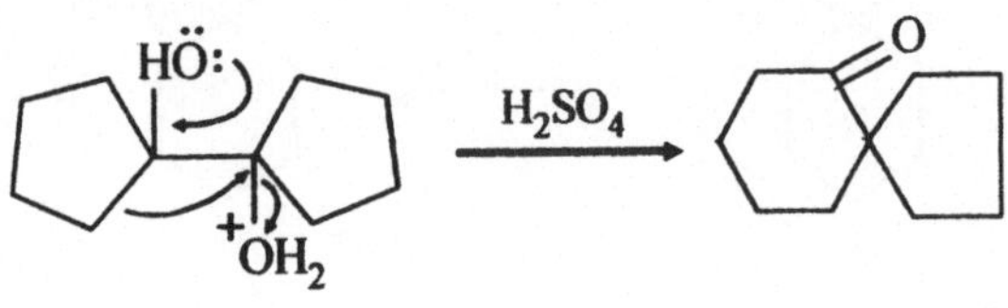

Problem 1 :

Identify **A, B, C, & D**

Cyclobutadiene + **Ethylene** $\xrightarrow{AlCl_3}$ C_6H_{10} **[A]** $\xrightarrow{Br_2}$ $C_6H_{10}Br_2$ **[B]**

$C_6H_{12}O_2$ **[C]** $\xrightarrow{H_2SO_4 \text{ conc.}}$ $C_6H_{12}O$ **[D]**

Solution :

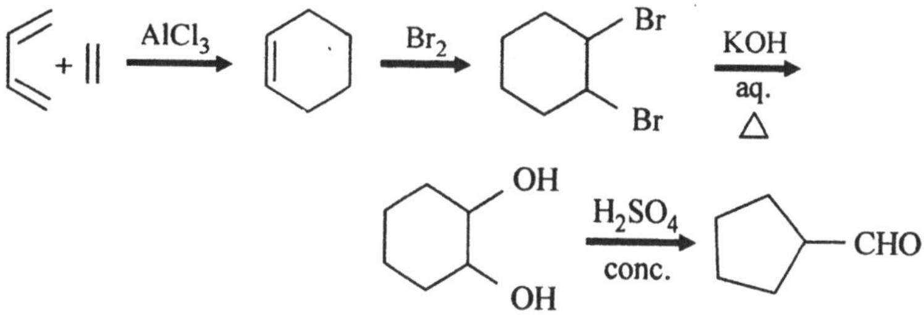

A,,	Cyclohexene;	B,,	1:2-Dibromocyclohexane
C,,	Cyclohex-1:2-diol;	D,,	Cyclopentaldehyde

Problem 2 :

Work out the following

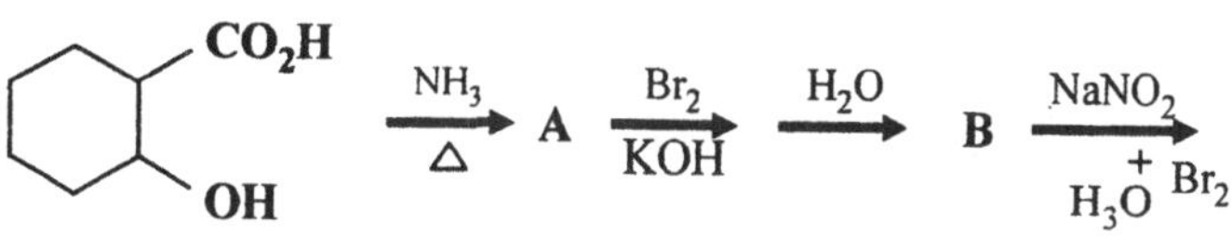

Answer :

A,, 1-Hydroxycyclohexyl amide

B,, 1-Hydroxycyclohexyl amine

C,, Cyclopentaldehyde

Problem 3 :

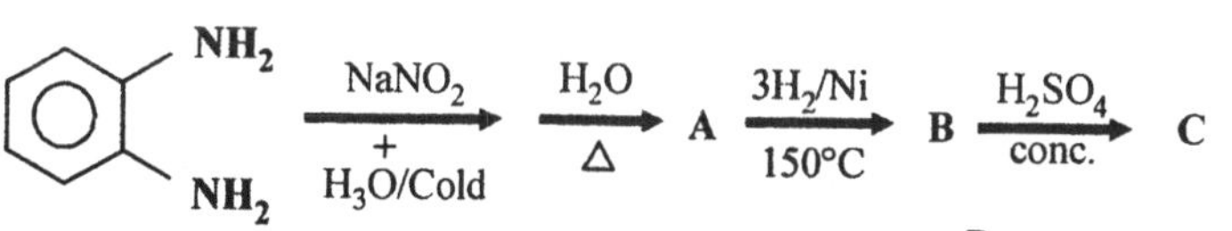

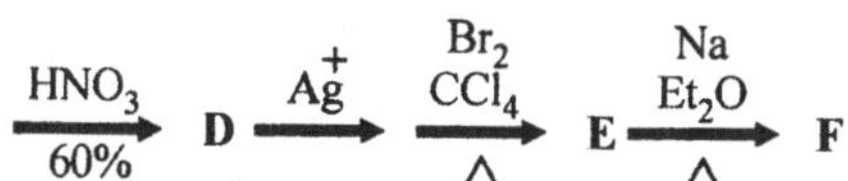

Solution :

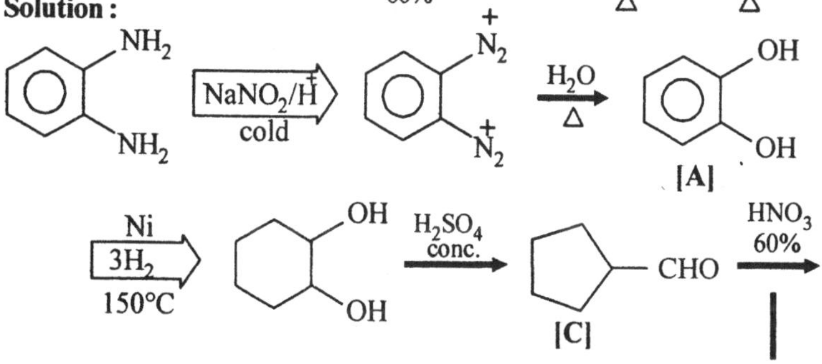

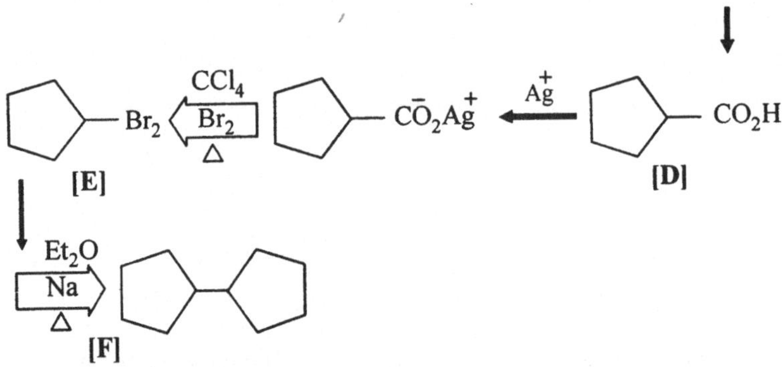

A,, Catechol; B,, 1:2-Dihydroxycyclohexane
C,, Cyclopentaldehyde; D,, Cyclopent-l-carboxylic acid
E,, Cyclopentyl bromide; F,, Dicyclopentane

Problem 4 :

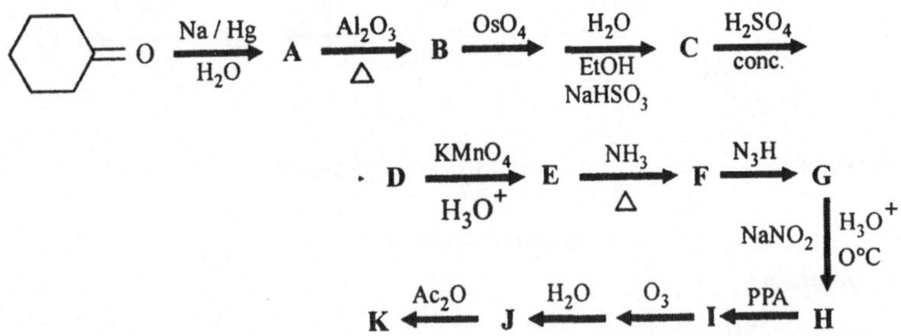

Solution :

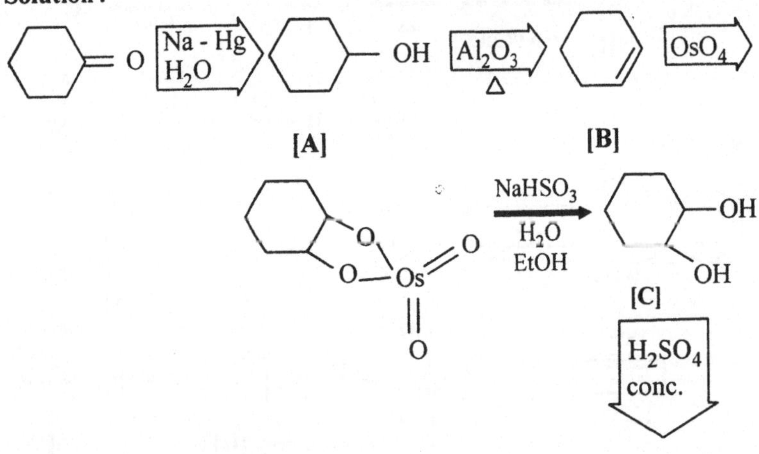

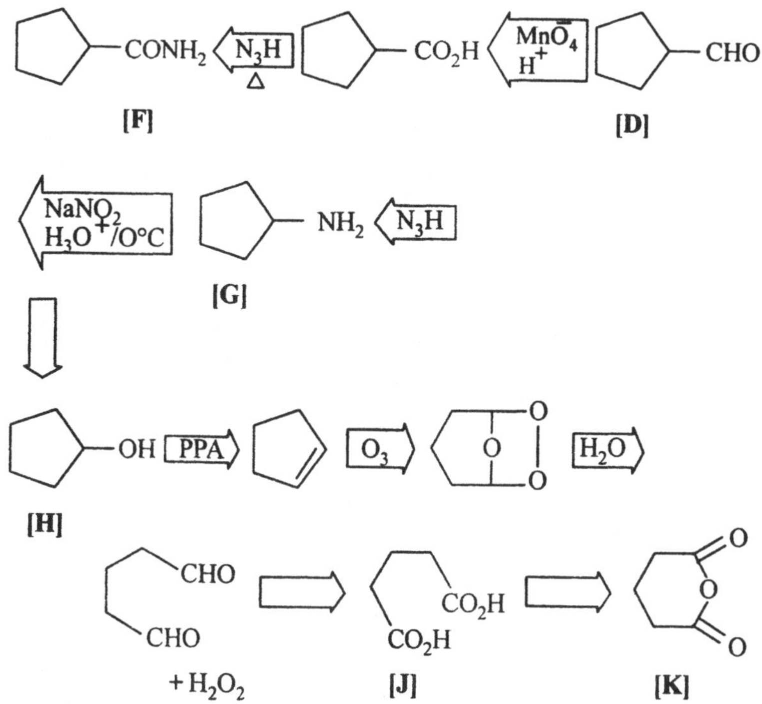

A,, Cyclohexanol;	**B**,, Cyclohexene;
C,, 1:2-Dihydroxycyclohexane;	**D**,, Cyclopentaldehyde;
E,, Cyclopent-l-carboxylic acid	**F**,, Cyclopent-1-carbamide;
G,, Cyclopentylamine;	**H**,, Cyclopentanol;
I,, Cyclopentene;	**J**,, Glutaric acid;
K,, Glutaric anhydride	

Problem 5 :

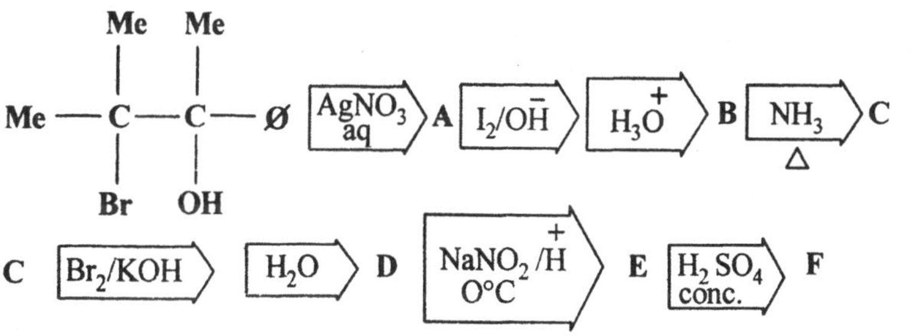

Answer :

A,, 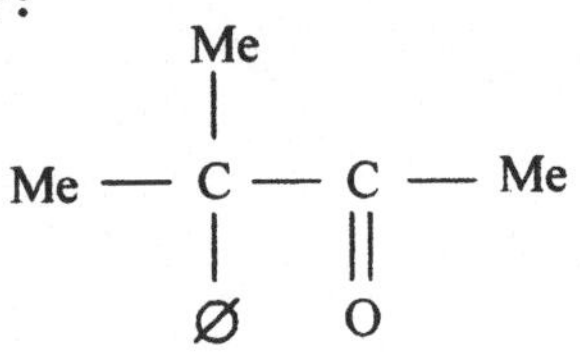 B,, 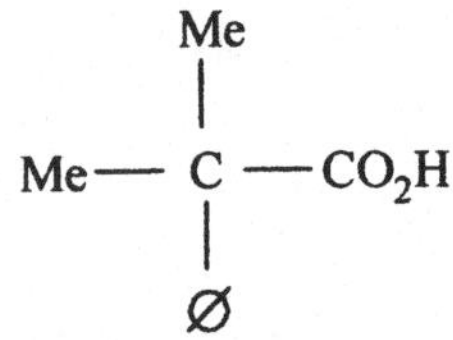

3-Methyl-3-phenyl but-2-one

2-Methyl-2-phenylpropionic acid

C,, 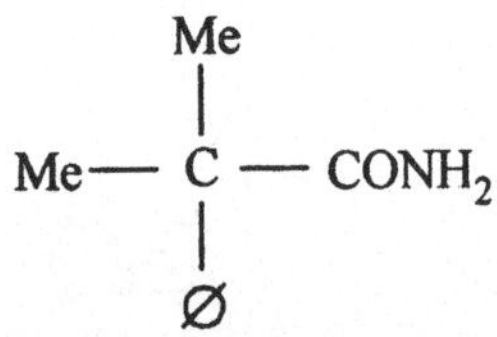 D,, 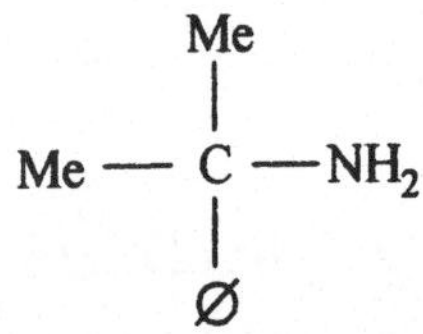

2-Methyl-2-phenylpropamide

2-Phenyl-2-aminopropane

E,, 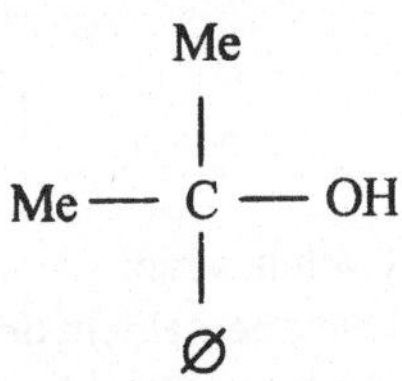 F,, 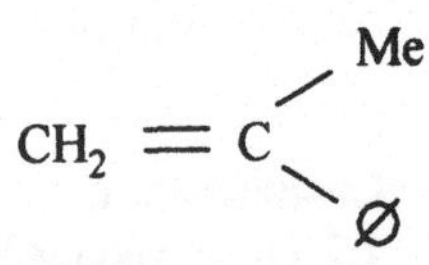

2-Phenyl-2-hydroxypropane

1-Methyl-1-phenyl ethylene

Problem 6 :

Work out

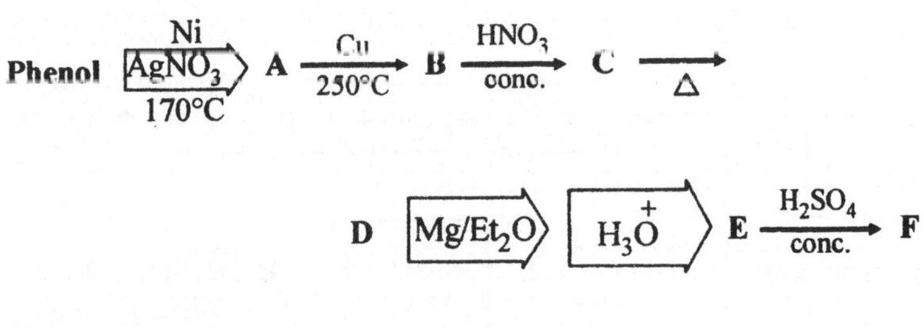

Solution :

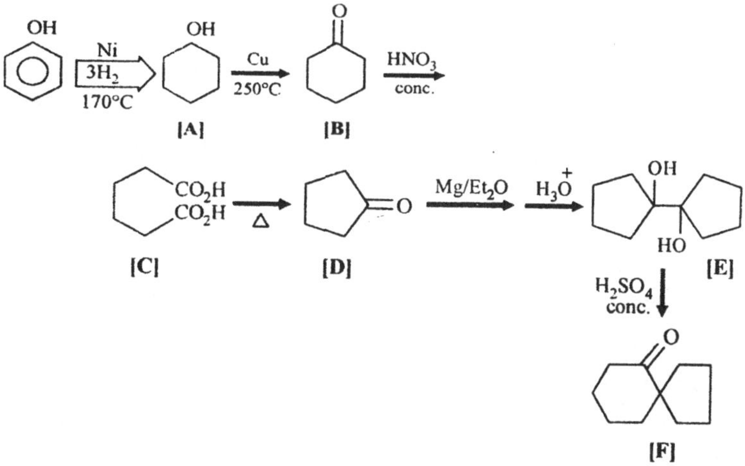
OH
Ni
3H2
170°C
OH
[A]
Cu
250°C
O
[B]
HNO3
conc.
CO2H
CO2H
Δ
[C]
O
[D]
Mg/Et2O
H3O+
OH
HO
[E]
H2SO4
conc.
O
[F]

79

PINNER SYNTHESIS

Addition of alcohol with cyanide in the presence of HCl is known as Pinner synthesis.

$$\overset{+\delta}{R-C}\equiv\overset{-\delta}{N} + R'-\overset{-\delta}{O}-\overset{+\delta}{H} \xrightarrow{HCl} \underset{\substack{|\\ OR'}}{R-C}=NH \cdot HCl$$

Salt of Imino ester

This salt on hydrolysis gives ester.

$$\underset{\substack{|\\ OR_1}}{R-C}=NH \cdot HCl \xrightarrow{HOH} R-\overset{\substack{O\\ \|}}{C}-OR_1$$

$$CH_3-CH_2-C\equiv N + CH_3-OH \xrightarrow[\text{(ii) } H_2O]{\text{(i) HCl}} CH_3-CH_2-\overset{\substack{O\\ \|}}{C}-OCH_3$$

$$C_6H_5-C\equiv N + CH_3-OH \xrightarrow[\text{(ii) } H_2O]{\text{(i) HCl}} C_6H_5-\overset{\substack{O\\ \|}}{C}-OCH_3$$

$$CH_3-\underset{\substack{|\\ CH_3}}{CH}-C\equiv N + C_2H_5OH \xrightarrow[\text{(ii) } H_2O]{\text{(i) HCl}} CH_3-\underset{\substack{|\\ CH_3}}{CH_2}-\overset{\substack{O\\ \|}}{C}-OC_2H_5$$

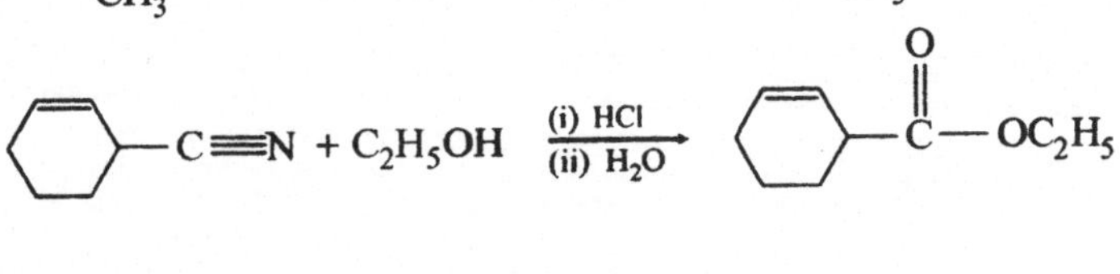

80

PREVOST REACTION

Preparation of trans 1, 2-diols by Prevost reagent is known as Prevost reaction. Prevost reagent is a solution of Iodine in carbon tetrachloride together with an equivalent of dry silver acetate or silver benzoate. Under anhydrous condition this reagent (oxidant) directly yields the diacetyl or dibenzoyl derivative of the trans diol. This on hydrolysis gives trans diol.

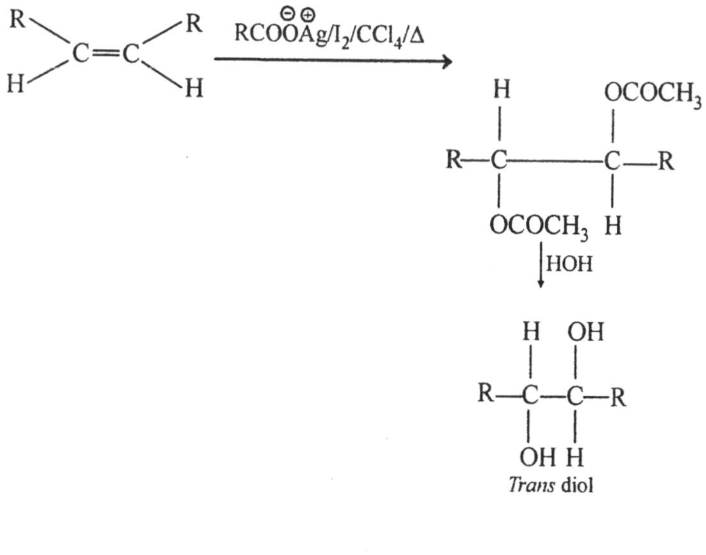

Trans diol

MECHANISM :

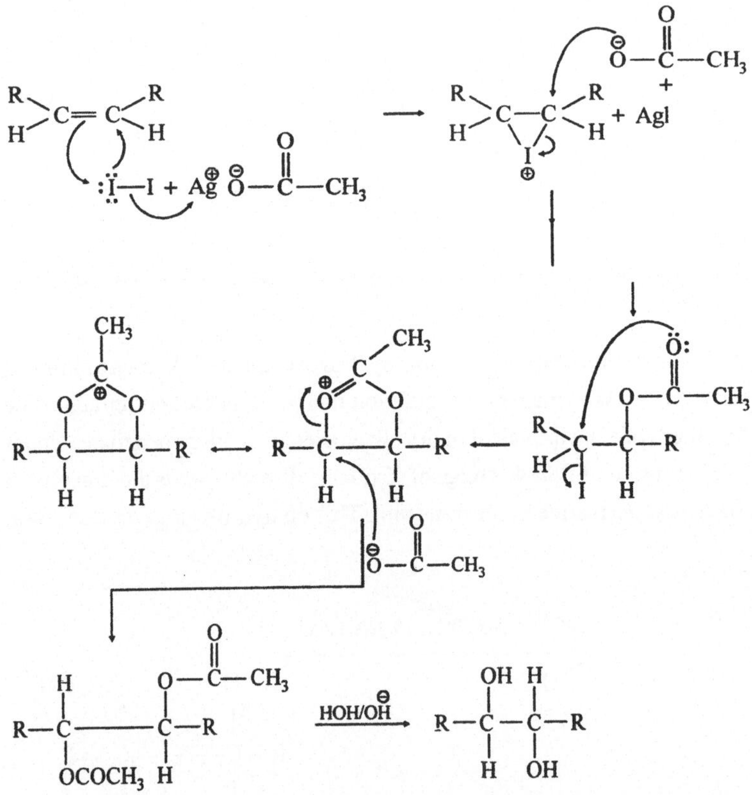

The value of this reagent is due to their specificity and due to the mildness of the reaction condition, free iodine under the conditions used, hardly affects other sensitive groups in the molecule.

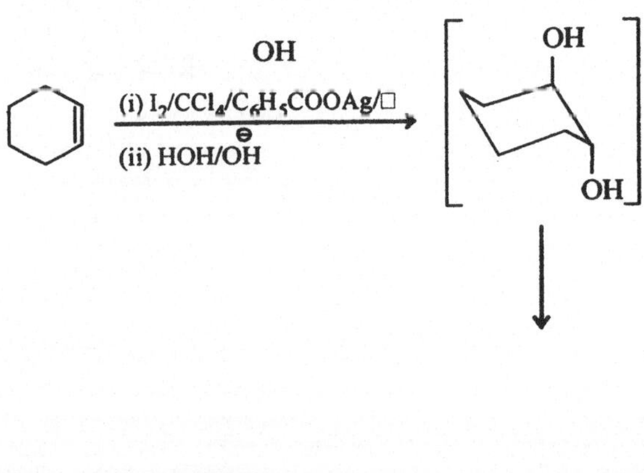

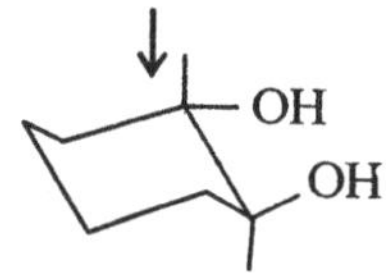

Trans diol

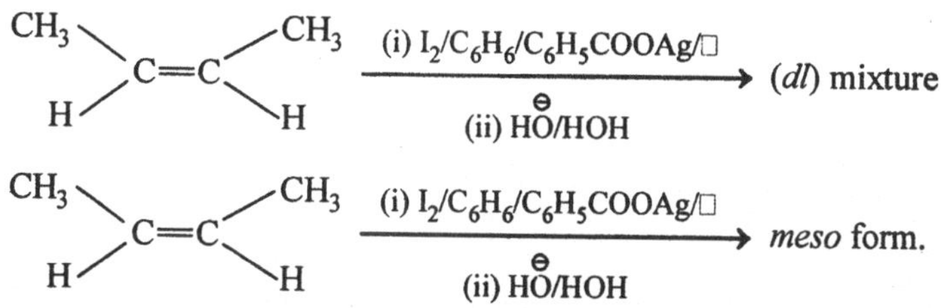

Note : If the above reaction is carried out in the presence of wet silver acetate or silver benzoate then the reagent in known as Woodward reagent and the product is, *cis* diol and the reaction is known as **Woodward reaction or Woodward hydroxylation.**

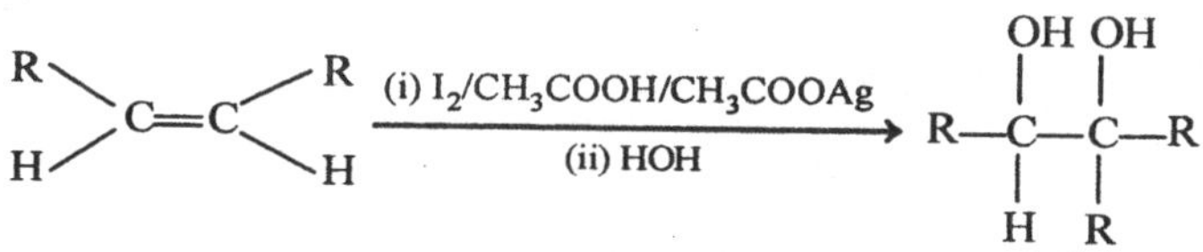

cis diol

MECHANISM OF WOODWARD REACTION

MECHANISM :

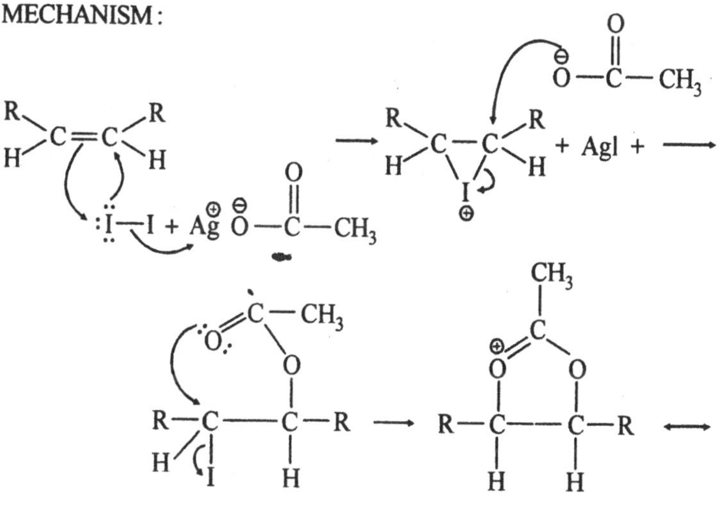

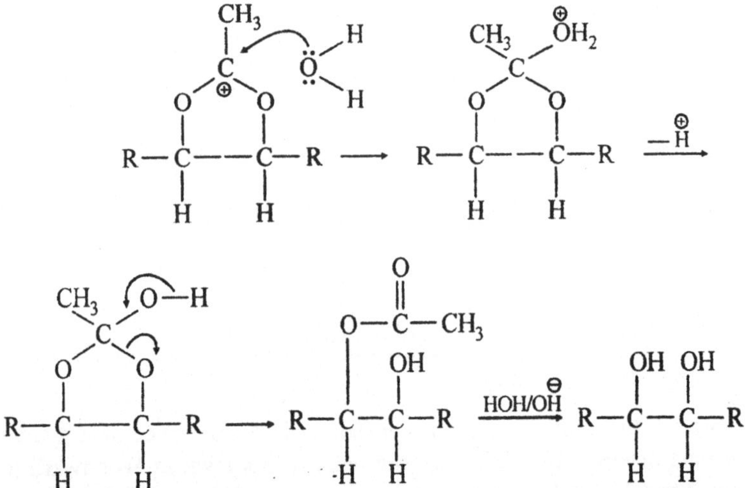

cis diol

Examples:

(*i*)

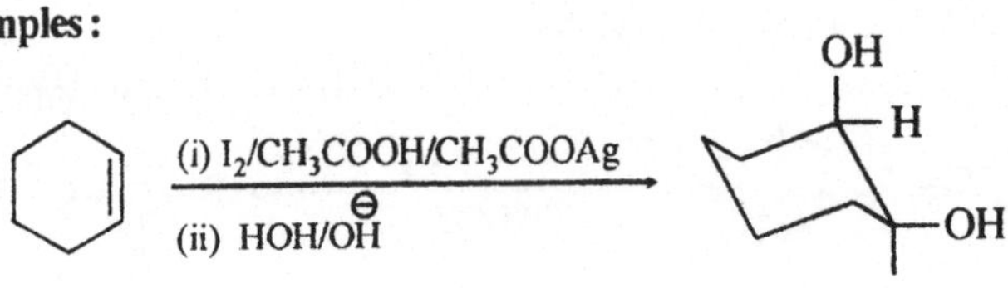

cis diol

(*ii*)

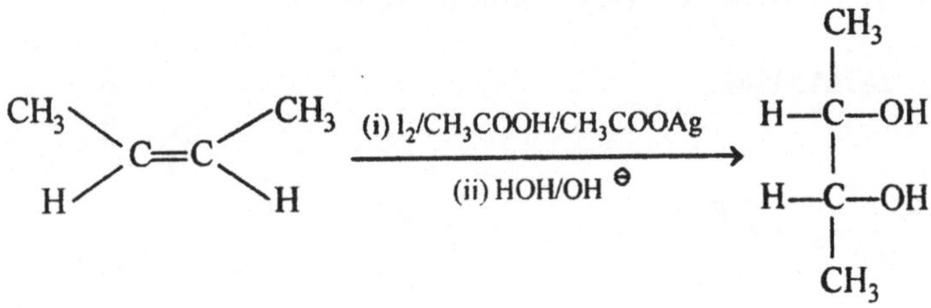

(*iii*)

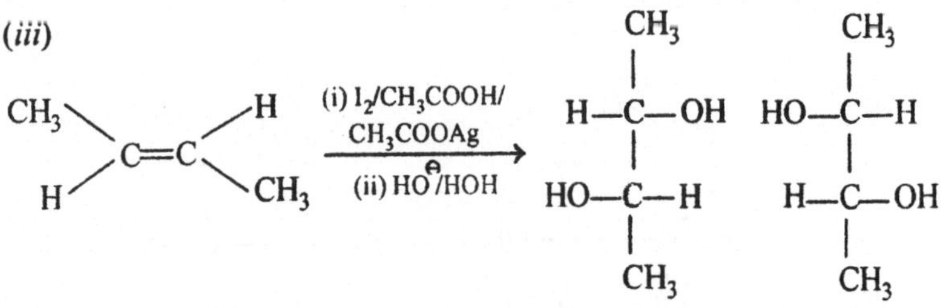

81

PRILEZHAEV REACTION

Treatment of alkene with peroxy acid in an inert solvent, such as $CHCl_3$ or CCl_4 yields an epoxide or oxirane. This reaction is known as Prilezhaev reaction. Peroxy acids may be aliphatic or aromatic acids such as peroxy acetic acid, peroxy benzoic acid, *m*-bromo peroxy benzoic acid and 3, 5-dinitro peroxy benzoic acid. The overall reaction amounts to the transfer of an oxygen from the oxidising agent to the alkene.

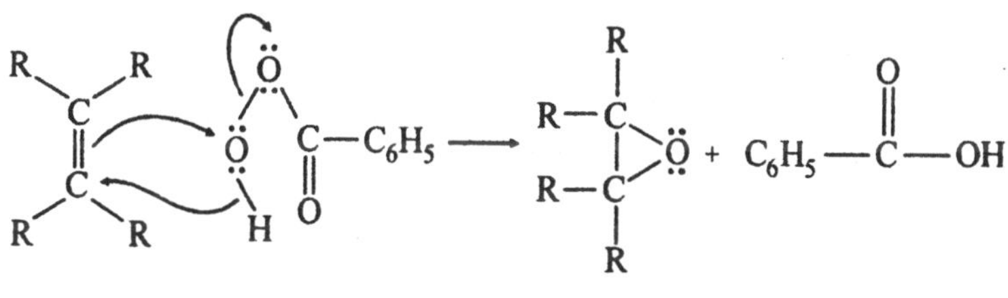

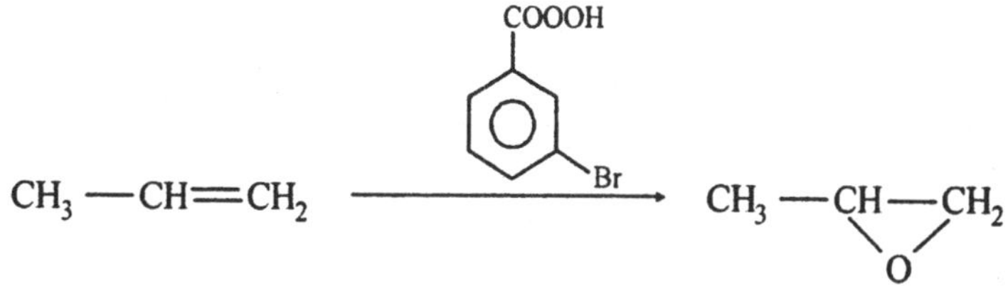

Alkene contains alkyl, aryl, hydroxy, ester but not amino and carbonyl groups because these two groups react with peroxy acids. Currently most widely used reagent for epoxidation is MMPP. (Magnesium monoperoxy phthalte).

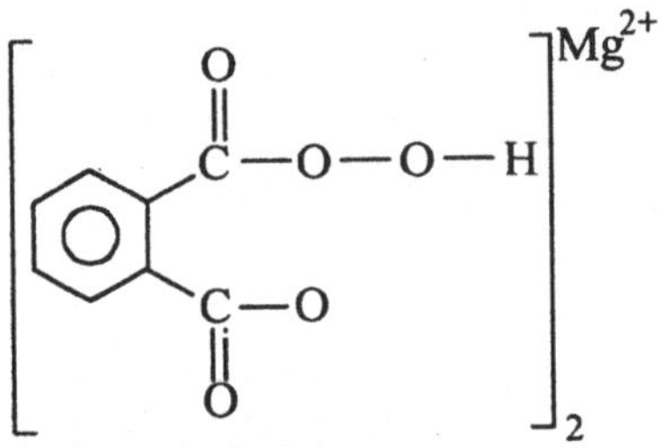

MMPP

The reaction is a concerted reaction and hence it is a stereospecific reaction *i.e.* the geometry of the alkene and the product is same in this reaction.

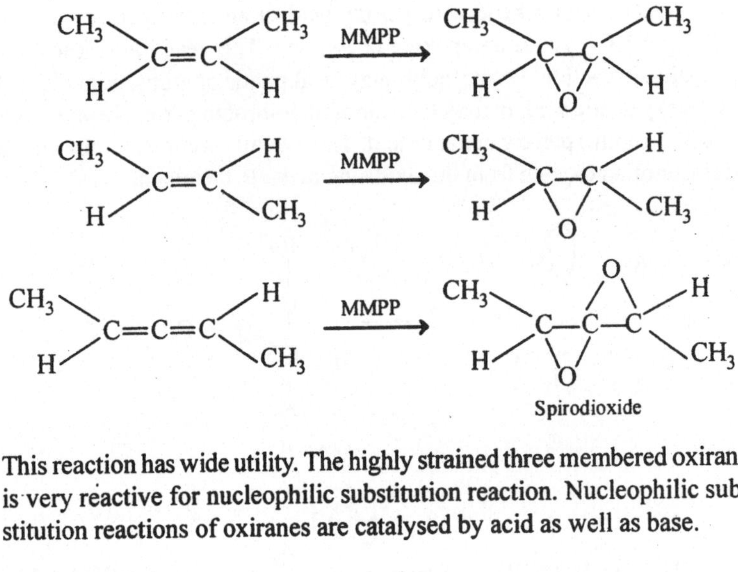

Spirodioxide

This reaction has wide utility. The highly strained three membered oxirane is very reactive for nucleophilic substitution reaction. Nucleophilic substitution reactions of oxiranes are catalysed by acid as well as base.

Acid-catalysed

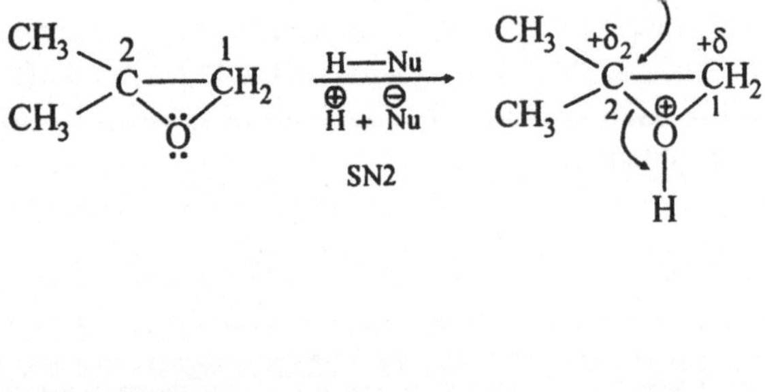

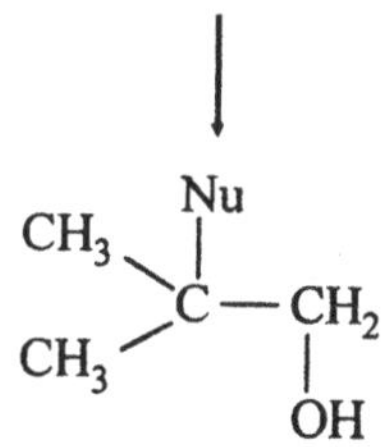

Due to presence of positive charge on oxygen, carbon-1 and carbon-2 both acquire carbocation character; hence nucleophile attacks on the carbon which is more hindred.

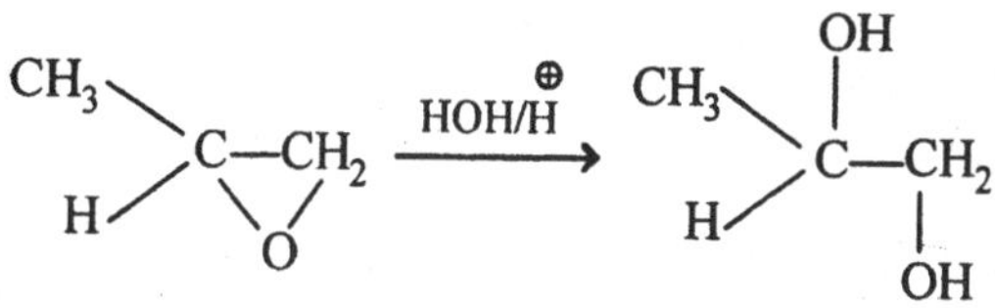

Trans-1, 2-diol

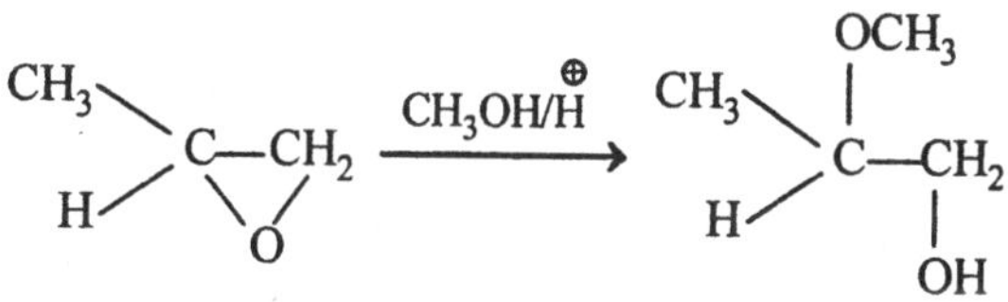

Base-catalysed

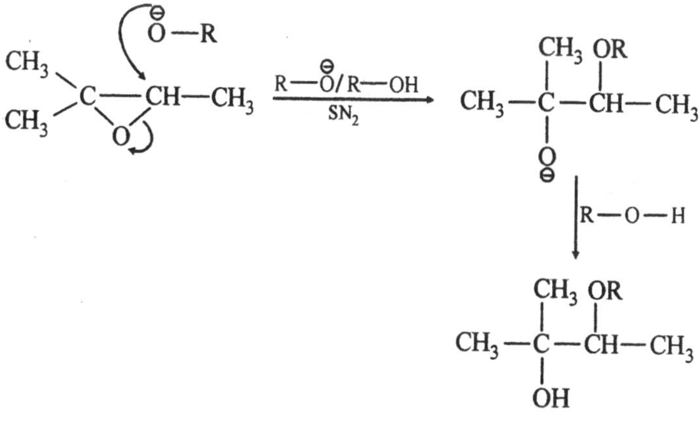

In presence of base, nucleophile attacks on less hindred carbon-atom because reaction is SN[2].

82

PRINS REACTION

Acid catalysed addition of aldehyde usually formaldehyde with alkene is known as Prins reactions.

1, 3-diol

MECHANISM :

Step I.

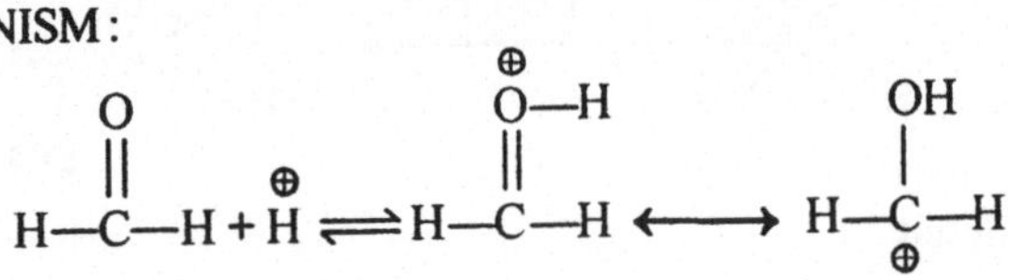

Step II.

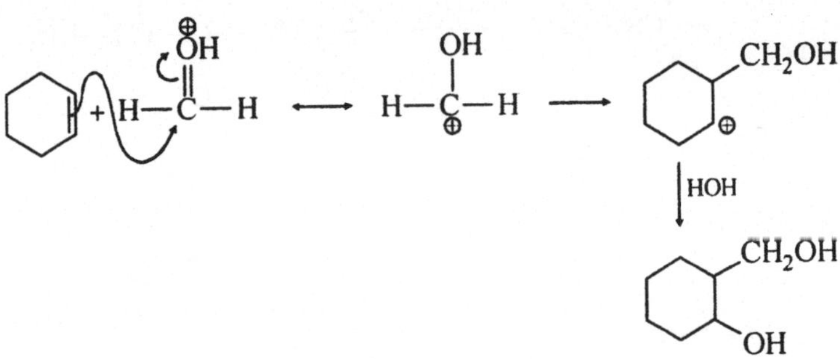

Example :

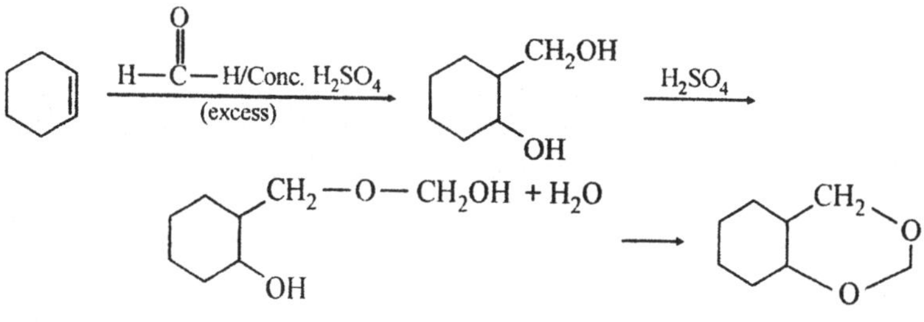

1, 3-Dioxane

83

RITTER REACTION

This reaction is given by alcohol (ter-alcohol) and alkene which forms ter carbocation in presence of conc. H_2SO_4. In this reaction, ter-alcohol or alkene is treated with conc. H_2SO_4 and nitrile to give amide which on acidic hydrolysis give *p*-amine.

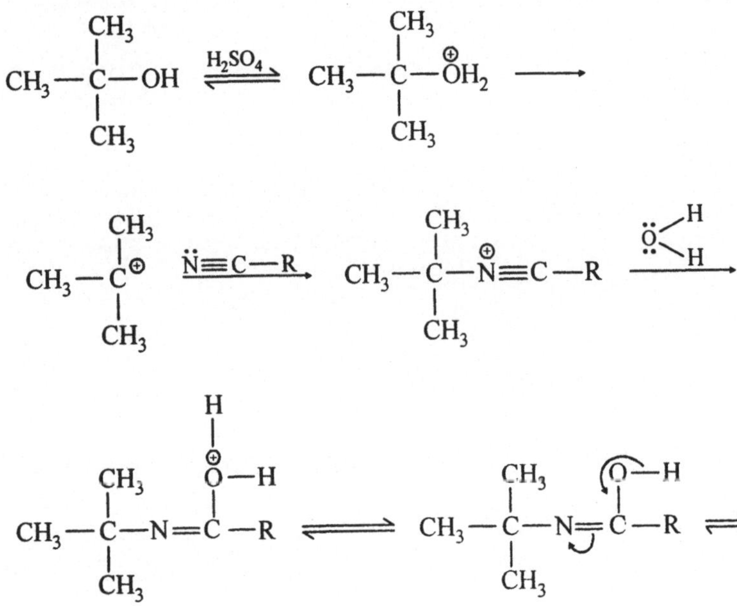

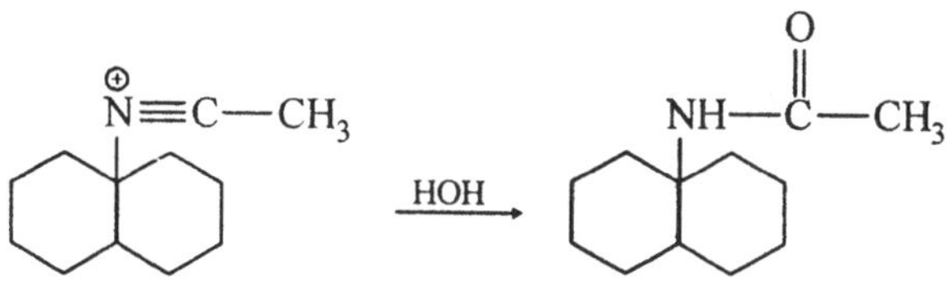

(C) (D)

Examples:

(i) 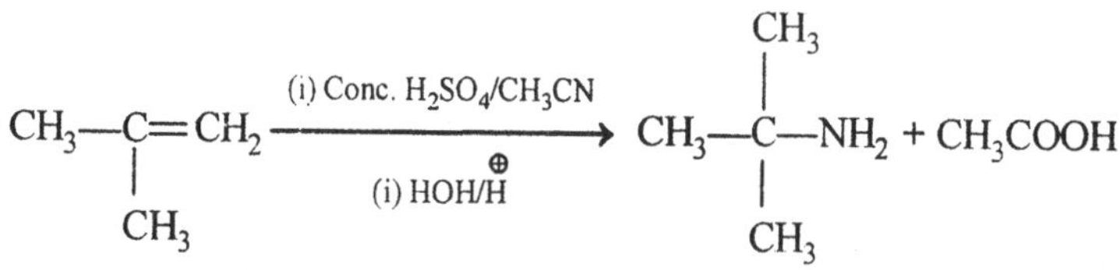

(ii) 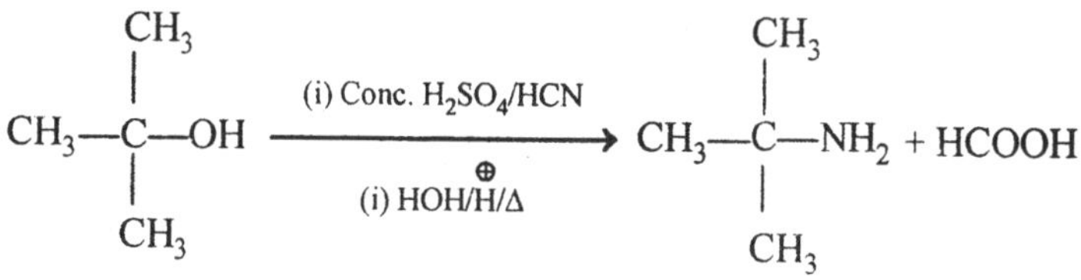

(iii) 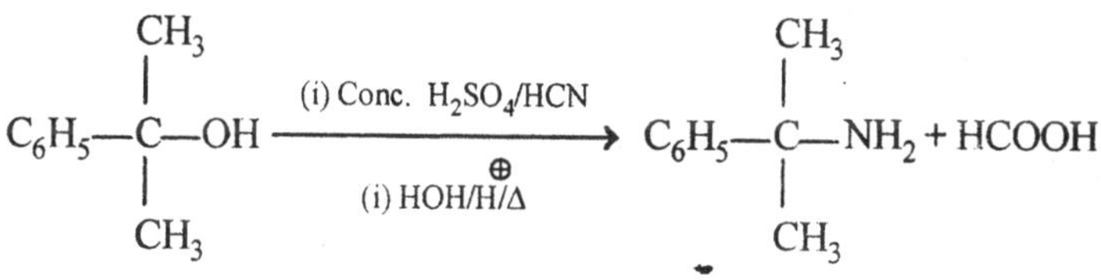

(iv) 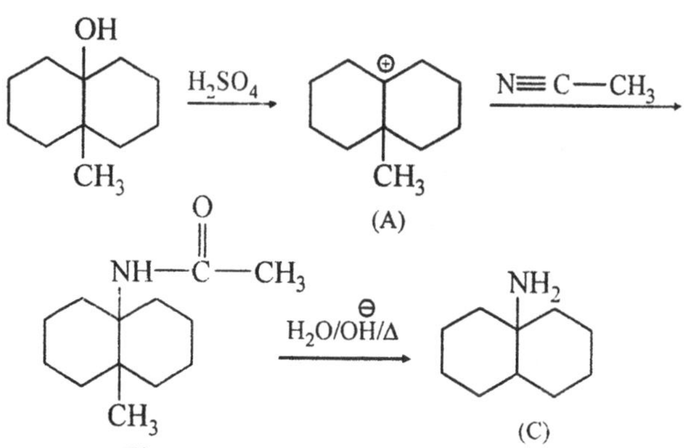

(v)

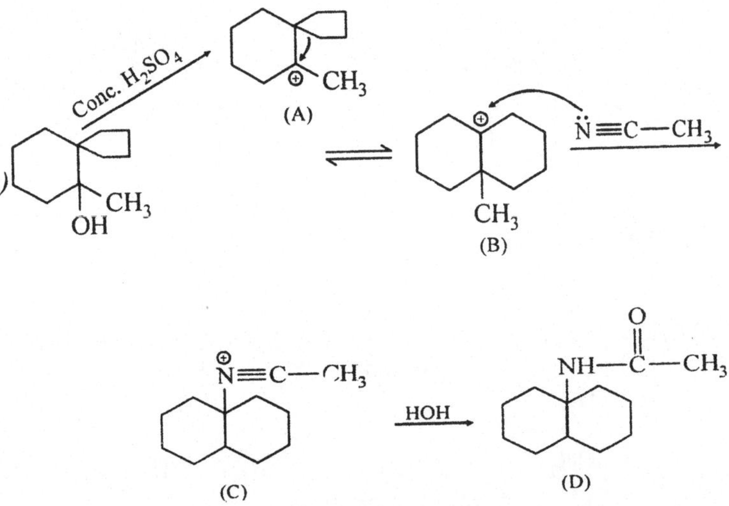
Conc. H_2SO_4
CH3
OH
CH3
(A)
$\ddot{N}\equiv C-CH_3$
CH3
(B)
$\overset{\oplus}{N}\equiv C-CH_3$
(C)
HOH
O
NH—C—CH3
(D)

84

REIMER-TIEMANN REACTION

The treatment of phenols with chloroform in the presence of aqueous base followed by treatment with aqueous acid give aldehydes.

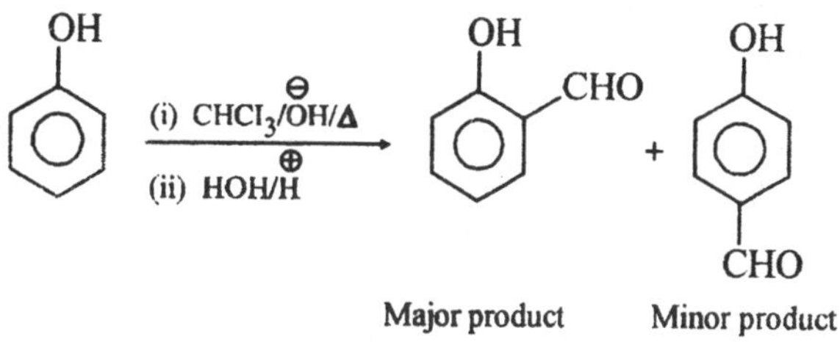

The reaction is known as Reimer-Tiemann formylation.

MECHANISM:

The first step is α-elimination of chloroform which gives a very reactive intermediate, dichlorocarbene as electrophile.

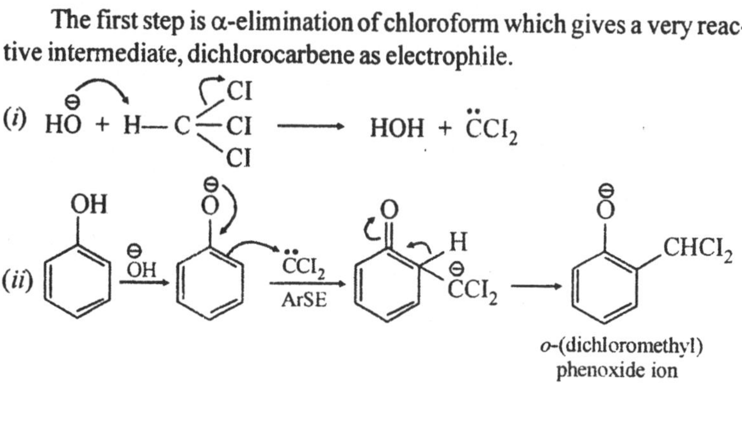

(*iii*)

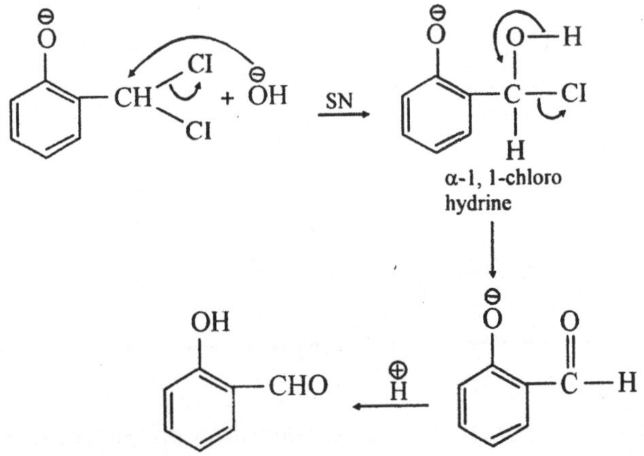

Pyrrole which resembles phenol in its reactivity for ArSE reaction also gives this reaction.

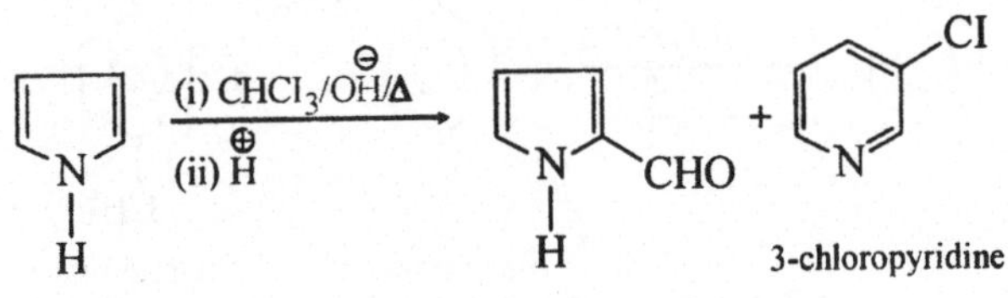

In case of Reimer-Tiemann reaction, salicyclic acid is obtained if CCl_4 is used in place of $CHCl_3$ and then the reaction is known as Reimer-Tiemann carboxylation. Carbontetrachloride is a neutral electrophile and it give ArSE reaction with phenoxide ion which is highly reactive for this reaction.

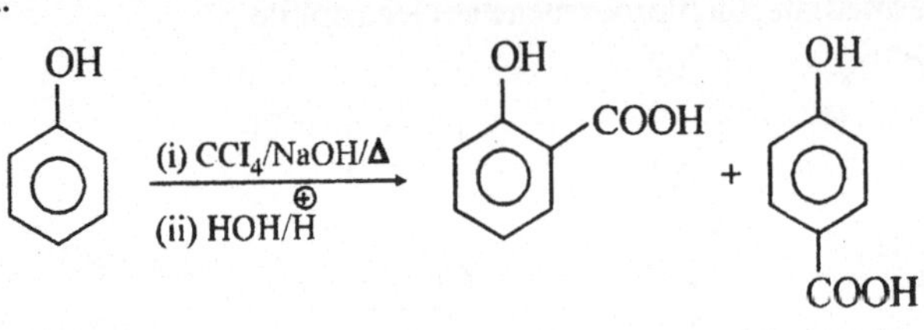

MECHANISM :

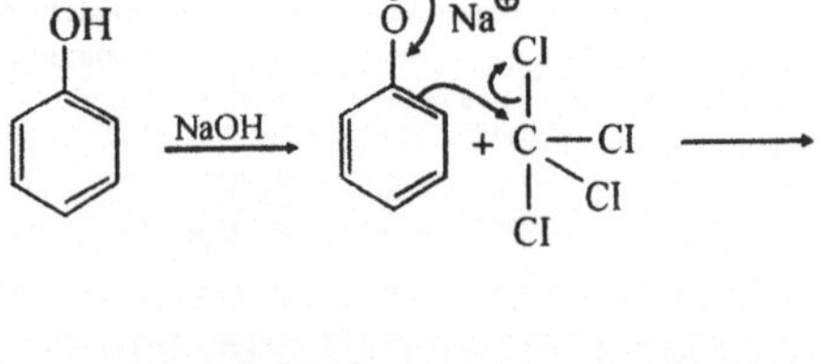

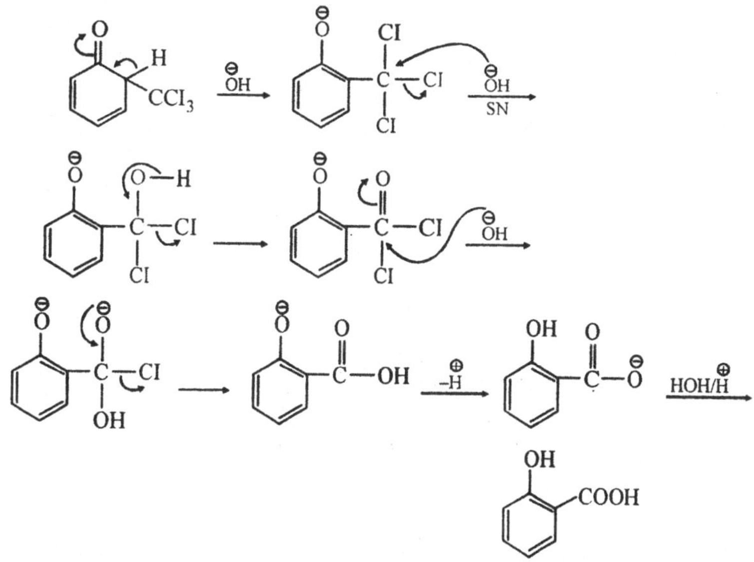

Although the product of Reimer-Tiemann reaction is the mixture of *o*- and *p*-aldehydes, yet the *o*-product predominates. If one of the *o*-positions is occupied, then the aldehyde group tends to enter *p*-position. Thus, guaiacol on refluxing with chloroform sodium hydroxide (or potassium hydroxide) give vanillin.

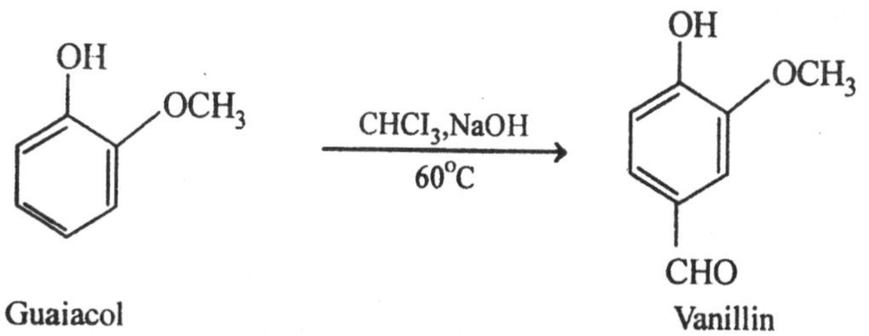

Similarly, catechol give 3, 4-dihydroxybenzaldehyde.

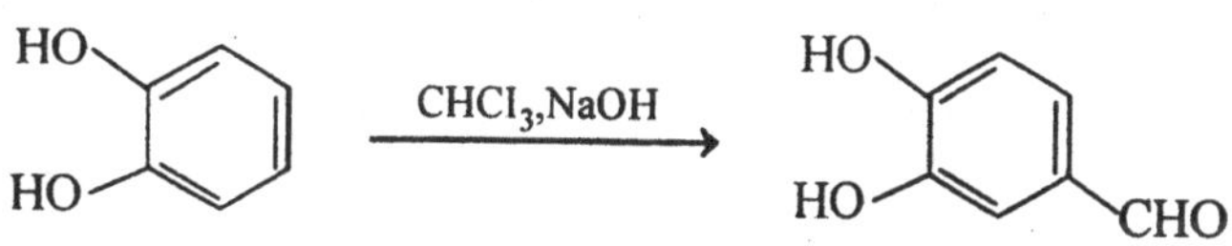

Phenol, with block p-position gives scyclohexadienone in which the two chlorine substituents, being in the neopentyl type environment are resistant to hydrolysis is basic conditions and remain intact.

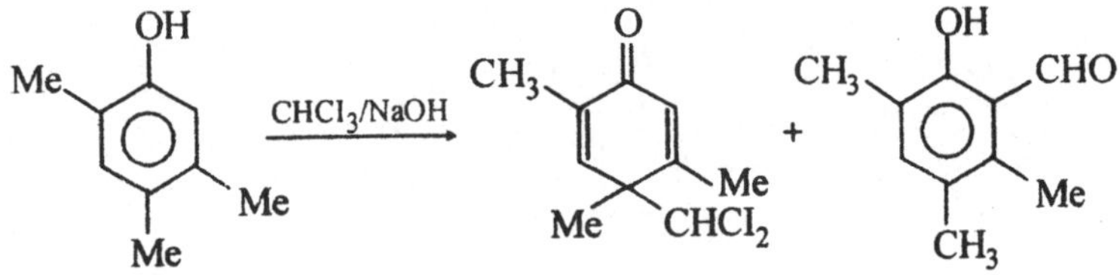

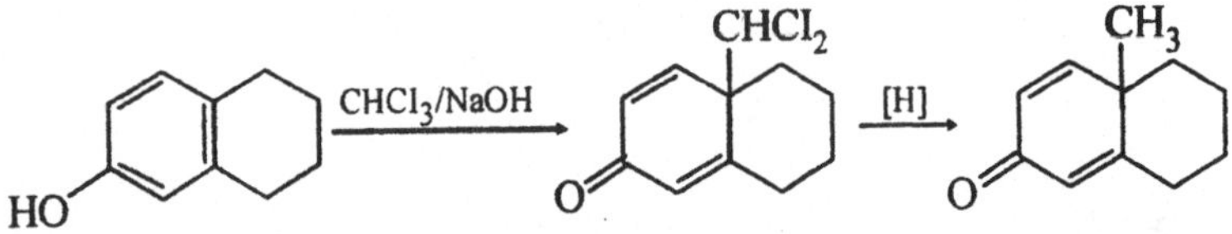

(*i*) Dichlorocarbene causes ring expansion which is seen in the reaction of pyrrole with chloroform and alkali.

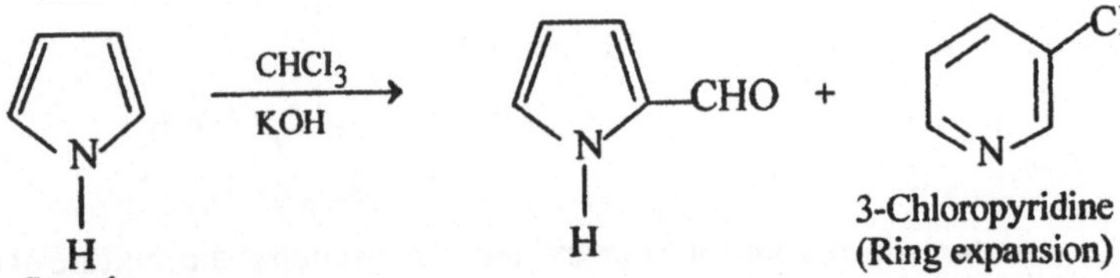

Pyrrole

3-Chloropyridine (Ring expansion)

(*ii*) If *p*-cresol, instead of phenol is used a ketone (I) with dichloromethyl group is obtained as a by-product.

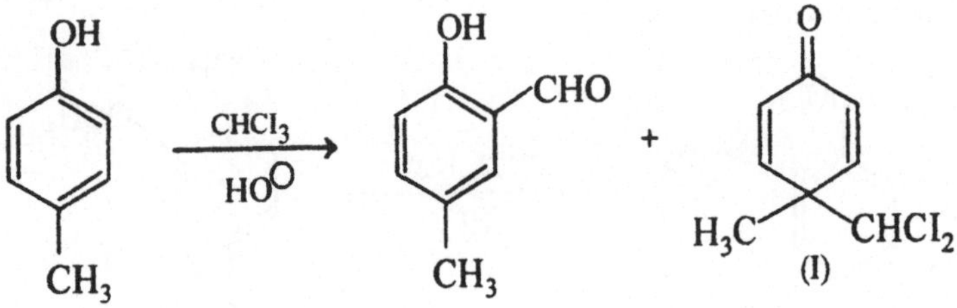

(I)

The compound (I) gets formed by the attack of dichlorocarbene at the para position.

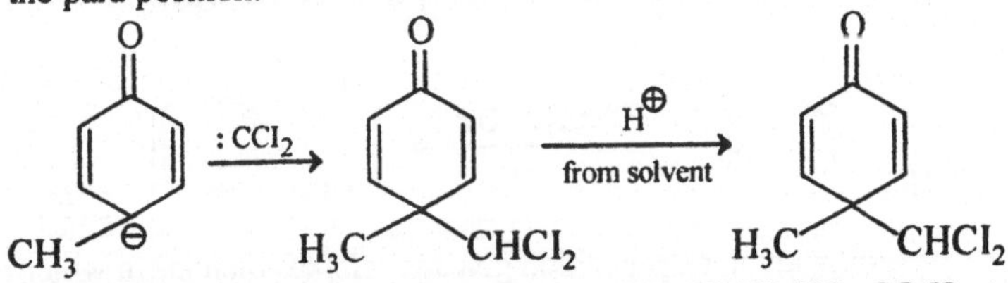

The presence of negative groups such as CN, COOH, NO_2, SO_3H, etc. tends to decrease the yield.

The yield is almost 50% and a large amount of phenol is left unreacted. This occurs probably due to the formation of diphenylacetal derivative.

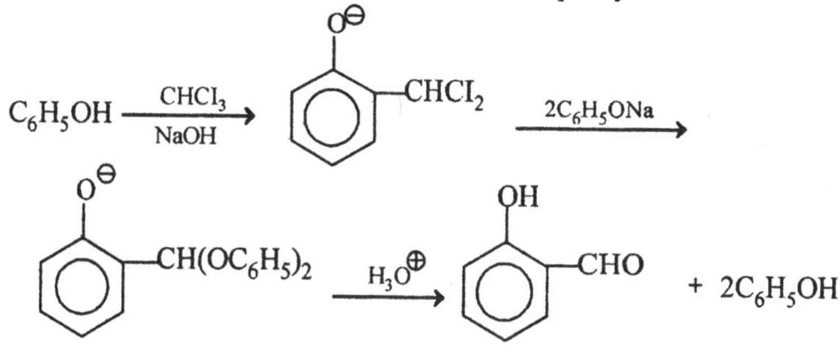

Carbon tetrachloride in place of chloroform under RTR condition yields salicyclic acid.

APPLICATIONS

Reimer-Tiemann's reaction is good method for introducing aldehyde or carbonyl group in phenols.

(*i*) *Preparation of vanillin*

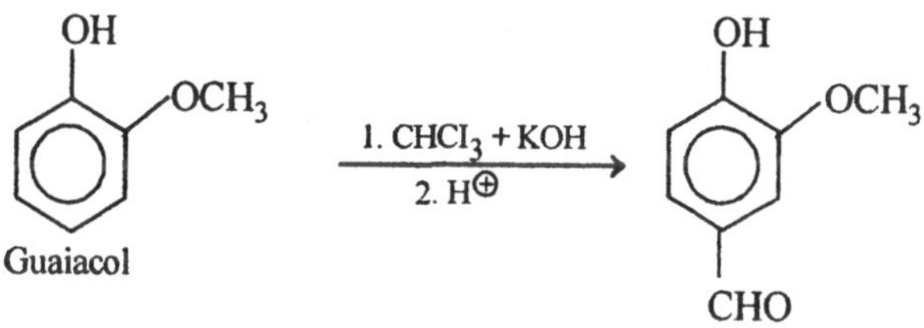

Guaiacol

Vanillin

(*ii*) *Preparation of piperonal*

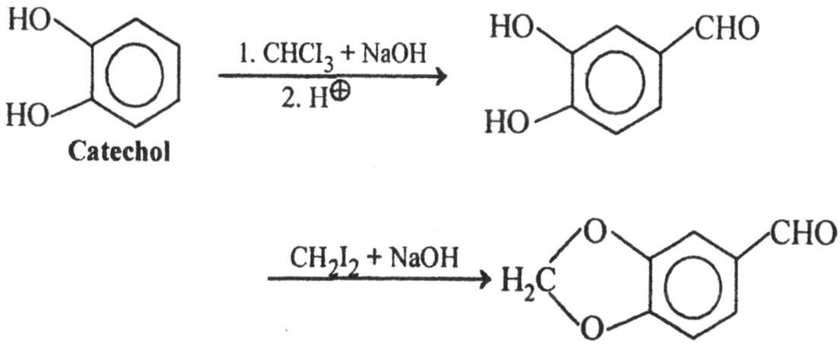

Catechol

Piperonal

(*iii*) *Formylation of heterocyclic compounds*

(a) 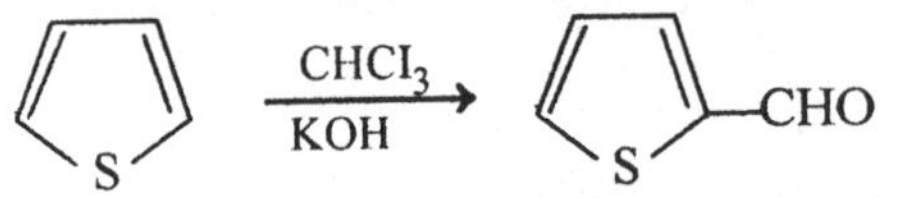

Thiophene-2-aldehyde

(b) 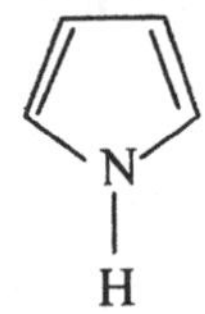

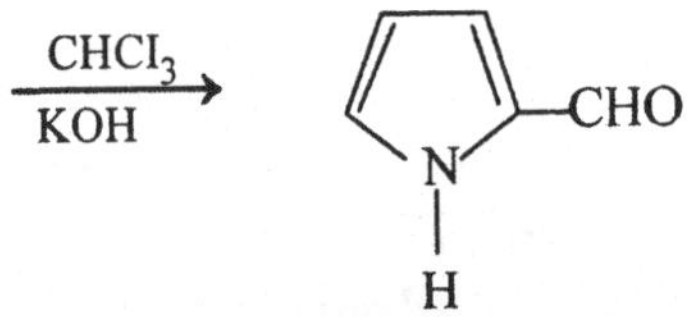

Pyrrole-2-aldehyde

The compound has been to synthesize the amino acid, tryptophan.

(*iv*) *Formylation of naphthol*

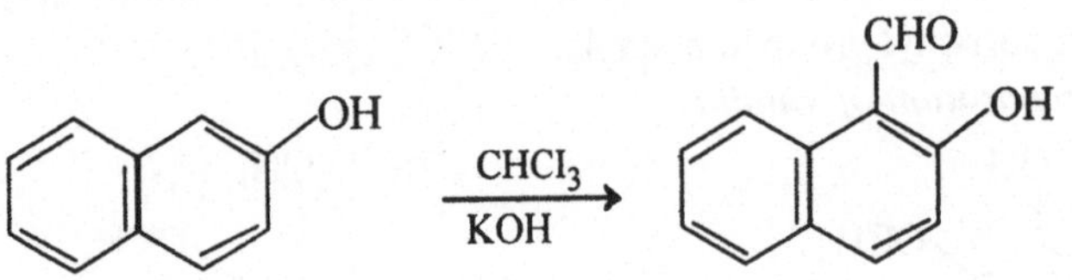

2-Naphthol **2-Hydroxy-1-naphthaldehyde**

(*v*) *Preparation of acid*

If carbon tetrachloride in place of chloroform is employed, a carboxyl group is introduced.

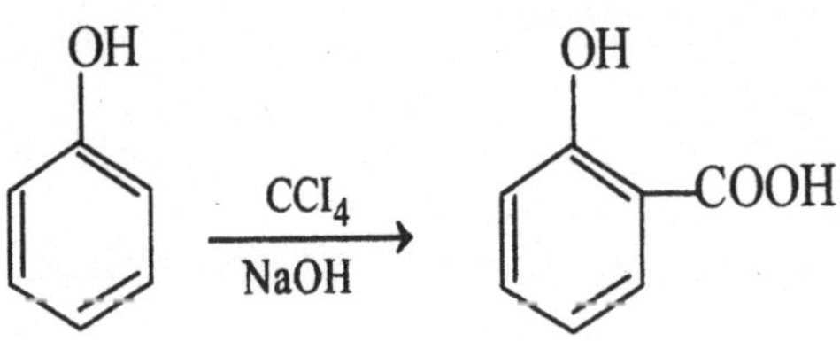

Salicylic acid

(*vi*) *Abnormal Reimer-Tiemann reaction*

1. There are many compounds which undergo normal and abnormal Reimer and Tiemann's reaction. For example, pyrrole undergoes normal and abnormal RTR.

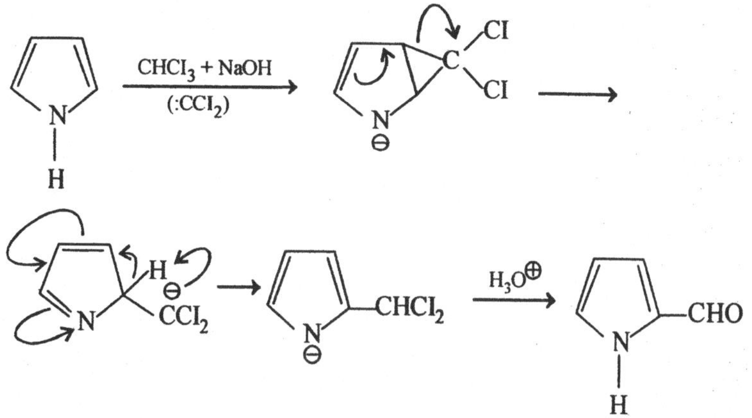

Pyrrole-2-aldehyde

Abnormal reaction (Ring expansion)

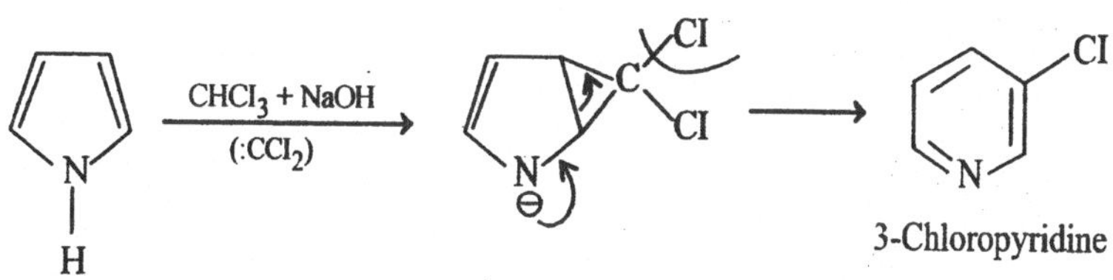

3-Chloropyridine

The reaction is being used to prepare pyridine and its derivatives.

(*vii*) **Certain phenolic compounds are also known which exhibit normal and abnormal** Reimer-Tiemann's reaction.

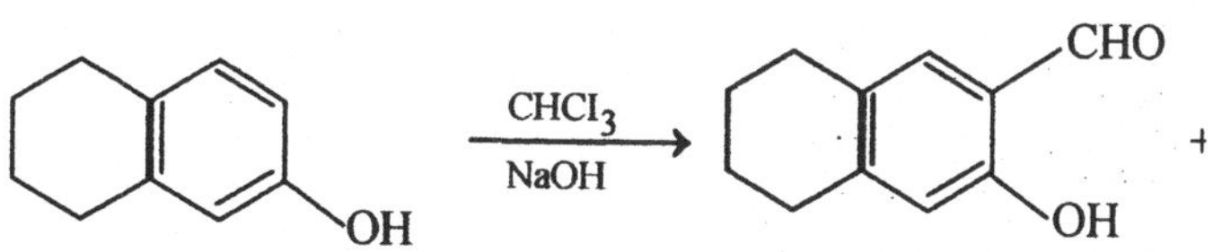

ar-2-Tetralol

Normal product

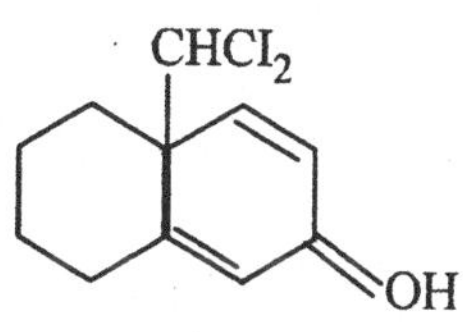

Abnormal product

The abnormal product may be reduced and then oxidized to yield a product with angular methyl group. The process finds use in steroid chemistry.

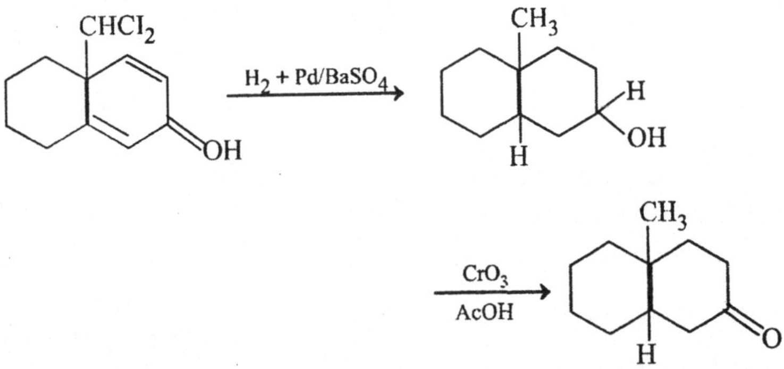

10-Methyl decal-2-one

85

REFORMATSKY REACTION

Condensation reaction of carbonyl compounds with α-haloester in presence of zinc-metal is known as Reformatsky reaction. The solvent most often used in this reaction is benzene or ether or a benzene ether mixture.

MECHANISM :

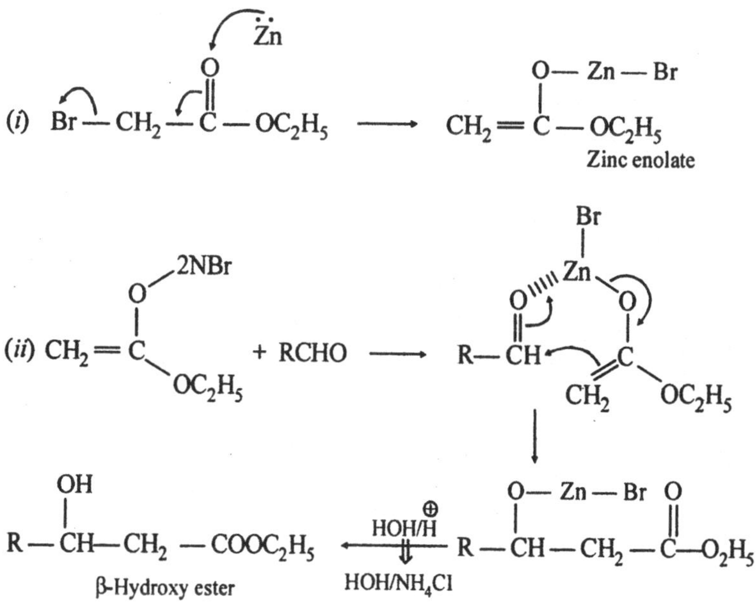

Zinc enolate does not react with ester, hence self condensation reaction does not take place between ester and ester. Zinc enolate reacts with carbonyl compounds to give zinc alkoxide which on hydrolysis gives β-hydroxy ester. The β-hydroxy ester produced in this reaction is easily dehydrated to α, β-unsaturated ester because dehydration yields conjugated system which is stable due to delocalisation.

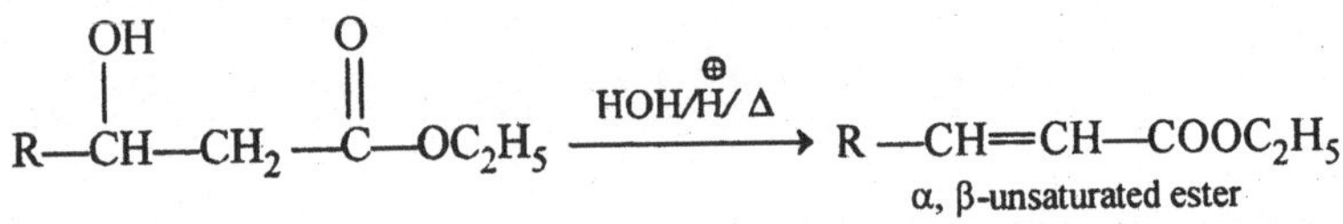

Example:

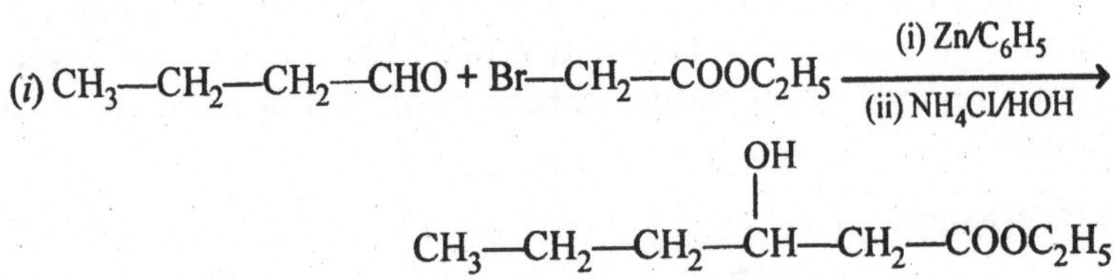

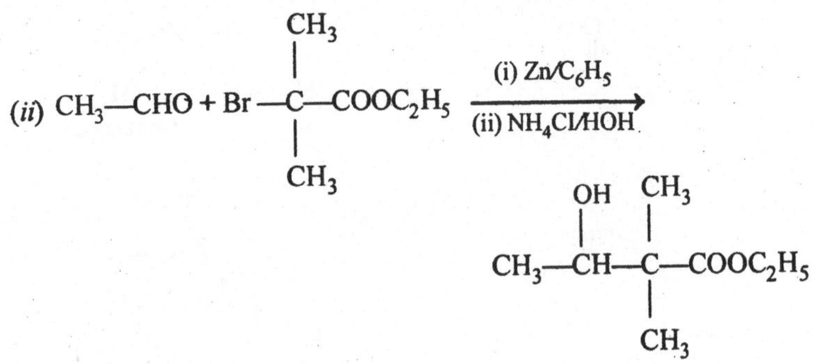

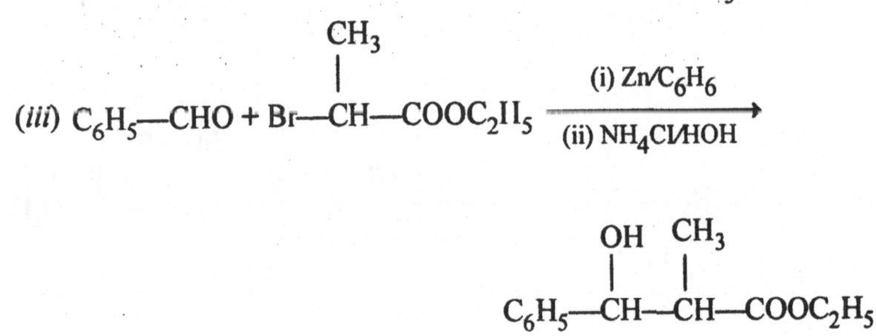

(*iv*)

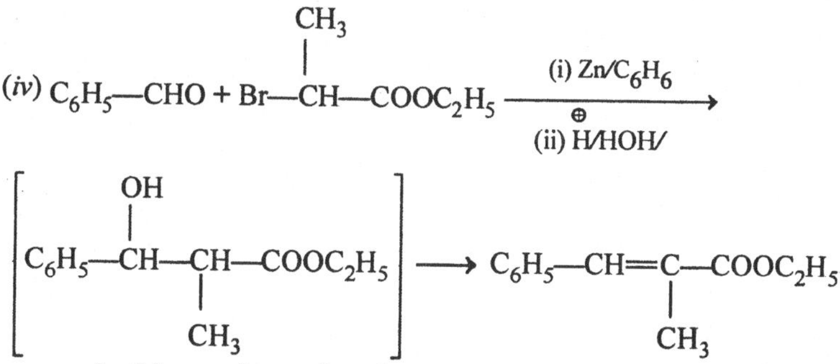

In this reaction carbonyl compounds may be aldehydes or ketones.

The haloester of the reaction in most of the cases is α-bromoester or its vinylog *i.e.*

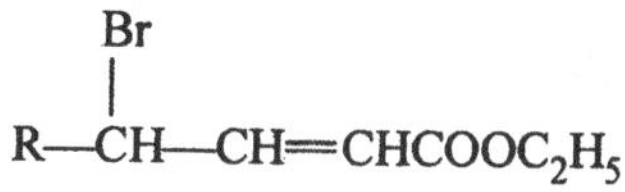

However, the reaction may be carried out with α-halo nitrites and α-halo N, N-disubstituted amides in place of α-halo ester. Similarly, nitrile can be used in place of carbonyl compound.

(*v*)

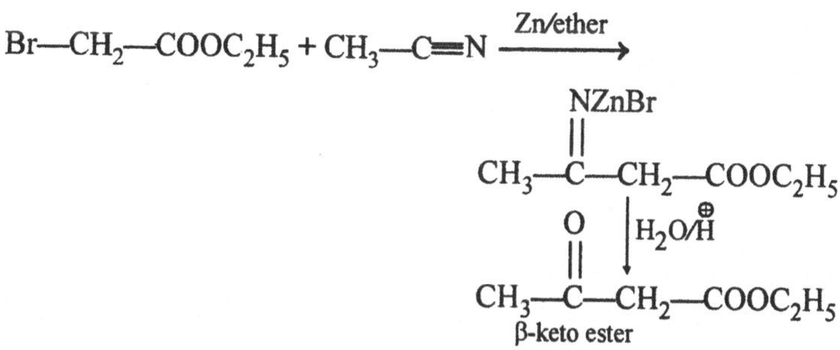

(*vi*)

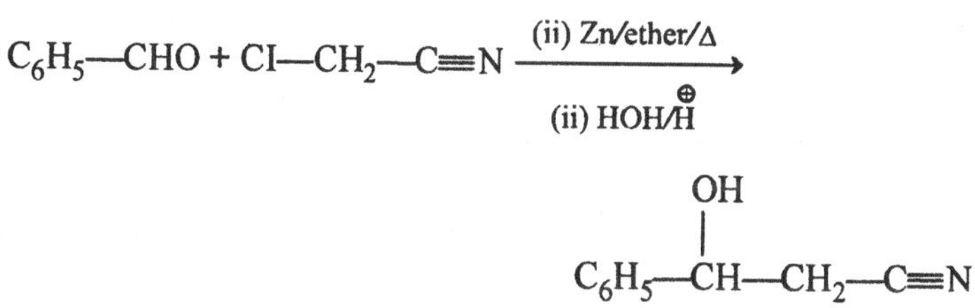

(*vii*)

$$CH_3—CH_2—CHO + Br—CH_2—\overset{\overset{\displaystyle O}{||}}{C}—N(CH_3)_2 \xrightarrow[\text{(ii) HOH/NH}_4\text{Cl}]{\text{(i) Zn/ether/C}_6\text{H}_5/\Delta} CH_3—CH_2—\overset{\overset{\displaystyle OH}{|}}{C}H—CH_2—\overset{\overset{\displaystyle O}{||}}{C}—N(CH_3)_2$$

SYNTHETIC APPLICATIONS

1. This reaction is mainly used for the synthesis of β-hydroxyesters and unsaturated esters and hence their corresponding acids.

(*i*)

$$C_6H_5CHO + BrCH_2COOC_2H_5 \xrightarrow{Zn} C_6H_5\overset{\overset{\displaystyle OZnBr}{|}}{C}H.COOC_2H_5$$

$$\xrightarrow{H_2O} C_6H_5\overset{\overset{\displaystyle OH}{|}}{C}H.CH_2COOC_2H_5 \xrightarrow{-H_2O} \underset{\text{Ethyl cinnamate}}{C_6H_5CH{=}CHCOOC_2H_5}$$

(*ii*)

$$C_6H_5.CO.CH_3 + BrCH_2COOC_2H_5 \xrightarrow{Zn} C_6H_5\overset{\overset{\displaystyle OH}{|}}{\underset{\underset{\displaystyle CH_3}{|}}{C}}.CH_2COOC_2H_5$$

β-Hydroxy-β-phenylbutyrate

(*iii*)

$$\underset{\text{Oxalacetic ester}}{\begin{matrix} COOEt \\ | \\ CO \\ | \\ CH_2—COOEt \end{matrix}} + BrCH_2COOEt \xrightarrow[I_2]{\text{Zn. Benzene}} \begin{matrix} COOEt\ \ CH_2—COOEt \\ | \\ BrZnO—C—COOEt \\ | \\ CH_2—COOEt \end{matrix}$$

$$\xrightarrow{H_3O^{\oplus}} \underset{\text{Citric acid}}{\begin{matrix} CH_2—COOEt \\ | \\ HO—C—COOH \\ | \\ CH_2—COOEt \end{matrix}}$$

(*iv*) $R{-}CHO + R'{-}CH(Br)COOEt \xrightarrow[2.\ H_3O^{\oplus}]{1.\ Zn} R{-}CH(OH)CHR'COOEt$ (β-Hydroxyester)

$\xrightarrow[-H_2O]{H_2SO_4,\ \Delta} R{-}CH{=}CR'COOEt$ (α, β-Unsaturated ester)

(*v*) 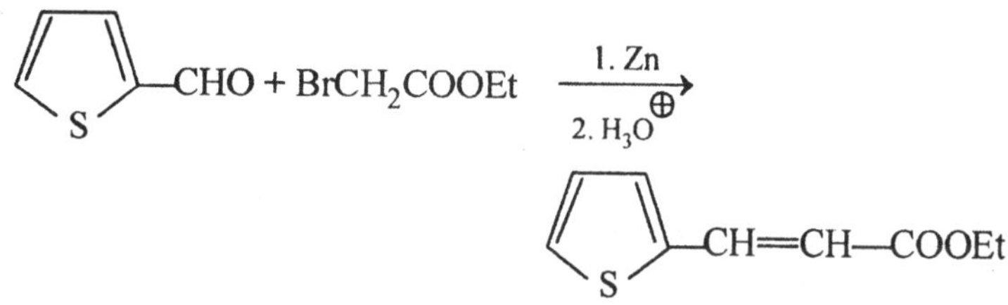

Ethyl-β-(2-thienyl)-acrylate

2. Synthesis of natural products : Reformatsky reaction has been found to be useful for synthesising various natural products, such as geranic ester and therefore citral, camphoronic acid, fenchane, cadalene (terpenoids), vitamin A, etc.

(*i*) **Geranic ester** is synthesised as follows :

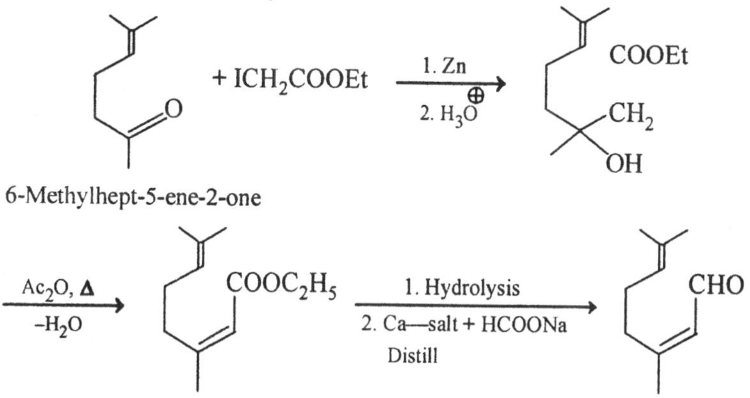

6-Methylhept-5-ene-2-one

Citral

(*ii*) **Camphoronic acid** is synthesised as follow :

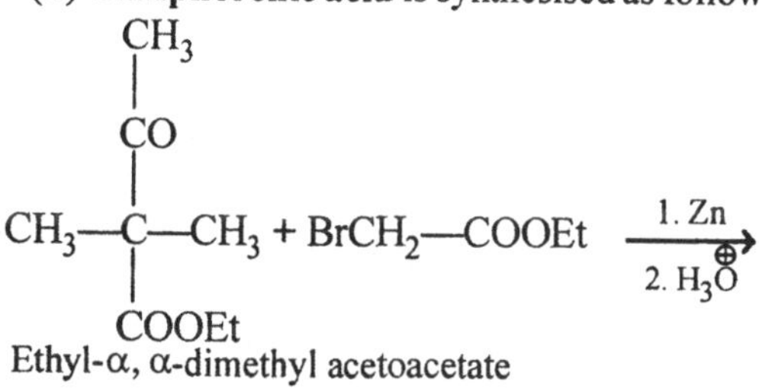

Ethyl-α, α-dimethyl acetoacetate

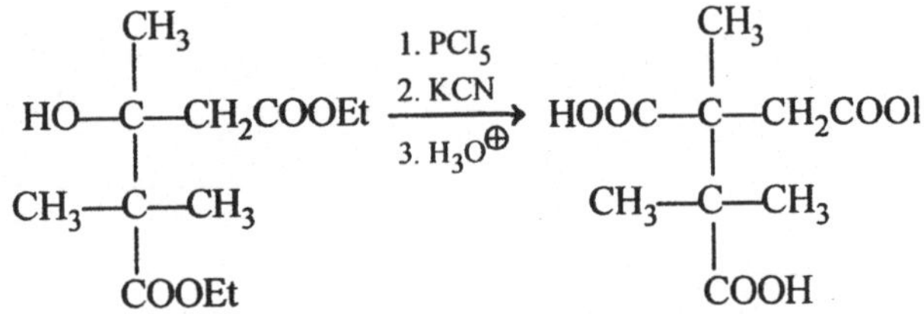

Camphoronic acid

(*iii*) **Vitamin A** is synthesised as follows :

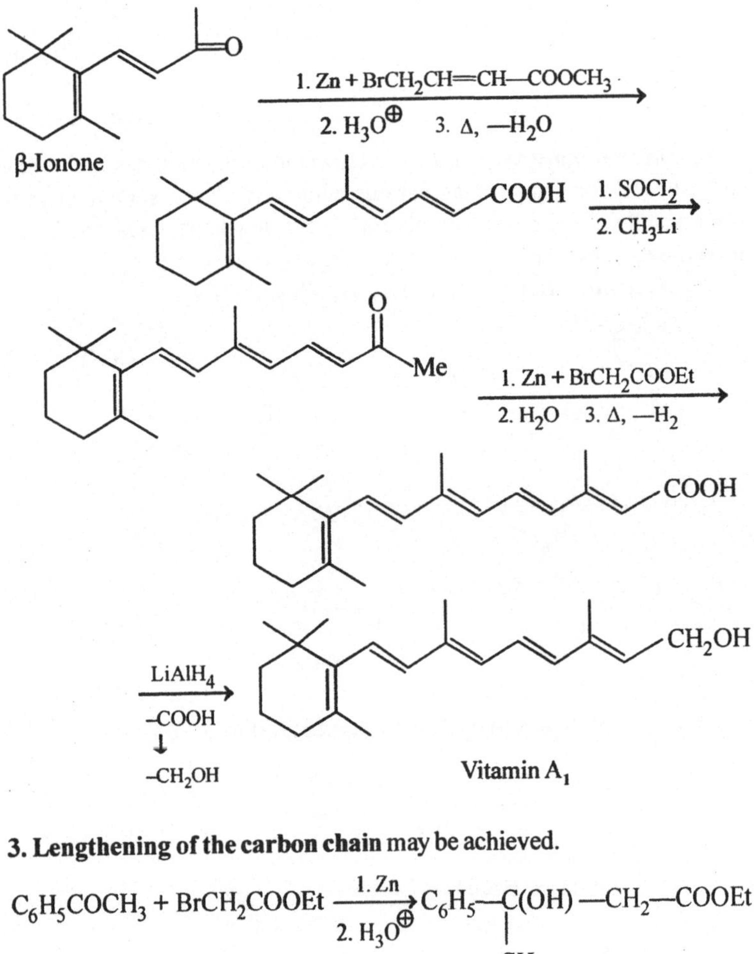

β-Ionone

Vitamin A_1

3. Lengthening of the carbon chain may be achieved.

$$C_6H_5COCH_3 + BrCH_2COOEt \xrightarrow[2.\ H_3O^{\oplus}]{1.\ Zn} C_6H_5-\underset{\underset{CH_3}{|}}{C}(OH)-CH_2-COOEt$$

$$\xrightarrow[-H_2O]{H_2SO_4,\ \Delta} C_6H_5-\underset{\underset{CH_3}{|}}{C}=CH-COOEt \xrightarrow{H_2, Pt}$$

$$C_6H_5CH(CH_3)-CH_2-COOEt \xrightarrow{H_2O} C_6H_5CH(CH_3)-CH_2-COOH$$

β-Phenylbutyric acid

By selecting suitable reactants, unsaturated or saturated acids or esters with branching at (α-and β-carbons can be synthesized.

4. **Synthesis of arylacetic acid** may be achieved.

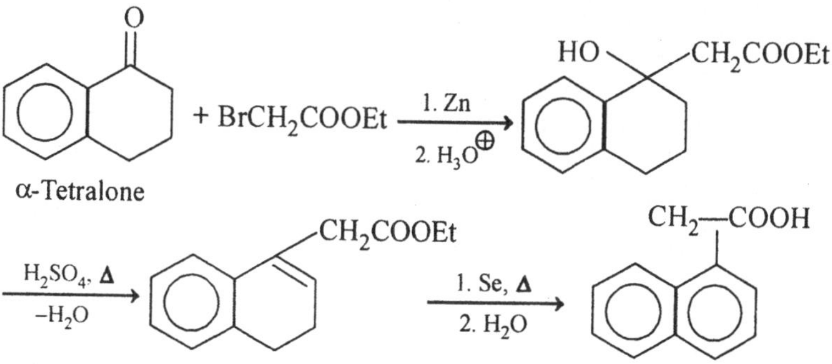

α-Naphthylacetic acid

If finds an important application in agriculture.

86

ROSENMUND'S REDUCTION

A landmark in the history of organic chemistry wascreated with the introduction of Rosenmund Reduction.

REACTION:

$$R \cdot CO \cdot Cl \xrightarrow[BaSO_4]{Pd \quad H_2} RCHO + HCl$$

The reduction of acid halides into aldehydes by molecular hydrogen in the presence of palladium-barium sulfate catalyst bed is known as Rosenmund Reaction.

MECHANISM:

The reaction mechanism follows via surface-adsorption phenomenon details of which is yet to be investigated. One possible mechanism being

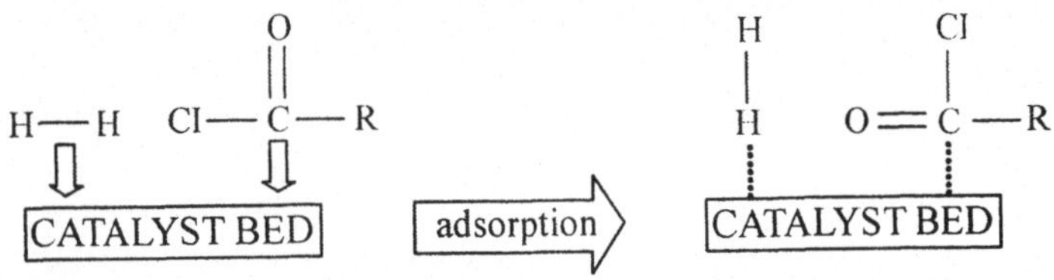

The catalyst bed of Pd-$BaSO_4$ provides on its surface a host of adsorption pockets—one pocket adsorbing one reactant molecule & no more at a time. And as a result a unimolecular layer of adsorbed molecules in produced. Following it there take place the successive processes like activation, reaction & desorption.

Note :

(I)This is the commonest way to reduce acids into aldehydes, since it is practically very difficult to cease the reduction of acids into aldehyde stage.

(II) Aldehydes being more susceptible than acid halides to reduction-process, so it is necessary to poison the catalyst to prevent reduction of aldehydes into alcohols. It is the barium sulfate that acts as a catalyst poison.

(III) The best effective catalyst poison is quinaline-sulfur prepared by refluxing quinoline with a small amount of sulphur.

(IV) Careful control of temperature has been proved to be successful.

(V) Usually the Rosenmund's Reaction is carried out in boiling xylene as a solvent containing suspended catalyst of Pd-$BaSO_4$. Improvements along this line have been reported by Sakurai by using $Pd(OH)_2$-$BaSO_4$ as catalyst and anhydrous acetone or ethyl acetate is solvent.

YIELDS

	CATALYST	*SOLVENT*	*YIELDS*
1.	Pd-$BaSO_4$	Xylene	70-90%
2.	$Pd(OH)_2$-$BaSO_4$	Acetone or Ethyl acetate (anhydrous)	80-90%

APPLICATIONS

(*i*) Synthesis of ***para*-anisaldehyde**

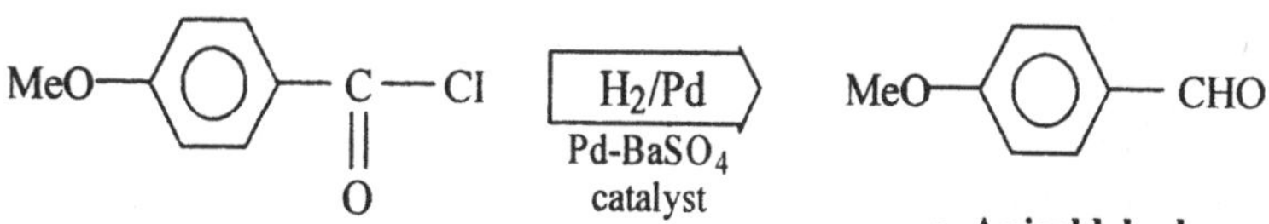

***p* - Aninsoyl chloride**

***p* -Anisaldehyde**
Yield :80-81%

(*ii*)

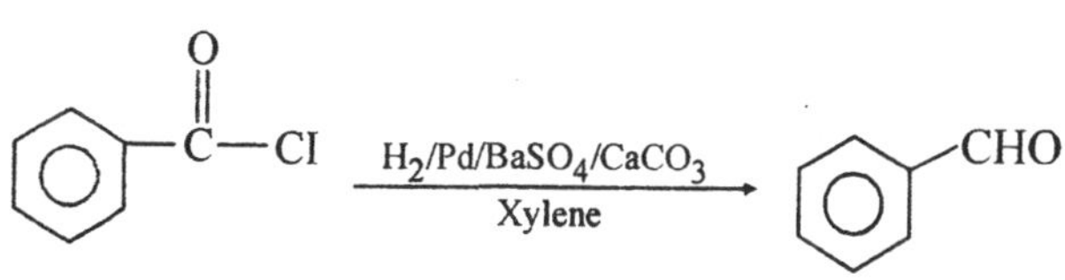

(*iii*)

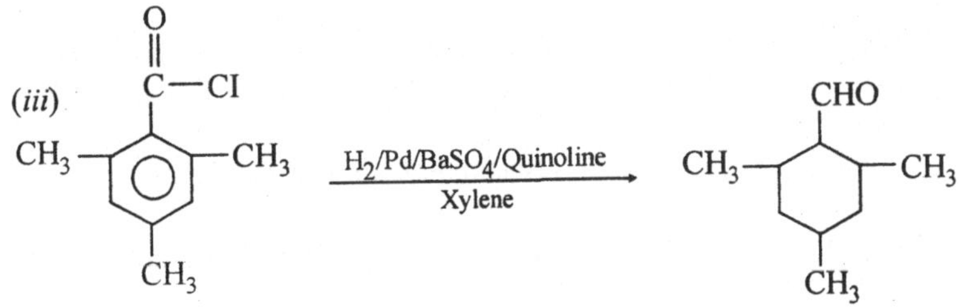

80% yield

Problem 1 :

$$\underset{[A]}{C_4H_6O_4} \longrightarrow \underset{[B]}{C_4H_4O_2Cl_2} \xrightarrow[\text{Xylene}, \ \Delta]{\text{Pd - BaSO}_4, \ H_2} \underset{[C]}{C_4H_6O_2}$$

$$\underset{[C]}{C_4H_6O_2} \xrightarrow{NH_3} \text{Pyrrole}$$

C undergoes tautomerism. Identify **A, B, C.**

Solution :

C on heating with ammonia produces pyrrole.

Evidently **C** is succinaldehyde which undergoes tautomerism

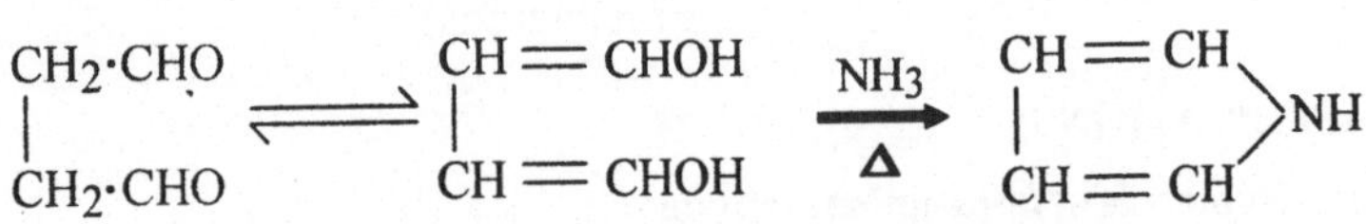

C is a dialdehyde which is produced from B by hydrogen-reduction carried over Pd-$BaSO_4$ catalyst bed—a process which cannot be dubbed anything other than Rosenmund reaction.

Hence B is succinyl chloride

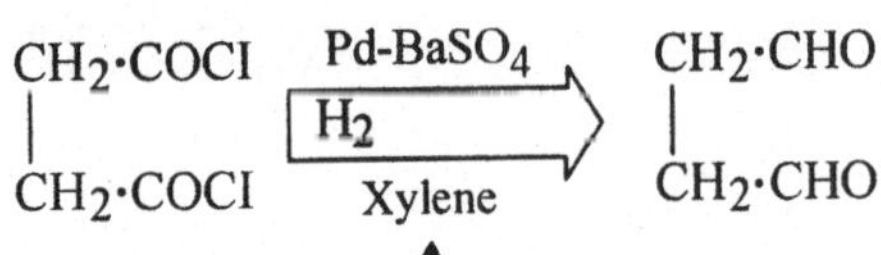

[B] [C]

B is produced from **A** on treating the latter with phosphorus pentachloride. Since **B** is the diacyl halide of **A, A** is succinic acid

$$\begin{array}{c} CH_2{\cdot}CO_2H \\ | \\ CH_2{\cdot}CO_2H \end{array} \xrightarrow{PCl_5} \begin{array}{c} CH_2{\cdot}COCl \\ | \\ CH_2{\cdot}COCl \end{array}$$

[A] [B]

Conclusion : **A** Succinic acid **B** Succinyl chloride
C Succinaldehyde

Problem 2 :

$$A \xrightarrow{KCN} B \xrightarrow{H_3O^+} C \xrightarrow{SO_2Cl_2} D \xrightarrow[\text{Xylene}\ \Delta]{Pd - BaSO_4,\ H_2} E$$

Compound	*%Composition*	*Mol. Wt.*
E	C 66.66	72
	O 22.22	
	H 11.11	

E has one side chain. **A** is a bromide. Identify **A, B, C, D, & E**

Answer : **A** *iso*-Propyl bromide; **B** *iso*-Propyl cyanide;
C 2-Methyl propanoic acid
D Propane-2-methyl-1-carbonyl chloride;
E 2-Methyl propanal

Problem 3 :

$$\text{Succinic anhydride} \xrightarrow{EtOH} A \xrightarrow{SO_2Cl_2} B \xrightarrow[\text{Quinoline sulfur}]{H_2/Pd} C$$

Identify **A, B, C.**

Answer :

A ▶ Monoethyl ester of succinic acid 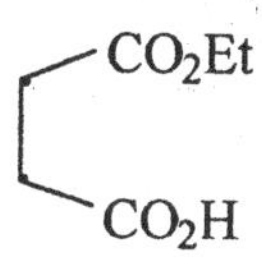

B ▶ Monoacyl chloride of ethyl succinate 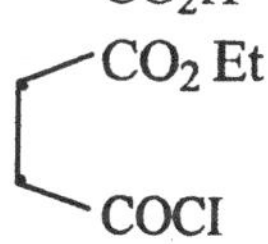

C ▶ 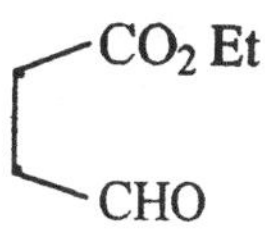

87

ROSENMUND-VON BRAUN REACTION

Aromatic nucleophilic substitution of aryl halide with anhydrous cuprous cyanide at higher temp. is known as Rosenmund-Von Braun Reaction.

$$C_6H_5X + CuCN \xrightarrow[200°C]{\Delta} C_6H_5CN$$

Presence of —R group/groups at *o*- and *p*-position increases the reactivity of substrate for this reaction because—R group is an activating group for ArSN reaction.

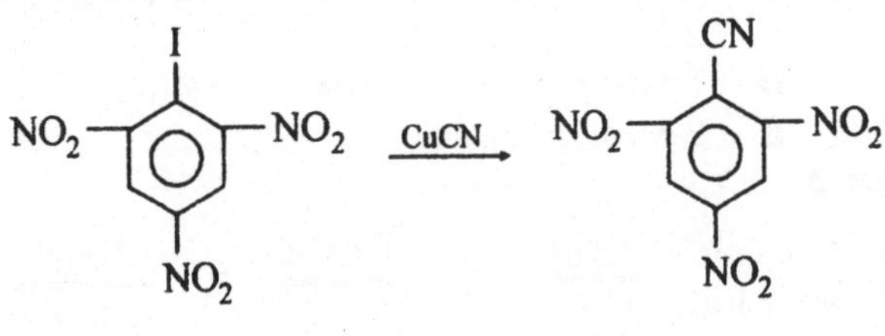

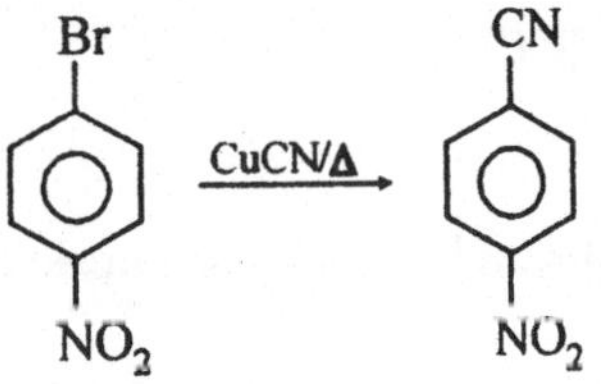

Presence of +R group/groups at *o*- and *p*-position retard the reaction because +R group is a deactivating group for ArSN reaction.

88

RUZICKA REACTION

Vapours of carboxylic acid when passed over heated ThO_2 or MnO above 300°C give aldehyde or ketone depending upon the nature of the acid. In this process, catalytic decomposition of acids takes place.

$$\mathrm{RCOOH + HCOOH \xrightarrow{ThO_2/\Delta} R-\overset{\overset{\displaystyle O}{\|}}{C}-H + CO_2 + H_2O}$$

$$\mathrm{RCOOH + HCOOH \xrightarrow{ThO_2/\Delta} R-\overset{\overset{\displaystyle O}{\|}}{C}-R + CO_2 + H_2O}$$

$$\mathrm{R-COOH + R'-COOH \xrightarrow{MnO/\Delta} R-\overset{\overset{\displaystyle O}{\|}}{C}-R' + CO_2 + H_2O}$$

If the acid is a dibasic acid having more than seven carbons, then intramolecular Ruzika reaction takes place which is also known as Ruzika cyclisation.

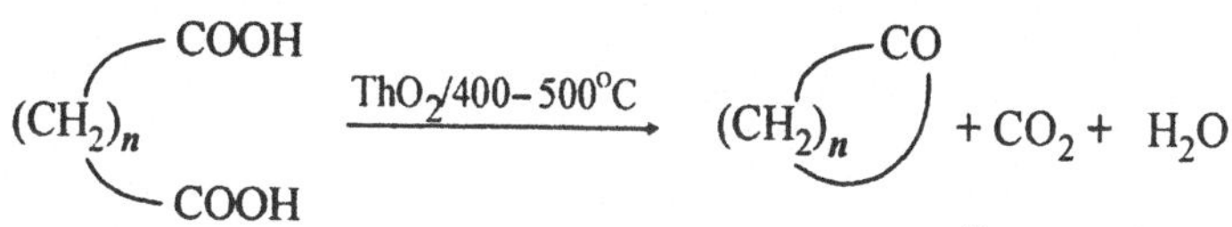

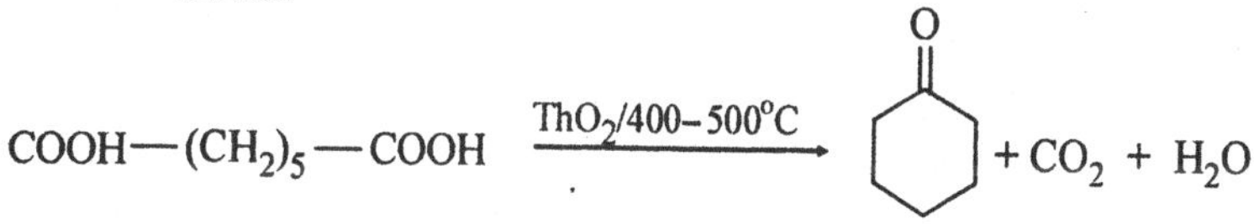

89

ROBINSON ANNELATION REACTION

Michael addition reaction followed by aldol condensation reaction is known as Robinson Annelation reaction

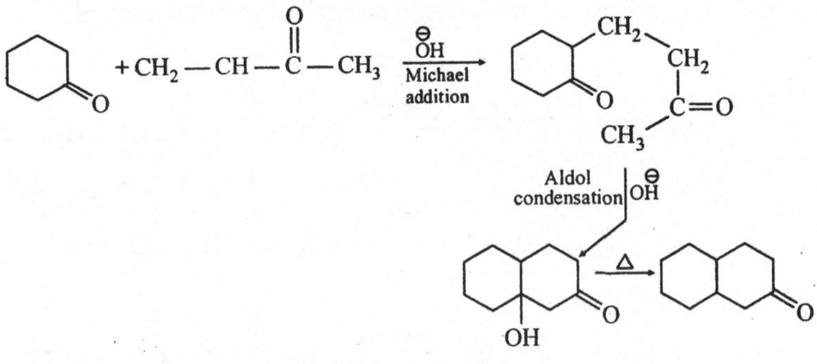

90

SANDMEYER REACTION

Replacement of diazo group of aryl diazonium cation by cyano group or halo group in presence of CuCN/HCN or CuX/HX is known as Sandmeyer reaction.

$$C_6H_5-\overset{\oplus}{N}\equiv N\overset{\ominus}{Cl} \xrightarrow{CuCN/HCN} C_6H_5-CN + N_2$$

$$C_6H_5-\overset{\oplus}{N}\equiv N\overset{\ominus}{Cl} \xrightarrow{Cu_2Cl_2/HCl} C_6H_5-Cl + N_2$$

$$C_6H_5-\overset{\oplus}{N}\equiv N\overset{\ominus}{Cl} \xrightarrow{Cu_2Cl_2/HBr} C_6H_5-Br + N_2$$

91

SCHOLL REACTION

This reaction is a coupling reaction of aromatic hydrocarbons which takes place in the presence of anhy $AlCl_3$ and proton acid. Reaction takes place at high temperature. This reaction is an example of dehydro coupling reaction.

$$C_6H_5—H + C_6H_5—H \xrightarrow{\text{Anhy } AlCl_3/\overset{\oplus}{H}/\Delta} C_6H_5—C_6H_5 + H_2$$

92

SCHOTTEN-BAUMANN REACTION

Benzoylation of alcohols and primary or sec amine by benzoyl chloride in aqueous basic medium is known as Schotten-Baumann reaction.

$$R{-}O{-}H + C_6H_5{-}\overset{O}{\overset{\|}{C}}{-}Cl \xrightarrow{\text{aq. NaOH}} R{-}O{-}\overset{O}{\overset{\|}{C}}{-}C_6H_5$$

$$RNH_2 + C_6H_5{-}\overset{O}{\overset{\|}{C}}{-}Cl \xrightarrow{\text{aq. NaOH}} R{-}NH{-}\overset{O}{\overset{\|}{C}}{-}C_6H_5$$

$$R_2NH + C_6H_5{-}\overset{O}{\overset{\|}{C}}{-}Cl \xrightarrow{\text{aq. NaOH}} R_2N{-}\overset{O}{\overset{\|}{C}}{-}C_6H_5$$

$$C_6H_5NH_2 \xrightarrow{C_6H_5\overset{O}{\overset{\|}{C}}{-}Cl/\text{aq. NaOH}} C_6H_5{-}NH{-}\overset{O}{\overset{\|}{C}}{-}C_6H_5$$

$$(CH_3)_2NH \xrightarrow{C_6H_5COCl/\text{aq. NaOH}} (CH_3)_2N{-}\overset{O}{\overset{\|}{C}}{-}C_6H_5$$

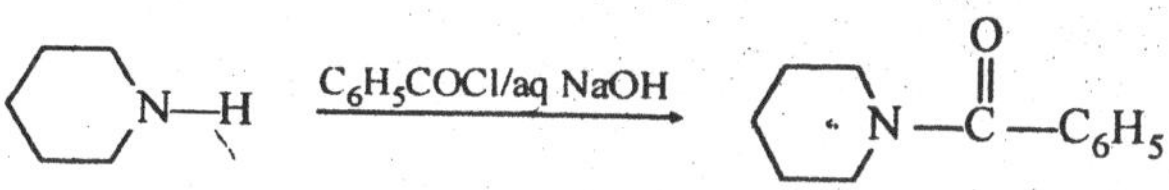

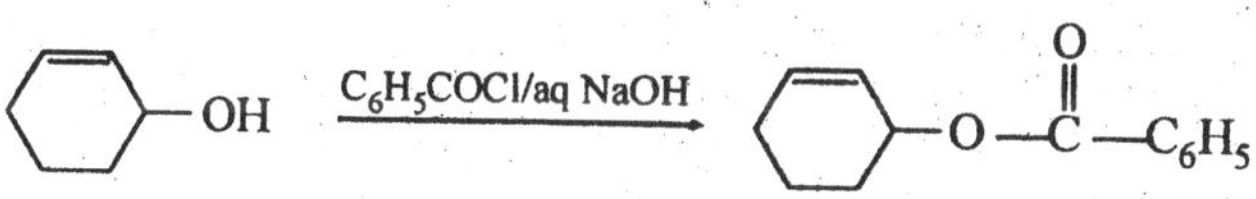

93

SANDMEYER REACTION

Sandmeyer Reaction opened the vistas for the synthesis of a great variety of aromatic compounds.

REACTION

The reaction is the conversion of aromatic amines into aromatic halides—chlorides & bromides—by substituting the amino group with a halogen atom. It is carried out by adding a freshly prepared diazonium salt solution to a solution of cuprous chloride (or bromide) dissolved in HCl (or HBr).

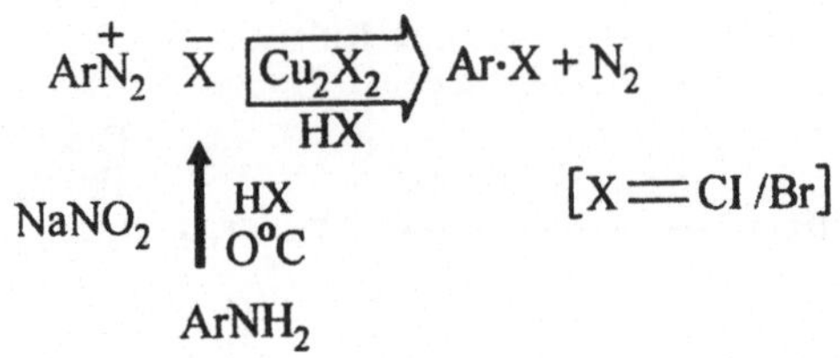

The reaction is carried out at room temperature or occassionally at elevated temperatures. Nitrogen bubbles off and aryl chloride or bromide can be isolated after several hours.

MECHANISM :

The reaction mechanism is by no means certain; one possibility being

$$Ar\overset{+}{N}_2 + \overset{I}{Cu}_2X_2 \longrightarrow Ar\cdot + N_2 + \overset{II}{Cu}_2X_2$$

$$Ar + X—Cu—X \longrightarrow Ar—X + CuX$$

$$CuX \xrightarrow{\text{dimerization}} Cu_2X_2$$

Cu_2X_2 acts as a catalyst

Note :

(I) For the synthesis of aryl bromide, diazotization can be effected by cold, dilute sulfuric acid since it is cheaper than hydrobromic acid. The yield is also very slightly affected.

(II) It is the halogen atom joined to the copper that substitutes the diazonium group.

(III) The cupric chloride is very effective if the diazotized amines contain a strong I group

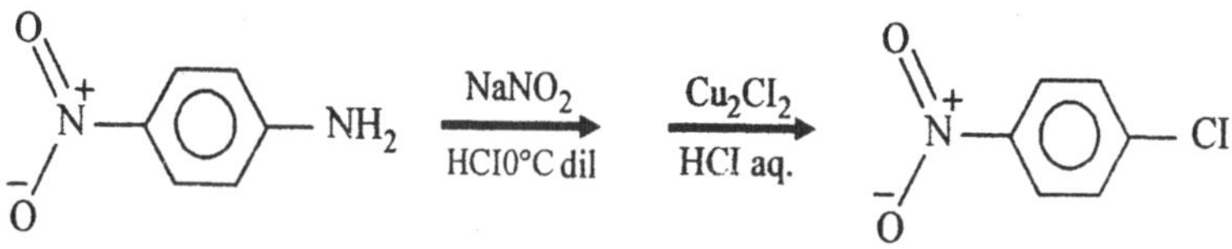

Yield : 98%

(IV) Though the best way of introducing a bromine or chloride into an aromatic ring, SANDMEYER REACTION is much less successful for the synthesis of aryl fluorides and iodides.

Gattermann modification : According to Gattermann, copper powder or copper bronze in place of cuprous chloride acts similarly to yield the aryl halides. This may take place because of the formation of cuprous copper on the surface of the copper metal.

APPLICATIONS AND EXTENSION

Sandmeyer reaction which is used for the preparation of arylhalides from diazonium salt is more useful than direct halogenation for several reasons.

(*i*) Direct halogenation of aromatic compounds yields a mixture of isomers so that the yield of the desired isomer becomes low and difficult to separate. On the other hand, the diazonium salts yield only one halogen derivative because the halogen enters at the position which is previously occupied by the diazonium group.

(*ii*) Aryl iodides and fluorides are seldom prepared by direct halogenation but they can be prepared by Sandmeyer reaction and Schiemann reaction respectively.

(*iii*) In some cases direct halogenation at a particular position is not possible because of electronic factors but this becomes possible through Sandmeyer reaction, e.g.,

(*a*) *Preparation of m-dichlorobenzene* : It can be done as follows :

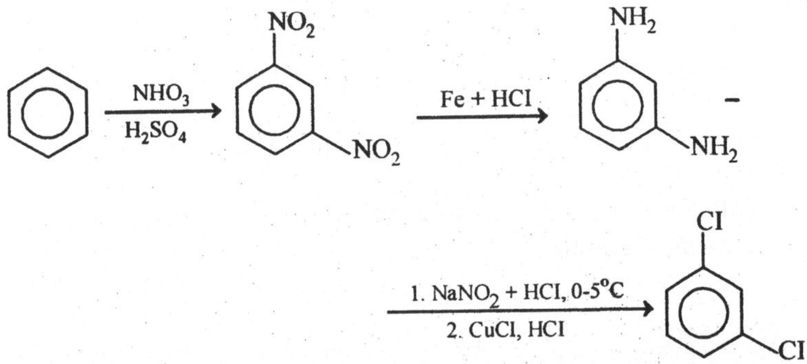

(*b*) *Preparation of 2-chloronaphthalene* : It may be done as follows :

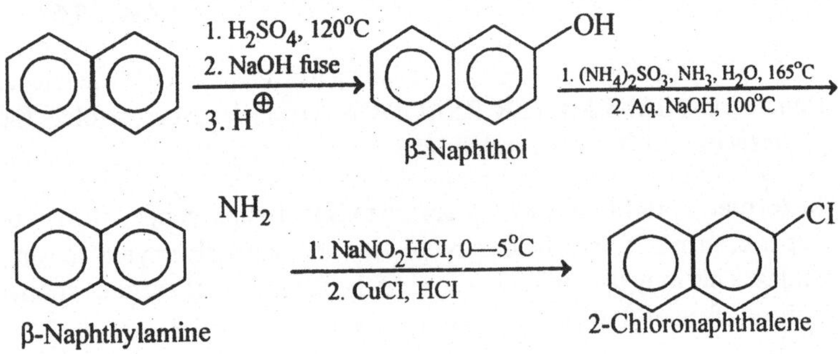

(*c*) *Preparation of acids* : Diazonium salts when treated with cuprous cyanide and potassium cyanide give nitriles which on hydrolysis yields acids.

$$C_6H_6 \xrightarrow[H_2SO_4]{HNO_3} C_6H_5NO_2 \xrightarrow{[H]} C_6H_5NH_2 \xrightarrow[0-5^\circ C]{NaNO_2 + HCl}$$

$$C_6H_5\overset{\oplus}{N}\overset{\ominus}{H_2} \xrightarrow[KCN]{CuCN} C_6H_5CN \xrightarrow{H_3O^{\oplus}} C_6H_5COOH$$

(*d*) *Removal of amino group from aromatic rings :* Amino groups of aromatic amines on conversion to the corresponding diazonium salt and treatment with hypophosphorus acid are removed from the aromatic rings. The reaction has been successfully used for preparing compounds which cannot be prepared by direct substitution. Thus, *m*-nitrotoluene may be prepared from toluene. Toluene on nitration yields *o*- and *p*-nitrotoluenes. The para-isomer after separation is treated as follows :

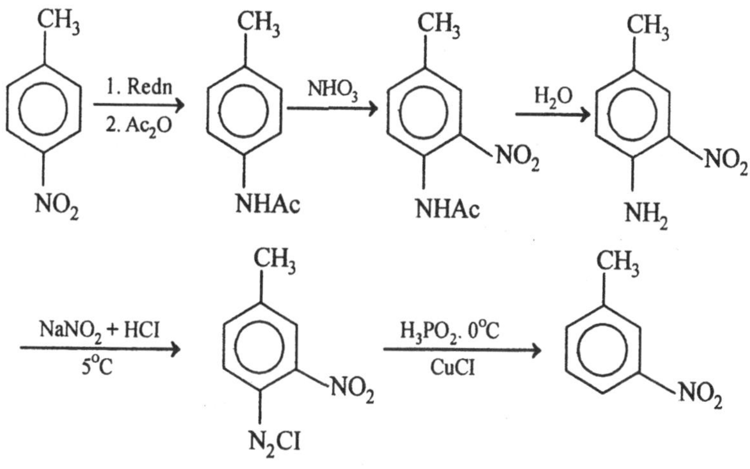

(*e*) Synthesis of *meta*-Nitrobenzene

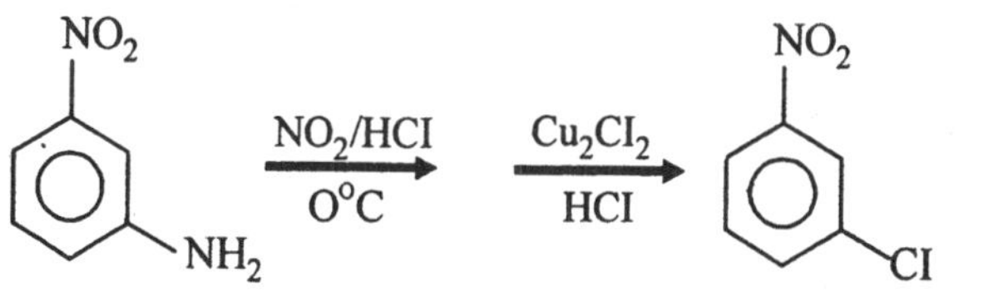

YIELD : 68-71%

(*f*) Synthesis of *para*-Bromotoluene

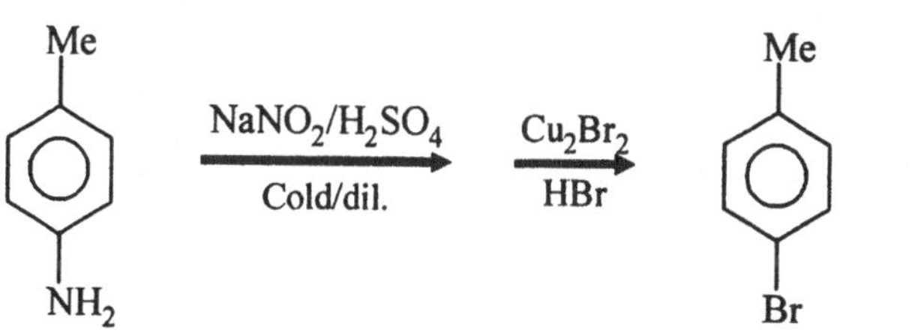

YIELD : 70-73%

Problem 1 :

[I] $$\text{Succinic anhydride} \xrightarrow[\Delta]{\text{EtOH}} \underset{[A]}{C_6H_{10}O_4} \xrightarrow[\Delta]{SO_2Cl_2} \underset{[B]}{C_6H_9O_3Cl}$$

[II] $$\underset{[C]}{C_7H_9ON} \xrightarrow[\text{HCl aq., } 0^{\circ}C]{NaNO_2} \xrightarrow[\text{HCl}]{Cu_2Cl_2} \underset{[D]}{C_7H_7OCl}$$

[III] $$B+D \xrightarrow[\Delta]{\text{Na}} \xrightarrow[\Delta]{\text{HCl conc.}} \xrightarrow{\text{PPA}} \underset{[E]}{C_{10}H_8O_3} \xrightarrow{\text{LAH}}$$

$$\xrightarrow{H_3\overset{+}{O}} \underset{[F]}{C_{10}H_{10}O_3} \xrightarrow[\text{Conc.}]{H_2SO_4} \underset{[G]}{C_{10}H_8O}$$

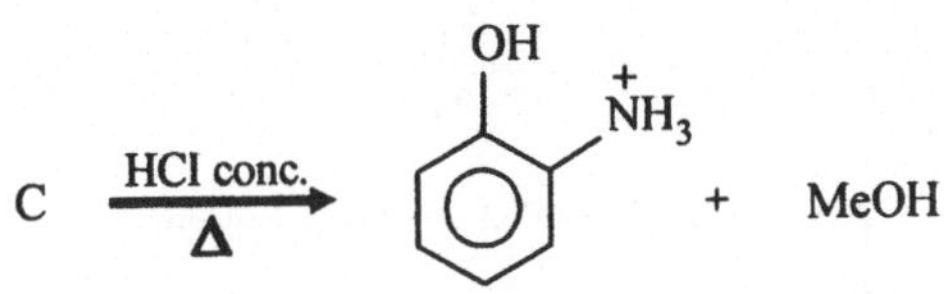

Identify **A, B, C, D, E, F, & G.**
PPA : Polyphosphoric acid; LAH : Lithium aluminium hydride

Answer :

A 1-Carboxy-2-carboethoxyethane
[3-Carboethoxypropanoyl chloride]

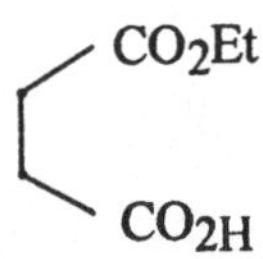

B 3-Carboethoxypropanoyl chloride

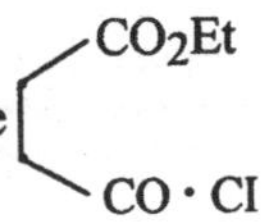

C *ortho*-Methoxy aniline

D *ortho*-Methoxy phenyl chloride

E 5:8-Diketo-6:7-dihydro-α-naphthol

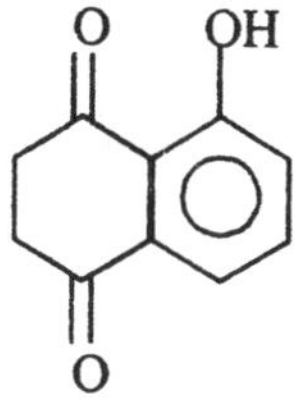

F 5:6:7:8-Tetrahydro-5:8-dihydroxy-α-naphthol

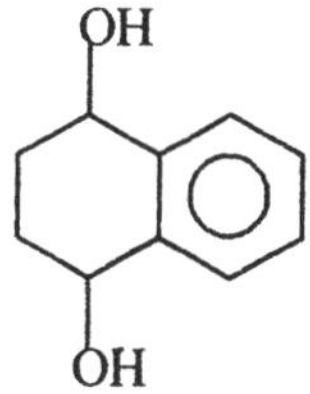

G α-Naphthol

Problem 2 : Work out

$$\textbf{\textit{ortho}-Diphenylenediamine} \xrightarrow[\overset{+}{H_3O}/O°C]{NO_2} \mathbf{A} \xrightarrow[HCl]{Cu_2Cl_2} \mathbf{B}$$

$$\mathbf{A} \xrightarrow{CuCN} \mathbf{C} \xrightarrow{H_3\overset{+}{O}} \mathbf{D} \xrightarrow[HCl]{SO_2Cl_2} \mathbf{E}$$

$$\mathbf{E} + \mathbf{B} \xrightarrow[\Delta]{Na} \mathbf{F} \xrightarrow[\Delta]{Zn} \mathbf{G}$$

Solution :

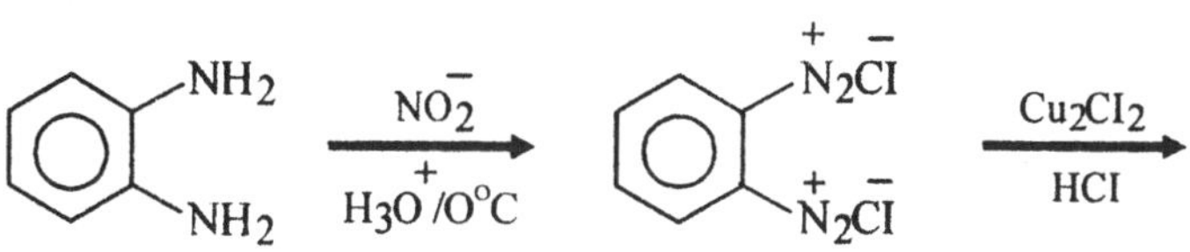

[A]

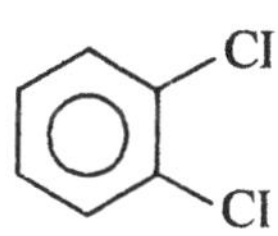

[B]

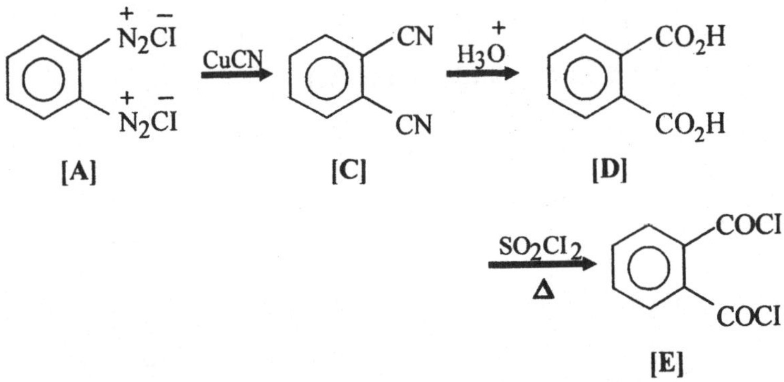

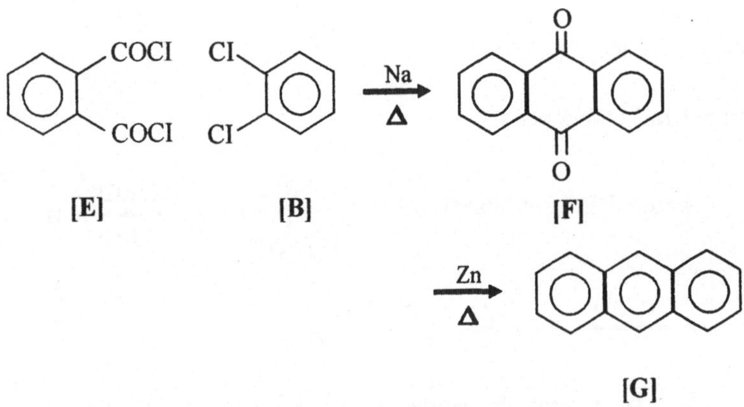

A	*ortho*-Tetrazo benzene	**B**	*ortho*-Dichloro benzene;
C	*ortho*-Dicyano benzene;	**D**	Phthalic acid;
E	Phthaloyl chloride	**F**	9:10-Diketo anthracene;
G	Anthracene		

94

SCHMIDT REACTION

Schmidt reaction is closely related to Hofmann and Curtius reactions, all of which accomplish the some overall process-conversion of a primary amine with less of the carboxy carbon of the acid.

$$R\text{—}\overset{\overset{\displaystyle O}{\|}}{C}OH \longrightarrow R\text{—}NH_2 + CO_2$$

In all these reactions, the common point is the formation of isocyanate (acyl nitrene) intermediate through the migration of group from carbon to nitrogen. Hofmann rearrangement begins with an amide, curtius reaction with an acyl azide and Schmidt reaction with an acid. Examples of the three reactions are as follows:

Hofmann

$$\underset{\text{Decanamide}}{CH_3(CH_2)_7CH_2CONH_2} + Cl_2 + OH^- \longrightarrow \underset{\substack{\text{Nonamine} \\ (66\%)}}{CH_3(CH_2)_7CH_2NH_2}$$

Curtius

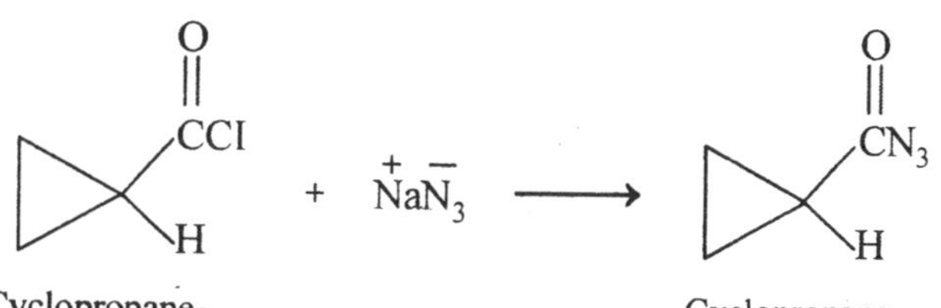

Cyclopropane--carbonyl chloride

Cyclopropanecarbonyl azide

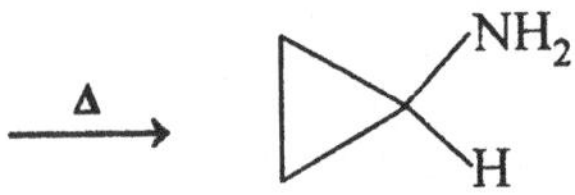

Cyclopropanamine

Schmidt Reaction

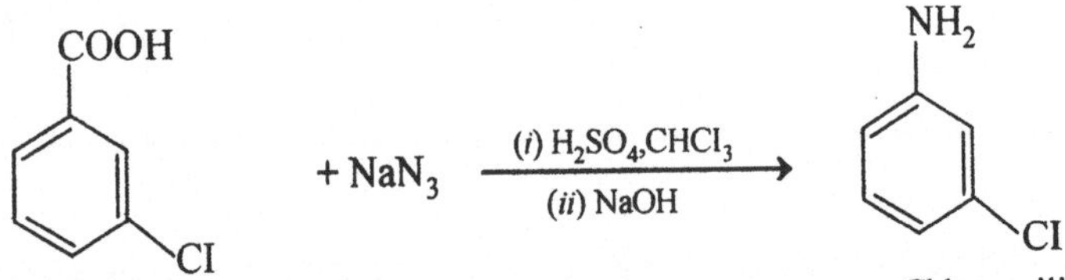

m-Chlorobenzoic acid — *m*-Chloroaniline (75%)

Schmidt reaction is a modification of Curtius reaction.

In Schmidt reaction, the acyl azide is obtained by condensing a carboxylic acid with hydrazoic acid (HN_3) in the presence of sulphuric acid. No attempt is made to isolate the acyl azide but it is allowed to decompose in the reaction mixture itself.

$$RCOOH + HN_3 \xrightarrow{H_2SO_4} R{-}NH_2 + CO_2 + N_2$$

When treated with hydrazoic acid in sulphuric acid, carbonyl compounds also undergo Schmidt reaction. Ketones form amides while aldehydes generally form a mixture of the corresponding nitrile and *N*-formyl derivative.

$$\underset{\text{Aldehyde}}{RCHO + HN_3} \xrightarrow{H_2SO_4} \underset{\text{Cyanide}}{R.CN} + \underset{\text{Formyl derivative of amine}}{RNH.CHO}$$

$$R.CO.R + HN_3 \xrightarrow{H_2SO_4} \underset{\text{Amide}}{RCONHR} + N_3$$

MECHANISM :

The mechanism of Schmidt reaction with acids has been found to be similar to that the Hofmann and Curtius reactions. The mechanism which is consistent with the experiment observations is as follows:

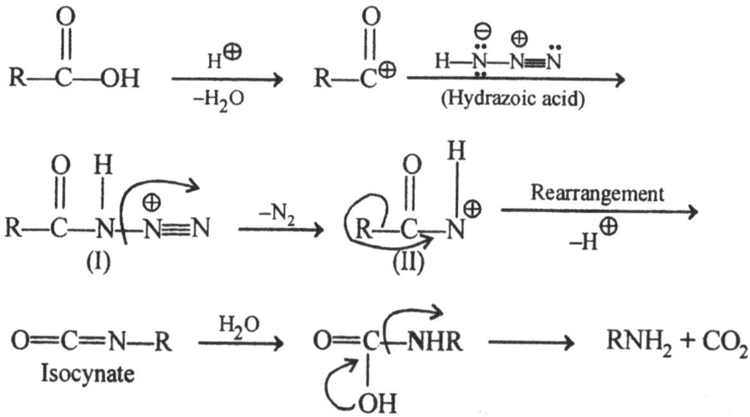

When N-alkylated cations derived from the condensation of alkyl azides (RN_3) and carboxylic acids are subjected to Schmidt rearrangement, they do not undergo Schmidt rearrangement. From this it is evident that the removal of water molecule precedes the rate determining step.

Acids which do not form acyl cation react through the protonated acid on heating as follows:

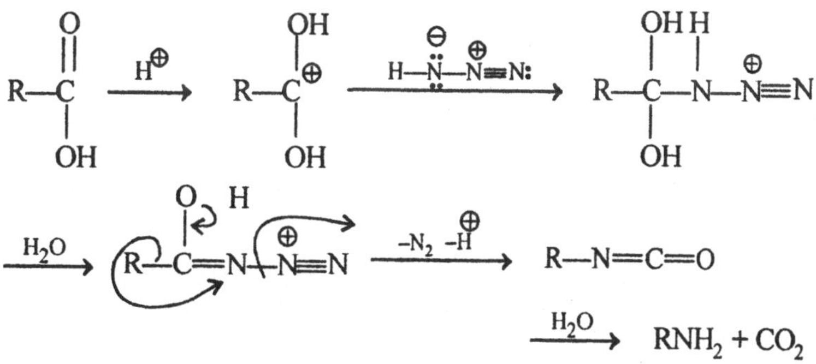

Isocyanate is not isolated under the acid condition although it has been isolated under anhydrous condition. The rearrangement is intramolecular since the migrating group never becomes free. This has been concluded from the crossover experiment and from the fact that the migrating group retains its configuration.

A probable mechanism of carbonyl compounds undergoing. Schmidt rearrangement may be written as follows:

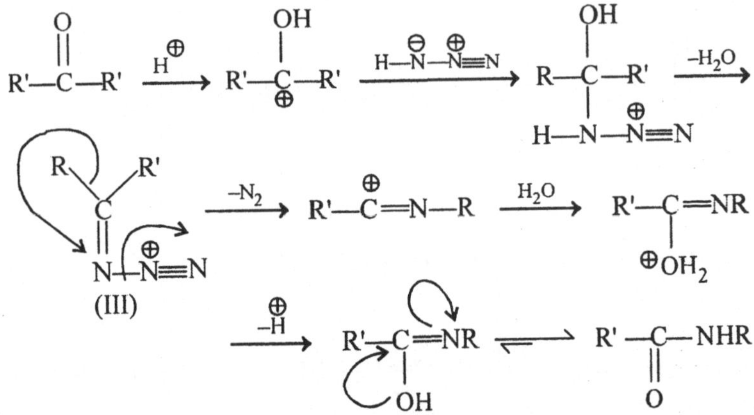

In the case of aldehydes, there occurs the preferential loss of a proton from the carbonium ion, thus forming nitrile.

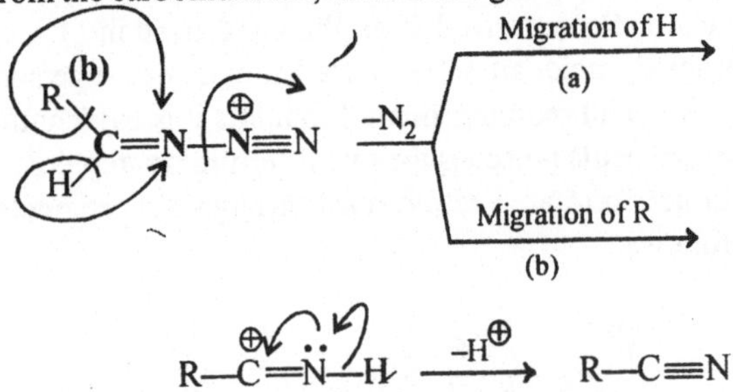

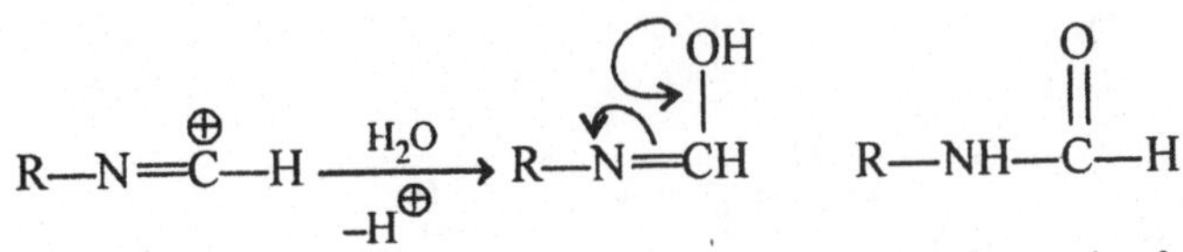

Sulphuric acid is the most common catalyst for Schmidt reaction but Lewis acids have also been used.

APPLICATIONS

(*i*) *Preparation of primary amines* : Schmidt reaction is a direct method for preparing primary amines from carboxylic acids and its yield is somewhat better than the Hofmann or Curtius reaction. However, this reaction must be carried out very cautiously due to the explosive and poisonous nature of the hydrazoic acid. Also, the reaction is applicable when the carboxylic acid used is not having any group which is sensitive to concentrated sulphuric acid.

$$(a)\ \underset{\textbf{Phenylacetic acid}}{C_6H_5CH_2COOH} + N_3H \xrightarrow[CHCl_3]{H_2SO_4} \underset{\textbf{Benzyl amine}}{C_6H_5CH_2NH_2} + N_2 + CO_2$$

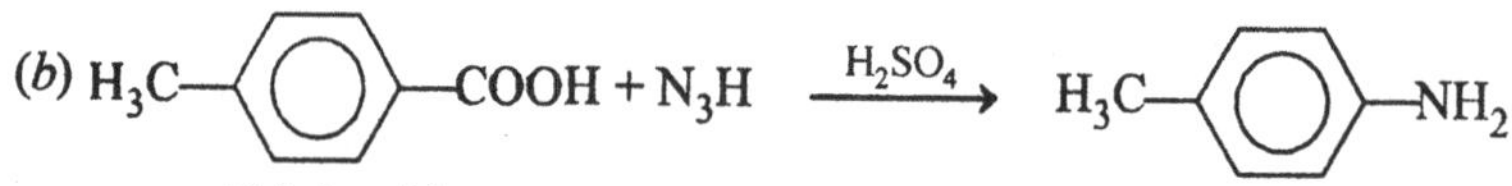

p-Toluic acid *p*-Toluidine

(*ii*) *Preparation of α-amino acids :* Schmidt reaction has been used for synthesizing α-amino acids from acetoacetic ester.

$$CH_3CO.\overset{\overset{R}{|}}{C}H.COOC_2H_5 \xrightarrow[(-N_2)]{HN_3/H_2SO_4} CH_3CONH.\overset{\overset{R}{|}}{C}H.COOC_2H_5$$

$$\xrightarrow[\text{hydrolysis}]{\text{acidic}} \underset{\alpha\text{-Amino acid}}{H_2N.\overset{\overset{R}{|}}{C}H.COOH} + CH_3COOH + C_2H_5OH$$

(*iii*) *Preparation of lactams :* Cyclic ketones react to yield lactams. Cyclohexanone yield Î-caprolactam.

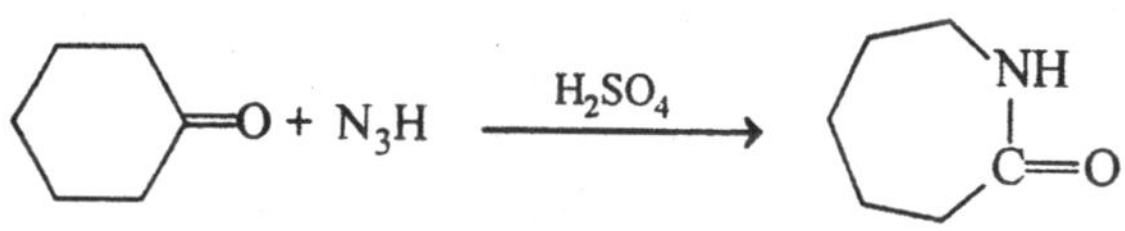

With excess of N_3H substituted tetrazoles are obtained from ketones. Cardiazole, a heart stimulant, is prepared from cyclohexanone and N_3H.

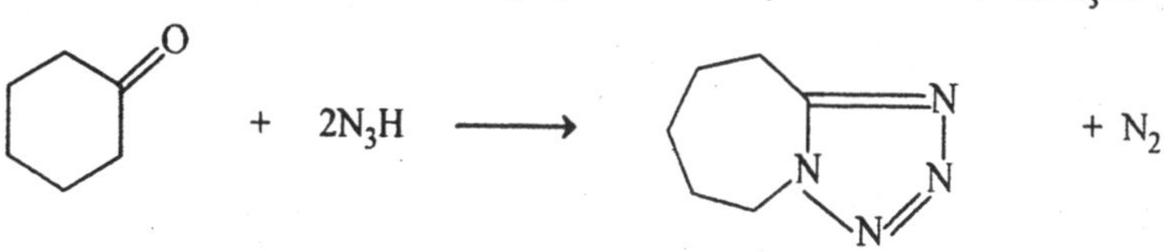

95

SCHIEMANN REACTIONS

SCHIEMANN REACTION is by far the best way of introducing a fluorine atom to an aromatic ring.

REACTION

The reaction sequence is represented schematically as follows:

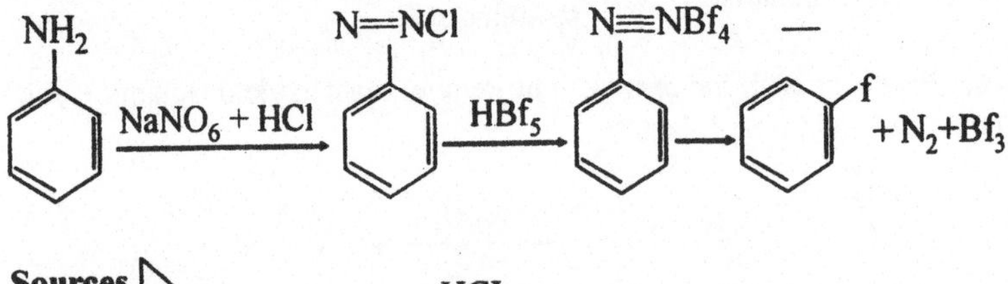

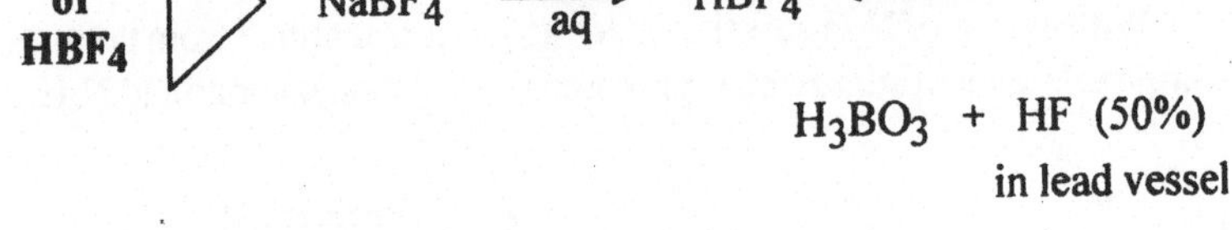

An improved technique is precipitation of diazonium hexafluorophosphate and heating it after drying.

$$\text{Ø}-\overset{+}{N_2} \xrightarrow{HPF_6} \text{Ø}-\overset{+}{N_2}\,[PF_6] \xrightarrow[\Delta]{} \text{Ø}-F + N_2 + PF_5$$

MECHANISM:

The mechanism is not free-radical one but ionic. Phenyl cation is the intermediate and the mechanism is S_N1 type.

$$\text{Ø}—\overset{+}{N_2} + [BF_4]^{-} \longrightarrow \text{Ø}—\cdot\overset{+}{N_2}\,[BF_4]^{-} \longrightarrow$$

$$\cdot\overset{+}{\text{Ø}} + \overset{-}{B}F_4 + N_2 \longrightarrow \text{Ø}—BF_4 \longrightarrow \text{Ø}\cdot F + BF_3$$

SUPPORT OF THE MECHANISM

(1) The mechanism is not a free radical one; since in that event a biaryl derivative would always result.

$$Ar\cdot \xrightarrow{Ar\cdot} Ar—Ar$$

No biaryl compound has been found in practice.

(2) If the intermediate is a phenyl cation, directive influence of *ortho-*, *meta-*, & *para*-orienting groups present in the phenyl ring will occur. Actually this has been observed experimentally [Makarova *et. al.*, **Bull. Acd. Sc. USSR**, Div. Chm-Sc., vol. 1958]

Note:

(I)Fluorination in the aromatic ring cannot be effect in the usual way chlorination or bromination is effected.

(II) The diazonium salt of fluoroborate is quite stable.

(III) This method is very successful and the yield is very high.

96

SKRAUP'S SYNTHESIS

INTRODUCTION

This is the commercial method for the preparation of quinoline. Aniline is heated with glycerol and conc. sulfuric acid in the presence of an oxidising agent usually nitrobenzene or arsenic pentoxide. The reaction is exothermic and tends to become very violent. Ferrous sulphate or boric acid is generally added to make the reaction less violent.

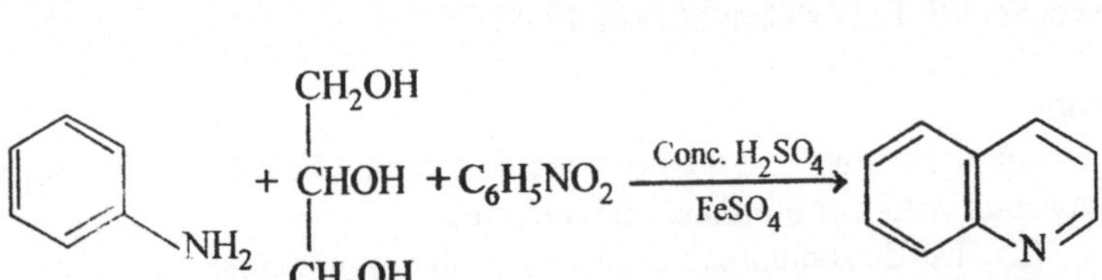

Aniline Quinoline

MECHANISM :

The mechanism of this reaction is not understood completely. However, it is believed to proceed as follows:

The first step involves the dehydration of glycerol to acrolein which condenses with aniline to form 1, 2-dihydroquinoline which is oxidized to quinoline. Ferrous sulfate is used to slow down the otherwise vigorous reaction. 8-Hydroxyquinoline (oxine) is available by this method using aniline.

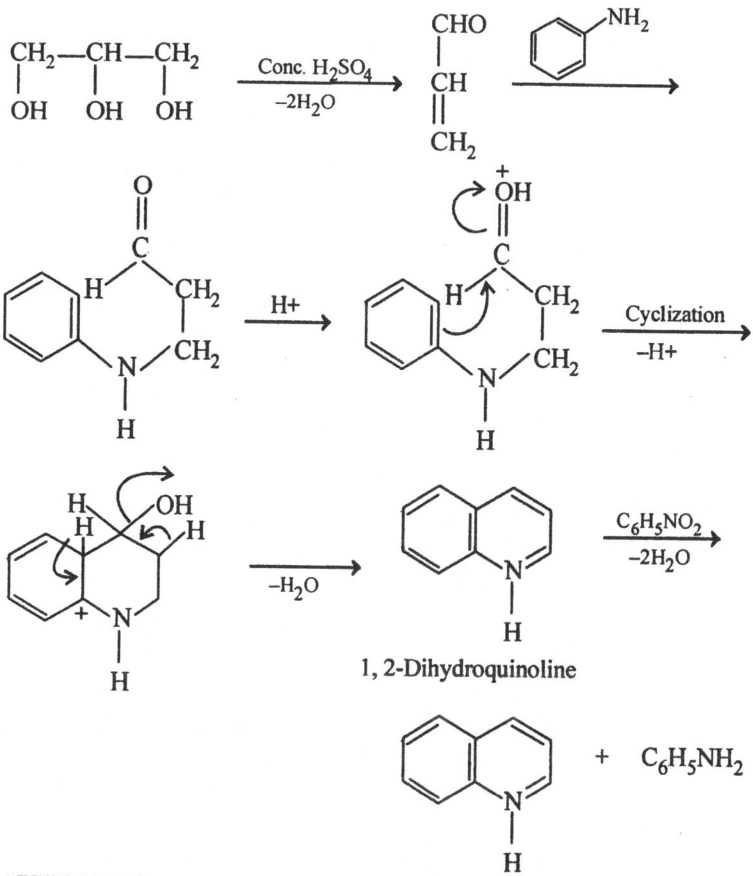

Quinoline

APPLICATION

Substituted quinolines can be prepared by taking suitable aldehydes and ketones.

Problem :

Two mechanisms for the skraup synthesis are :

1. 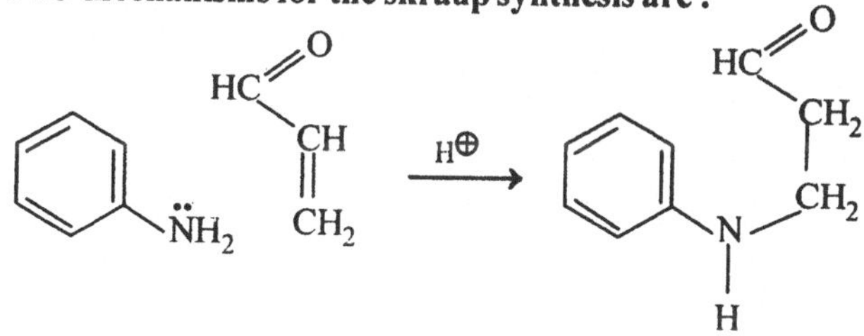

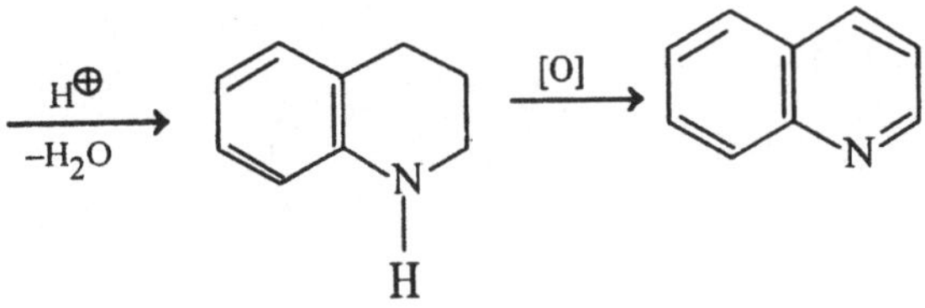

2. 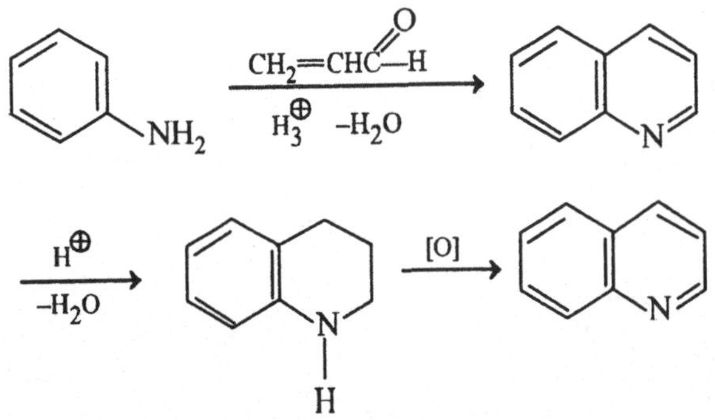

Devise an experiment whereby we can tell which is the actual mechanism.

Solution : The difference between the two mechanisms is the place of attack by the nucleophilic nitrogen atom. The first mechanism involves attack on the β-carbon of the α, β-unsaturated aldehyde. The second mechanism involves attack at aldehydic carbon. By using ^{14}C labled acrolein, we can deduce the actual mechanism by the position of the label in the product :

1. 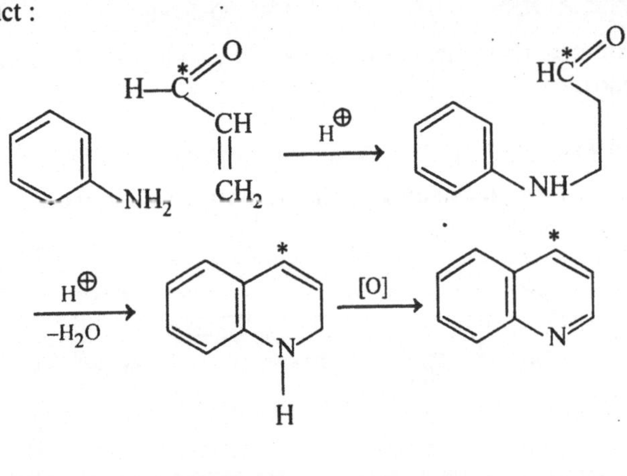

2.

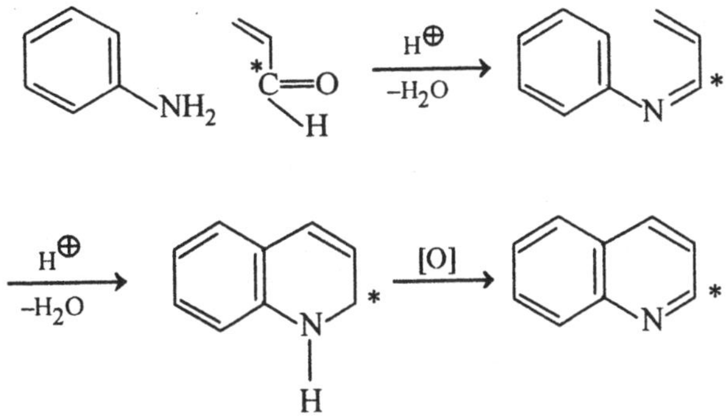

97

SONN-MULLER ALDEHYDE SYNTHESIS

Conversion of an aroyl halide to the corresponding aldehyde via reduction of an imino-chloride and hydrolysis of resulting Schiff's base is known as Sonn-Muller aldehyde synthesis.

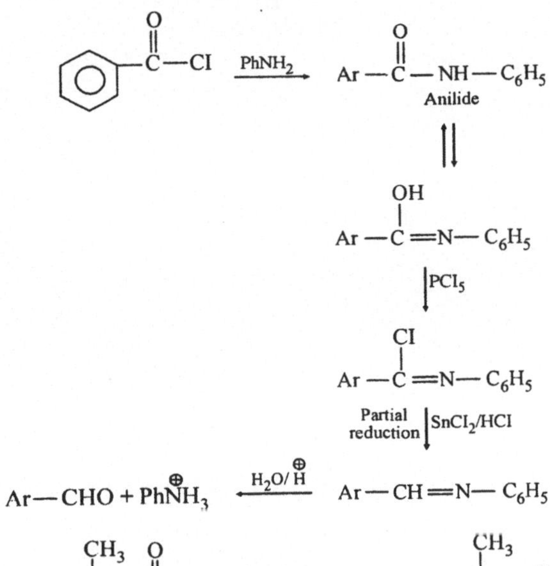

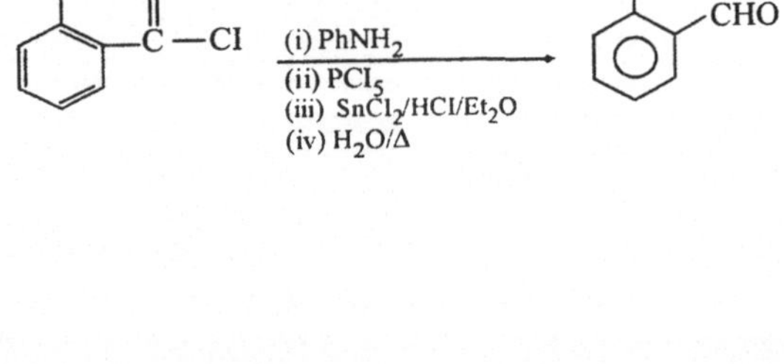

98

STAUDINGER REACTION

Reduction of azides into *p*-amines by $LiAlH_4$, $NaBH_4$ and $(C_6H_5)_3P/H_2$ is called Staudinger reaction.

$$RN_3 \longrightarrow R{-}NH_2$$

$$R{-}X + N_3Na \longrightarrow RN_3 \xrightarrow{LiAlH_4} R{-}NH_2$$

99

SOMMELET REACTION

Oxidation of —CH_2Cl to to —CHO group by means of hexamethylenetetramine (urotropine) is known as **Sommelet reaction**.

$$\underset{\text{Benzyl chloride}}{C_6H_5.CH_2Cl} \xrightarrow[60\text{-}70\%]{(CH_2)_6N_4} \underset{\text{Benzyldehyde}}{C_6H_5.CHO}$$

MECHANISM :

The mechanism of this reaction is uncertain. According to Angyl *et. al.* (1953) its mechanism is outlined as follows :

The halide on reaction with hexamethylenetetramine yields a quaternary salt (I) which is hydrolysed to amine (II). Hexamethylenetetramine also gets hydrolysed to form formaldehyde and ammonia. Both these react to form methylene amine (III). Then, the amine (II) transfers a hydride ion to methylene amine (III), thereby forming an imine (IV) which gets finally hydrolysed to yield aldehyde.

$$(i)\ C_6H_5CH_2Cl + C_6H_{12}N_4 \longrightarrow \underset{\text{(I)}}{C_6H_5CH_2(C_6H_{12}\overset{\oplus}{N}_4)\overset{\ominus}{Cl}}$$

$$\xrightarrow{H_2O} \underset{\text{(II)}}{C_6H_5CH_2NH_2}$$

$$(ii)\ C_6H_{12}N_4 \xrightarrow{H_3O^{\oplus}} CH_2O + NH_3 \longrightarrow \underset{\text{(III)}}{CH_2{=}\overset{\oplus}{N}H_2}$$

$$(iii)\quad \underset{\textbf{(II)}}{C_6H_5-\underset{|}{\overset{}{C}}H=\ddot{N}H_2} + \underset{\textbf{(III)}}{CH_2=\overset{\oplus}{N}H_2} \longrightarrow$$

$$\longrightarrow \underset{\textbf{(IV)}}{C_6H_5-CH_2=\overset{\oplus}{N}H_2} + CH_3NH_2$$

$$(iv)\quad C_6H_5-CH=\overset{\oplus}{N}H_2 \xrightarrow{H_3\overset{\oplus}{O}} C_6H_5CHO + NH_3$$

The above mechanism has been supported by the conversion of benzylamine into corresponding aldehyde by using hexamine.

$$C_6H_5CH_2NH_2.HCI \xrightarrow{(CH_2)_6N_4} C_6H_5.CHO$$

It is to noted that the electronegative groups lowers the yield and *o*-substituents hinder the reaction. When both the *o*-positions are *o*-substituent, the reaction fails completely, *e.g.* 2, 4-dinitro- and 2, 6-dimethyl benzyl chloride do not undergo this reaction.

With phenolic compounds, the yield is much poor.

Electron withdrawing substituents decrease the yield and ortho substituents hinder the reaction.

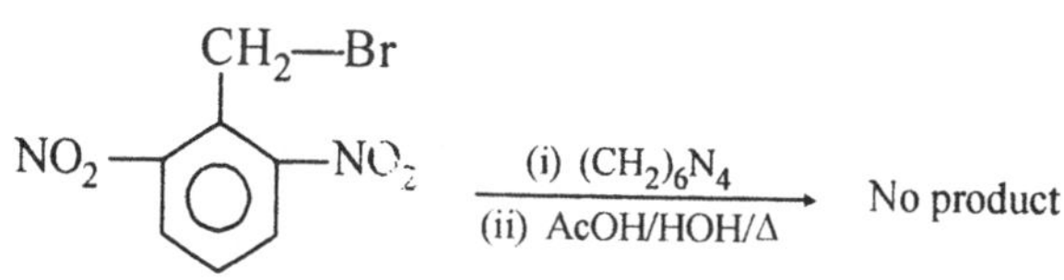

Electron withdrawing groups are present in the substrate

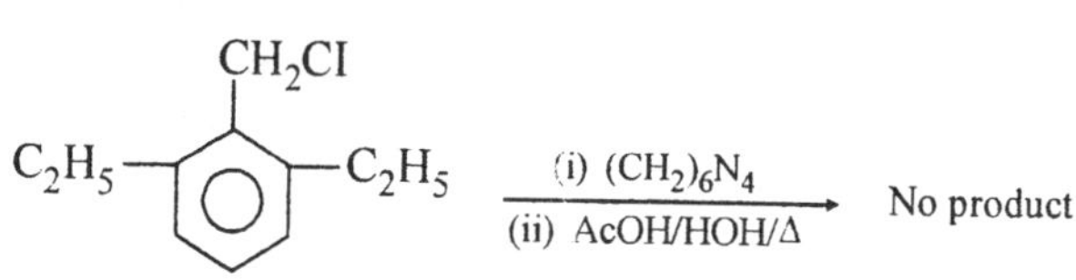

Sterric hinderence in the molecule

Phenolic compounds do not give this reaction.

This reaction is also given by aliphatic compounds.

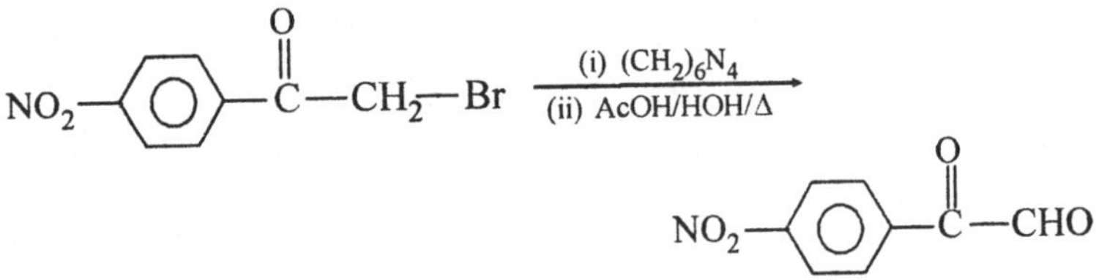

If the reaction is carried out in the presence of excess ammonia then product is *p*-amine instead of aldehyde.

$$C_6H_5\text{—}CH_2\text{—}Cl \xrightarrow[\text{(ii) } NH_3/\Delta]{\text{(i) } (CH_2)_6N_4} C_6H_5\text{—}CH_2\text{—}NH_2$$

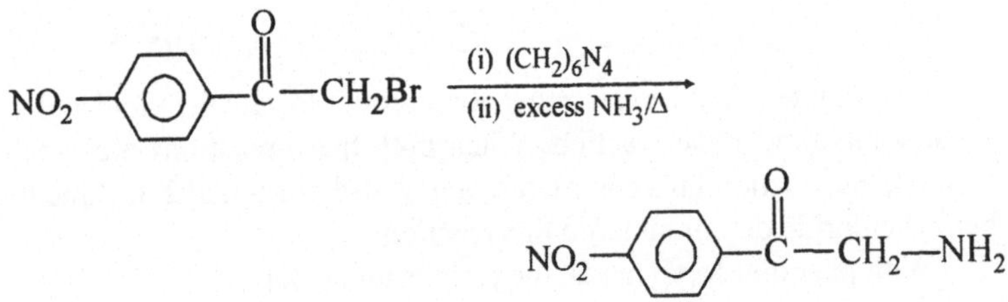

Halomethyl aromatics can also be oxidised into aldehyde by other nitrogen containing compounds.

$$Ar\text{—}CH_2\text{—}X \xrightarrow[\text{(ii) } H_2O/\overset{\oplus}{H}]{\text{(i) Pyridine}} Ar\text{—}CHO$$

The reaction is known as **Krohnk reaction**.
This reaction occurs under very mild condition (Pyridine is a very weak base) so that it is suitable for the preparation of sensitive aldehydes. This reaction is facilitated by electron withdrawing group.

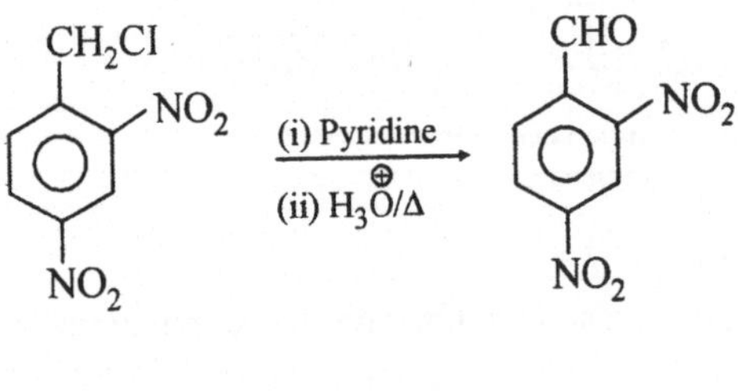

APPLICATIONS

1. The reaction has been found to be very useful for the synthesis of aldehydes from halides and amines, *e.g.*

(*i*) $$CH_3{-}(CH_2)_4{-}CH_2NH_2 + (CH_2)_6N_4 \xrightarrow[\Delta]{HCl} CH_3{-}(CH_2)_4{-}CHO$$

n-Hexanal

(*ii*)

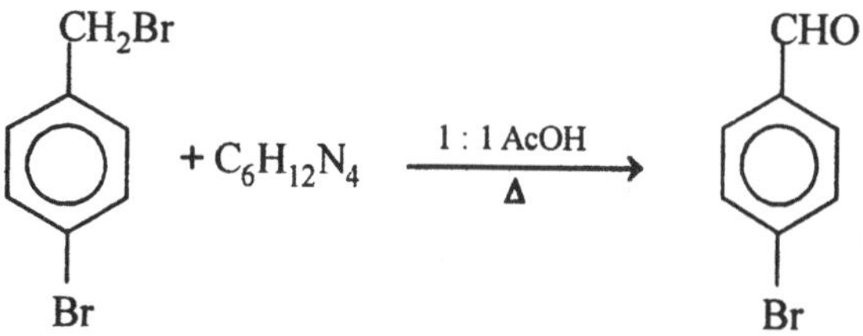

***p*-Bromobenzaldehyde**

(*iii*)

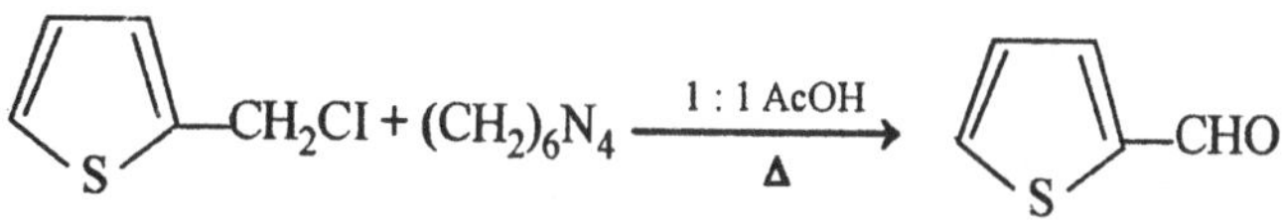

Thiophene-2-aldehyde

(*iv*)

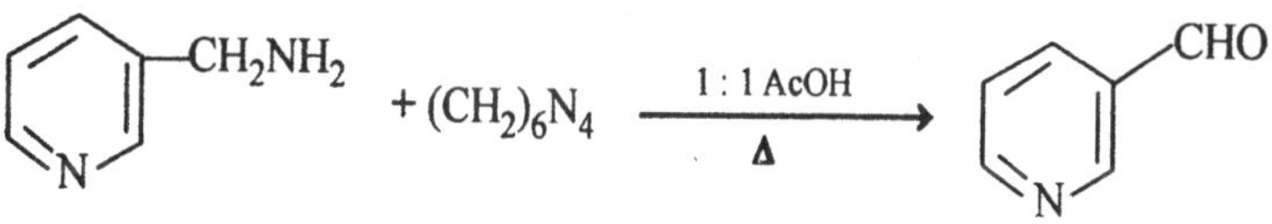

Pyridine-3-aldehyde

(*v*)

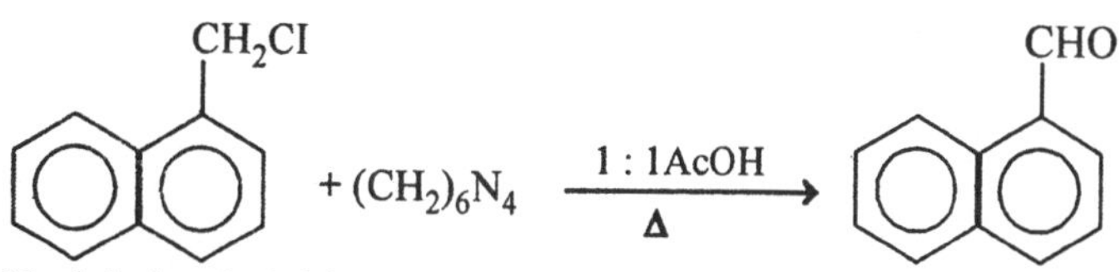

1-Naphthylmethyl chloride 1-Naphthaldehyde

2. *Preparation of amines :* When warmed with excess of ammonia, the hexamine-alkyl halides form amines in satisfactory yield. For example, benzylamine may be prepared from benzyl chloride.

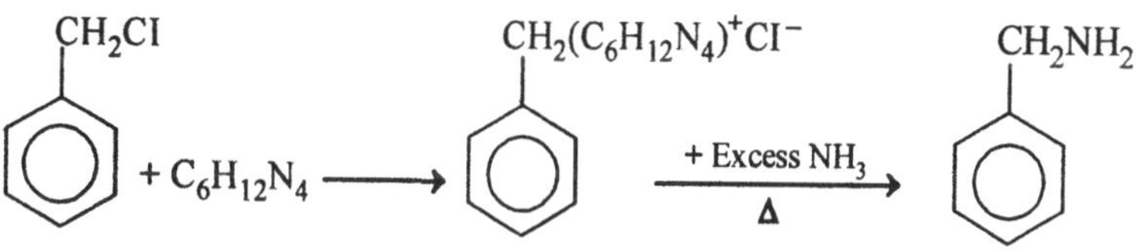

The reaction (2) is used for the preparation of an amino ketone intermediate which is required in the synthesis of chloromycetion.

p-Nitro-ω-bromoacetophenone

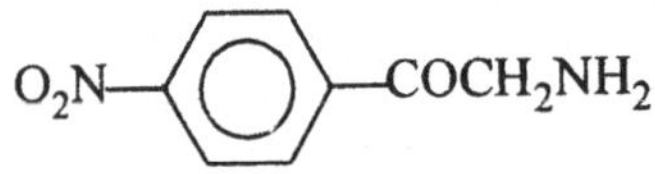

Intermediate for chloromycetin

Another way to oxidize benzilic halides to aldehydes is the Krohnke reaction. The reaction consists of treatment with pyridine followed by *p*-nitrosoaimethylaniline and subsequent hydrolysis.

$$C_6H_5CH_2CI + C_5H_5N \longrightarrow [C_6H_5CH_2\overset{\oplus}{N}C_5H_5]\overset{\ominus}{C}I$$

$$\xrightarrow[2.\ H_3O^{\oplus}]{1.\ p\text{-}ONC_6H_4—N(CH_3)_2,\ \overset{\ominus}{O}H} C_6H_5CHO$$

100

SIMMONS-SMITH REACTION

There are some compounds that behave like carbene but, in fact, are not free carbene. These compounds are referred to as carbenoids. The best known example of a carbenoid is Simmons-Smith reagent. This reagent is made from methylene iodide and a zinc-copper couple or better, zinc silver couple.

$$CH_2I_2 \xrightarrow[\text{Zn - Ag}]{\text{Zn - Cu or}} \underset{\text{Simmon-Smith reagent}}{I—CH_2—Zn—I}$$

This reagent is unable and is usually generated in situ.

This reagent is the most useful reagent in the synthesis of cyclopropanes from alkene and its derivative.

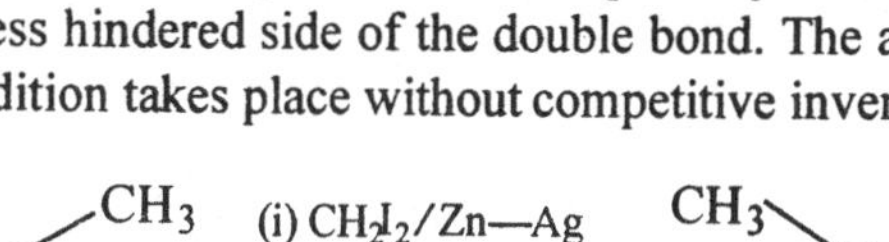

This reaction is methylene transfer reaction.

The reaction is stereospecific and takes place by *cis* addition of methylene to the less hindered side of the double bond. The addition is concerted and addition takes place without competitive inversion.

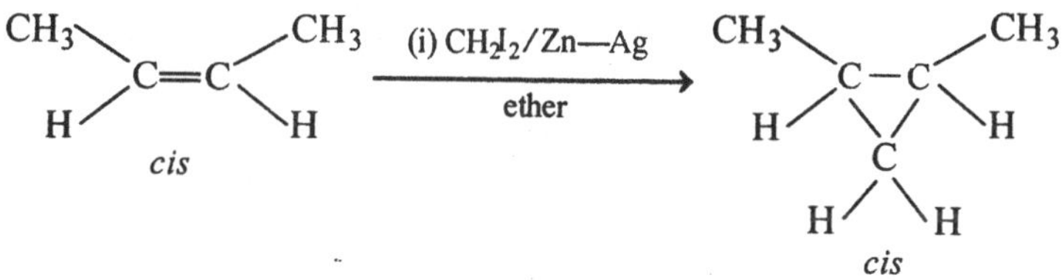

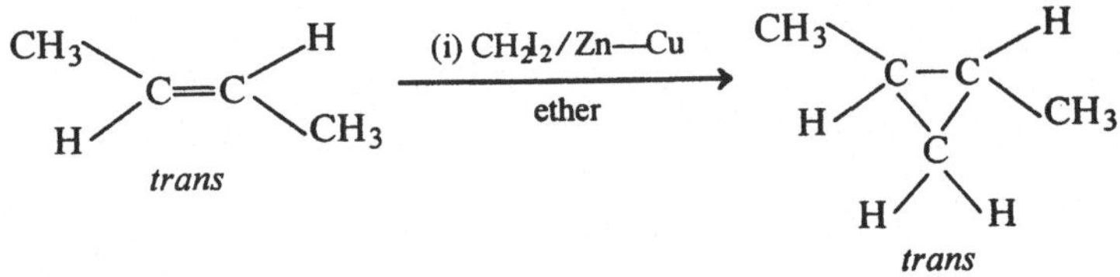

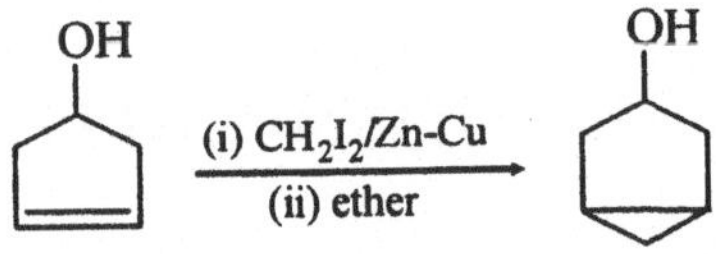

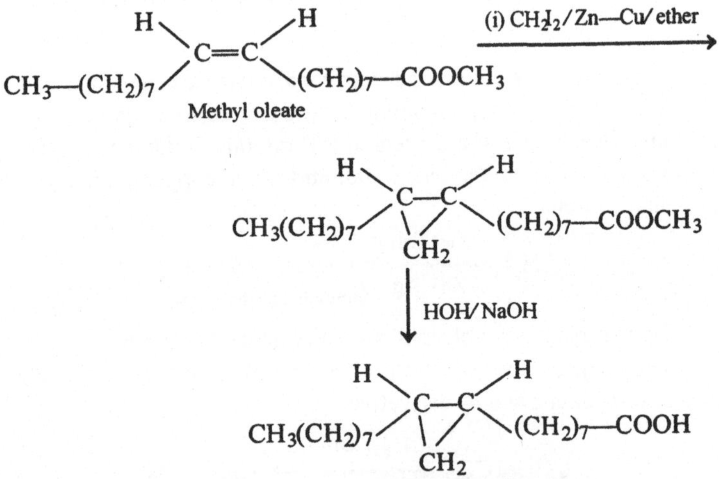

MECHANISM :

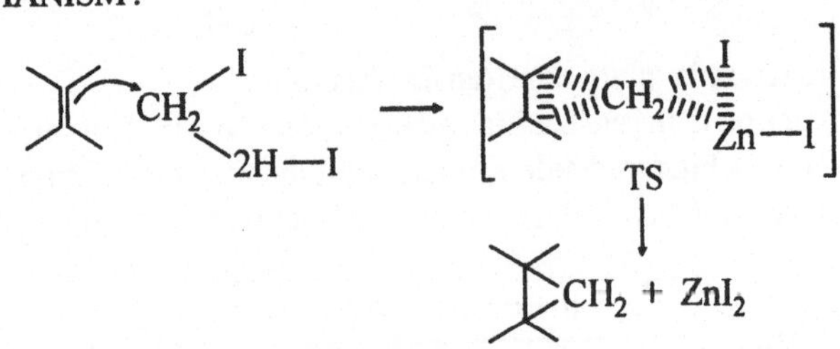

In case of allylic alcohol, OH and CH_2 groups are always in the same plane.

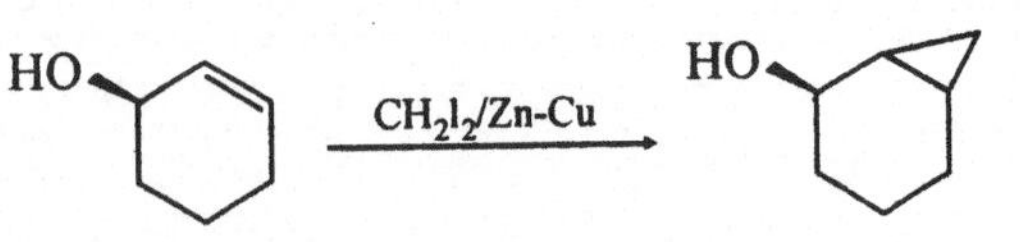

101

STOBBE CONDENSATION

Aldehydes and ketones react with dialkyl succinate in the presence of strong base such as NaH or sodium *t*-butoxide to form monoesters of an α-alkylidene or arylidene succinic acid. This reaction is known as the Stobbe condensation.

(I)

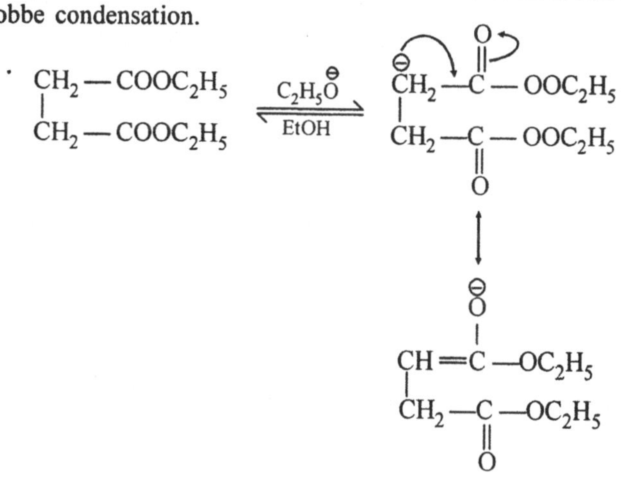

(II)

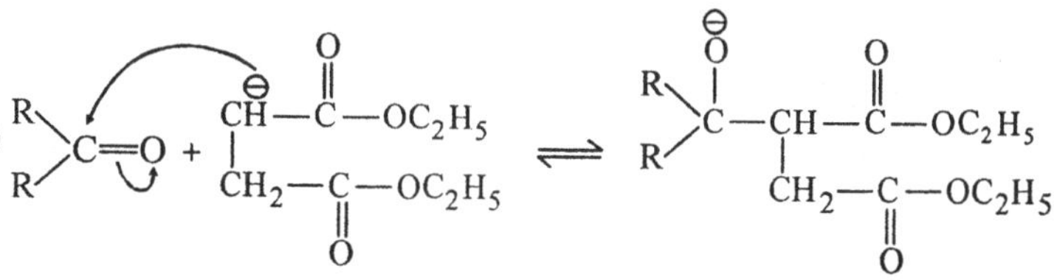

The resulting cyclises to γ-lactone which then undergoes base-catalysed ring opening to give carboxylate salt.

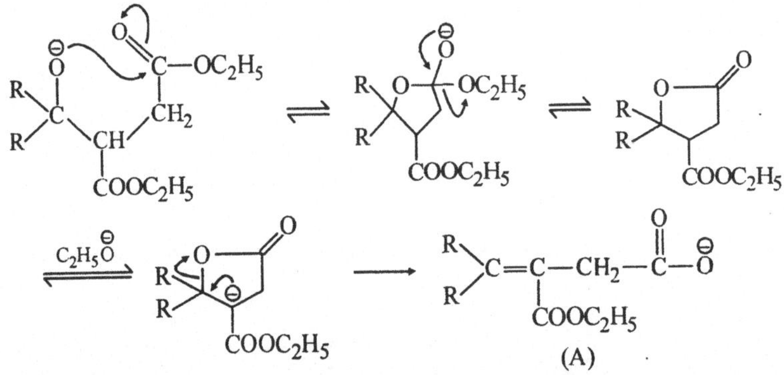

Product of the reaction can be written as follows :

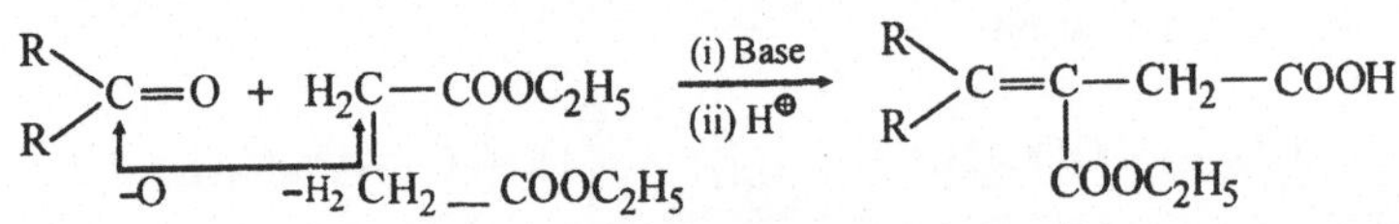

In this reaction, dialkyl succinate behaves differently from other esters. The enolate is formed from ester and not from the ketone or aldehyde and then enolate ester adds with carbonyl compound.

Aldehydes containing α-hydrogen undergo self condensation in the presence of strong base in this reaction, but those with no α-hydrogen react successfully.

In this reaction, formation of the carboxylate ion (A) is irreversible, hence this condensation may be applied successfully even to relatively hindered ketones, where other types of Aldol condensation fail.

Examples

(*i*)

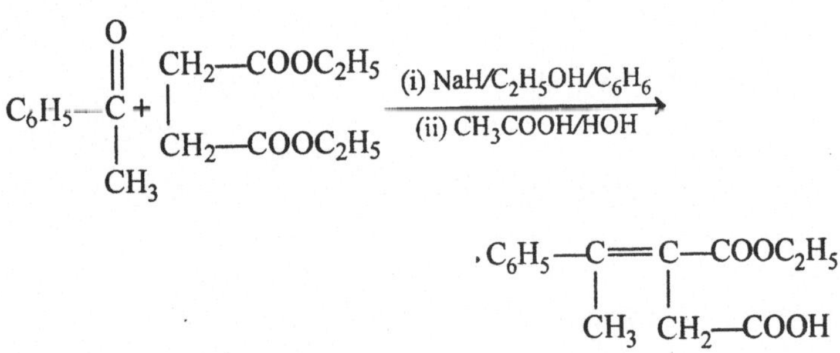

(*ii*)

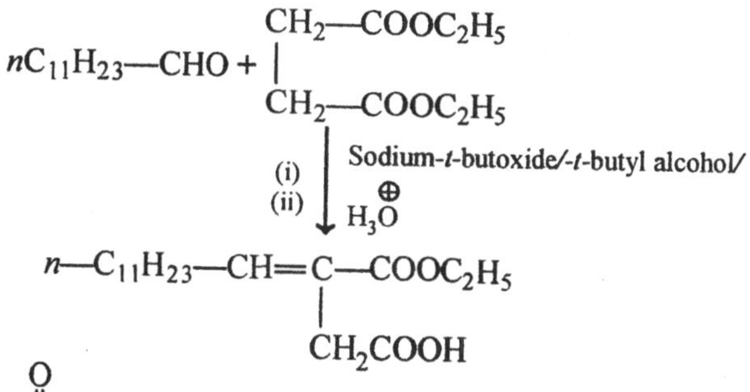
Sodium-*t*-butoxide/-*t*-butyl alcohol/
(i)
(ii)
$H_3\overset{\oplus}{O}$

(*iii*)

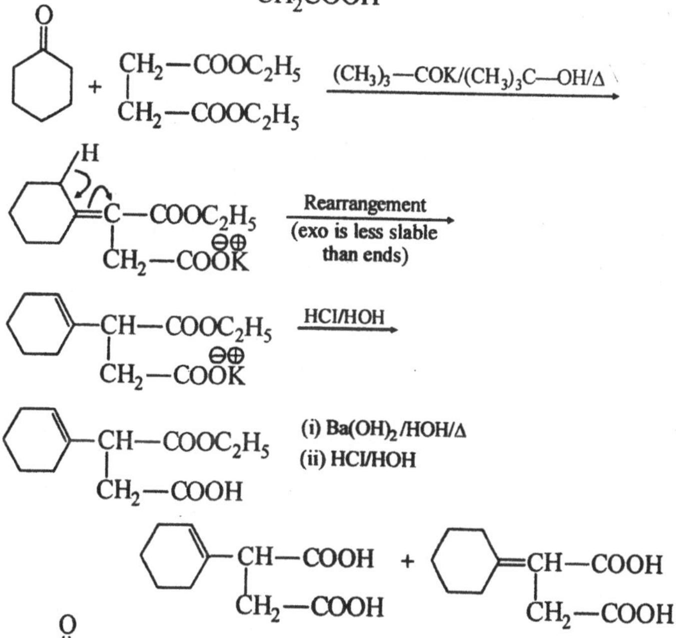
$(CH_3)_3$—COK/$(CH_3)_3$C—OH/Δ
Rearrangement
(exo is less slable
than ends)
HCl/HOH
(i) $Ba(OH)_2$/HOH/Δ
(ii) HCl/HOH

(*iv*)

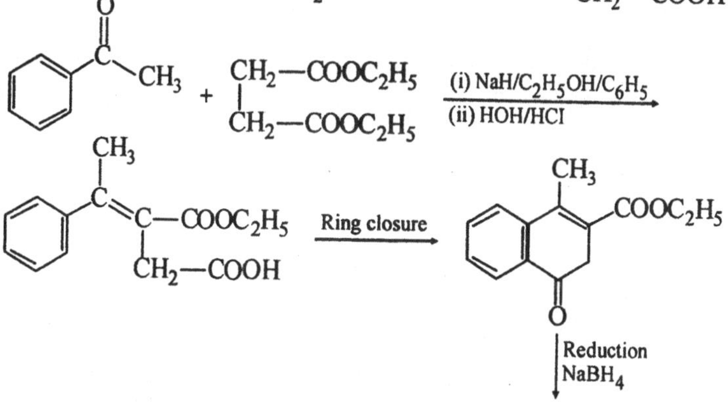
(i) NaH/C_2H_5OH/C_6H_5
(ii) HOH/HCl
Ring closure
Reduction
$NaBH_4$

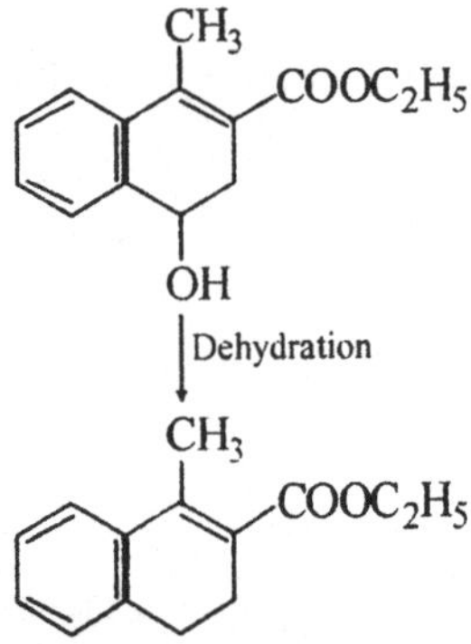

Scope of the reaction :

Numerous variety of carbonyl compounds, such as aliphatic, aromatic and α, β-unsaturated aldehydes, aliphatic and aromatic ketones and diketones, keto-esters and cyanoketones have been used.

The succinic esters employed are dimethyl, diethyl and di-butyl succinate, α-Substituted aryl, alkyl and alkylidene succinic esters have also been used.

Sodium or potassium alkoxides are used as condensing agents. When sodium ethoxide is used Stobbe reaction is sometimes accompanied by reaction of some ketone to the corresponding alcohol. This can be eliminated by use of *t*-butyl alcohol. Under these conditions the yields are better and the reaction time is also shorter than with sodium ethoxide. Other condensing agents such as sodium hydride, sodium triphenylmethyl have also been used.

APPLICATIONS

(*i*) Preparation of unsaturated and saturated (by hydrogenation) substituted succinic acids

(*a*) *Synthesis of tetraconic acid :*

$$\underset{\text{Acetone}}{(CH_3)_2C{=}O} + \underset{\text{Diethyl succinate}}{\begin{array}{l} CH_2COOC_2H_5 \\ | \\ CH_2COOC_2H_5 \end{array}} \xrightarrow{C_2H_5ONa}$$

$$\begin{array}{l} (CH_3)_2C{=}CCOOC_2H_5 \\ \qquad\quad\;\; | \\ \qquad\quad CH_2COOH \end{array} \xrightarrow[(ii)\ H^{\oplus}]{(i)\ KOH} \underset{\text{Tetraconic acid}}{\begin{array}{l} (CH_3)_2C{=}C{-}COOH \\ \qquad\qquad\; | \\ \qquad\quad\;\; CH_2COOH \end{array}}$$

(*b*) *Synthesis of β-carboxy-γ, γ-diphenylvinylacetic acid*

$$\underset{\text{Benzophenone}}{(C_6H_5)_2C{=}O} + \begin{array}{l} CH_2COOC_2H_5 \\ | \\ CH_2COOC_2H_5 \end{array} \xrightarrow[(CH_3)_3COH]{(CH_3)_3COK}$$

$$\underset{(90\text{-}94\%)}{(C_6H_5)_2C{=}\underset{\underset{CH_2COOH}{|}}{C}{-}COOC_2H_5} \xrightarrow[(ii)\ H^{\oplus}]{(i)\ KOH} \underset{\text{β-Carboxy-γ, γ-diephenylvinyl acetic acid}}{(C_6H_5)_2C{=}\underset{\underset{CH_2COOH}{|}}{C}{-}COOH}$$

The acid obtained can be decarboxylated to yield monobasic acid as follows :

$$(C_6H_5)_2C{=}\underset{\underset{CH_2COOH}{|}}{C}{-}COOH \xrightarrow[\substack{H_2O \\ \text{Reflux}}]{\substack{HBr \\ CH_3COOH}}$$

$$\underset{(73\%)}{(C_6H_5)_2C{=}CHCH_2COOH} + \underset{\text{A γ-lactone (22\%)}}{(C_6H_5)_2C\begin{array}{l} \diagup O{-}C{=}O \\ \diagdown CH_2{-}CH_2 \end{array}}$$

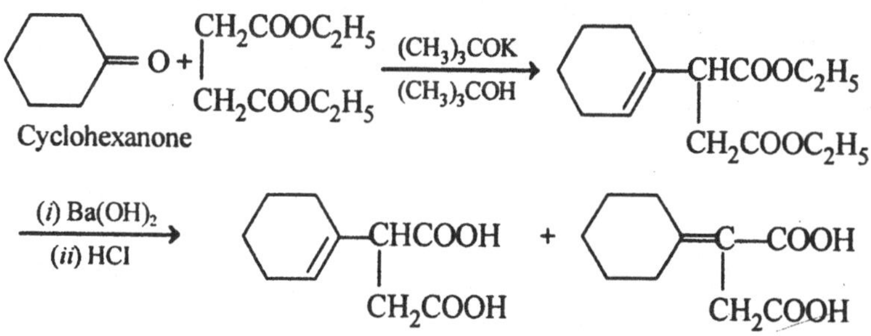

The dibasic and monobasic acids can be hydrogenated to yield the corresponding saturated acids.

(*ii*) Synthesis of γ-keto acids :

$$\underset{\textbf{Heptanal}}{n\text{–}C_6H_{13}CHO} + \underset{\text{Diethyl succinate}}{\begin{array}{l} CH_2COOC_2H_5 \\ | \\ CH_2COOC_2H_5 \end{array}} \xrightarrow[(CH_3)_3COH]{(CH_3)_3COK}$$

$$n\text{–}C_6H_{13}CH{=}\underset{\underset{CH_2COOC_2H_5}{|}}{C}COOC_2H_5 \xrightarrow[(ii)\ H^{\oplus}]{(i)\ KOH} n\text{–}C_6H_{13}CH{=}\underset{\underset{CH_2COOH}{|}}{C}{-}COOH$$

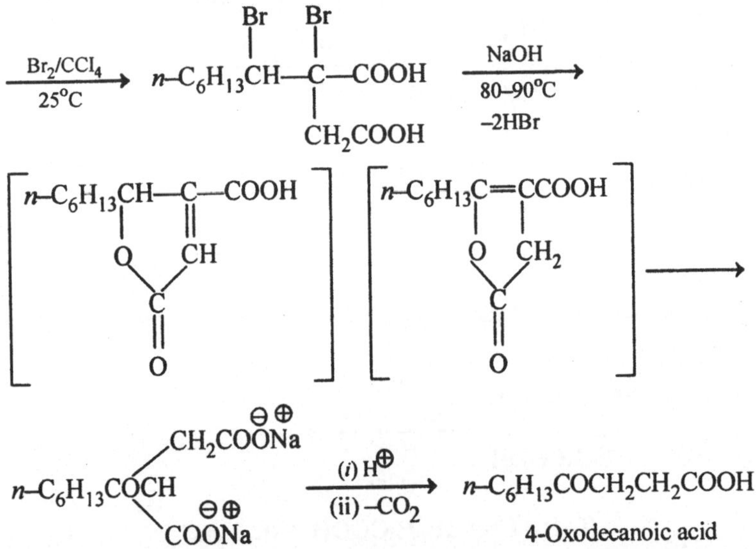

(*iii*) **Synthesis of naphthols :** The half-ester having appropriate stereochemical configuration, *viz.*, an aryl group *cis* to the —CH_2COOH group, can undergo cyclodehydration and enolisation to yield substituted 1-naphthol-3-carboxylic acids.

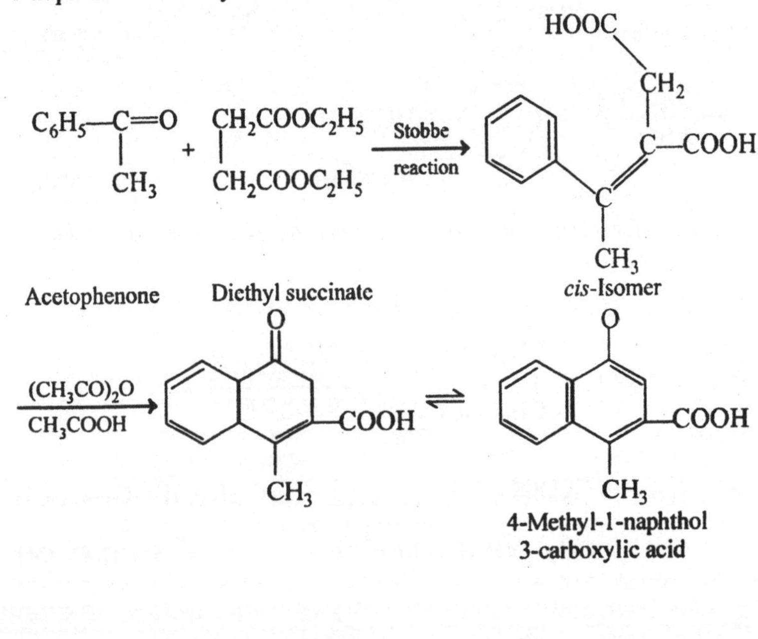

(*iv*) **Synthesis of tetralone derivatives :** The γ-lactone obtained in (i-b) can be reduced to substituted γ-arylbutyric acid which on cyclodehydration yields substituted 1-tetralones.

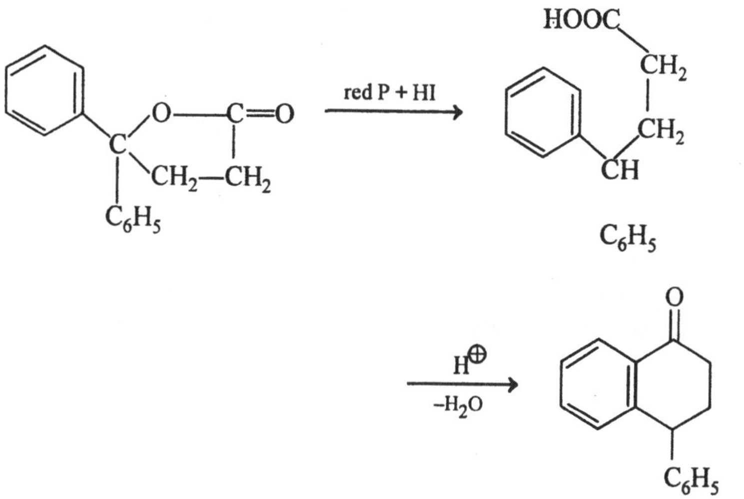

4-Phenyl-1-tetralone

STEPHEN ALDEHYDE SYNTHESIS

Conversion of cyanide into aldehyde by partial reduction with $SnCl_2$/HCl, followed by hydrolysis, is known as Stephens aldehyde synthesis.

$$R{-}C{\equiv}N \xrightarrow[\text{(ii) } H_2O/\Delta \text{ or Steam distillation}]{\text{(i) } SnCl_2/HCl/\text{ether}} R{-}CHO$$

MECHANISM :

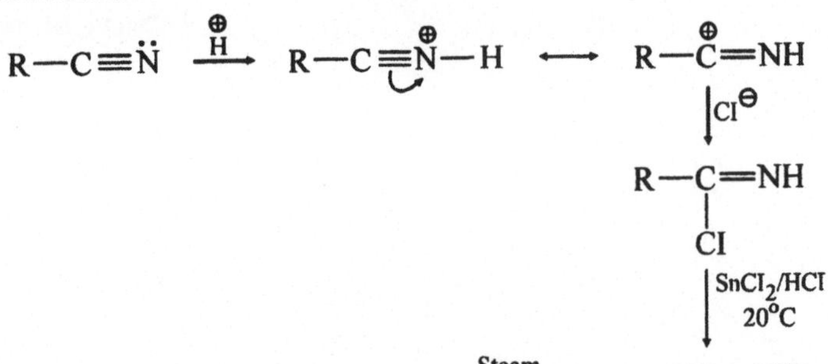

EXAMPLES :

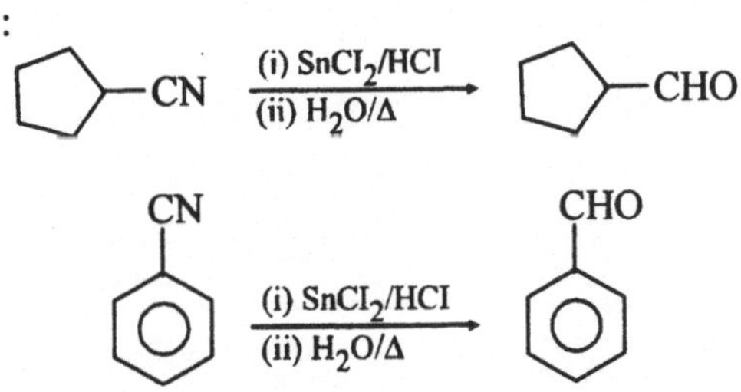

103

STORK ENAMINE REACTION

Carbonyl compounds react with secondary amines to form compounds called enamines. The reaction is catalysed by acid. The secondary amines most commonly used to prepare enamines are cyclic, such as pyrrolidine, piperidine and morpholine. For this reaction carbonyl compound should have at least one α-hydrogen.

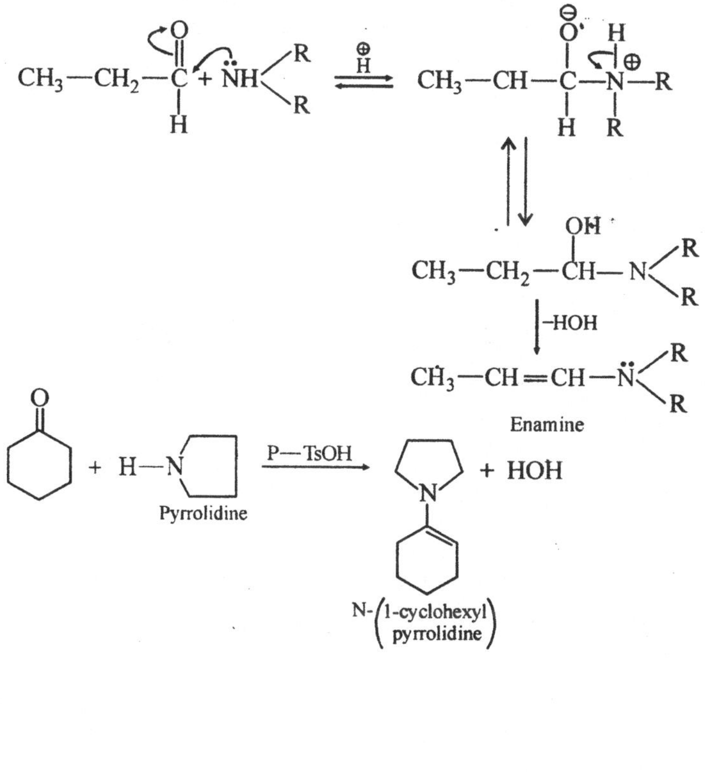

Enamines are good nucleophiles in which nitrogen and carbon both are nucleophilic centres.

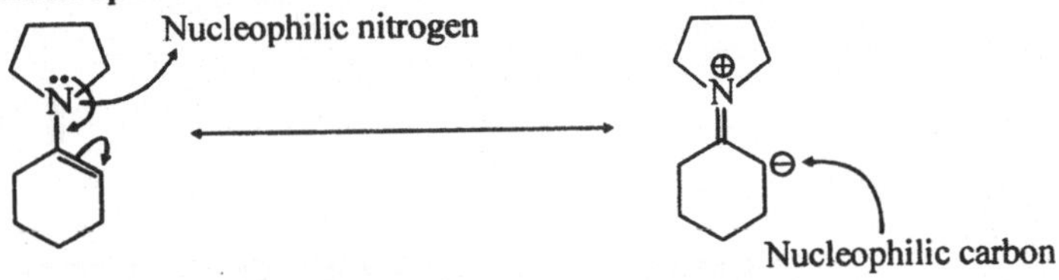

The nucleophilicity of the carbon of enamine makes them particularly useful reagents in organic synthesis because they can be acylated, alkylated and used in Michael addition.

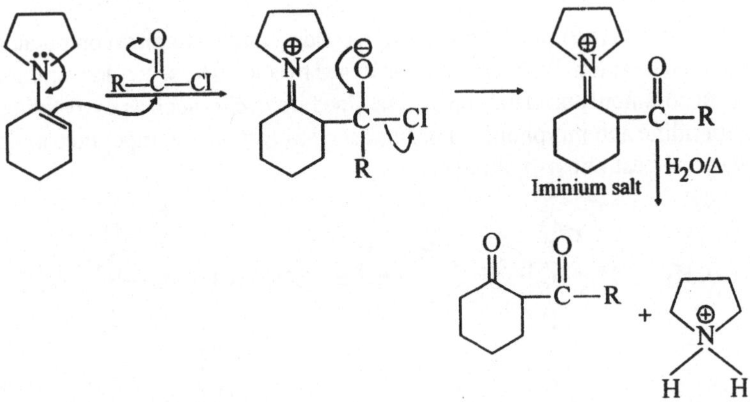

Examples :

(1) 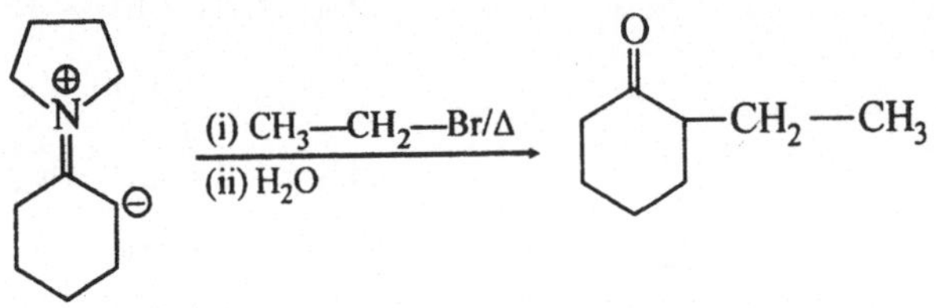

(2) 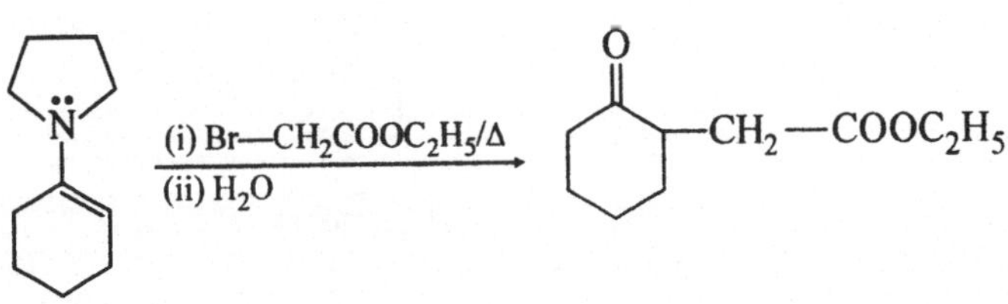

(3)

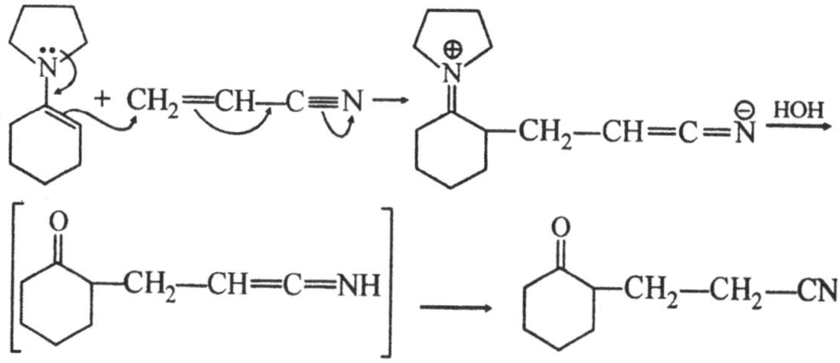
+ CH₂═CH—C≡N →
CH₂—CH═C═N⊖
HOH
CH₂—CH═C═NH
CH₂—CH₂—CN

(4)

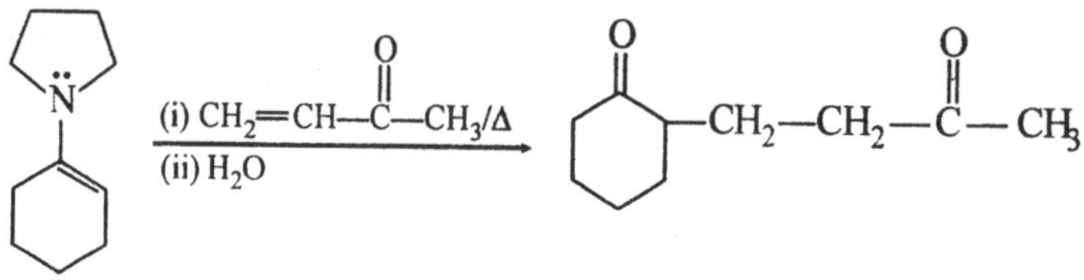
(i) CH₂═CH—C(═O)—CH₃/Δ
(ii) H₂O
CH₂—CH₂—C(═O)—CH₃

(5)

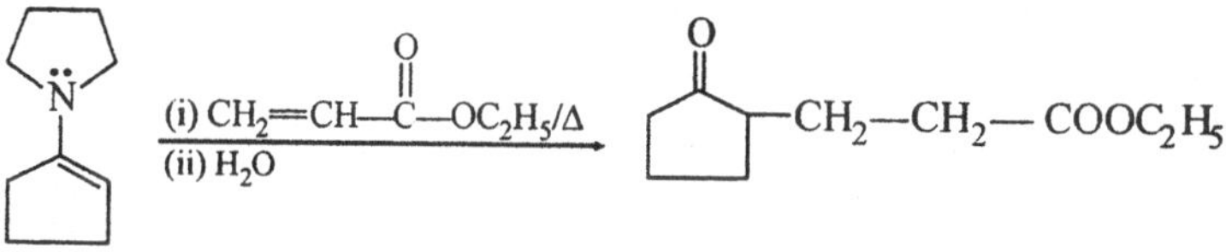
(i) CH₂═CH—C(═O)—OC₂H₅/Δ
(ii) H₂O
CH₂—CH₂— COOC₂H₅

(6)

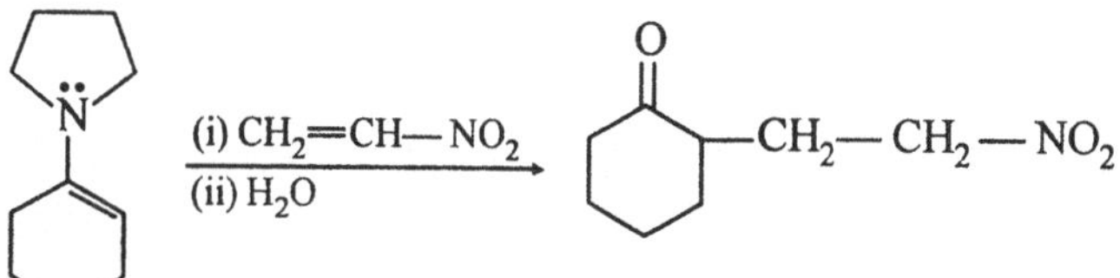
(i) CH₂═CH— NO₂
(ii) H₂O
CH₂— CH₂— NO₂

104

STRECKER AMINO ACID SYNTHESIS

Conversion of aldehydes into α-aminoacids by NH_3 and CN followed by hydrolysis is known as Strecker amino acid synthesis.

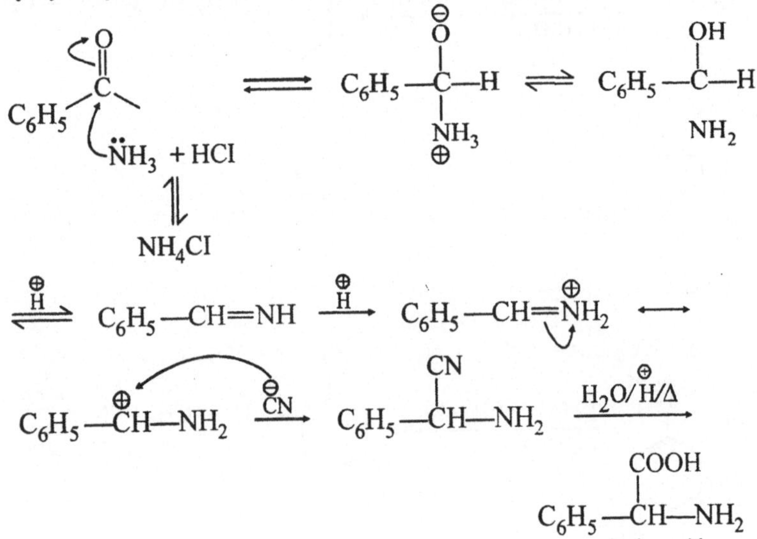

α-Amino acid

It is convenient to employ aqueous ammonium chloride and NaOH as the reagents, because ammonia is generated in situ by hydrolysis.

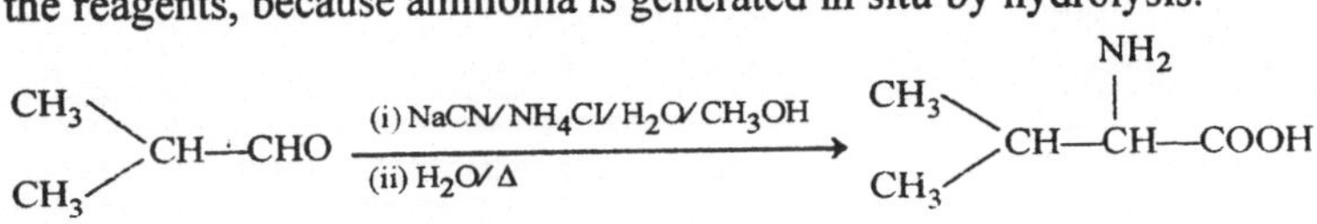

105

SWERN OXIDATION

Oxidation of primary and sec alkyl halides into aldehyde and ketone, respectively by dimethyl sulphoxide is known as Swern oxidation. In this reaction, the yield of the product is good.

$$R{-}CH_2{-}X \xrightarrow{\text{DMSO}} R{-}CHO + HX + CH_3{-}S{-}CH_3$$

$$R{-}\overset{\overset{\large X}{|}}{C}H{-}R \xrightarrow{\text{DMSO}} R{-}\overset{\overset{\large O}{\|}}{C}{-}R + HX + CH_3{-}S{-}CH_3$$

MECHANISM :

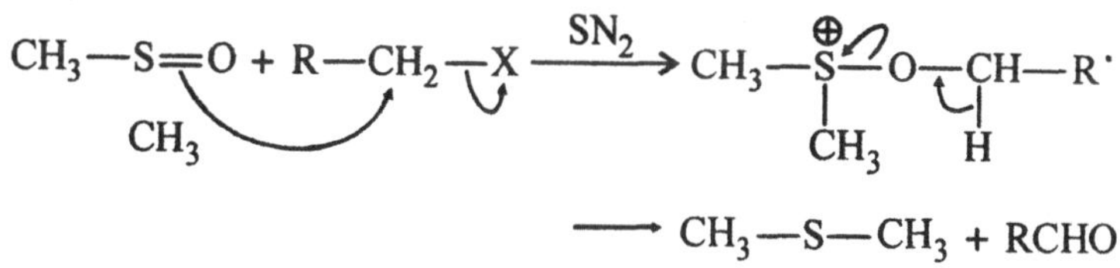

106

THORPE REACTION

Condensation reaction between two molecules of nitrile in the presence of strong base is known as Thorpe reaction. In this reaction the two nitriles may be same or different but one should have at least one α-hydrogen. Sodium ethoxide is used as a strong base.

$$CH_3—CN + CH_3CN \xrightarrow{\text{(i) } C_2H_5ONa} \underset{\substack{\| \\ NH}}{CH_3—C—CH_2—CN}$$

β-Iminonitrile

MECHANISM:

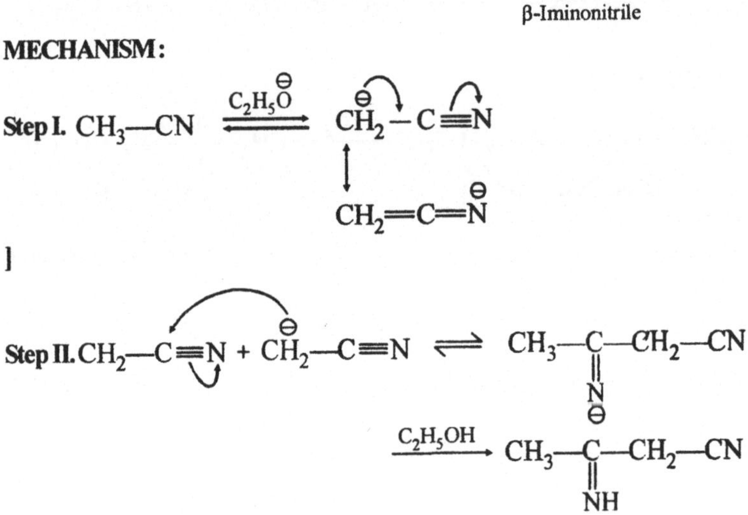

The reaction is a very useful reaction for the preparation of β-ketonitrile.

$$CH_3-\underset{\underset{NH}{||}}{C}-CH_2-CN \xrightarrow{H_2O/\Delta} CH_3-\underset{\underset{O}{||}}{C}-CH_2-CN$$

Cyano group is hydrolysed in the presence of catalyst, acid or base, but imino group is easily hydrolysed by water.

107

THORPE -ZIEGLER CYCLISATION

A base catalysed condensation of a dinitrile to form a β-keto nitrile is known as Thorpe-Ziegler cyclisation. This reaction is carried out at high dilution so that only intramolecular reaction takes place. Base catalyst is the lithium salt of secondary amine.

First Step :

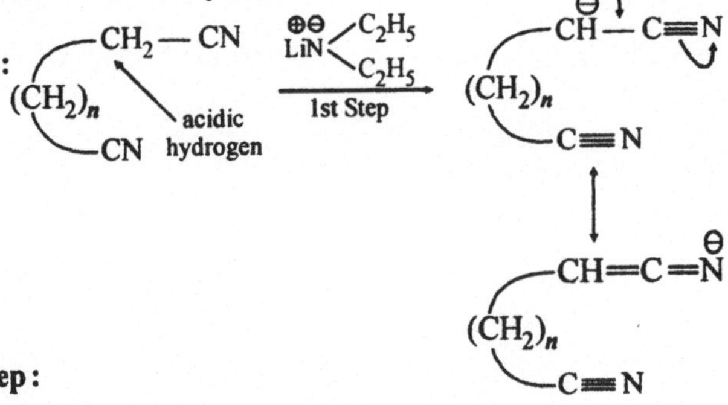

Second Step :

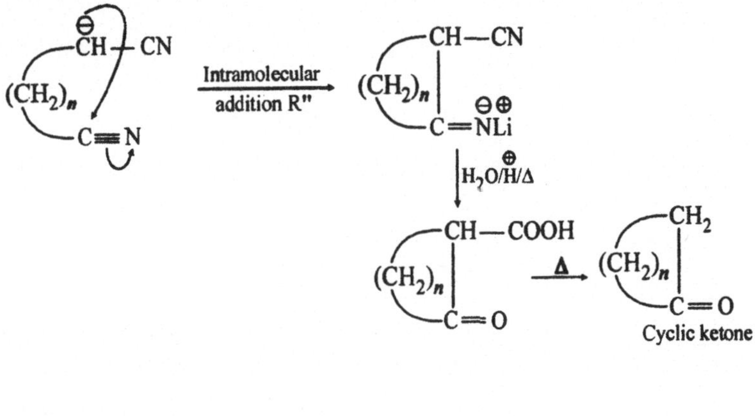

108

TOLLENS REACTION

Cross aldol condensation between aldehydes or ketones having at least one α-hydrogen with excess of formaldehyde followed by Cannizzaro reaction is known as Tollens reaction. This reaction is catalysed by a base like $Ca(OH)_2$, $Ba(OH)_2$ or $Sr(OH)_2$.

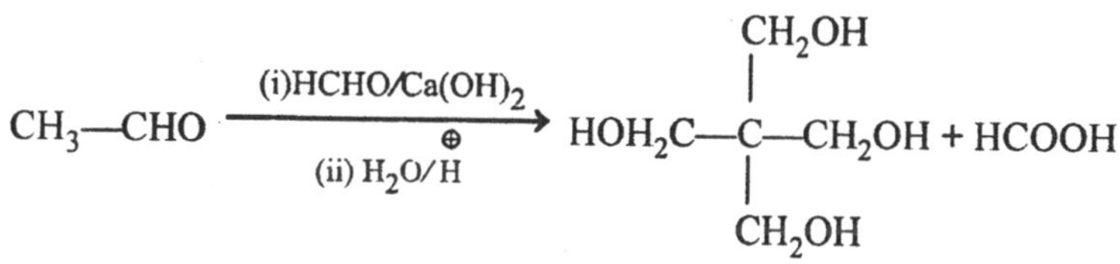

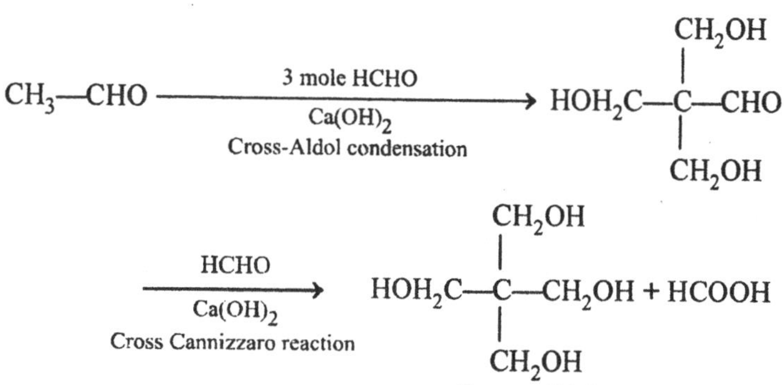

Pentaerythriol

This reaction also takes place between nitroalkanes having at least one α-hydrogen and excess of formaldehyde. In this case, nitro group is not reduced as in the case of aldehydes or ketones.

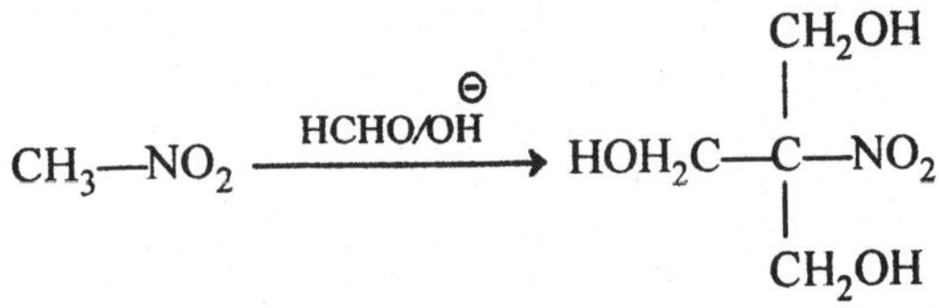

109

TISCHENKO REACTION OR TISCHENKO-CLAISEN REACTION

The alkoxide-catalysed disproportionation of aldehydes to produce ester is known as Tischenko reaction. Alkoxide of the reaction is Aluminium tri ethoxide, aluminium tri isopropoxide. The reaction is also catalysed by Al_2O_3. In case of aromatic aldehydes reaction is catalysed by disodium tetra carbonyl ferrate.

$$CH_3—CHO + CH_3CHO \xrightarrow{(C_2H_5O)_3Al/\Delta} CH_3—\overset{\overset{O}{||}}{C}—O—CH_2—CH_3$$

MECHANISM :

Step I.

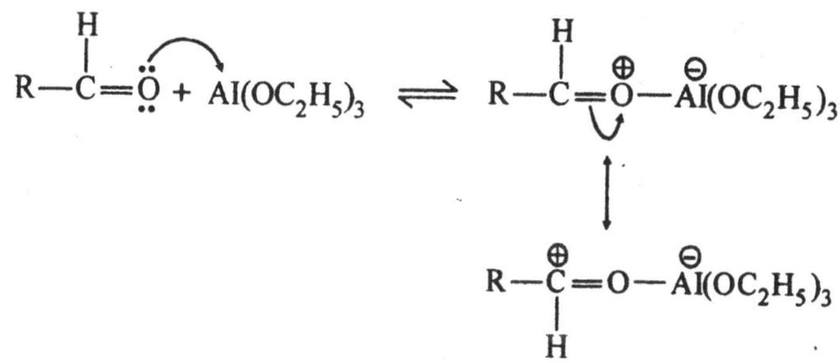

Step II.

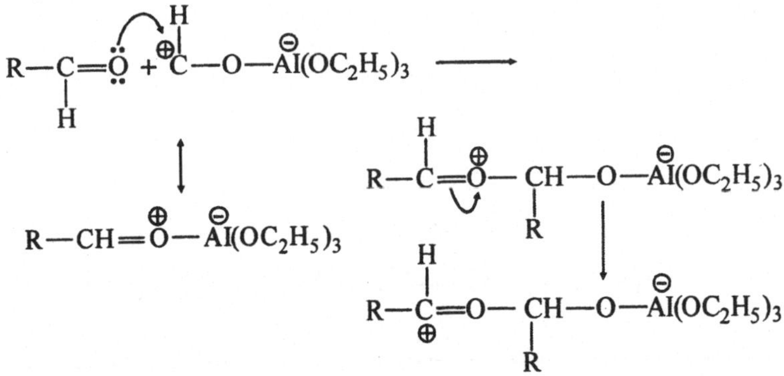

Step III.

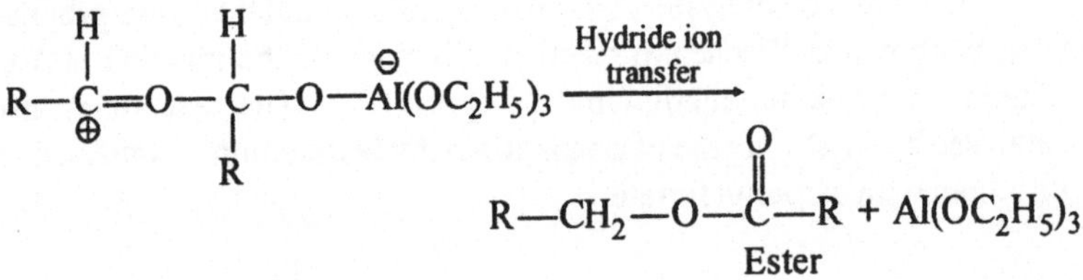

110

ULLMANN REACTION

It is a coupling reaction between two molecules of aryl halides in the presence of Cu or copper bronze. The reaction is carried out in the absence of a solvent at high temperatures (> 200°C). This reaction is not given by fluorobenzene. Product of the reaction is biaryl.

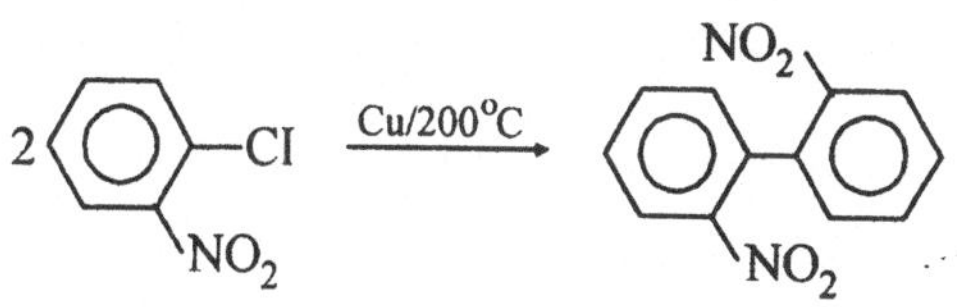

When dimethylformamide (DMF) is used as solvent, a better yield is obtained. Nitro group strongly activates the halides but only from the ortho position whereas alkyl (R) and alkoxy (OR) groups tend to activate from all positions. However, electron-releasing groups tend to inhibit the reaction.

Ullmann reaction is quite important because it is simple and straight forward and the structure of the product is known from its synthesis.

A method of coupling aryl halides that is complementary to the Ullmann reaction involves formation of the diarylcuprate which is oxidised by oxygen at low temperature.

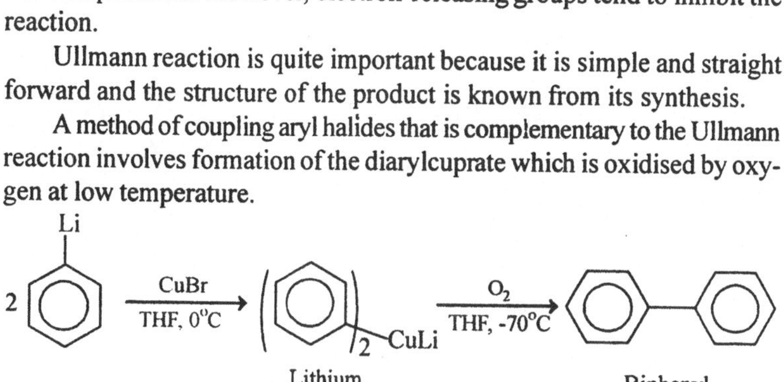

Lithium diphenylcuprate

Biphenyl (75%)

The yield of this method is 75%. The cuprate method avoids the high temperatures but it is only applicable with arenes not having functional groups that react with aryllithium intermediate.

MECHANISM :

Step I. Aryl halide forms organocopper compound which is unstable.

$$C_6H_5—X + Cu \longrightarrow C_6H_5—Cu—X$$

This organomatic copper reacts with Aryl halide to give product

$$\overset{+\delta}{Cu}X + \overset{+\delta}{C_6H_5}—\overset{-\delta}{X} \longrightarrow C_6H_5—C_6H_5 + CuX_2$$

APPLICATIONS

1. *Synthesis of biphenyls :* Ullmann reaction is used for synthesising biphenyls. An important advantage of this reaction is that its yield is higher than that of Wurtz reaction. Also, Ullmann reaction is quite valuable in synthesising symmetrical as well as unsymmetrical biphenyls.

(*a*) *Symmetrical biaryls :* Ullmann reaction is quite useful in synthesising many symmetrical biaryls, *e.g.*,

(*i*) 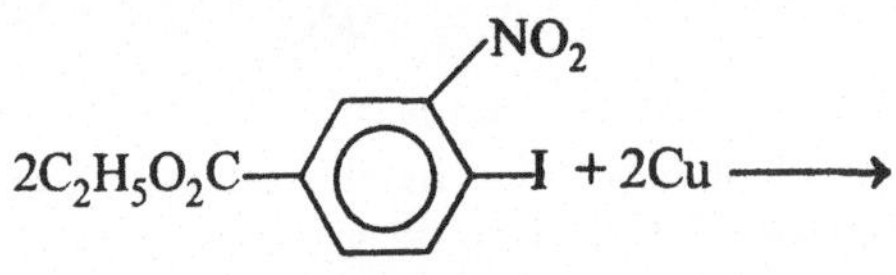

4-Iodo-3-nitrobenzoate

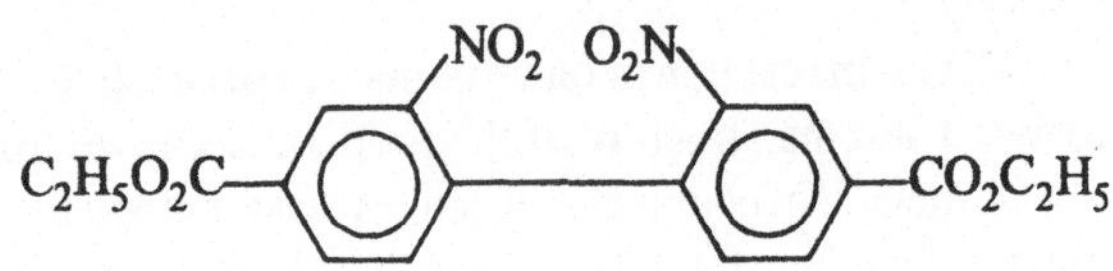

2 : 2'-Dinitro-4 : 4'-biphenyldicarboxylate

(*ii*) 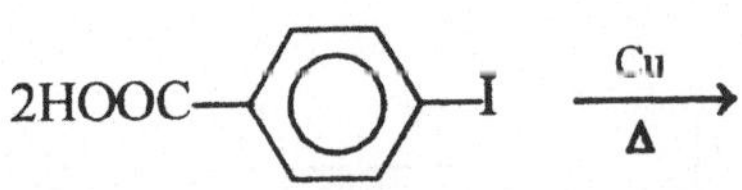

p-Iodobenzoic acid

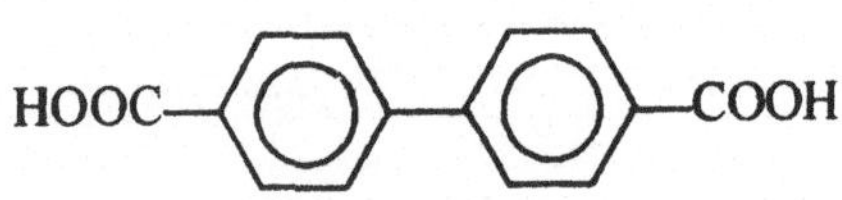

4, 4'-Diphenic acid

(*iii*) 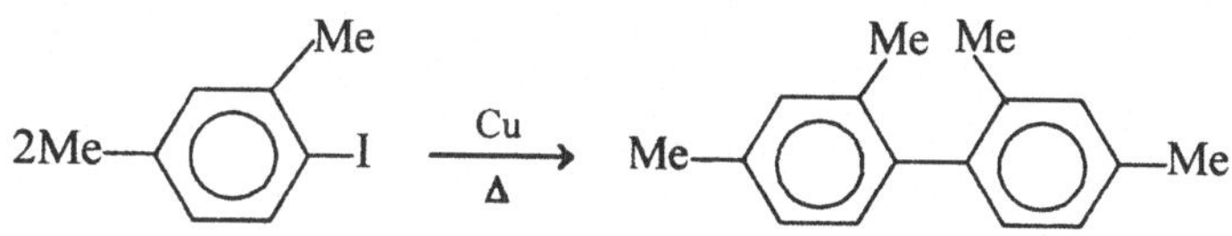

2,4-Dimethyl iodobenzene 2, 2', 4, 4'-Tetramethyl biphenyl

(*b*) *Unsymmetrical biaryls :* Ullmann reaction is used for synthesising unsymmetrical biaryls by heating a mixture of equal amount of different aryl halides with copper. The symmetrical biaryls are also formed along with unsymmetrical biaryls. Thus, the yield of unsymmetrical biaryl is not always high. Further, it is difficult to isolate.

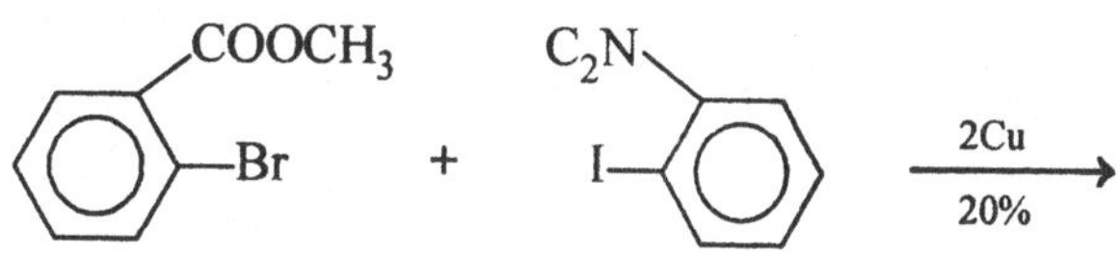

Methyl-
o-bromobenzoate 2-nitro-iodobenzene

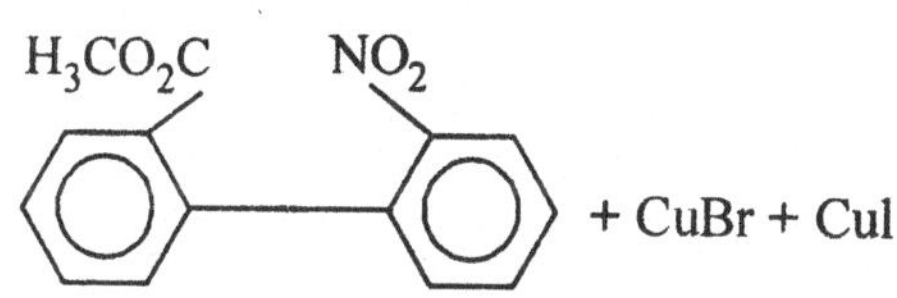

2-Nitro-2'-diphenylcarboxylate

However, unsymmetrical biaryl is the main product in some reactions, *e.g.*,

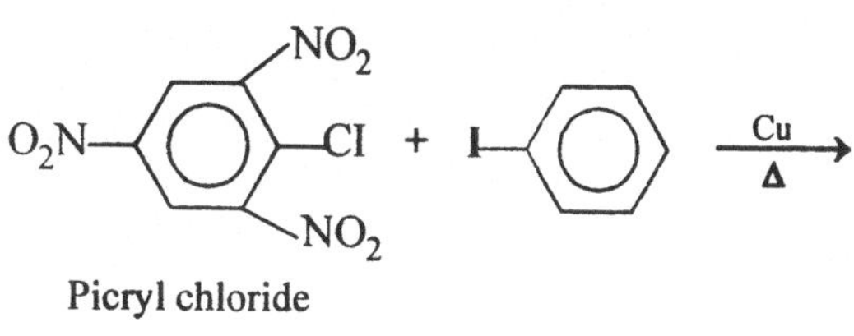

Picryl chloride

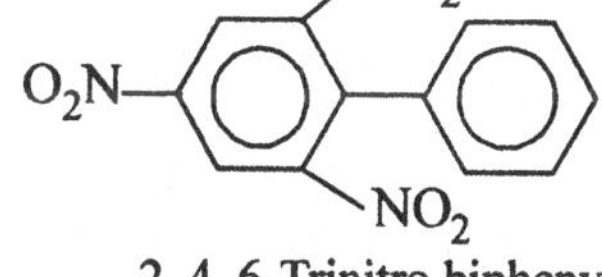

2, 4, 6-Trinitro biphenyl

2. *Preparation of polycyclic hydrocarbons :* Ullmann reaction has been used in synthesising many polycyclic hydrocarbons, e.g.,

(*i*) 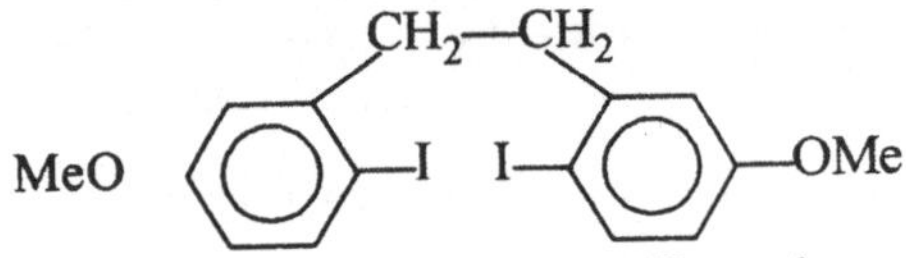

2.2'-Diiodo-5, 5'-dimethoxydibenzyl

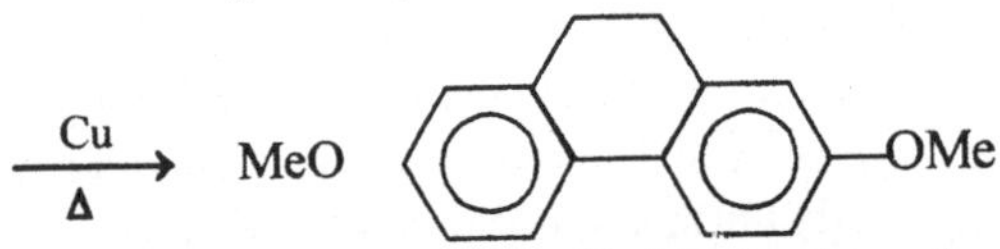

2, 7-Dimethoxy-9, 10-dehydrophenanthrene

(*ii*) 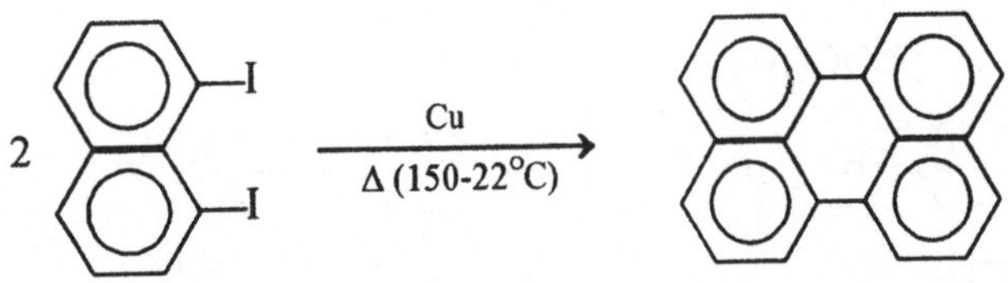

1, 8-Diiodonaphthalene　　　　Perylene

(*iii*) 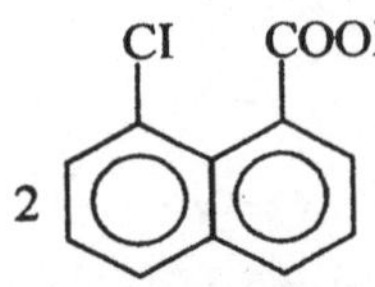

Ethyl-8-chloro-1-naphthoate

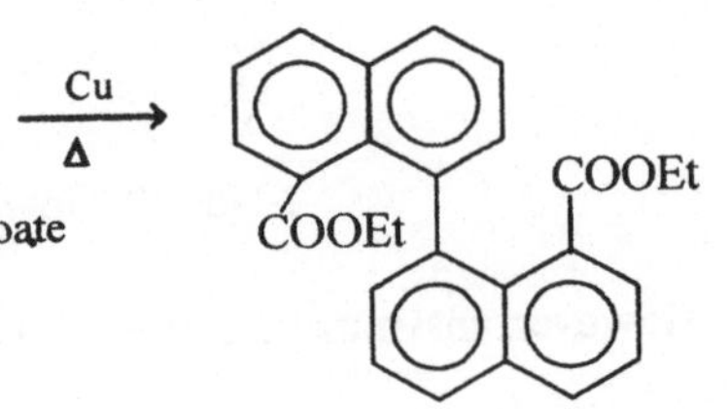

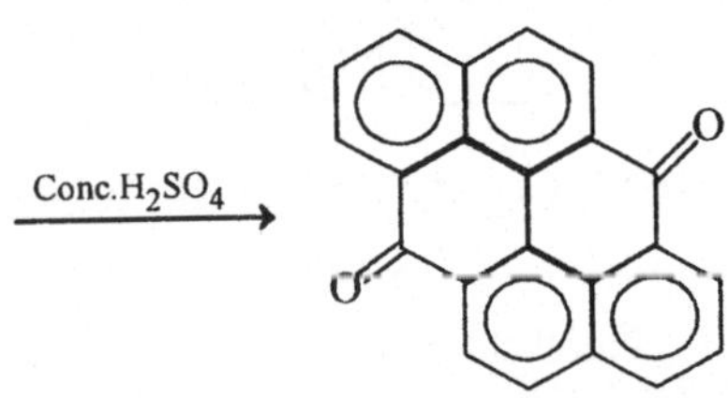

Anthanthrone

Anthanthrone and its halogen derivatives e.g., dichloroanthanthrone (indanthrene brilliant orange) and dibromoanthanthrone have been used as dyesheff.

3. *Preparation of diaryl-and triarylamines :* Ullmann reaction is used for synthesising diaryl-and triarylamines by refluxing an aryl halide, K_2CO_3 or KOH and another aromatic compound with a small amount of copper.

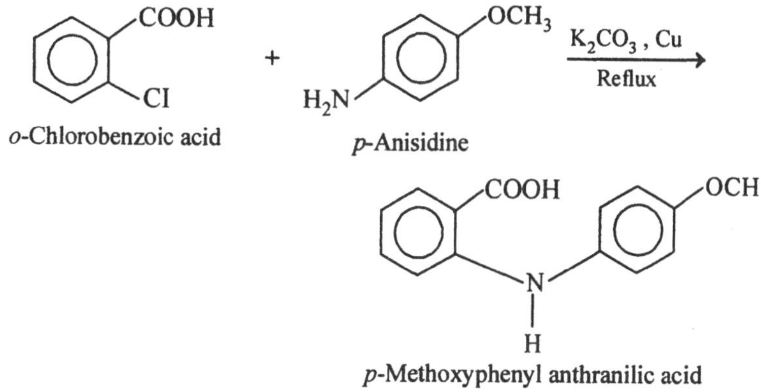

o-Chlorobenzoic acid *p*-Anisidine

p-Methoxyphenyl anthranilic acid

Other examples are as follows :

(*i*) $C_6H_5NHCOCH_3 + C_6H_5Br + K_2CO_3 \xrightarrow{Cu} (C_6H_5)_2NH + CH_3COOK + CO_2 + KBr$

(*ii*) $2(C_6H_5)_2NH + 2C_6H_5I + K_2CO_3 \xrightarrow{Cu} 2(C_6H_5)_3N + 2KI + CO_2 + H_2O$

(*iii*) $C_6H_5Br + C_6H_5OH + KOH \xrightarrow{Cu} (C_6H_5)_2O + K\,Br + H_2O$

4. *Preparation of diaryl ethers :* These are obtained by refluxing compounds and aryl halides in the presence of potassium hydroxide or potassium carbonate and copper. This reaction has been used in the synthesis of thyroxine.

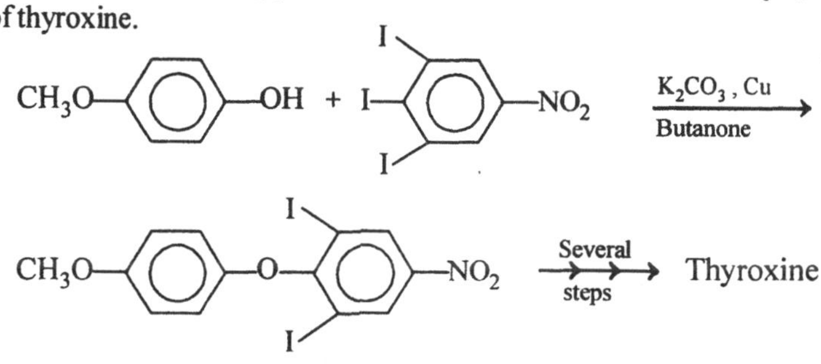

A diaryl ether

EXTENSION IN ULLMANN REACTION

(1) Ullmann ether synthesis : Phenol and phenol derivatives form diaryl ether with Aryl halides in the presence of $K_2CO_3/Cu/\Delta$.

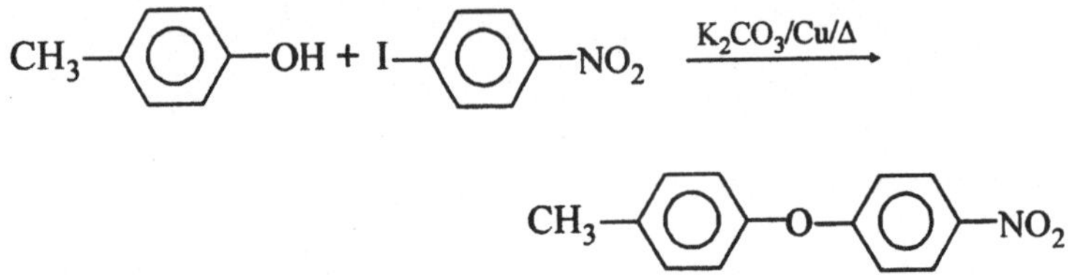

111

VILSMEIER REACTION OR VILSMEIER-HAACK

The reaction of an *N*-formyl derivative of secondary amine (DMF) and $POCl_3$ with a reactive aromatic compound affords a convenient method of directly introducing an aldehydic group into the aromatic compound.

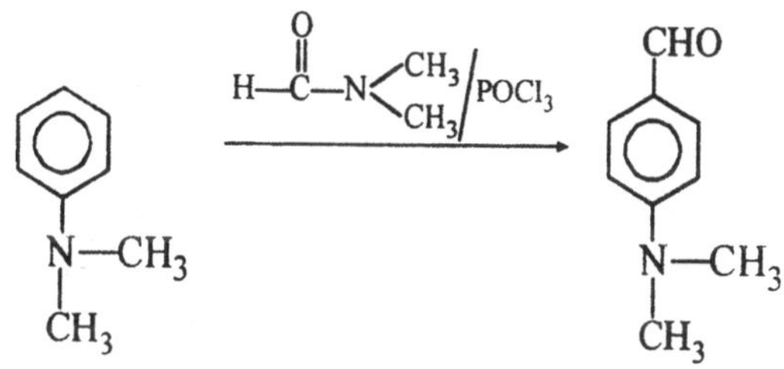

MECHANISM :

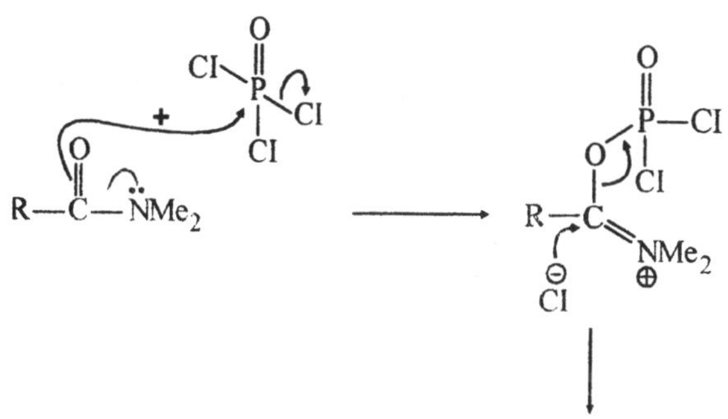

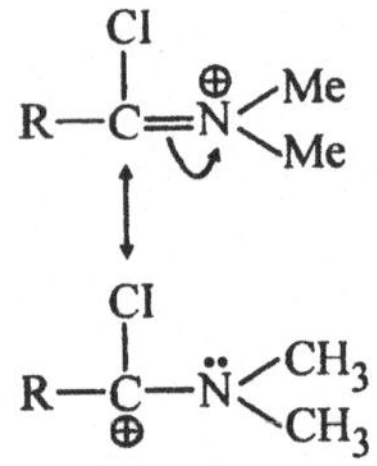

This species is an electrophile and hence can give electrophilic substitution reaction with aromatic compounds.

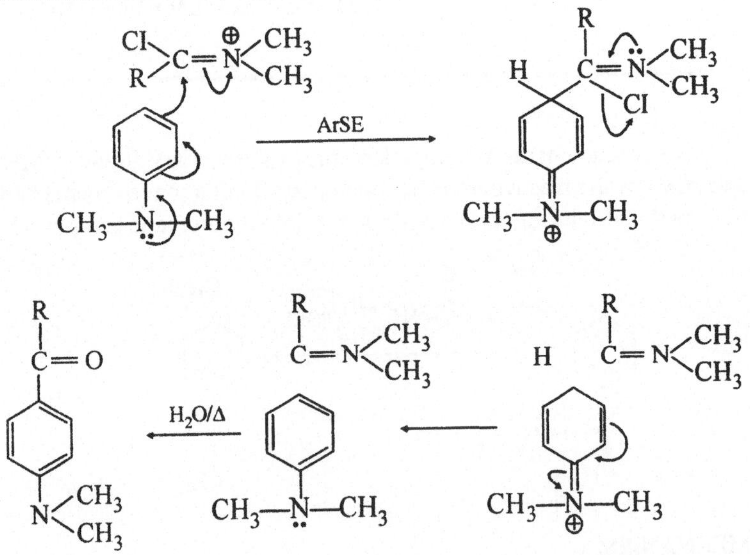

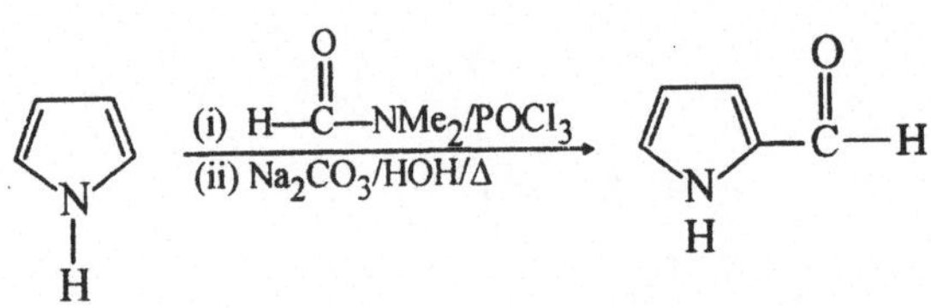

It is acid sensitive, hence hydrolysis is carried out in basic medium.

In this reaction the species, 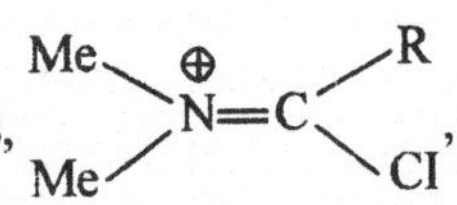, is known as Vilsmeier reagent. Vilsmeier reaction also takes place in the presence of other halides such as $SOCl_2$ and $COCl_2$. Similarly amides may be simple amides and substituted amides such as

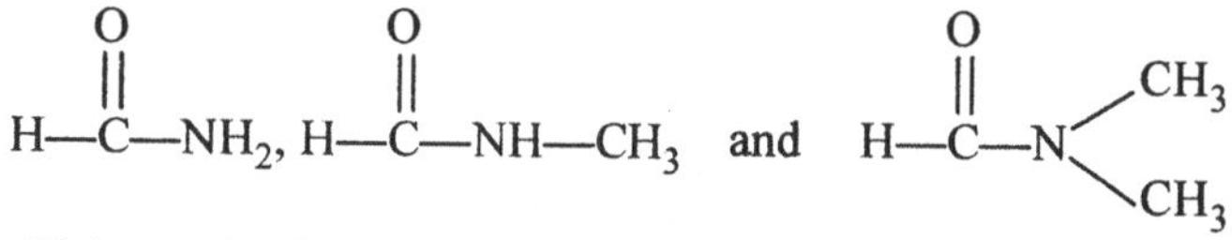

This reaction is an example of aromatic substitution reaction in which Vilsmeier reagent behaves as electrophile. The reaction is given by reactive aromatic compounds such as *N*, *N*-dialkyl aniline, phenols, pyrrole, thiophene, phenolic ethers and anthracene.

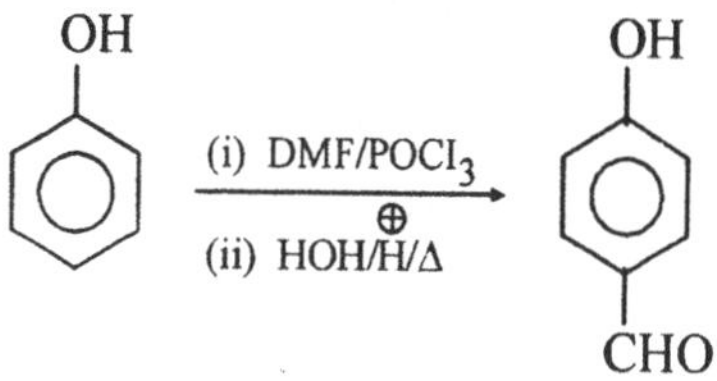

85%

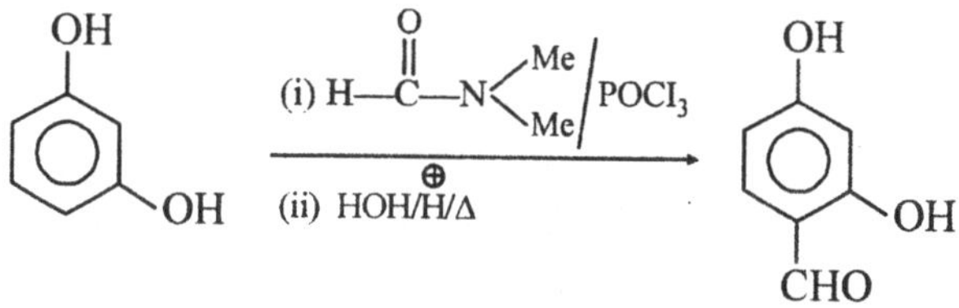

60%

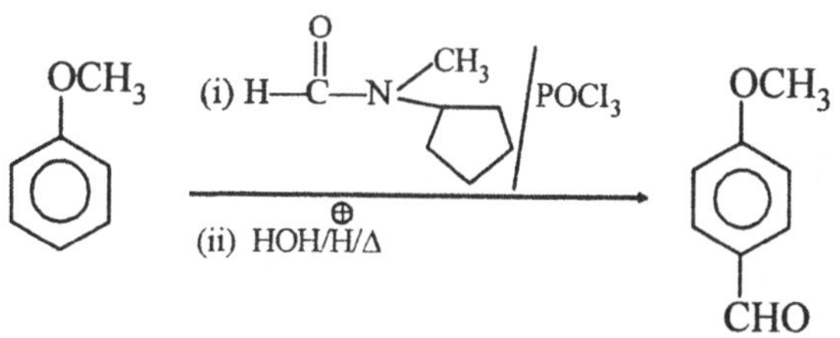

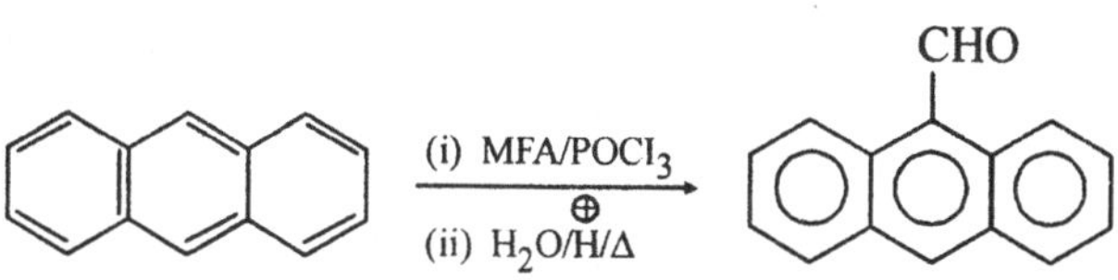

If substrate is sensitive to acid like pyrrole or thiophene, then this reaction is a very useful reaction. In such cases, hydrolysis is carried out in basic medium.

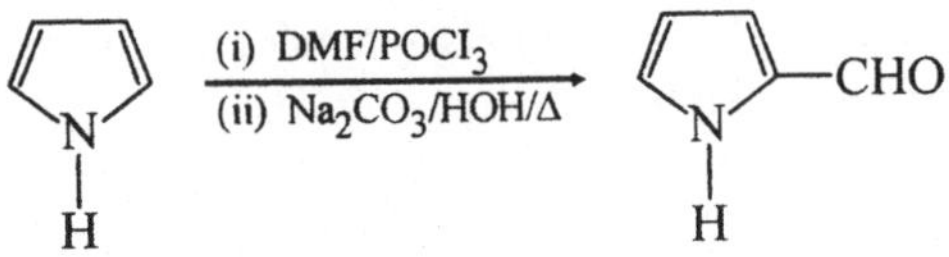

APPLICATIONS

It is used to prepare the aldehydes derived from phenols, ethers, anthracene etc.,

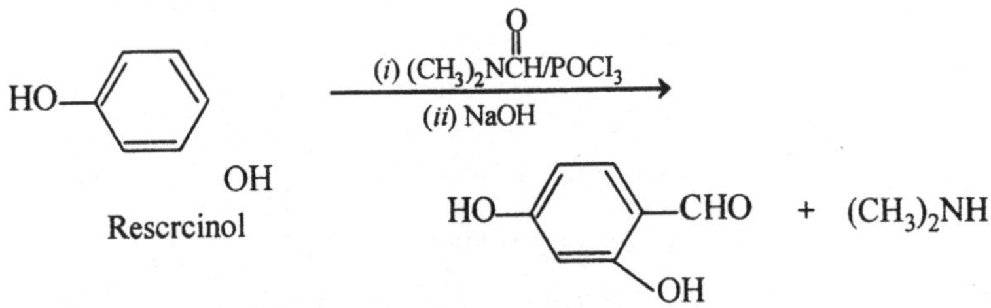

Rescrcinol

2,4-Dihydroxy benzaldehyde

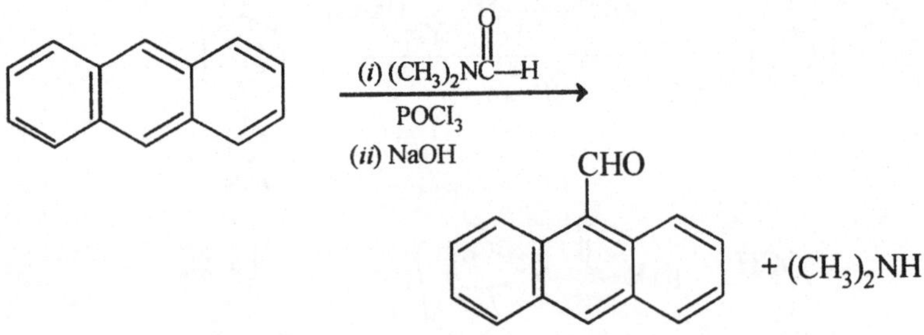

Anthracene-9-aldehyde
77-84%

112

VON BRAUN REACTION

Very familiar in the domain of alkaloid chemistry, Von Braun Reaction is frequently employed to cleave *tertiary* cyclic amines.

REACTION

It is essentially the reaction of a *tertiary* amine with cyanogen bromide—the products being an alkyl bromide and a disubstituted cyanamide

$$R_3N + Br—C \equiv N \longrightarrow R — Br + R_2N — C \equiv N$$

If the *tertiary* amine is cyclic, cleavage takes place opening up the chain

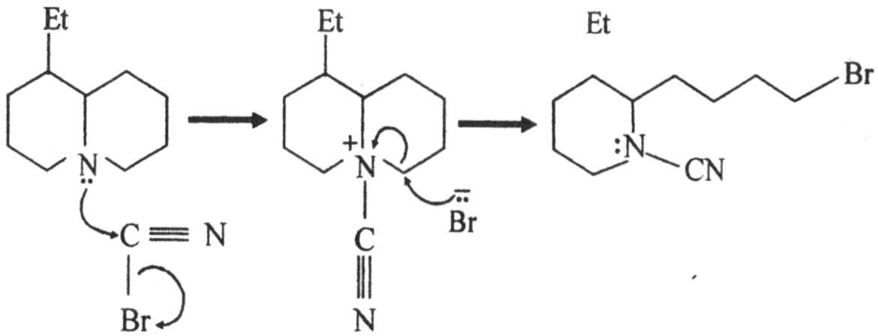

MECHANISM:

The mechanism is ionic and composed of two successive nucleophilic substitutions.

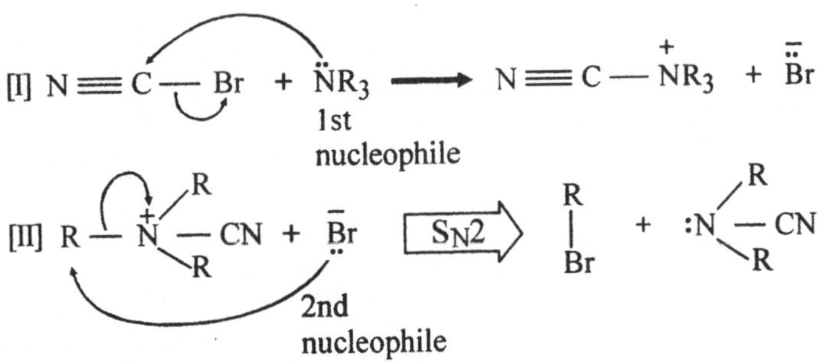

Note :

(I)For the simple alkyl groups, the least bulky groups are the ones most readily cleaved.

(II) One or two groups on the amines may be phenyl; but they do not cleave.

(III) This reaction is chiefly used to cleave cyclic amines.

(IV) *Secondary* amines also give this reaction but these are seldom met with success.

Problem 1 :

Put the missing figures

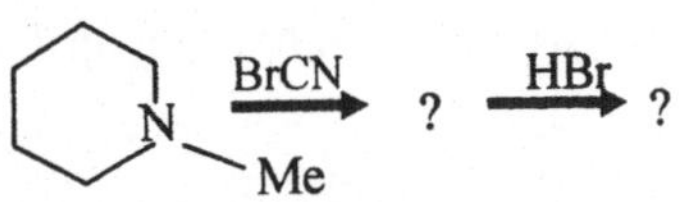

Solution :

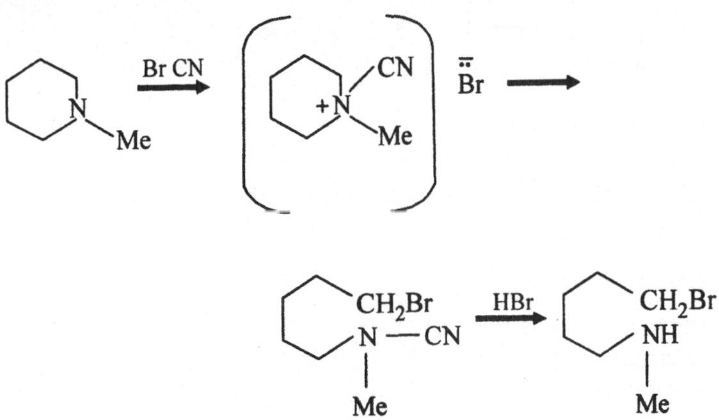

Problem 2 :

Predict the products

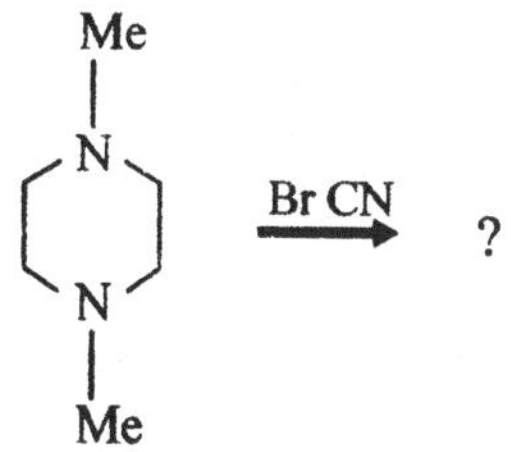

Solution :

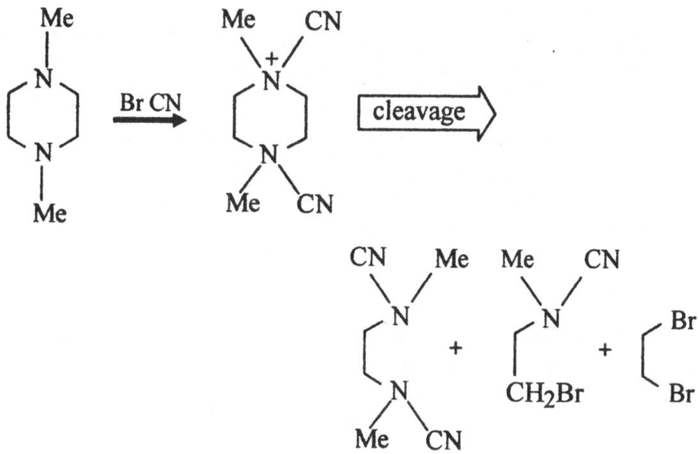

113

WAKCER PROCESS

Oxidation of ethylene, monosubstituted and 1, 2-disubstituted alkene into carbonyl compounds in the presence of $PdCl_2$, HOH is known as Wacker process for the synthesis of carbonyl compounds. The oxidation is usually carried out with oxygen in the presence of CuCl.

$$CH_2{=}\underset{\underset{H}{|}}{CH} + \frac{1}{2}O_2 \xrightarrow[CuCl]{PdCl_2} CH_3{-}\overset{\overset{O}{\|}}{C}{-}H$$

With longer chain monosubstituted alkenes more substituted carbon converts into carbonyl group.

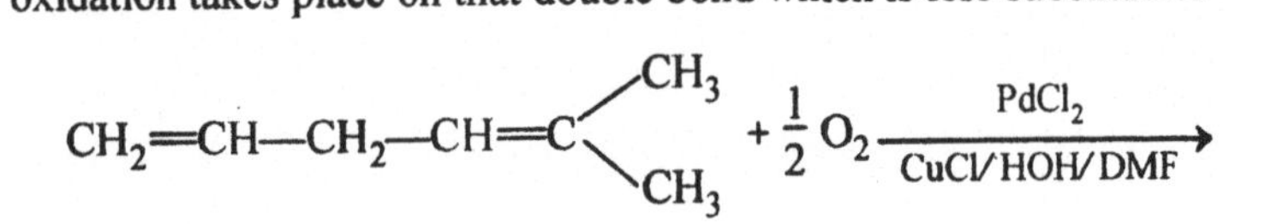

If compound has more than one carbon, carbon double bond then oxidation takes place on that double bond which is less substituted.

$$CH_2{=}CH{-}CH_2{-}CH{=}C\begin{matrix} \diagup CH_3 \\ \diagdown CH_3 \end{matrix} + \frac{1}{2}O_2 \xrightarrow[CuCl/HOH/DMF]{PdCl_2}$$

$$CH_3{-}\overset{\overset{O}{\|}}{C}{-}CH_2{-}CH{=}C\begin{matrix} \diagup CH_3 \\ \diagdown CH_3 \end{matrix}$$

114

WAGNER-MEERWEIN REARRANGEMENTS

These rearrangements take place in tertiary-amyl and neo-pentyl type of compounds, *i.e.*, highly branched compounds. When these compounds are subjected to reactions such as substitution, addition, elimination (dehydration, dehydrohalogenation), they give rise to products which are having different carbon skeleton from the parent skeleton. Such reactions which involve a change in carbon skeleton are called **Wagner-Meerwein rearrangements**. In these rearrangements, the change in the carbon skeleton occurs through the rearrangement of carbocation intermediate. Some typical examples of Wagner-Meerwein rearrangements are as follows :

(*i*) When neopentyl bromide is hydrolysed under S_N1 condition, the expected product is neopentyl alcohol. But it is not formed but instead of this 2-methyl-butan-2-ol and 2-methyl but-2-ene are formed.

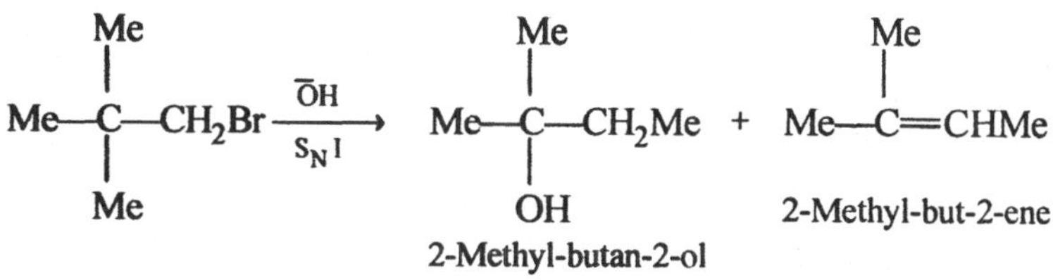

.......(i)

In this example, the change in the carbon skeleton is observed.

(*ii*) When the acid catalysed dehydration of neopentyl alcohol is carried out, it yield 2-methyl-but-2-ene and 2-methyl-but-1-ene.

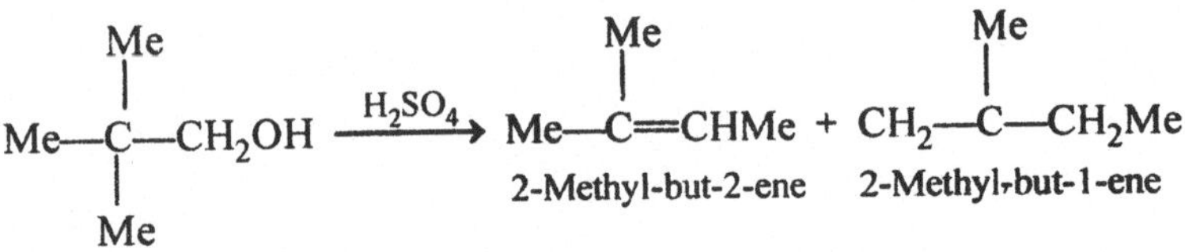

In this example, there also occurs the change in carbon-skeleton.

(*iii*) Wagner-Meerwein rearrangement was first observed in bicyclic terpenes.

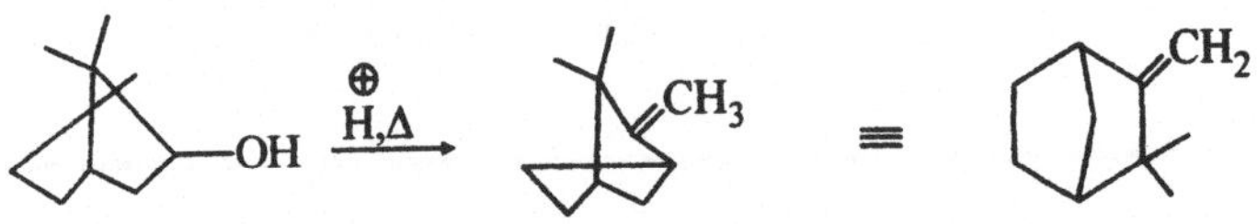

Isoborneol　　　　Camphene

(*iv*) Some more simpler examples of Wagner-Meerwein rearrangement are as follows :

(*a*) 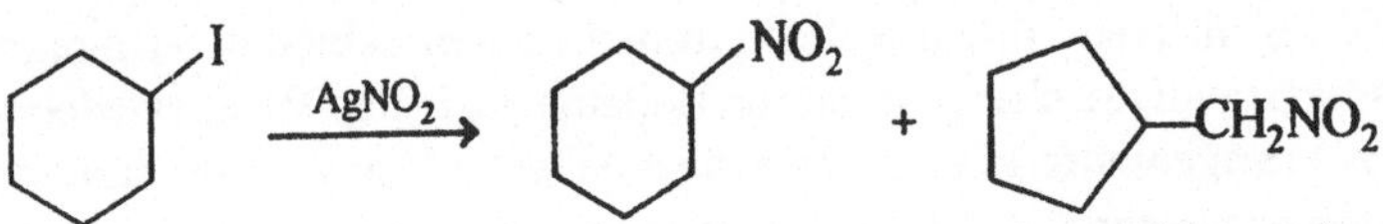

Iodocyclohexane　　Nitrocyclohexane　　Cyclopentylnitromethane

(*b*) 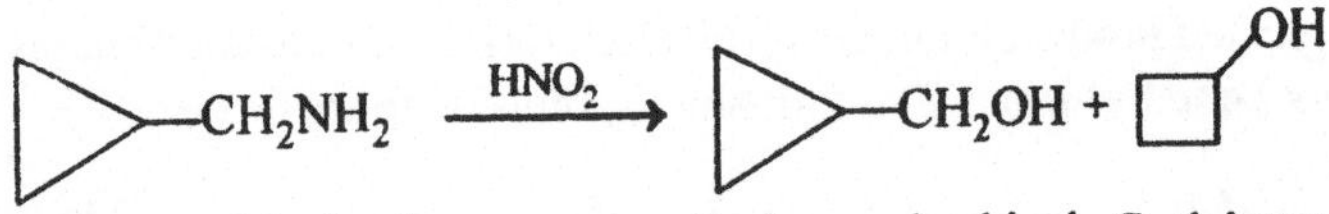

Cyclopropylcarbinylamine　　Cyclopropylcarbinol　Cyclobutanol

MECHANISM :

The protonation of the hydroxyl group of the alcohol followed by the loss of water molecule gives rise to the formation of 1° carbocation. This less stable 1° carbocation rearranges to the relatively more stable 3° carbocation by involving 1, 2-methyl shift before the product gets formed. The greater stability of 3° carbocation gives the necessary force for the migration of the methyl group from the adjacent carbon. The rearrangement carbocation then attains stability by losing a β-hydrogen to form olefin. Which of the two β-hydrogens will be eliminated will be governed by Zaitsev's rule.

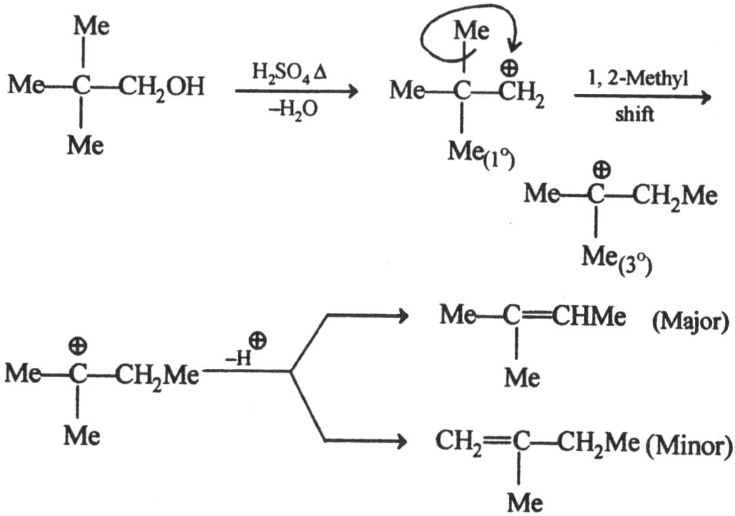

Similarly, hydrolysis of neopentyl halide first of all yields a rearranged 3° carbocation which can accept a nucleophilic from the solvent to form alcohol or eliminate a proton to form olefin (See Eq. (*i*)

If the initially formed carbocation gets stabilised by other factors, rearrangement does not take place. For example, 3, 3-dimethyl-2-bromobutane (neopentyl type) undergoes S_N 1 hydrolysis along with re-arrangement.

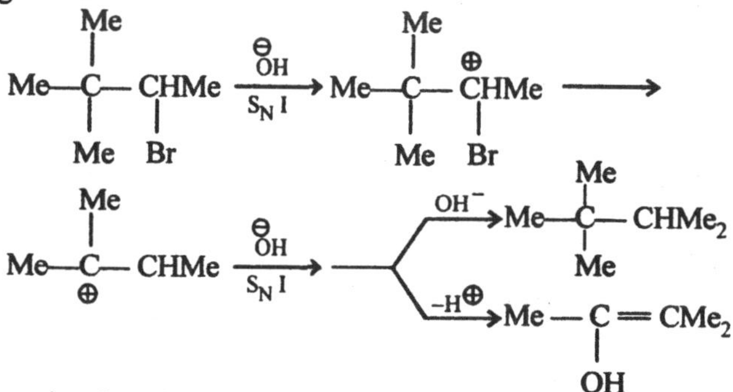

Whereas its phenyl analogue yields product without rearrangement.

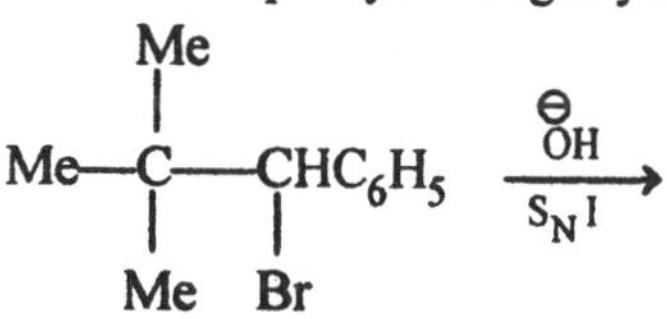

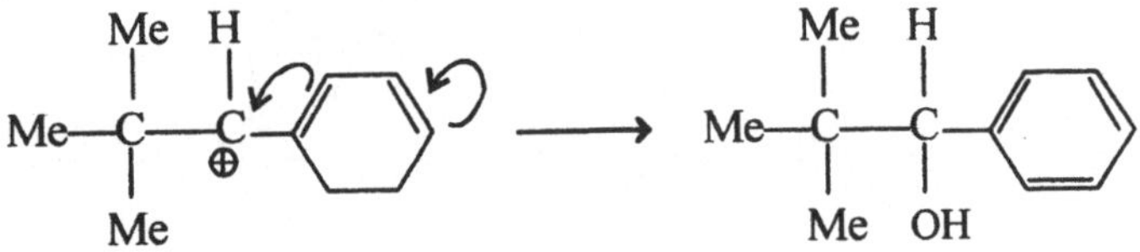

Benzilic cation
(stable)

The benzylic cation gets stabilised by the phenyl group through delocalisation and does not undergo rearrangements. Rearrangement in the alicyclic system may take place by migration of ring methylene group, giving rise to the ring expansion and contraction.

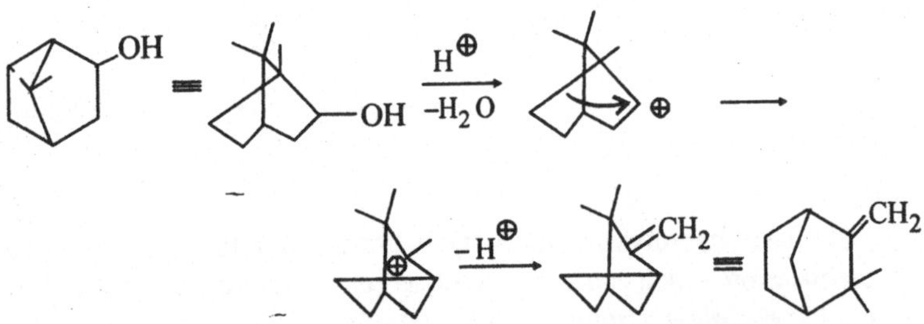

Camphene

Other examples include the conversions of camphene hydrochloride to isobornyl chloride, bornyl chloride to camphene, pinene hydrochloride to boryl chloride, etc.

In addition to alkyl and ring methylene migrating groups, there are also hydride ion or aryl group.

$$CH_3—CH_2—CH_2—Br + AlBr_3 \xrightarrow{\Delta} CH_3—\underset{H}{CH}—\overset{\oplus}{C}H_2 + \overset{\ominus}{A}lBr_4$$

$$\longrightarrow CH_3—\overset{\oplus}{C}H—CH_3 + \overset{\ominus}{A}lBr_4 \longrightarrow CH_3—CHBr—CH_3 + AlBr_3$$

The migration of the aryl group takes place at a much faster rate than alkyl group. The reason is that the electron-releasing group in the aryl group tends to increase the rate of migration whereas electron-withdrawing group tends to decrease the rate of migration. Therefore, elimination becomes many times faster in neophyl bromide than in neopentyl bromide. This is attributed to the simple fact that the rearranged phenonium ion becomes relatively more stable because of delocalisation than the rearranged carbocation from neopentyl bromide. Thus, the former is having lower energy and more readily formed than the latter.

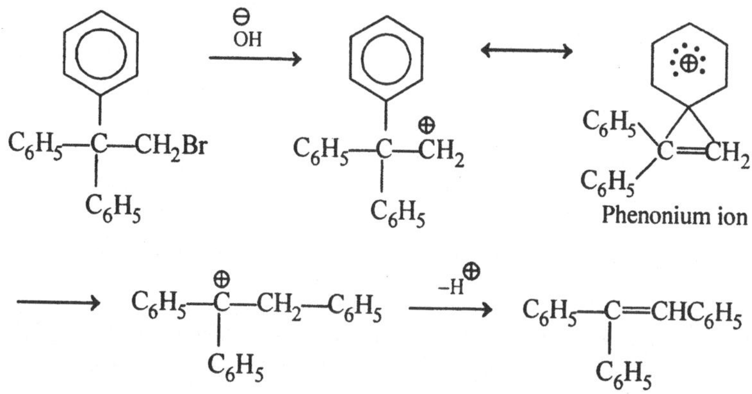

APPLICATIONS

Wanger-Meerwein rearrangements find many extensive applications. They are used for ring expansion or contraction. They are used in commercial synthesis of camphor from α-pinene, and the preparation of santene, present in sandalwood oil, from camphene.

115

WILLIAMSON ETHER SYNTHESIS

Preparation of ether (symmetrical as well as unsymmetrical) by SN2 reaction between alkoxide and alkyl halide (for alkyl sulphonate or alkyl sulphate) is known as Williamson ether synthesis.

$$R{-}\overset{\ominus\ \oplus}{ONa} + R'{-}L \xrightarrow{SN2} R{-}O{-}R' + \overset{\oplus\ominus}{Na\,L}$$

$$\downarrow$$

$$L{=}X, {-}OSO_2R'', {-}O{-}SO_2{-}OR''$$

The solution alkoxide can be prepared from alcohol and NaH.

$$R{-}O{-}H + NaH \longrightarrow R{-}\overset{\ominus\ \oplus}{ONa} + H_2$$

The reaction only works for alkyl halides (alkyl sulphonate or alkyl sulphate) that are active in the SN2 reaction. Best result is obtained when the alkyl halide, sulphonate or sulphate is primary (or methyl). If the substrate is tertiary, elimination is the exclusive result. Substitution is also favoured over elimination at lower temperature. Even secondary alkyl halides are of little use in this reaction. Alkoxides are strong bases and the elimination reaction is a prominent side reaction when see halide is used.

$$CH_3{-}CH_2{-}CH_2{-}\overset{\ominus\ \oplus}{ONa} + CH_3{-}CH_2{-}I \xrightarrow{SN2}$$

$$\underset{70\%}{CH_2{-}CH_2{-}CH_2{-}O{-}CH_2{-}CH_3}$$

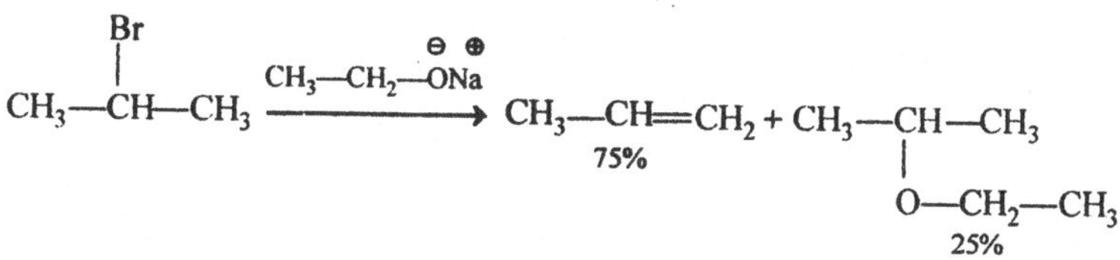

If alkoxide is a ter-alkoxide then all types of alkyl halides (except methyl halides) always give elimination reaction because the alkoxide is a very strong base.

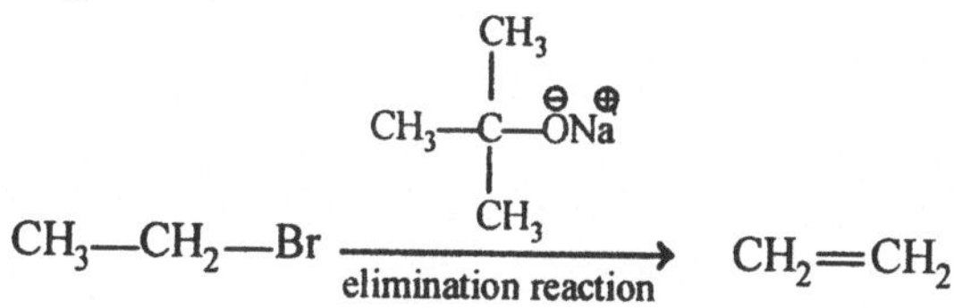

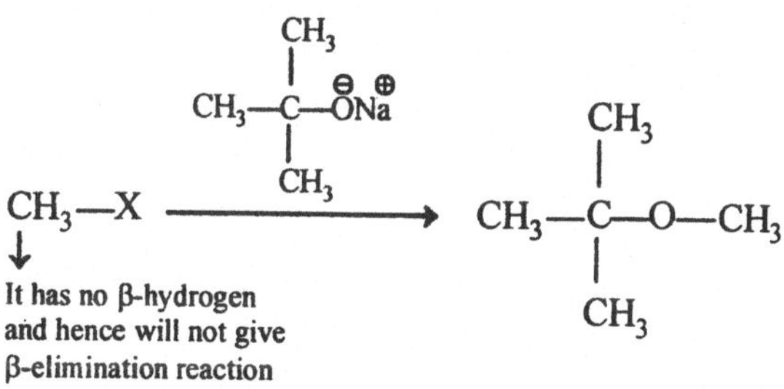

Halo alcohol undergoes intramolecular Williamson ether synthesis.

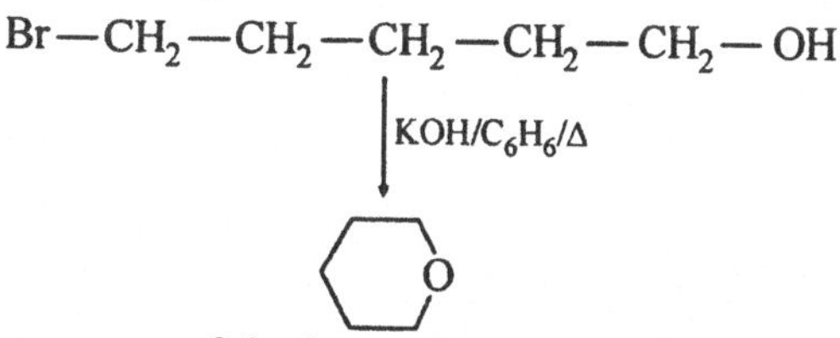

An important use of the intramolecular Williamson synthesis is in the making of oxiranes.

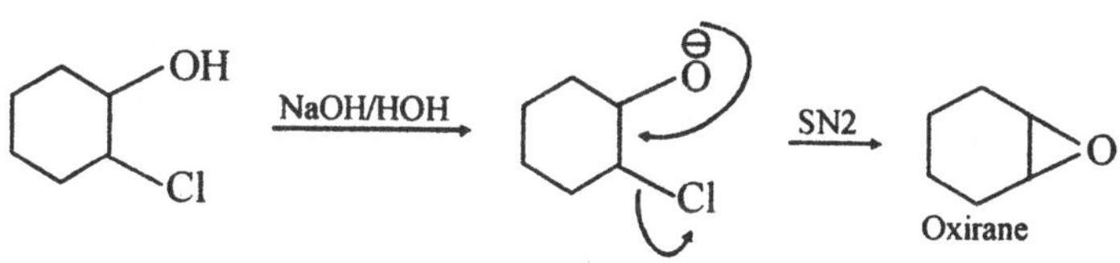

116

WITTIG REACTION

The reaction of aldehydes and ketones with phosphorus yield to produce an alkene is known as Wittig reaction.

An yield is a neutral species with a positive and negative charge on adjacent atoms and a phosphonium ylide carries its positive charge on phosphorus. Phosphorus ylides are prepared from triphenyl phosphine and alkyl halides having at least one α-hydrogen (*i.e.*, primary and secondary alkyl halides).

The first step is a nucleophilic substitution reaction. Triphenylphosphine is the excellent nucleophile and a weak base. It reacts readily with primary and secondary alkyl halide by an SN2 mechanism to give an alkyl phosphonium salt.

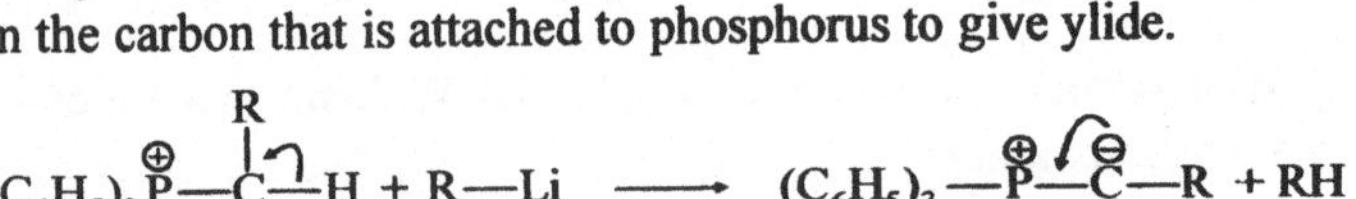

An alkyl triphenyl phosphonium halide

A strong base (usually R—Li, C_6H_5—Li, NaH or $NaNH_2$) removes a proton from the carbon that is attached to phosphorus to give ylide.

Phosphorous ylide

The carbon of the ylide is a nucleophile and gives nucleophilic addition with carbonyl compound to form betaine which leads to the formation of a four membered ring, an oxaphosphetane. Oxaphosphetane is strained and unstable and give alkene and triphenylphospheneoxide.

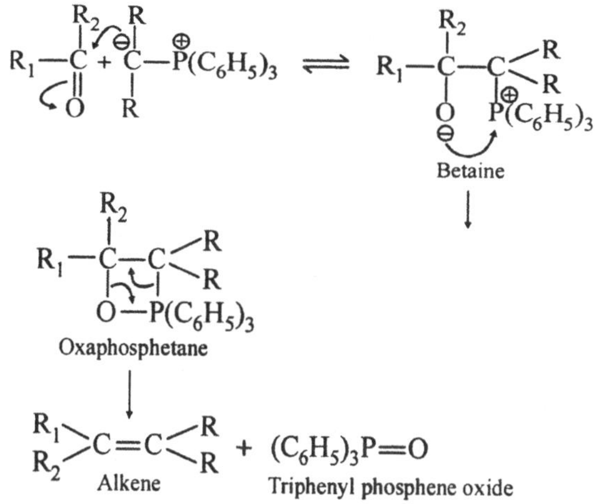

The Wittig reaction is applicable to a wide variety of compounds and although a mixture of (E) and (Z) isomers may result, the Wittig reaction gives a great advantage over most alkene synthesis in that no ambiguity exists as to the location of the double bond in the product. This is contrast to β-elimination reaction of alkyl halides which give multiple alkene products either by rearrangement or by β—C'S.

$$C_6H_5-CH_2-Cl \xrightarrow[\text{(ii) RLi} \atop \text{(iii) } C_6H_5CHO]{\text{(i) } (C_6H_5)_3P} C_6H_5-CH=CH-C_6H_5$$

$$CH_3-Cl \xrightarrow[\text{(ii) RLi} \atop \text{(iii) } C_6H_5CHO]{\text{(i) } (C_6H_5)_3P} C_6H_5-CH=CH_2$$

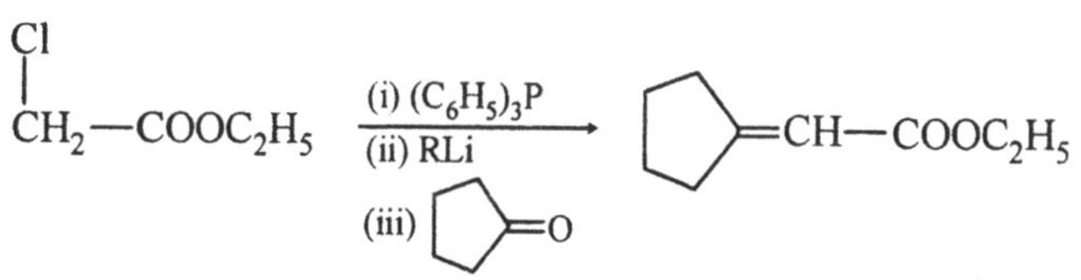

A widely used variation of the Wittig reaction is the Horner-Wadsworth Emmons modification. The **Horner Wadsworth-Emmons reaction** involves the use of phosphonate ester instead of triphenylphosphonium salt. In this reaction the major product is *E*-alkenes. Formation of phosphonate is known as Arbuzov reaction. Bases used to form the phosphonate ester carbonium includes NaH, potassium-*t*-butoxide and alkyl and phenyl lithium.

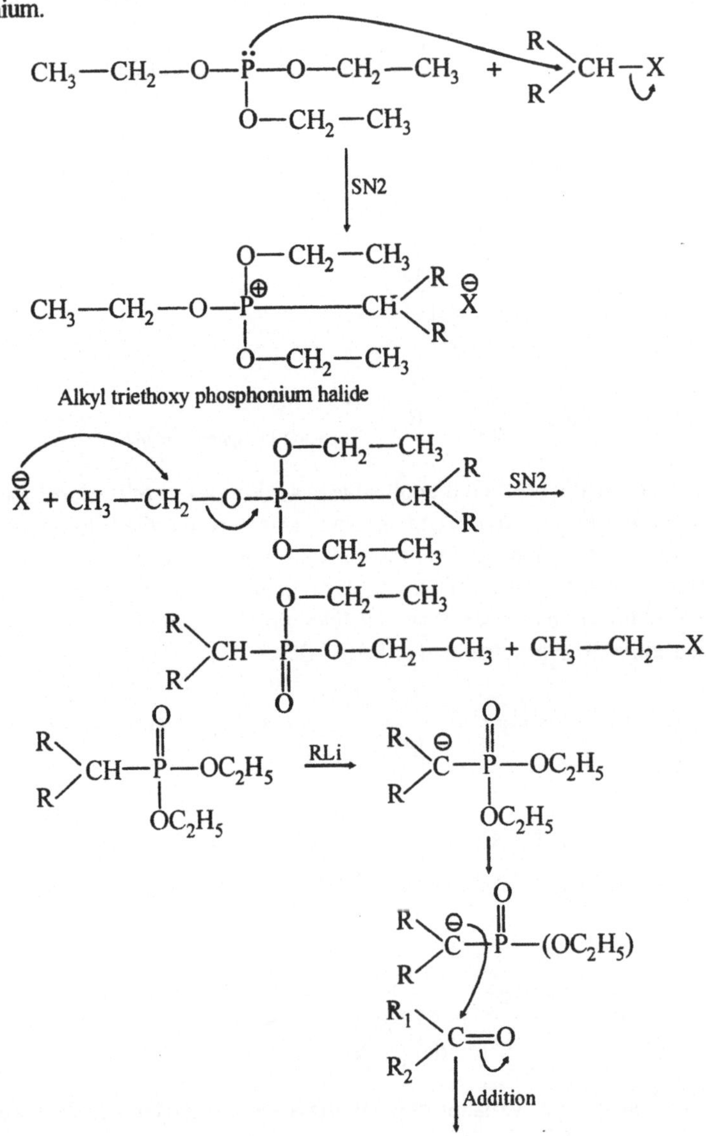

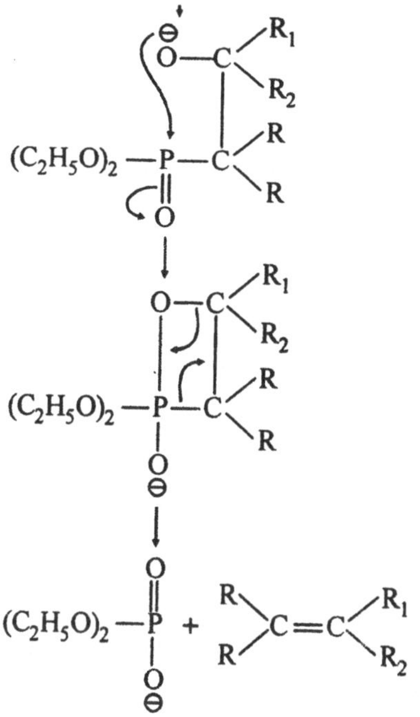

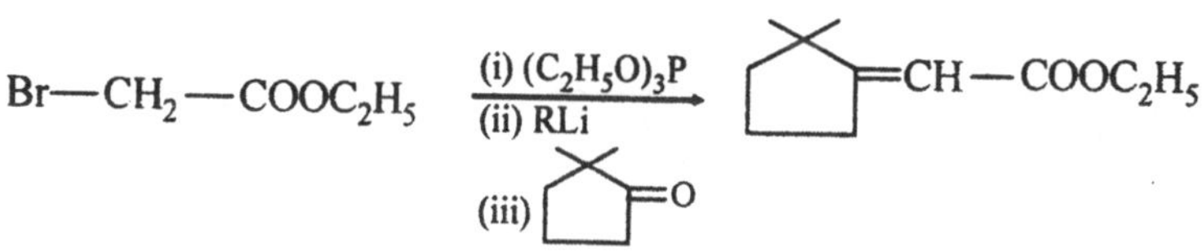

Note: Triphenylphosphine reacts with epoxide to give alkene because phosphorus has great affinity with oxygen.

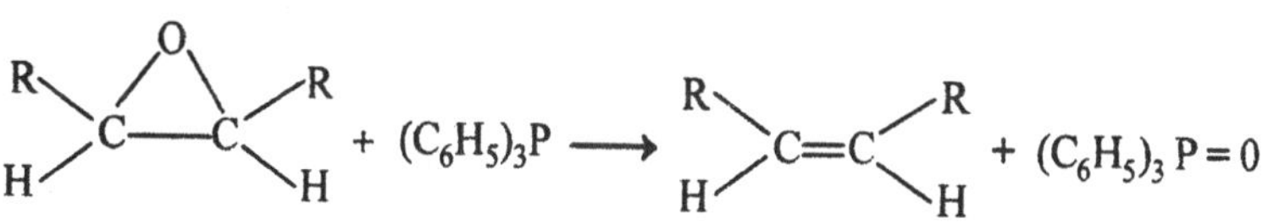

A modified Wittig reaction was introduced to overcome the difficulty in preparing alkenes having electron-withdrawing group in the alpha position. The modified Wittig reaction involves the attack of phosphonate ylide on the carbonyl carbon as carbanion.

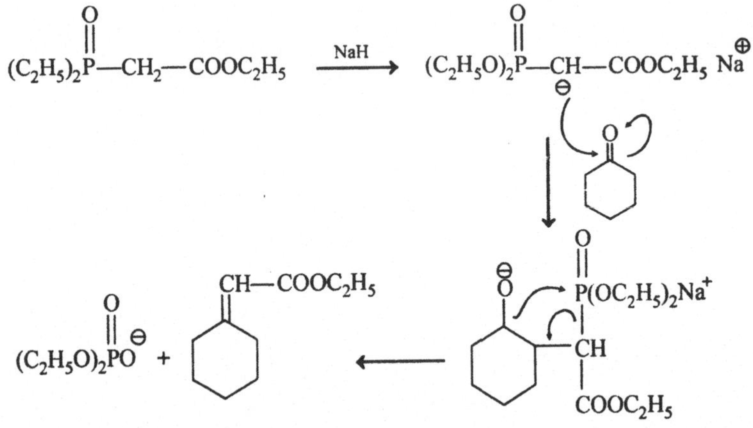

APPLICATIONS:

Wittig reaction is of great synthetic importance. The following examples illustrate the importance of this reaction:

1. Preparation of olefins : Wittig reaction is used for the preparation of olefins in which the position of the double bond is definite, *e.g.*

(a) $\underset{\text{Acetophenone}}{(C_6H_5)(CH_3)C{=}O} \xrightarrow{CH_2{=}PPh_3} \underset{\alpha\text{-Methylstyrene}}{(C_6H_5)(CH_3)C{=}CH_2}$

(b) $C_6H_5.CH{=}O \xrightarrow{CH_2{=}PPh_3} \underset{\text{Styrene}}{C_6H_5.CH{=}CH_2}$

(c) $C_6H_5CHO \xrightarrow{Ph_3P{=}CH.COOC_6H_5} \underset{\text{Ethyl cinnamate}}{C_6H_5.CH{=}CH.COOC_2H_5}$

(d) $C_6H_5.CH{=}CH.CHO \xrightarrow{Ph_3P{=}CH_2} \underset{\text{1-Phenylbutadiene}}{C_6H_5.CH{=}CH.CH{=}CH_2}$

(e) $C_6H_5CHO \xrightarrow[\text{(Vinylmethyleneylide)}]{Ph_3P{=}CH.CH{=}CH_2} \underset{\text{1-Phenylbutadiene}}{C_6H_5.CH{=}CH.CH{=}CH_2}$

2. Formation of exocyclic double bond : It is exemplified as follows :

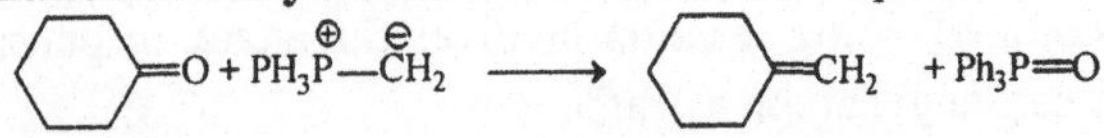

Cyclohexanone   Methylene cyclohexane

This method of introducing exocyclic double bond has been widely used in the preparation of methylene steroids.

3. Stepping up of aldehydes : By using Wittig reaction Kochetkov *et. al.* (1965) stepped up and aldose by two carbon atoms.

$$\underset{\text{Carbethoxymethylenephosphorane}}{RCHOH.CHO + Ph_3P{=}CH.COOC_2H_5} \longrightarrow$$

$$RCHOH.CH{=}CHCOOC_2H_5 \xrightarrow[HClO_4]{O_sO_4}$$

$$[R.CHOH.CHOH.CHOH.COOH] \longrightarrow$$

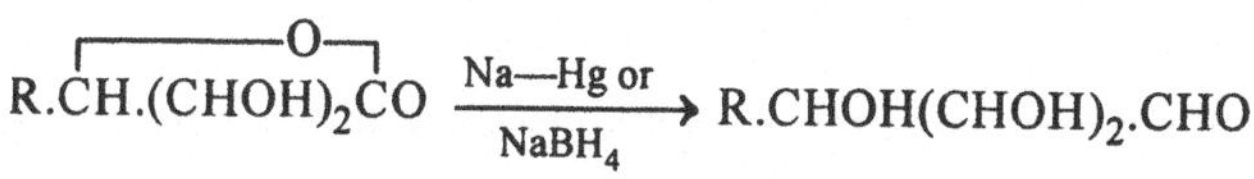

4. Preparation of carbodi-imides : Carbodi-imides, the important reagents used during the synthesis of polypeptides and nucleic acids are prepared from isocyanates in the following way.

$$RN{=}C{=}O + R'_3\,P{=}O \longrightarrow RN{=}PR'_3 + CO_2$$

$$RN{=}C{=}O + RN{=}PR'_3 \longrightarrow \underset{\text{Carbodi-imide}}{RN{=}C{=}NR} + R'_3\,P{=}O$$

Here R is alkyl, alicyclic or aromatic group *e.g.* in dicyclohexyl carbodi-imide, $R{=}C_6H_{11}$(cyclohexyl).

5. Preparation of β, γ-unsaturated acids : It is exemplified as follows :

$$R_2C{=}O + PH_3\,\overset{\oplus}{P}{-}\overset{\ominus}{C}H{-}CH_2CO\overset{\ominus}{O} \longrightarrow$$

$$R_2C{=}CHCH_2CO\overset{\ominus}{O} + PH_3P{=}O$$

In all other methods isomerization to α, β-unsaturated acids occurs.

6. Formation of large rings containing 5 to 16 carbons : It is exemplified as follows :

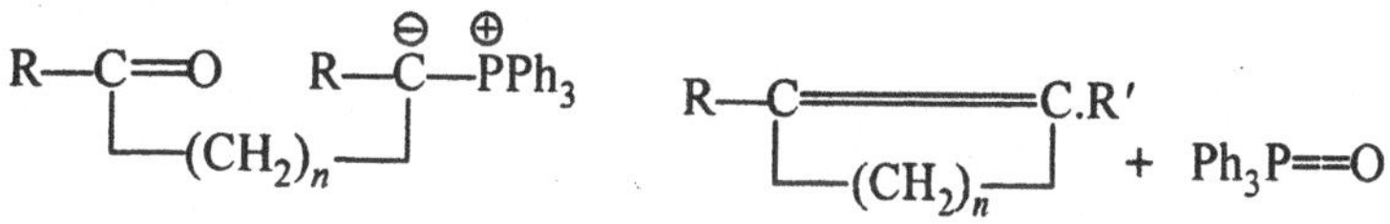

7. Synthesis of vinyl halides : Chloromethylenetriphenylphosphorane ($Ph_3P{=}CHCl$) needed for this synthesis is formed by reacting chlorocarbene with triphenylphosphine.

$$:CHCl + Ph_3P \longrightarrow \overset{\ominus}{Ph_3}\overset{\oplus}{P}-\overset{\ominus}{\ddot{C}}HCl$$

(*i*)

$$C_6H_5CHO + Ph_3\overset{\oplus}{P}-\overset{\ominus}{C}HCl \longrightarrow C_6H_5CH{-}CHCl + Ph_3P{=}O$$

(*ii*)

8. Synthesis of ethers : Methoxymethylenetriphenylphosphorane undergoes reaction with carbonyl compounds to yield diphenyl substituted vinyl methyl ethers.

$$Ph_2C{=}O + Ph_3P{=}CHOCH_3 \longrightarrow Ph_2C{=}CHOC_3$$

Hydrolysis of the product yields aldehyde.

$$Ph_2C{=}CHOCH_3 \xrightarrow{H_2O} \underset{\text{Diphenyl acetaldehyde}}{Ph_2CH{-}CHO}$$

9. Synthesis of natural products : Wittig reaction has been used for synthesising a number of naturally occurring compounds *e.g.* bisabolene, squalene (terpenoids), β-carotene, lycopene., crocetin (pigments), bixin esters, vitamin A_1, vitamin D_3, corticoids, etc.

(i) β-carotene (pro vitamin A)

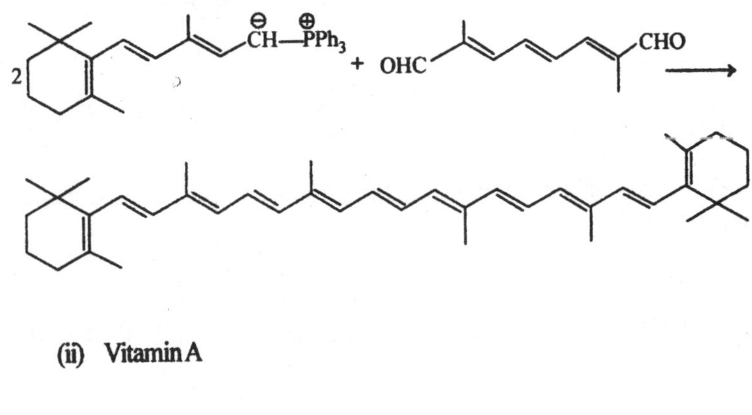

(ii) Vitamin A

β-Carotene

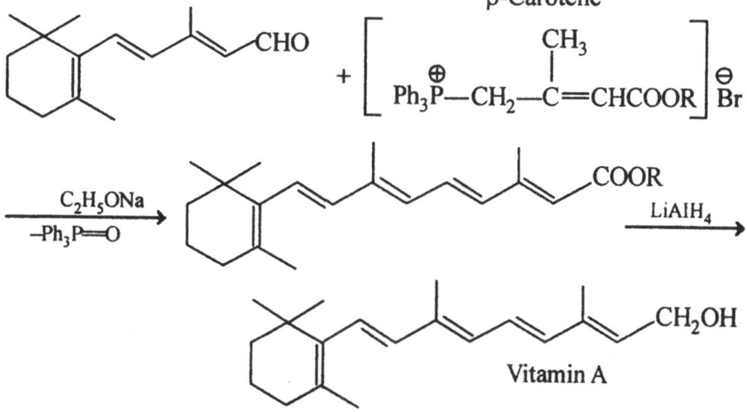

Vitamin A

Problem 1 : Show how you would use a Witting reaction to synthesize 1-phenyl-1, 3-butadiene.

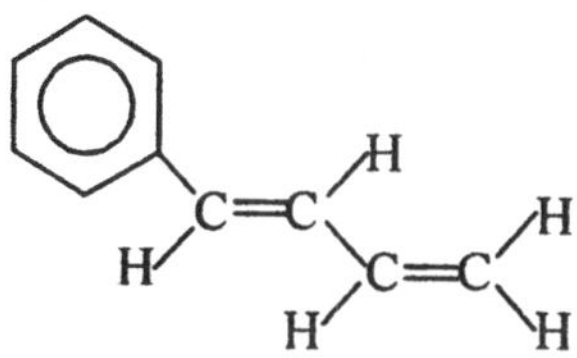

1-Phenyl-1, 3-butadiene

Hint : This molecule has two double bonds that might be formed by Witting reactions. The central double bond could be formed by a Witting reaction in either of two ways; both of these syntheses will probably work, and both will produce a mixture of *cis* and trans isomers.

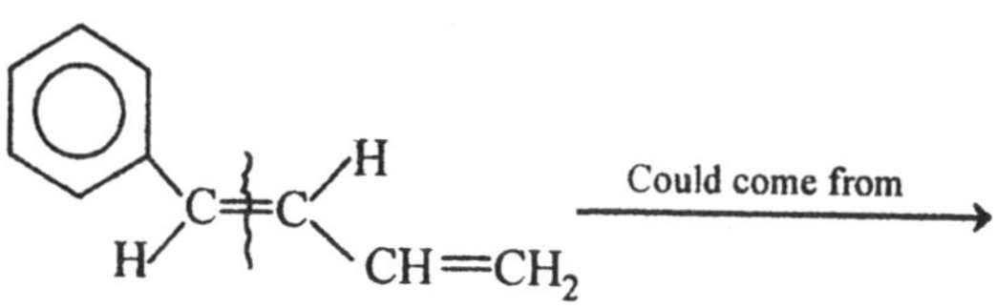

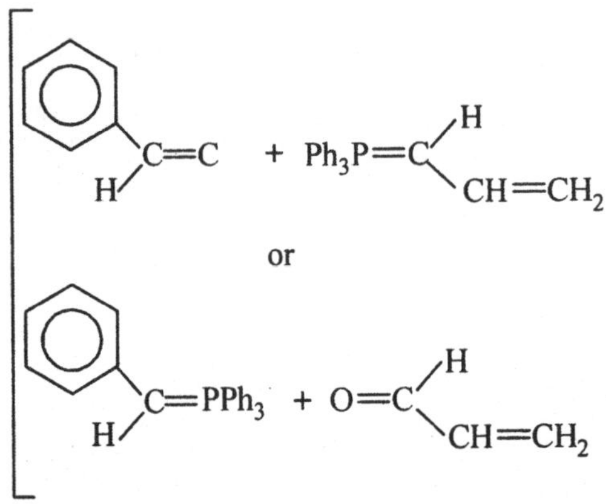

Problem 2 : Show how a Wittig reaction might be used to synthesize each of the following compounds. In each case, start with an alkyl halide and a ketone or an aldehyde.

(a) $Ph-CH=C(CH_3)_2$

(b) $Ph-C(CH_3)=CH_2$

(c) $Ph-CH=CH-CH=CH-Ph$

(d) $(CH_3)_2C=CH-(CH_2)_3-CH_3$

(e) 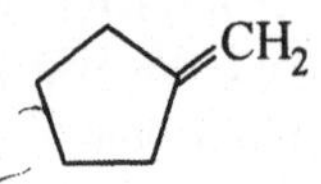 (f) 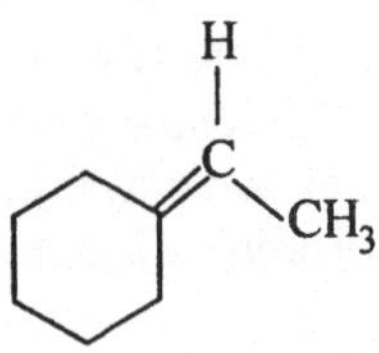

Solution : (a) $PhCH_2Br + PH_3P$

(b) $CH_3I + Ph_3P$, then acetophenone

(c) $PhCH_2Br + Ph_3P$, then cinnamaldehyde

(d) 1-Bromopentance + Ph_3P, then acetone

(e) $CH_3I + Ph_3P$, cyclopentanone

(f) Ethyl bromide + Ph_3P, then cyclohexanone

117

WOLFF-KISHNER REDUCTION

Reduction of $-\overset{\overset{\large O}{\|}}{C}-$ group of aldehydes and ketones into $-CH_2-$ group in the presence of excess hydrazine and strong base at elevated temperature is known as Wolff-Kishner reduction. Diethylene glycol (b.p. 245°C) is a convenient solvent for the reaction. The water formed during the reaction lowers the boiling out of water.

The W.K. reduction takes place is strongly basic solution and can be used for those compounds that are sensitive to acid. This reaction is not suitable for those compounds which are sensitive to alkali.

$$R-\overset{\overset{O}{\|}}{C}-R' \xrightarrow[\text{HO CH}_2\text{ CH}_2\text{ O CH}_2\text{ CH}_2\text{OH}]{NH_2-NH_2/\text{strong base}/\Delta} R-CH_2-R' + N_2 + HOH$$

Unlike Clemmensen reduction, the reaction does not fail with acid-sensitive or high molecular weight substrates.

Some typical examples of Wolff-Kishner's reduction are as follows :

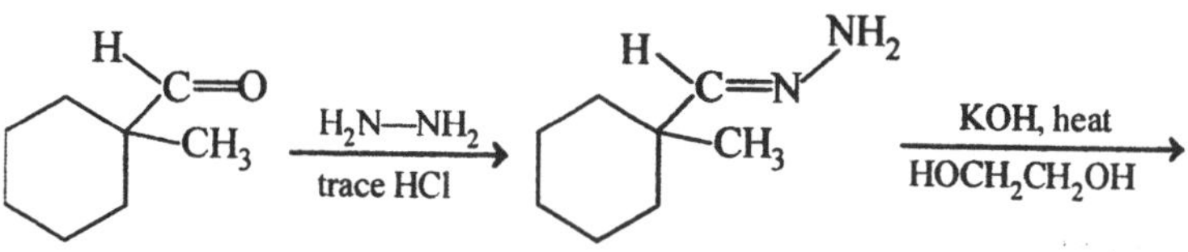

Hydrazone

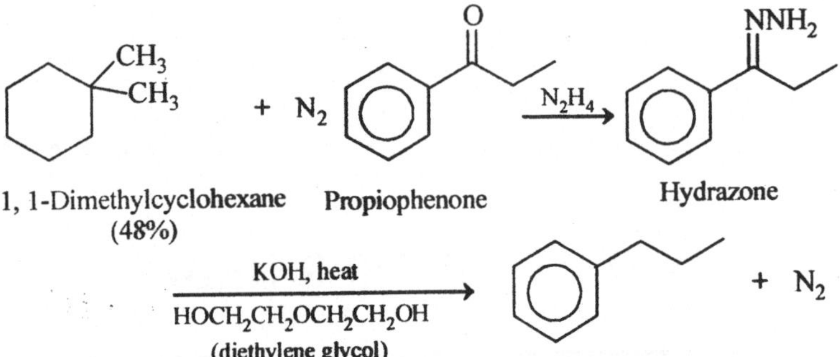

1, 1-Dimethylcyclohexane (48%) Propiophenone Hydrazone

n-Propylbenzene (82%)

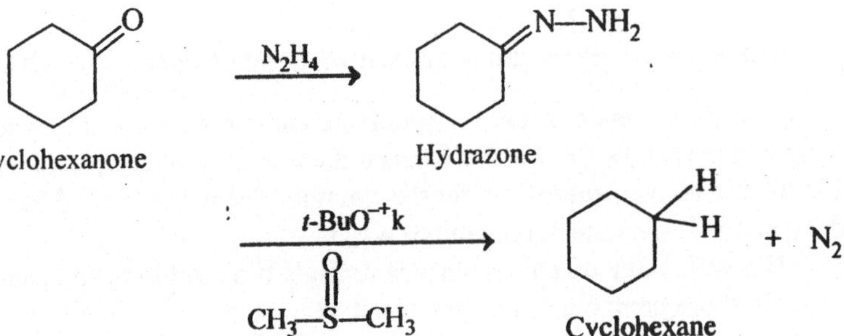

Cyclohexanone Hydrazone

Cyclohexane (80%)

It is to be noted that during the reduction of semicarbazones or azines the derivatives are first converted to hydrazones, e.g.,

$$\begin{matrix} R \\ R \end{matrix}\!\!>C{=}N.NH.CONH_2 + H_2O \longrightarrow \begin{matrix} R \\ R \end{matrix}\!\!>C{=}N.NH_2$$

Semicarbazone Hydrazone

$$\begin{matrix} R \\ R \end{matrix}\!\!>C{=}N{-}N{=}C\!\!<\begin{matrix} R \\ R \end{matrix} + H_2N.NH_2 \longrightarrow \begin{matrix} R \\ R \end{matrix}\!\!>C{=}N.NH_2 + N_2$$

Azine Hydrazine Hydrazone

MECHANISM:

Step I.

$$R{-}\underset{R}{\underset{|}{C}}{=}O + NH_2{-}NH_2 \underset{}{\overset{\text{Slow}}{\rightleftharpoons}} \begin{matrix} R \\ R \end{matrix}\!\!>C{=}\ddot{N}{-}NH_2 + H_2O$$

In this reaction hydrazone is first isolated in the free state by removal of water.

Step II. In this step tautomerisation and elimination of nitrogen takes place in the presence of strong base.

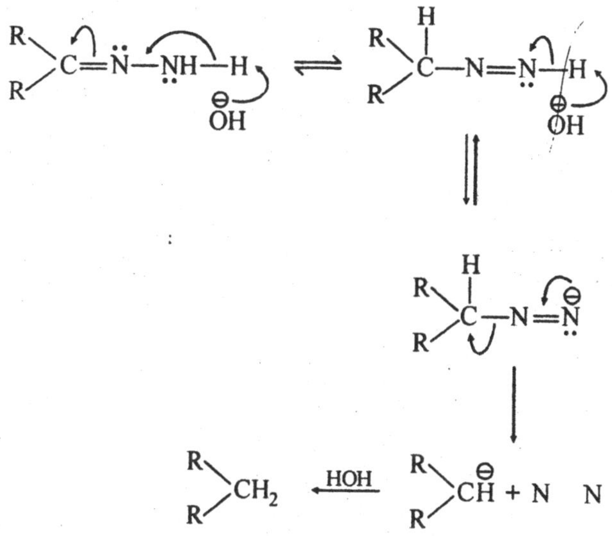

W.K. Reduction takes place at very high temperature but it is now more conveniently effected by heating a mixture of carbonyl compound hydrazine hydrate and sodium or potassium hydroxide in a high-boiling solvent (diethylene glycol) at 180-200°C. Excellent yield of the production is often obtained. With potassium *t*-butoxide in dimethyl sulphoxide, reduction takes place at room temperature. This modification is known as **Huang-Minlon** modification and the entire reaction takes place in one pot.

$$C_6H_5{-}\overset{\overset{\displaystyle O}{\|}}{C}{-}C_6H_5 \xrightarrow{NH_2{-}NH_2.H_2O/K\text{-}t\text{-buloxide/DMSO}} \underset{\text{90\% yield}}{C_6H_5{-}CH_2{-}C_6H_5}$$

$$C_6H_5{-}\overset{\overset{\displaystyle O}{\|}}{C}{-}CH_2{-}CH_3 \xrightarrow[\text{(ii) NaOH/diethylene glycol/200°C}]{\text{(i) }NH_2NH_2} \underset{82\%}{C_6H_5{-}CH_2{-}CH_2{-}CH_3}$$

Reduction of conjugated, unsaturated carbonyl compounds is sometimes accompanied by a shift in the position of double bond.

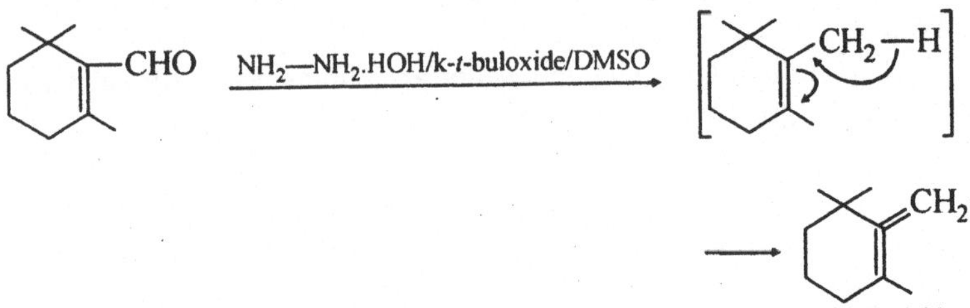

80% yield

Allylic hydrogen is acidic; hence in the presence of base, rearrangement takes place.

With many carbonyl compounds having a leaving group at α-position, elimination accompanies reduction.

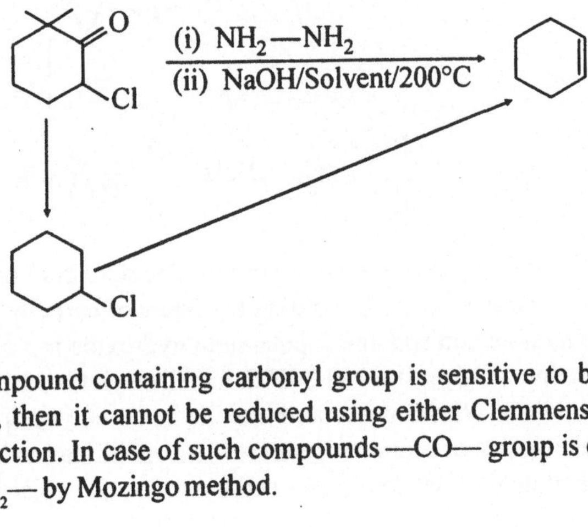

If compound containing carbonyl group is sensitive to both acids and bases, then it cannot be reduced using either Clemmenson or the W.K. reduction. In case of such compounds —CO— group is converted into —CH_2— by Mozingo method.

$$R{-}\overset{\overset{\displaystyle O}{\|}}{C}{-}R \xrightarrow[\text{(ii) } H_2\text{—Ni}]{\text{(i) SH—}CH_2\text{—}CH_2\text{—SH/}Et_2O/BF_3} R{-}CH_2{-}R$$

$$R{-}\overset{\overset{\displaystyle O}{\|}}{C}{-}R \xrightarrow[\text{(ii) } NH_2\text{—}NH_2/\Delta]{\text{(i) SH—}CH_2\text{—}CH_2\text{—SH/}Et_2O/BF_3} R{-}CH_2{-}R$$

The above sequence of reaction is known as **Mozingo reaction**.

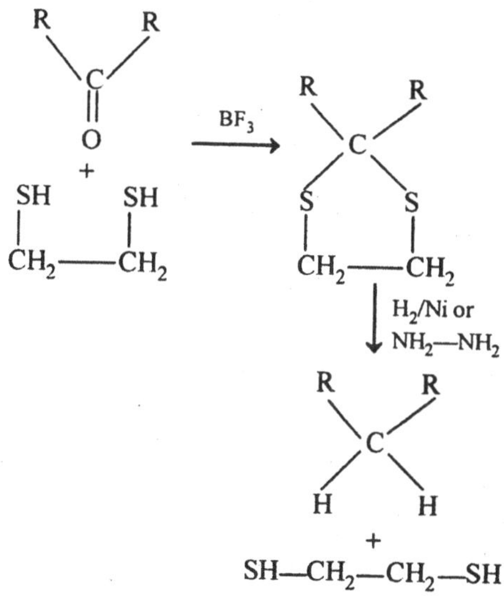

APPLICATIONS :

Wolff-Kishner reduction is used mainly for the reduction of carbonyl group (>C=O) to methylene (>CH_2) during the synthesis of many natural products. This method unlike Clemmenson reduction is applicable for the reduction for the reduction of high molecular weight and acid-sensitive compounds.

(a) *Reduction of camphor to camphane :*

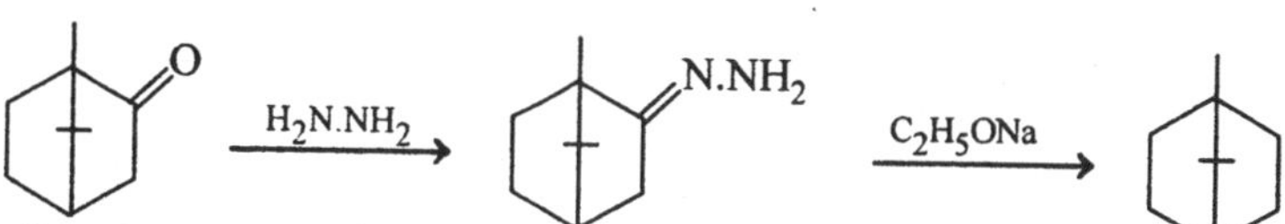

Camphor Camphane

(b) *Elucidating the structure of oestrone :* When oestrone methyl ether is subjected to Wolff-Kishner reduction followed by selenium dehydrogenation, it yields a compounds, 7-methoxy-1, 2-cyclopentenophenanthrene, of known structure.

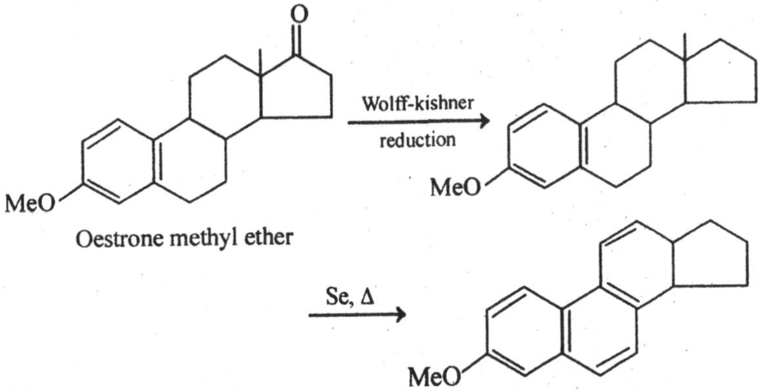

Oestrone methyl ether

7-Methoxy-1 ; 2-cyclopentan-phenanthrene

(I)

The known structure of degraded product, I clearly reveals the carbon skeleton of oestrone. Further, it also ascertains the position of the phonoelic group in oestrone.

(c) *Reduction of high molecule weight compounds* : This reaction has been successfully used for the reduction of high molecular weight compounds such as cholesterol.

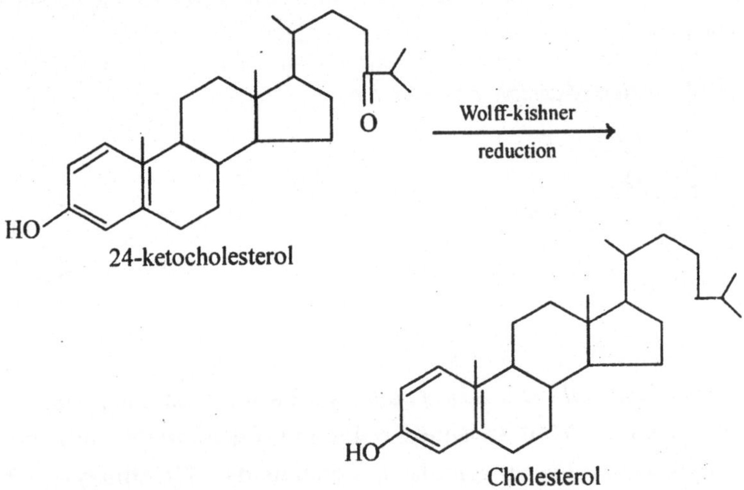

24-ketocholesterol

Cholesterol

(I)

(d) *Introduction of long straight-chain alkyl groups of aromatic rings :* By using Friedel-Crafts alkylation, it is not possible to introduce a straight chain alkyl group longer than ethyl group into an aromatic ring because of rearrangement of the alkyl group. However, this has been achieved by acylation followed by reduction.

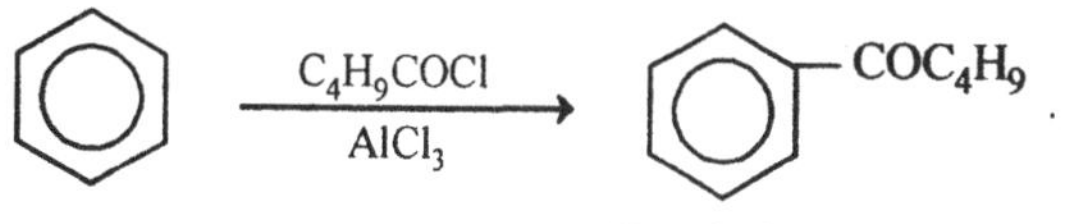

n-Butyl phenyl ketone

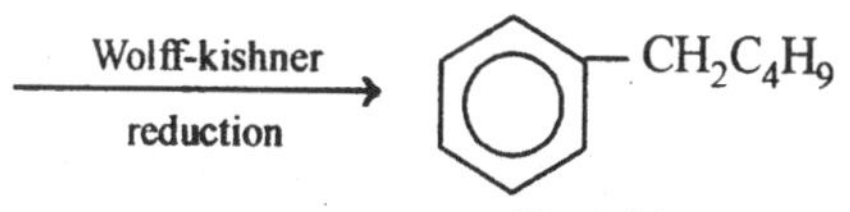

n-Pentyl benzene

(e) *Synthesis of pyrroles :* Structure of the pyrroles (degraded products of haemin and chlorophyll) has been ascertained by their synthesis which involve Wolff-Kishner reduction. An example in the synthesis of opsopyrrole.

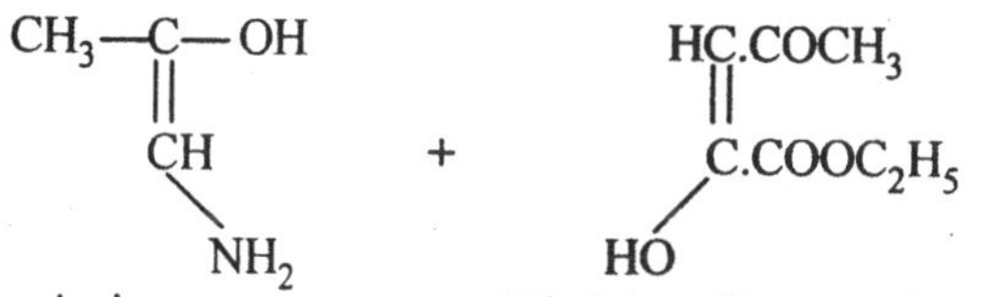

Amino acetone (enolic form) Ethyl-2, 4-diketopentanoate (enolic form)

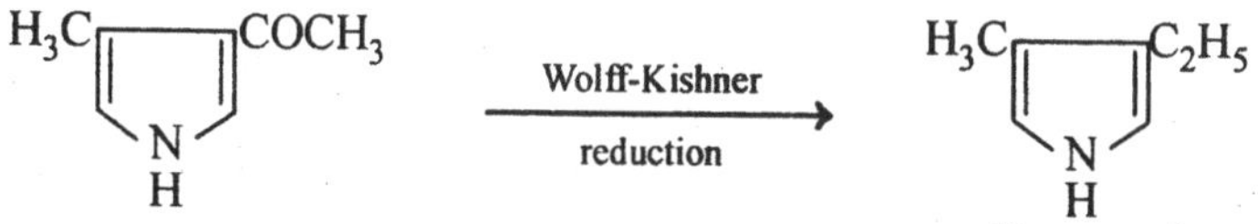

Opsopyrrole

Problem 1 : Predict the major products of the following reactions.

(a) 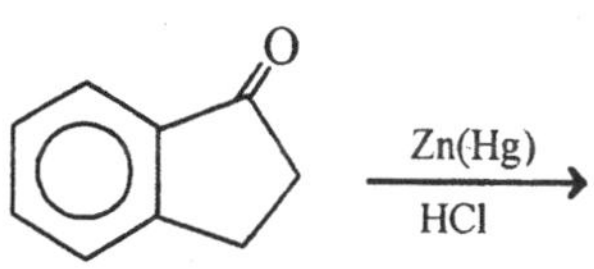

(b) 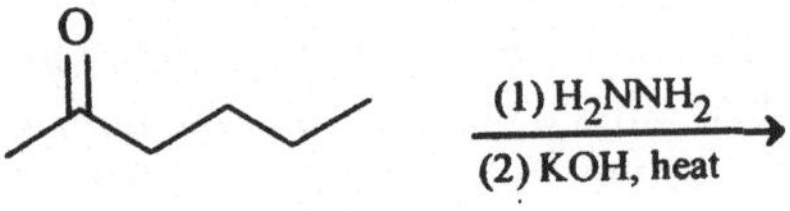

(c) 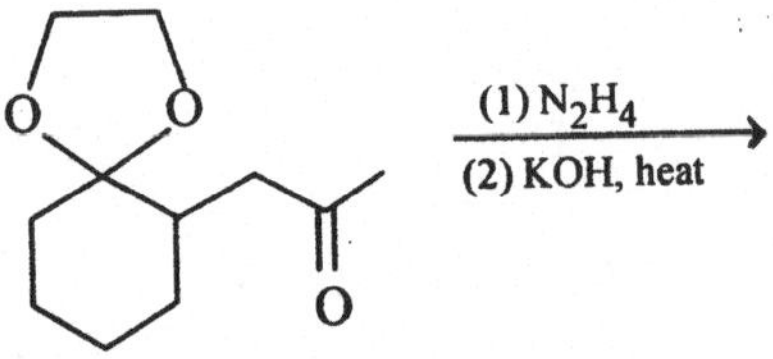

(d) 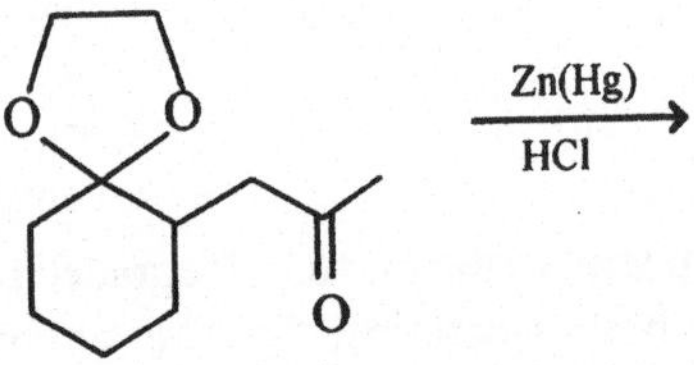

Ans : (a) Indane (b) Hexane
(c) Ethylene ketal of 2-propylcyclohexanone
(d) Propylcyclohexane.

118

WURTZ REACTION

An ethereal solution of an alkyl halide is treated with sodium which removes the halogen of alkyl halide and the two alkyl radicals join together to form an alkane.

$$\underset{\text{Alkyl halide}}{R}\,\vdots\, X + 2Na + X \,\vdots\, \underset{\text{Alkyl halide}}{R} \xrightarrow{\text{Dry ether}} \underset{\text{Alkane}}{R{-}R} + 2NaX$$

$$\underset{\text{Ethyl bromide}}{C_2H_5}\,\vdots\, Br + 2Na + Br \,\vdots\, \underset{\text{Ethyl bromide}}{C_2H_5} \xrightarrow{\text{Dry ether}} \underset{\text{Butane}}{C_2H_5{-}C_2H_5} + 2NaBr$$

In this method, we get alkane containing even number of carbon atoms. For preparing alkanes with odd number of carbon atoms, a mixture of two different alkyl halides is used, e.g.,

$$\underset{\text{Methyl bromide}}{CH_3}\,\vdots\, Br + 2Na + Br \,\vdots\, \underset{\text{Ethyl bromide}}{C_2H_5} \xrightarrow{\text{Dry ether}} \underset{\text{Propane}}{CH_3{-}C_2H_5} + 2NaBr$$

There will also be side reactions and we will get ethane ($CH_3.CH_3$) and butane ($C_2H_5.C_2H_5$) in addition to propane.

$$\underset{\text{Methyl bromide}}{2CH_3Br} + 2Na \xrightarrow{\text{Dry ether}} \underset{\text{Ethane}}{CH_3.CH_3} + 2NaBr$$

$$\underset{\text{Ethyl bromide}}{2C_2H_5Br} + 2Na \xrightarrow{\text{Dry ether}} \underset{\text{Butane}}{C_2H_5.C_2H_5} + 2NaBr$$

So Wurtz reaction is not suitable for the synthesis of alkanes containing odd number of carbon atoms.

For carrying out Wurtz reaction, the alkyl bromides and iodides may be used. The order of reactivity of alkyl halides is RCl < RBr < RI.

In place of sodium, other metals such as zinc, finely divided silver, etc. can also be used.

MECHANISM:

Mechanisms have been proposed :

FREE RADICAL MECHANISM which follows via generation of alkyl free-radical

Step 1. ***Generation of Free-Radical***

$$R-Br + Na. \longrightarrow R\cdot + NaBr$$

Step 2. ***Coupling between two Free-Radicals***

$$R\cdot \ \cdot R \longrightarrow R-R$$

[II] Ionic Mechanism

$$Na \longrightarrow \overset{+}{Na} + e$$

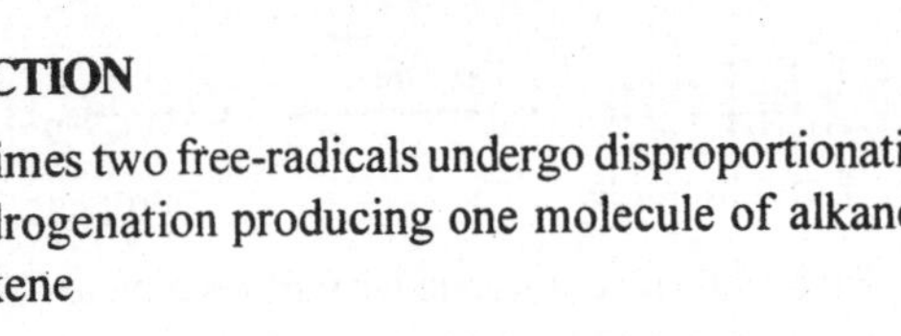

SIDE REACTION

Sometimes two free-radicals undergo disproportionation i.e. intermolecular hydrogenation producing one molecule of alkane and one molecule of alkene

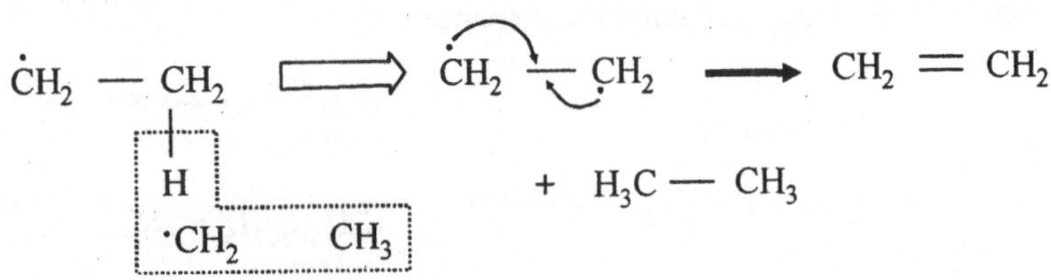

This can be represented ionically too

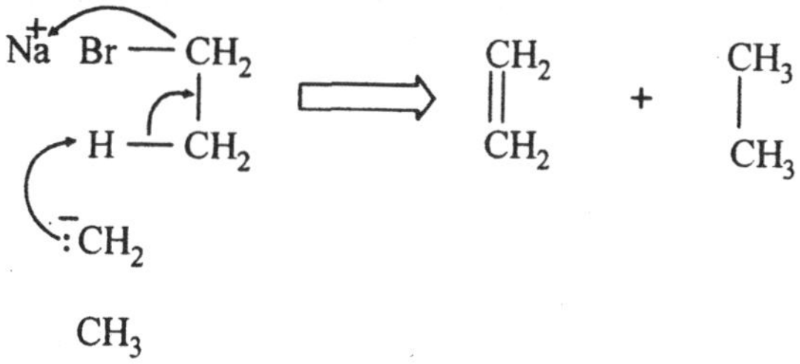

LIMITATIONS

The method is particularly useful in the synthesis of symmetrical alkanes having plane of symmetry e.g. ethane

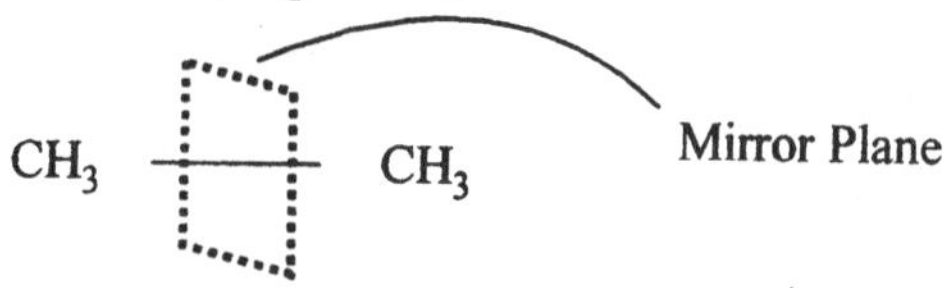

Applied to the synthesis of unsymmetrical alkanes, WURTZ REACTION results in poor yields.

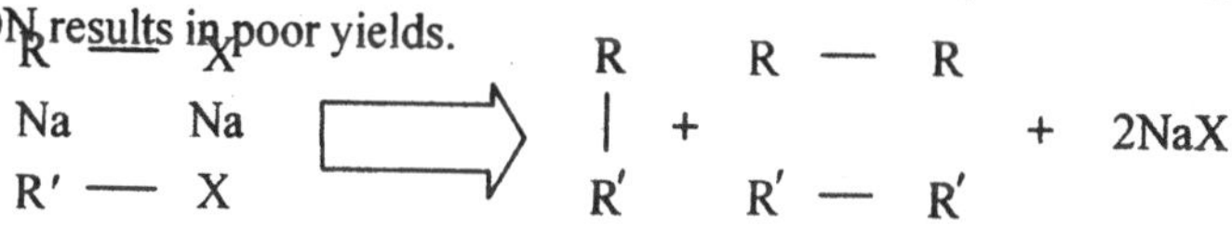

Owing to low reactivity of aryl halides towards nucleophilic substitution, ethereal sodium fails to bring about coupling between two aryl groups.

Finally, WURTZ REACTION will fail if the alkyl group in the halide bears an active center e.g. OH. In such cases, the metallic sodium will preferably replace hydrogen atom of the alkyl halide.

APPLICATION

WURTZ REACTION can be applied for the synthesis of cycloalkanes. The yield is very good for cyclopropane : but with higher cycloalkanes the yield gets poorer.

Internal Wurtz Reaction

A particularly interesting synthesis being the preparation of bicyclobutane

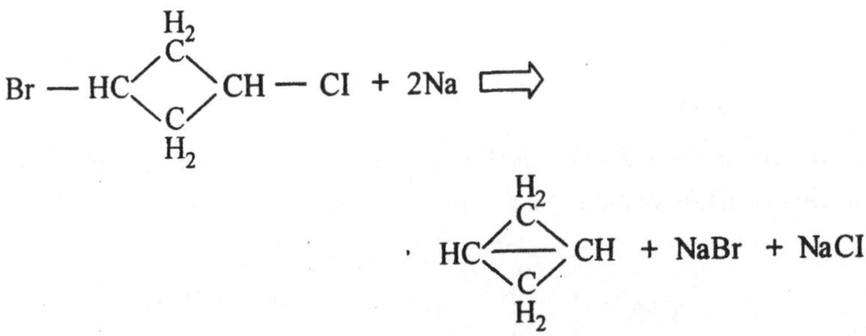

YIELD : 93 - 96%

[Wiberg & Lapman, 1963]

119

WURTZ-FITTIG REACTION

This reaction is a coupling reaction which takes place between aryl halide and alkyl halide in the presence of sodium metal and dry ether. Mechanism of the reaction is same area the case of Wurtz reaction.

$$Ar-X + R-X \xrightarrow{\text{Na/dry ether}} Ar-R + Ar-Ar + R-R$$

The main product of the reaction is always cross product.

$$C_6H_5-I + CH_3-CH_2-I \xrightarrow{\text{Na/dry ether}} \underset{\text{Ethyl benzene}}{C_6H_5-CH_2-CH_2}$$

FITTIG REACTION

This reaction is also coupling reaction which takes place between two molecules of aryl halides in the presence of Na/dry ether.

$$C_6H_5-X + C_6H_5-X \xrightarrow{\text{Na/dry ether}/\Delta} \underset{\text{Diphenyl}}{C_6H_5-C_6H_5}$$

Note : Halides give coupling reaction in the presence of Na, Mg, Ag and Zn.

(1) Coupling of mono halides given alkanes

$$R-X + R-X \longrightarrow R-R$$

(2) Coupling of gem dihalides give alkenes

$$R-X+R-X \longrightarrow R-R$$

(3) Coupling of gem trihalides gives alkynes

$$R-X+R-X \longrightarrow R-R$$

The above coupling reactions are examples of dehalogenation.